Current Progress in Biogeophysics and Biogeochemistry

Current Progress in Biogeophysics and Biogeochemistry

Edited by **Karl Seibert**

SYRAWOOD
PUBLISHING HOUSE
New York

Published by Syrawood Publishing House,
750 Third Avenue, 9th Floor,
New York, NY 10017, USA
www.syrawoodpublishinghouse.com

Current Progress in Biogeophysics and Biogeochemistry
Edited by Karl Seibert

International Standard Book Number: 978-1-68286-297-1 (Hardback)

Printed in China.

Contents

Preface

Biogeochemistry and biogeophysics are branches of biogeosciences which focus on the study of diverse processes occurring within the environment. This book includes some of the vital pieces of work being conducted across the world on various topics related to biogeophysics and biogeochemistry. Some of the topics that have been elucidated in this book are interactions between microbes and organic matter sediments, study of gas exchange and gas cycles, global elemental cycles, etc. This book caters to graduate and postgraduate students along with scholars and researchers in the field of biogeosciences and associated fields of study.

After months of intensive research and writing, this book is the end result of all who devoted their time and efforts in the initiation and progress of this book. It will surely be a source of reference in enhancing the required knowledge of the new developments in the area. During the course of developing this book, certain measures such as accuracy, authenticity and research focused analytical studies were given preference in order to produce a comprehensive book in the area of study.

This book would not have been possible without the efforts of the authors and the publisher. I extend my sincere thanks to them. Secondly, I express my gratitude to my family and well-wishers. And most importantly, I thank my students for constantly expressing their willingness and curiosity in enhancing their knowledge in the field, which encourages me to take up further research projects for the advancement of the area.

Editor

A novel reflectance-based model for evaluating chlorophyll concentrations of fresh and water-stressed leaves

C. Lin[1]**, S. C. Popescu**[2]**, S. C. Huang**[1]**, P. T. Chang**[3]**, and H. L. Wen**[1]

[1]Department of Forestry and Natural Resources, National Chiayi University, Chiayi, Taiwan
[2]Department of Ecosystem Science and Management, Texas A&M University, College Station, Texas, USA
[3]Department of Horticultural Science, National Chiayi University, Chiayi, Taiwan

Correspondence to: C. Lin (chinsu@mail.ncyu.edu.tw)

Abstract. Water deficits can cause chlorophyll degradation which decreases the total concentration of chlorophyll *a* and *b* (Chls). Few studies have investigated the effectiveness of spectral indices under water-stressed conditions. Chlorophyll meters have been extensively used for a wide variety of leaf chlorophyll and nitrogen estimations. Since a chlorophyll meter works by sensing leaves absorptance and transmittance, the reading of chlorophyll concentration will be affected by changes in transmittance as if there were a water deficit in the leaves. The overall objective of this paper was to develop a novel and reliable reflectance-based model for estimating Chls of fresh and water-stressed leaves using the reflectance at the absorption bands of chlorophyll *a* and *b* and the red edge spectrum.

Three independent experiments were designed to collect data from three leaf sample sets for the construction and validation of Chls estimation models. First, a reflectance experiment was conducted to collect foliar Chls and reflectance of leaves with varying water stress using the ASD FieldSpec spectroradiometer. Second, a chlorophyll meter (SPAD-502) experiment was carried out to collect foliar Chls and meter readings. These two data sets were separately used for developing reflectance-based or absorptance-based Chls estimation models using linear and nonlinear regression analysis. Suitable models were suggested mainly based on the coefficient of determination (R^2). Finally, an experiment was conducted to collect the third data set for the validation of Chls models using the root mean squared error (RMSE) and the mean absolute error (MAE). In all of the experiments, the observations (real values) of the foliar Chls were extracted from acetone solution and determined by using a Hitachi U-2000 spectrophotometer.

The spectral indices in the form of reflectance ratio/difference/slope derived from the Chl *b* absorption bands (ρ_{645} and ρ_{455}) provided Chls estimates with RMSE around 0.40–0.55 mg g^{-1} for both fresh and water-stressed samples. We improved Chls prediction accuracy by incorporating the reflectance at red edge position (ρ_{REP}) in regression models. An effective chlorophyll indicator with the form of ($\rho_{645}-\rho_{455}$) / ρ_{REP} proved to be the most accurate and stable predictor for foliar Chls concentration. This model was derived with an R^2 of 0.90 ($P < 0.01$) from the training samples and evaluated with RMSE 0.35 and 0.38 mg g^{-1} for the validation samples of fresh and water-stressed leaves, respectively. The average prediction error was within 14 % of the mean absolute error.

1 Introduction

Photosynthesis is the largest-scale synthetic process on earth. There are many kinds of photosynthetic pigments, i.e., chlorophylls, carotenoids, and phycobilins in plant leaves, but chlorophylls are considered to be the key factor because the photochemical reactions take place only at the trapped chlorophyll molecules. Light absorbed by chlorophyll excites electrons in the molecules, enabling them to be transferred to other molecules for glucose production and thus enabling vegetation growth. Chlorophyll content can directly determine photosynthetic potential and primary production (e.g., Whittaker and Marks, 1975). It is believed that about

Table 1. Previously developed spectral indices for foliar chlorophyll estimation.

Index	Formula	Source
Vogelmann red edge index 1	$\text{Vog1} = \rho_{740}/\rho_{720}$	Vogelmann et al. (1993)
Vogelmann red edge index 2	$\text{Vog2} = (\rho_{734} - \rho_{747})/(\rho_{715} + \rho_{726})$	Vogelmann et al. (1993)
Vogelmann red edge index 3	$\text{Vog3} = (\rho_{734} - \rho_{747})/(\rho_{715} + \rho_{720})$	Vogelmann et al. (1993)
Red edge NDVI	$\text{NDVI}_{705} = (\rho_{750} - \rho_{705})/(\rho_{750} + \rho_{705})$	Gitelson and Merzlyak (1994a, b)
Normalized difference vegetation index	$\text{NDVI}_{700} = (\rho_{800} - \rho_{700})/(\rho_{800} + \rho_{700})$	Gitelson and Merzlyak (1994a, b)
Simple ratio index	$\text{SR}_{700} = \rho_{750}/\rho_{700}$	Gitelson and Merzlyak (1996, 1997), Boegh et al. (2012)
Weighted simple ratio	$\text{wSR} = \rho_{860}/(\rho_{708} \times \rho_{550})$	Datt (1998), Gitelson et al. (2003)
Normalized difference vegetation index	$\text{NDVI}_{680} = (\rho_{800} - \rho_{680})/(\rho_{800} + \rho_{680})$	Blackburn (1998)
Modified chlorophyll absorption reflectance index	$\text{MCARI} = [(\rho_{700} - \rho_{670}) - 0.2(\rho_{700} - \rho_{550})][\rho_{700}/\rho_{670}]$	Daughtry et al. (2000)
Modified red edge simple ratio	$\text{mSR} = (\rho_{750} - \rho_{445})/(\rho_{705} - \rho_{445})$	Sims and Gamon (2002)
Modified red edge NDVI	$\text{mNDVI} = (\rho_{750} - \rho_{705})/(\rho_{750} + \rho_{705} - 2\rho_{445})$	Sims and Gamon (2002)
MERIS total chlorophyll index	$\text{MTCI} = (\rho_{750} - \rho_{710})/(\rho_{710} - \rho_{680})$	Dash and Curran (2004), Rossini et al. (2012)
Reciprocal-based simple ratio index	$\text{rSR}_{705} = \left(\frac{1}{\rho_{705}} - \frac{1}{\rho_{780}}\right) \times \rho_{780} = (\rho_{780}/\rho_{705}) - 1$	Gitelson et al. (2006)
Triangular chlorophyll index	$\text{TCI} = 1.2(\rho_{700} - \rho_{550}) - 1.5(\rho_{670} - \rho_{550})(\rho_{700}/\rho_{670})^{0.5}$	Haboudane et al. (2008)
Simple ratio index	$\text{SR}_{775} = \rho_{708}/\rho_{775}$	Féret et al. (2011)
Normalized difference vegetation index	$\text{NDVI}_{712} = (\rho_{780} - \rho_{712})/(\rho_{780} + \rho_{712})$	Féret et al. (2011)
Triangular greenness index	$\text{TGI} = -0.5[190(\rho_{670} - \rho_{550}) - 120(\rho_{670} - \rho_{480})]$	Hunt et al. (2011, 2013)
Simple ratio stress index	$\text{SR}_{760} = \rho_{695}/\rho_{760}$	Carter (1994)
Simple ratio stress index	$\text{SR}_{420} = \rho_{695}/\rho_{420}$	Carter (1994)
Simple ratio index broadband red edge	$\text{CI}_{\text{red edge}} = (\rho_{760\text{–}800}/\rho_{690\text{–}710}) - 1$	Gitelson et al. (2009)

100 billion tons of carbon is fixed annually into organic compounds by photosynthetic organisms (Nobel, 2005). The forest, a complex ecosystem of numerous trees, shrubs, savanna, and lichens, uptakes carbons for vegetation metabolism, and thus it makes a positive net primary production of biomass carbon.

Foliar chlorophyll concentration (shortened as Chls) has always been one of the important issues of research using vegetation remote sensing techniques in last 2 decades. As a consequence, a number of spectral indices were developed for foliar Chls estimation. Table 1 lists some of the chlorophyll indicators that have been examined by Vogelmann et al. (1993), Elvidge and Chen (1995), Blackburn and Ferwerda (2008), Ustin et al. (2009), Féret et al. (2011), and Hunt et al. (2013). Those indices integrate a couple of specific signatures of visible and near-infrared bands, for example, the reflectance at 445, 550, 680, 700, 705, 710, 720, 750, 780, 800, 860 nm for foliar Chls estimation.

Typical reflectance of vegetation in the visible–infrared region will increase as water deficit occurs (Knipling, 1970; Gausman and Allen, 1973; Gausman et al., 1982; Hunt and Rock, 1989; Carter, 1991 and 1993; Ceccato et al., 2001; Zygielbaum et al., 2009; Lin et al., 2012). As leaves dehydrate or vegetation suffers water stress, leaf water potential becomes increasingly negative and the rate of photosynthesis is reduced (Nilsen and Orcutt, 1996; Montagu and Woo, 1999; Keenan et al., 2009; Lavoir et al., 2009) because water deficit can cause chlorophyll degradation and thus significantly decreases foliar chlorophyll concentration (Kirnak et al., 2001; Pirzad et al., 2011; Desotgiu et al., 2012; Ghorbanli et al., 2013). Specifically, the magnesium ion (Mg^{2+}) of the chlorophyll will be removed. As a result, chlorophyll becomes pheophytin (chlorophyll without Mg^{2+}) and inactivates the photochemical reaction (Kaoau et al., 2007; Schelbert et al., 2009; Weber et al., 2009); furthermore leaves decrease the absorptance of blue and red light while increasing the reflectance at the corresponding wavelength bands. This underpins the first basic assumption of this study that a spectral index that effectively integrates the reflectance at the blue and red bands, at which the light specifically absorbed by chlorophyll is only used for plant photosynthesis, is better for foliar Chls estimation than the indices (listed in Table 1) that use other than the blue and red bands.

The dynamics of pigment concentrations are diagnostic of a range of plant physiological properties and processes (Blackburn, 2007). A suitable chlorophyll index can offer useful information for estimating the gross productivity of a terrestrial ecosystem (Nave et al., 2011) and even for understanding dust storm events (Tan et al., 2011). In order to address the effects of global climate changes, it is necessary to continuously update the prediction of forest carbon sequestration and the net primary productivity of terrestrial ecosystem. Traditional methods of using a spectrophotometer and/or fluorometer in ground leaf liquids operate on the light absorption of leaf in a laboratory setting. Determining foliar Chls with this technique is a standard measurement. In recent years, chlorophyll meters have offered a fast and convenient alternative for foliar Chls measurement. The Minolta SPAD 502 is used for a wide variety of leaf chlorophyll and nitrogen estimation by measuring the amount of light transmittance and absorptance in an easy and non-destructive way (Takebe et al., 1990; Ma et al., 1995; Blackmer and Schepers, 1995; Cate and Perkins, 2003; Read et al., 2003; Rowland et al., 2004; Hawkins et al., 2009; Rascher et al., 2009; Boegh et al., 2012). Nevertheless, it will be extremely hard work to extend the result of traditional methods to field study because such a single leaf measurement will be useless (due to no connection) for up-scaling the Chls and/or the fraction of absorbed photosynthetically active radiation (Gond et al., 1999) for the levels at tree crown or stand canopy.

In addition, since a chlorophyll meter works based on the sensing of leaves absorptance and transmittance, the reading of chlorophyll concentration will be affected by changes in transmittance as if there is a water deficit in leaves. On the other hand, the use of the relationship between the chlorophyll concentration and the readings of the chlorophyll meter may not accurately observe the chlorophyll variations due to physiological stresses. Because remote sensing only uses the reflectance to differentiate materials and/or discern the properties of targets, it therefore can offer good opportunity to indirectly determine foliar Chls. Therefore, the second assumption of this study is based on the fact that foliar reflectance at the photosynthesis wavebands is a better representative of foliar biochemical spectra than single leaf transmittance (hand-held chlorophyll meter) in a non-destructive detection base.

Many studies indicated that the spectral characteristics of red edge (RE) and green peak (GP) are directly or indirectly correlated to the level of leaf chlorophyll (Horler et al., 1983; Curran et al., 1990; Filella and Peñuelas, 1994; Pinar and Curran, 1996; Jongschaap and Booij, 2004; Mutanga and Skidmore, 2007) and can provide a method to distinguish between water and nutrient stress (Estep and Carter, 2005); they should be helpful in the prediction of leaf chlorophyll concentration. However, little research has examined the effectiveness of remote sensing models in the estimation chlorophyll content of both fresh and water-stressed leaves. We therefore proposed as a major goal of our study to develop a spectral index which could effectively integrate the reflectance of the photosynthetic-related spectra for leaf chlorophyll determination in a reliable non-destructive way for field application.

Briefly, the null hypotheses specified as follows will be examined in this paper.

H01: The reflectance at the wavelengths (e.g., 663, 645, 455, and 426 nm, that is, ChlsPn variables) directly absorbed by chlorophyll for photosynthesis is significantly and negatively related to foliar Chls with respect to variation of leaf water content.

H02: Reflectance-based spectral indices derived from ChlsPn/RE/GP variables are closely related to foliar Chls and can make better estimations of Chls than other indices without ChlsPn variables in respect to different degrees of water-stressed conditions.

H03: The determination of foliar Chls using the transmittance-based meter (e.g., SPAD chlorophyll meter) is insensitive to leaf water content.

2 Materials and methods

There were two independent experiments adopted to develop the foliar chlorophyll concentration models. The first was the chlorophyll-reflectance experiment from which training leaf samples were collected for reflectance measurement and chlorophyll determination, and the second was the chlorophyll SPAD experiment in which a new set of leaf samples was collected independently for SPAD measurement and chlorophyll concentration determination. Finally, additional leaf samples were used as a test data set for further validation of those models developed based on foliar reflectance experiment or SPAD absorptance experiment.

A hardwood species, namely Camphor tree (*Cinnamomum camphora* (Linn.) Seib), was selected for experiments. Leaf samples sized around 6–8 cm long by 3–4 cm wide were collected from the campus of the National Chiayi University in Taiwan. The authors intended to have samples collected in a wide range of pigment concentrations to meet the needs of this study. According to the ground inventory, we collected samples to meet leaf colors including dark green, light green, yellowish green, red, and dark red for laboratory experiments. Leaf samples of the data sets for chlorophyll-reflectance experiment and chlorophyll SPAD experiment is

50 and 45, respectively, and the additional evaluation data set is 70 leaves.

2.1 Data acquisition

2.1.1 Determination of foliar chlorophyll contents

Wellburn (1994) demonstrated that the chlorophyll concentration of leaf pigments could be determined by acetone, chloroform, dimethylformamide, and dimethyl-sulfoxide with spectrophotometer analysis. Concentrations of the tested foliar chlorophyll were extracted from the 80 % acetone solution and determined spectrophotometrically using a Hitachi U-2000 spectrophotometer following the method of Arnon (1949). Concentrations of chlorophyll a (Chl a) and chlorophyll b (Chl b) are determined using Eqs. (1) and (2) where D_λ stands for the absorptance at the specific wavelength λ, and where V and W represent the volume of ground leaf-acetone liquid (ml) and the fresh weight (g) of the ground leaf, respectively. Total chlorophyll concentration, Chls, was expressed as milligrams of chlorophyll per gram of fresh leaf weight (mg g^{-1}) and can be derived by summing up the values of Chl a and Chl b.

$$\text{Chl } a = (12.7 \times D_{663} - 2.69 \times D_{645}) \times (V/1000W) \quad (1)$$

$$\text{Chl } b = (22.9 \times D_{645} - 4.68 \times D_{663}) \times (V/1000W) \quad (2)$$

2.1.2 Foliar reflectance measurement

Spectral data were obtained from the FieldSpec Pro FR spectroradiometer manufactured by Analytical Spectral Devices (ASD). This instrument measures spectra over a spectral range of 350–2500 nm and offers 1 nm wide narrowband spectral data. Specifically, the full-width at half-maximum (FWHM) spectral resolution of the FieldSpec Pro FR spectroradiometer is 3 nm for the region 350–1000 and 10 nm for the region 1000–2500 nm (Hatchell, 1999) which meets the nominal sampling and resolution requirements for hyperspectral remote sensing applications (Curtiss and Goetz, 1994).

Procedures to gathering spectra involve optimizing the integration time (typically set at 17 ms), providing foreoptic information, recording dark current, collecting Spectralon reference radiance, and then obtaining target radiance. A 25° field-of-view (FOV) foreoptic which connected ASD spectroradiometer and the computer control system was mounted 35 cm above and leveled at a tripod on the top of leaf samples. As a result, a pixel size of 1.5 cm was determined as optimal. A black cloth was used to cover the platform to avoid the influence of background reflection. Two light sources were face to face mounted at an elevation angle of 45° and 1 m away from the sample. The target reflectance is determined as the ratio of the energy reflected off the target (target radiance) to energy incident on the reference Spectralon (reference radiance). For each measurement, the radiance was taken with spectrum averaging set to 15 and then filtered using a median filter (Hatchell, 1999; Lin et al., 2012).

Leaves reflectance spectra were measured in a laboratory with an artificial illuminator (USHIO jc 14.5V-50WC) before the leaves grinding process for chlorophyll determination. The ASD spectroradiometer collects one nanometer resolution hyperspectral data. A first derivative transformation of the reflectance spectra (Novo et al., 1995; Dawson et al., 1998) was applied to calculate the slope values of the foliar reflectance spectra, also known as first derivative spectra (FDS), and to determine the red edge position. This position has the largest FDS value which indicates the maximum change in the slope of the reflectance spectra per unit change in wavelength. Red edge position generally moves toward the longer wavelength if the FDS become larger which is a result of high chlorophyll concentrations in leaves. Leaf color is generally applied to visually diagnose foliar chlorophyll or health status. Green peak is supposed to be the main syndrome of foliar greenness and probably could offer potential value in the foliar Chls estimation. Green peak position is determined if the FDS value equals to zero.

2.1.3 Implement of spectral and chlorophyll measurement of fresh and water-stressed leaves

Relative water content (RWC) of leaves is commonly used to assess the water status of plants in tree physiology research. It has been applied to describe the status of leaf water stress in remote sensing (Pu et al., 2003; Lin et al., 2012) and is therefore used in this study. Fresh leaf samples were first detached, measured for fresh weight (FW), and then spectral reflectance data were immediately collected. Leaf samples were then left to dry naturally in an air-conditioned room at 26 °C with air circulated with a fan. Measurements of leaf weight and reflectance were made every 2 h during the drying process for 24 h. After collecting the final drying leaf weight and spectra, the leaf samples were oven-dried and the absolute dry weight (WD) was recorded. Finally, the RWC of fresh leaves and drying leaves were determined using Eq. (3). We further refer to this experiment as the pilot experiment.

$$\text{RWC} = \frac{\text{FW} - \text{WD}}{\text{FW}} \times 100 \quad (3)$$

The first experiment is a chlorophyll-reflectance experiment in which a total of 50 leaves were first used for spectral measurement and then ground for chlorophyll concentration determination. The second experiment is the chlorophyll SPAD experiment which was designed for exploring the relationship of the SPAD readings and leaf chlorophyll contents. A chlorophyll meter SPAD-502 (Konica Minolta Sensing, Inc.) was used for quick measurements of the chlorophyll content. In this experiment, we had 45 samples which were first measured by SPAD readings then ground and dissolved in acetone solution for chlorophyll concentration determination. Data collected from these two experiments were used for correlation and regression analysis to derive the relationships of chlorophyll-reflectance spectra and chlorophyll SPAD readings. The third experiment is a validation experiment. An additional 70 leaf samples were first detached, and measurements of their weight were recorded along with SPAD and reflectance of fresh leaves, and then left to dry naturally in the same environmental conditions as the pilot experiment. 24 h later, spectral measurements were implemented immediately after obtaining the weight and SPAD measurements for every leaf samples. A small portion, a circle with a diameter of 1.5 cm (set W_s) of each of the leaf samples (set W_{total}) was taken for the determination of foliar Chls in the acetone solution, and the another part of each of the leaf samples was oven-dried to get the dry weight for the determination of leaf RWC based on the weight ratio of W_s and W_{total}.

2.2 Correlation analysis and regression analysis

A correlation analysis was applied to determine the correlation coefficient (r) between the foliar chlorophyll concentration and its reflectance. All of the coefficients were further tested using the Student's t statistic: $t = r\sqrt{(n-2)/(1-r^2)} \sim t_{\alpha/2,\,n-2}$ to examine whether it is statistical meaningful for diagnosing foliar chlorophyll status.

Arnon (1949) demonstrated that chlorophyll a and chlorophyll b have particular absorption features in the blue and red spectral region in the acetone solution. Specifically, Chl a and Chl b have two absorption peaks for photosynthesis at the wavelengths of 426 and 663 nm, and 455 and 645 nm, respectively. The absorption peaks of Chl a and Chl b in other solvents, such as chloroform, shift a little from the predefined wavelength (Wellburn, 1994). Since the chlorophyll was determined by using the acetone solvent, the spectral reflectance at those four specific wavelengths, i.e., ρ_{663}, ρ_{645}, ρ_{445}, and ρ_{426} are called ChlsPn variables; the chlorophyll-related spectra such as the position and reflectance of the green peak feature (λ_{Gmax} and ρ_{Gmax}) and the red edge feature (λ_{REP} and ρ_{REP}) are called GP variables and RE variables. A transformation of two key spectral features can be integrated by simple ratio (ρ_i / ρ_j), simple difference ($\rho_i - \rho_j$), and normalized difference $(\rho_i - \rho_j)/(\rho_i + \rho_j)$ methods to derive a new spectral index for remote sensing analysis. A new transformation, the slope index (SI), was defined as the ratio of the spectral difference and the distance of any two key features. That is $\mathrm{SI} = (\rho_i - \rho_j)/(|\lambda_i - \lambda_j|)$. This index integrates two spectral reflectance values based on their spectral curve (or spectral behavior) into a standardized index value and potentially can reduce the influence caused by background reflection and various albedo. The original form of the variables ChlsPn, GP, and RE and their derived spectral indices (Eqs. 6–21) were used as variables (shown in Table 2) in regression analysis. Reflectance-based empirical Chls models were than validated to examine the hypotheses of this study.

In the regression analysis of reflectance-based models, the dependent variable is the natural-logarithm-transformed foliar Chls, denoted as lnChls. The transformation is used to stabilize the constant variance of the predicted error term. Also, the independent variables are the ChlsPn variables, the GP variables, the RE variables, and/or their derived spectral indices. The statistics such as the coefficient of determination (R^2), the prediction error sum of squares (PRESS), and the standard error of estimates (SE(Y)) were used to measure the model adequacy. In the regression analysis of SPAD-based chlorophyll model, the SPAD readings and acetone-extracted chlorophyll was set to be the regressor variable and the dependent variable, respectively; the fitted model is named as the absorptance-based chlorophyll model.

2.3 Validation of reflectance-based and absorptance-based chlorophyll empirical models

The reflectance spectra and SPAD readings collected by the third experiment were input to the absorptance-based model and the reflectance-based models to get the estimates of the foliar chlorophyll concentration; each of the estimates was then assessed by the acetone-method-determined chlorophyll contents. In the prediction assessment, the formula of root mean squared error (RMSE) and mean absolute error (MAE) are listed in Eqs. (4) and (5) and applied to demonstrate how the estimator differs from the measured value of the quantity being estimated. RMSE has the same units (mg g^{-1}) as the quantity being estimated, and MAE is presented in percentage, indicating a relative degree of the estimation differs from the observation. In Eqs. (4) and (5), n is the number of samples, y and $\hat{y}$ represent the observed and predicted value, respectively.

$$\mathrm{RMSE} = \sqrt{\frac{\sum_{i=1}^{n} (y_i - \hat{y}_i)^2}{n}} \tag{4}$$

$$\mathrm{MAE} = \frac{1}{n}\left[\sum_{i=1}^{n} \frac{\mathrm{abs}(y_i - \hat{y}_i)}{y_i} \times 100\,\%\right] \tag{5}$$

3 Results

3.1 Reflectance spectra of fresh and water-stressed leaf

Figure 1a shows the spectral behavior of the fresh and water-stressed leaves of camphor trees. RWC of the leaf sample varied from 51 to 5 %. Four aspects could be pointed out about the difference of reflectance spectra between fresh and water-stressed leaves. First, reflectance spectra in the visible–infrared region behaved like a general reflectance curve of fresh green leaves while the curve lifts up as the RWC decreased. Second, the green peak of the reflectance curves was always clearly visible and the slope from the peak at

Table 2. The meaning and the formula of some extended spectral indices derived from the ChlsPn variables, red edge, and green peak variables.

Regressor names	Meaning and mathematical formula	
RDI group	Reflectance difference index of two spectral features from ChlsPn, red edge, and green peak.	
RDI_a	RDI derived based on the two absorption peaks of chlorophyll *a* at 663 nm and 426 nm. $RDI_a = \rho_{663} - \rho_{426}$	(6)
RDI_b	RDI derived based on the two absorption peaks of chlorophyll *b* at 645 nm and 455 nm. $RDI_b = \rho_{645} - \rho_{455}$	(7)
$RDI_{\rho REP - \rho Gmax}$	RDI derived based on the red edge and green peak position. $RDI_{\rho REP - \rho Gmax} = \rho REP - \rho Gmax$	(8)
SI group	Slope index of two spectral features from ChlsPn, red edge, and green peak.	
SI_a	SI determined using the two absorption peaks of chlorophyll *a*. $SI_a = (\rho_{663} - \rho_{426})/(\lambda_{663} - \lambda_{426})$	(9)
SI_b	SI determined using the two absorption peaks of chlorophyll *b*. $SI_b = (\rho_{645} - \rho_{455})/(\lambda_{645} - \lambda_{455})$	(10)
$SI_{\rho REP \vert \rho Gmax}$	SI determined using the red edge and green peak features. $SI_{\rho REP \vert \rho Gmax} = (\rho REP - \rho Gmax)/(\lambda REP - \lambda Gmax)$	(11)
NDI group	Normalized difference index of two spectral features from ChlsPn, red edge, and green peak.	
NDI_a	$NDI_a = (\rho_{663} - \rho_{426})/(\rho_{663} + \rho_{426})$	(12)
NDI_b	$NDI_b = (\rho_{645} - \rho_{455})/(\rho_{645} + \rho_{455})$	(13)
$NDI_{REPRDIb}$	$NDI_{REPRDIb} = (\rho_{REP} - RDI_b)/(\rho_{REP} + RDI_b)$	(14)
$NDI_{REPGmax}$	$NDI_{REPGmax} = (\rho_{REP} - \rho_{Gmax})/(\rho_{REP} + \rho_{Gmax})$	(15)
RI group	Ratio Index of two spectral features from ChlsPn, red edge, and green peak.	
RI_a	$RI_a = \rho_{663}/\rho_{426}$	(16)
RI_b	$RI_b = \rho_{645}/\rho_{455}$	(17)
$RI_{RDI_b/\rho REP}$	$RI_{RDI_b/\rho REP} = RDI_b/\rho REP$	(18)
$RI_{RDI_b/\lambda REP}$	$RI_{RDI_b/\lambda REP} = RDI_b/\lambda REP$	(19)
$RI_{\rho Gmax/\rho REP}$	$RI_{\rho Gmax/\rho REP} = \rho Gmax/\rho REP$	(20)
RI_{SPAD}	$RI_{SPAD} = \rho_{650}/\rho_{940}$	(21)

green region to the lowest point at red region was significantly decreased when RWC is less than 30 %. Third, the significant water absorption valleys could be seen at the spectral regions centered at 1450 and 1910 nm. The depth and the area of the absorption valley are negatively closely related to leaf RWC (Lin et al., 2012). Finally, a water-stressed leaf reflectance peaked at around 2000 nm as the RWC was less than or equal to 16 %. Although the absorption feature in the infrared region (2000–2200 nm) is possibly due to the dry matter constituents (such as protein, lignin, and cellulose) (Cheng et al., 2011), this particular phenomenon is probably related to physiological reactivity. It should be worthwhile to explore in further studies. Figure 1b shows the first derivative of reflectance in the visible region of a sample leaf. The red edge and green peak of the sample leaf were detected at around 701–697 and 540–535 nm respectively. It was observed that a small shift from longer to shorter wavelengths happened as RWC changed from 51 to 5 %. Even though the shifts were not very significant, it still indicated that the foliar Chls would decrease if the water content of leaves decreased. This phenomenon agree with the one addressed by Kirnak et al. (2001), Pirzad et al. (2011), Desotgiu et al. (2012), and Ghorbanli et al. (2013).

3.2 Response of leaf reflectance to variations in leaf chlorophyll concentration

Figure 2a demonstrates the reflectance spectra over the visible–infrared wavebands of the fresh leaves in the first

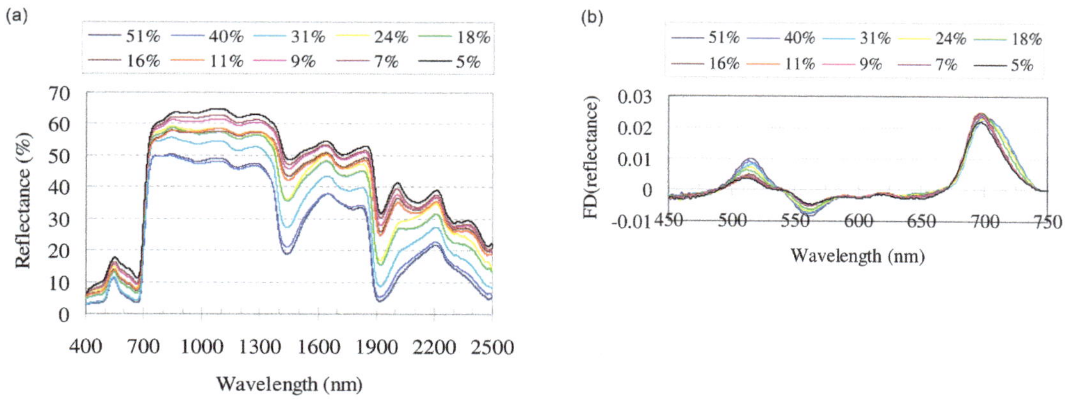

Figure 1. Foliar reflectance spectra of *Cinnamomum camphora*. **(a)** Spectral curves showed the reflectance variation of visible–infrared bands due to the changes of water content. **(b)** An example of the association between the relative water content in fresh and water-stressed leaves and the green peak and red edge spectra features.

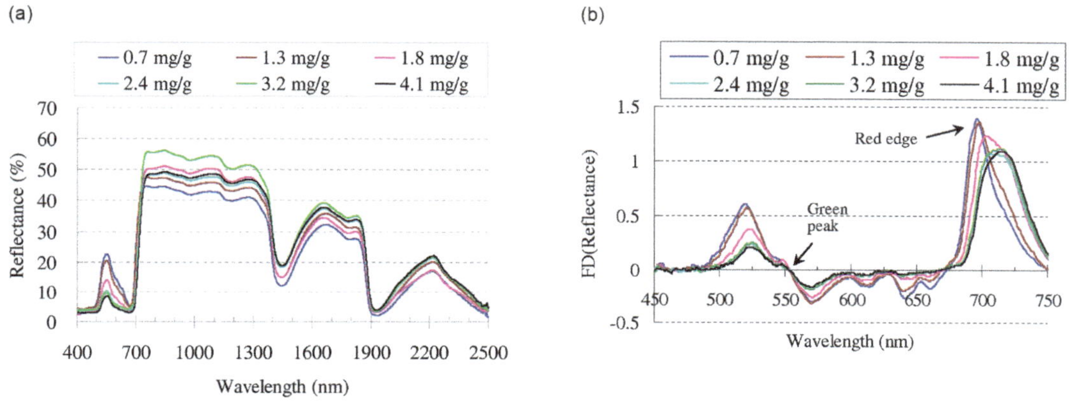

Figure 2. Foliar reflectance spectra of *Cinnamomum camphora*. **(a)** Spectral curves showed the reflectance variation of visible–infrared bands due to the changes of foliar chlorophyll concentration. **(b)** Association between the chlorophyll concentration and the green peak and red edge spectra of fresh leaves.

experiment. The Chls of these samples ranged from 0.7 to 4.1 $mg\,g^{-1}$. An important feature that was observed showed that the change of reflectance in the visible region behaves obviously different from the one in the infrared region due to changes of foliar Chls. Leaves with smaller Chls showed a higher reflectance in the visible portion of the spectrum. This is very similar to the increase of the reflectance curve as leaves are under water-stressed conditions. On the contrary, a leaf with higher Chls demonstrated higher reflectance in the infrared area, which is not always consistent with the increase of foliar Chls among all of the samples.

The blue drop and the red drop in the visible region are due to photons being absorbed by the chlorophylls *a* and *b* in the photosynthesis process (Emerson and Lewis, 1943; Hopkins and Hüner, 2004). It indicates that a sharp decrease happened to the gradient between these two points (that is, the gradient) for the leaf with higher value of Chls. In other words, a leaf whose reflectance in visible region will decrease (level down), the gradient will also decline as it is maturing. Associations between the green peak and the red edge features and values of foliar Chls are shown in Fig. 2b, which indicates that the green peaks and the red edge occur at wavelengths around 554–557 and 694–715 nm, respectively. As foliar Chls decreased, the red edge position moves toward shorter wavelengths, the same trend being observed for the blue edge.

3.3 Relationship of chlorophyll concentration and visible–infrared reflectance spectra

Leaf spectral reflectance (ρ_λ) is correlated to foliar Chls. Figure 3 demonstrates the generalized visible–infrared spectra of *Cinnamomum camphora* leaves and associated with the corresponding correlogram of Pearson's correlation coefficients, denoted as $r(\text{Chls}, \rho_\lambda)$ a correlation coefficient of the foliar chlorophyll concentration and the reflectance at wavelength λ. A negative coefficient $r(\text{Chls}, \rho_\lambda) < 0$ was found in the visible region while a positive coefficient $r(\text{Chls}, \rho_\lambda) > 0$ was found in the infrared region.

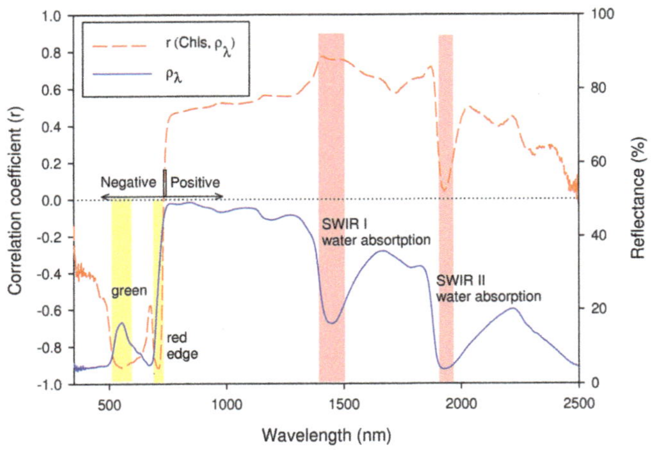

Figure 3. Generalized reflectance spectra of *Cinnamomum camphora* leaf (solid line) and trends of the corresponding correlation coefficients (dashed line) between the concentration of total chlorophyll and reflectance in the visible–infrared wavelength region. Results derived from the training data set, the 50 leaf samples.

Most of the reflectance values between 350–2500 nm have proven to be significantly linearly related to foliar Chls based on the Student's t statistic $t = r\sqrt{(n-2)/(1-r^2)} \sim t_{\alpha/2,\,n-2}$. Exceptions are the sub-regions 730–741, 1905–1970, and 2408–2500 nm. Reflectance of the bands at green sub-region 530–580 nm and red edge sub-region 700–716 nm are valuable because the absolute value of their $r(\text{Chls}, \rho_\lambda)$ are greater than 0.90 ($P < 0.01$). Although vegetation is proven to use the light energy in blue (PS I) and red (PS II) wavelengths for photosynthesis, the reflectance spectra appears different than the absorption spectra.

Infrared reflectance is positively related to foliar Chls. It is observed that in the first sub-region of shortwave infrared, a dramatic drop of reflectance curve happened in the water absorption area (1395–1504 nm, denoted as SWIR I), with a value of $r(\text{Chls}, \rho_\lambda)$ around 0.75. In the second sub-region of shortwave infrared (1905–1970 nm, denoted as SWIR II), the reflectance is almost independent of foliar Chls because the value of $r(\text{Chls}, \rho_\lambda)$ is almost identical to zero, as indicated by the Student's t test results.

4 Discussions

4.1 Reflectance-based empirical models for the estimation of total chlorophyll content

4.1.1 Adequacy comparison among the models with prototype variables

Recall that a measured value of foliar chlorophyll is determined by the spectrophotometrical method. This method uses the absorptance peaks of blue spectra at 426 and 455 nm and the absorptance peaks of red spectra at 645 and 663 nm. The reflectance of those specific wavelengths, i.e., ρ_{426}, ρ_{455}, ρ_{645}, and ρ_{663} are named as the ChlsPn variables, the red edge characteristics ρ_{REP} and λ_{REP} are named as red edge variables, and the green peak characteristics ρ_{Gmax} and λ_{Gmax} are named as green peak variables hereafter in this paper.

Based on the fundamentals of remote sensing, a target will reflect smaller amount of incident energy if it absorbs most of the incident energy. Figure 4a–d shows that foliar chlorophyll concentration is negatively related to the reflectance of the ChlsPn variables. It indicates that higher foliar chlorophyll concentration causes lower reflectance of the blue and red spectra. The models with one of the ChlsPn variables work like the spectrophotometric method. Figure 4e and f show that the chlorophyll concentration is positively linearly related to the red edge variables indicating a result similar to Curran et al. (1991). The adequacy statistics R^2, PRESS, and SE(Y) show that λ_{REP} is better than ρ_{REP} and even better than ρ_{645} in the prediction of chlorophyll concentration. There are 89 % of the Chls variation could be explained by λ_{REP} while only 30 % of the Chls variation could be explained by ρ_{REP}. This result agrees with the research of Mutanga and Skidmore (2007) and Reddy and Matcha (2010), who demonstrated that the red edge position is strongly negatively correlated with the foliar pigment concentration in plants. The reductions in Chls increased leaf reflectance at red spectral region and caused the red edge shift to shorter wavelengths. This agrees with conclusions of Carter (1993) and Carter and Knapp (2001). Though the Chls of training samples show a big variation, the green peak position (λ_{Gmax}) changes only in a very short span from 554 to 557 nm. There is only 10 % of Chls that could be explained by the variable λ_{Gmax} (Fig. 4g). The Chls is much better fitted by the reflectance of the green peak position (ρ_{Gmax}) with a negatively linear relationship (Fig. 4h). This model has adequacy very close to the model with the regressor λ_{REP}.

4.1.2 Adequacy comparison among the models using a derived spectral index

Figure 5 shows the relationship between the lnChls and the difference index or the slope index of red and blue absorption peaks by the prototype variables, and of the green peak and red edge variables. RDI_a and SI_a are negatively linearly related to lnChls while SI_a has a better model adequacy than RDI_a (Fig. 5a and b). It is observed that both predictors RDI_b and SI_b have an exponential decay relationship with the lnChls (Fig. 5c and d); the regression coefficient of the predictor is negative which means that a lower value of total chlorophyll concentration will be observed when having a higher value of RDI_b or SI_b.

The predictors $\text{RDI}_{\rho_{\text{REP}}-\rho_{\text{Gmax}}}$ and $\text{SI}_{\rho_{\text{REP}}|\rho_{\text{Gmax}}}$ work like a two-order polynomial function in the prediction of lnChls. These two models also have the same levels of the indicators of model adequacy (Fig. 5e and f). The coeffi-

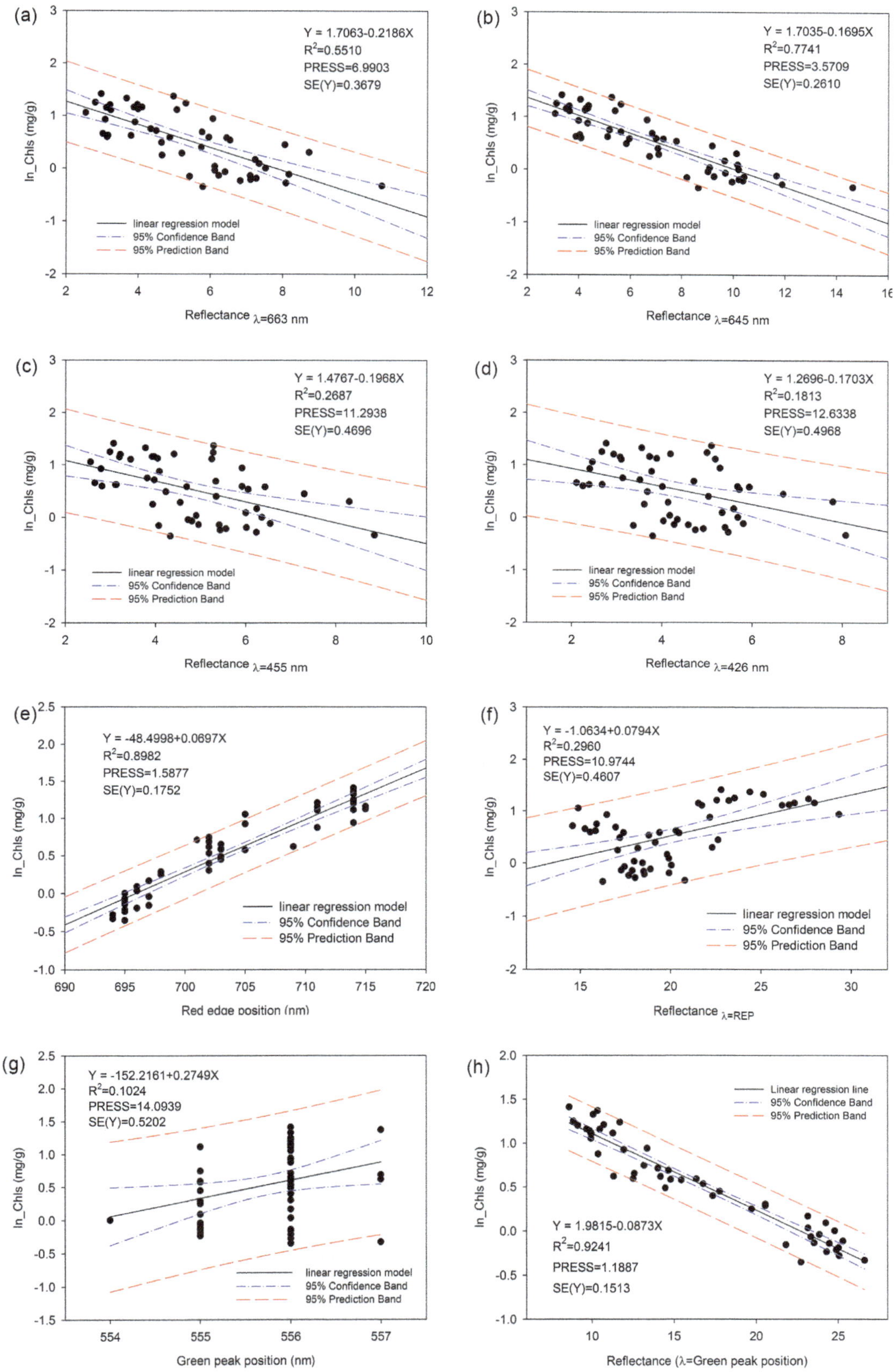

Figure 4. Reflectance-based empirical models with the ChlsPn/red edge/green peak variables for leaf total chlorophyll content estimation.

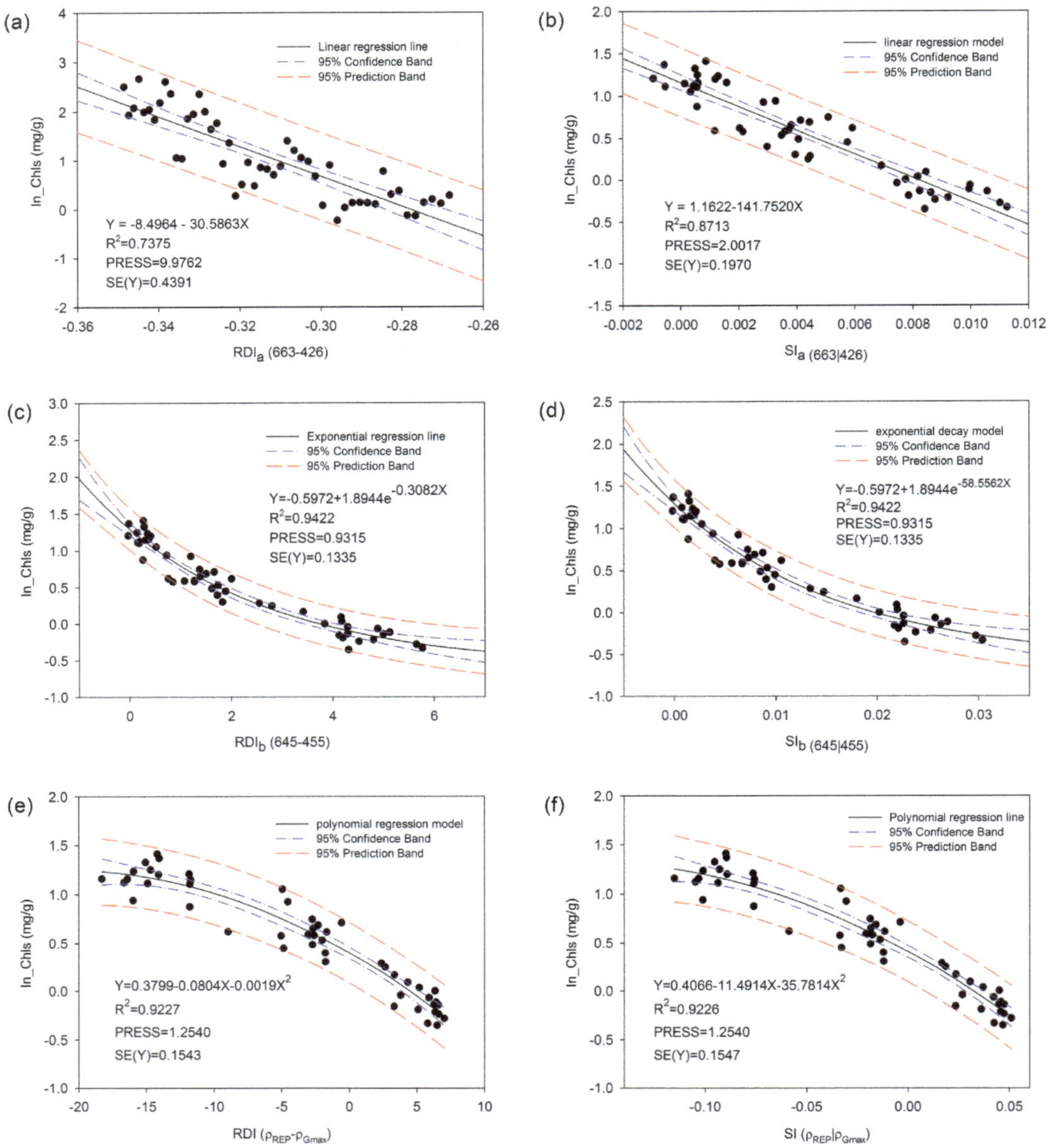

Figure 5. RDI- and SI-based empirical models for the estimation of foliar total chlorophyll concentration.

cients of first- and second-order variables are negative, indicating a higher value of foliar Chls has a lower value of $\mathrm{RDI}_{\rho\mathrm{REP}-\rho\mathrm{Gmax}}$ and $\mathrm{SI}_{\rho\mathrm{REP}|\rho\mathrm{Gmax}}$. In other words, a leaf with high chlorophyll concentration will have the reflectance at the red edge position far greater than the reflectance at the green peak position.

Figure 6 shows six empirical models with adequacy assessments for the estimation of foliar chlorophyll concentration using the ratio index of spectral features. Those models are all negatively related to the lnChls. Figure 7 shows the behavior of normalized difference indices with respect to the changes of foliar total chlorophyll concentration. The predictors NDI_a and NDI_b are linearly related to lnChls (Fig. 7a and b). These two indices display a negative relationship in the estimation of lnChls, with an R^2 value of 0.67 and 0.83 ($P < 0.01$).

4.1.3 Validation of empirical reflectance-based Chls models

The empirical reflectance-based Chls models developed using the training samples were validated using another data set by using samples which contain fresh and water-stressed leaves. The Chls prediction bias was presented with the indicators RMSE and MAE in response to the group of fresh and water-stressed samples, that is, RMSEf, RMSEw, MAEf, and MAEw. The prediction biases of both fresh and water-stressed groups were averaged to get the mean values of RMSEa and MAEa. Details of the model validation and per-

Table 3. Prediction accuracy assessment of the developed Chls empirical models.

Predictor	Model adequacy			Model validation					
	R^2	PRESS	SE(Y)	RMSEf	RMSEw	RMSEa	MAEf	MAEw	MAEa
$\mathrm{RI}_{\mathrm{RDI}_b/\rho_{\mathrm{REP}}}$	0.90	1.55	0.17	0.35	0.38	0.36	15.85	12.49	14.17
SI_b	0.94	0.93	0.13	0.46	0.43	0.44	19.07	14.63	16.85
$\mathrm{RI}_{\mathrm{RDI}_b/\lambda_{\mathrm{REP}}}$	0.94	0.93	0.13	0.46	0.42	0.44	19.33	14.47	16.90
RDI_b	0.94	0.93	0.13	0.44	0.49	0.46	17.11	18.03	17.57
$\mathrm{NDI}_{\mathrm{REPRDI}b}$	0.92	1.30	0.16	0.36	0.48	0.43	16.78	18.97	17.88
ρ_{REP}	0.30	10.97	0.46	0.52	0.44	0.48	20.45	20.79	20.62
NDI_b	0.84	2.49	0.22	0.40	0.55	0.48	21.11	24.36	22.74
RI_b	0.83	2.70	0.23	0.39	0.54	0.47	21.49	24.01	22.75
NDI_a	0.68	5.08	0.31	0.73	0.67	0.70	32.83	31.12	31.98
RI_a	0.66	5.36	0.32	0.64	0.70	0.67	32.05	33.05	32.55
SI_a	0.87	2.00	0.20	0.85	0.54	0.71	50.56	20.47	35.51
RDI_a	0.74	9.98	0.44	0.85	0.54	0.71	50.56	20.47	35.51
$\mathrm{RI}_{\mathrm{SPAD}}$	0.79	3.41	0.25	0.79	0.49	0.66	48.04	36.20	42.12
$\mathrm{RDI}_{\rho_{\mathrm{REP}}-\rho_{\mathrm{Gmax}}}$	0.92	1.25	0.15	1.20	0.78	1.01	55.63	39.73	47.68
$\mathrm{SI}_{\rho_{\mathrm{REP}}\vert\rho_{\mathrm{Gmax}}}$	0.92	1.25	0.25	1.15	0.78	0.98	54.55	39.34	49.95
λ_{REP}	0.90	1.59	0.18	1.18	0.64	0.95	56.05	48.43	52.24
$\mathrm{RI}_{\rho_{\mathrm{Gmax}}/\rho_{\mathrm{REP}}}$	0.93	1.06	0.14	0.40	1.17	0.88	16.33	89.52	52.93
ρ_{645}	0.77	3.57	0.26	0.98	0.63	0.82	60.38	46.41	53.39
$\mathrm{NDI}_{\mathrm{REPGmax}}$	0.93	1.08	0.14	0.40	1.19	0.89	16.49	90.71	53.60
ρ_{426}	0.19	12.63	0.50	0.84	0.72	0.78	54.73	54.99	54.86
ρ_{663}	0.55	6.99	0.37	1.12	0.60	0.90	70.54	43.91	57.23
ρ_{455}	0.27	11.29	0.47	1.02	0.85	0.94	66.31	64.24	65.28
ρ_{Gmax}	0.92	1.19	0.15	0.78	1.94	1.48	44.61	147.80	96.20
λ_{Gmax}	0.10	14.09	0.52	0.90	> 1000	> 1000	57.00	> 1000	> 1000

cent variances explained are listed in Table 3. Among the 24 models, there are 11 models whose R^2 is greater than 0.90 ($P < 0.01$), PRESS less than 1.60, and SE(Y) less than 0.25. Though those models have high R^2 values, the predicted Chls accuracy for the validation data set varies significantly. Some of the models have a big difference in prediction power for fresh leaves and water-stressed leaves samples. These results indicate that $\mathrm{RI}_{\rho\mathrm{Gmax}/\rho\mathrm{REP}}$ and $\mathrm{NDI}_{\mathrm{REPGmax}}$ models are only recommended for the Chls estimation of fresh leaves, while they failed to capture the changes caused by the water stress effect on spectral features variations.

Five models among those 24 models could be applied to estimate the foliar Chls of tree leaves because their predictor is able to capture accurately the Chls variation due to the changes of water content in leaves. The relatively high performance models have MAE ranges between 15–20 and 12–19 % for fresh leaves and water-stressed leaves, and have average MAEs between 14 and 18 %. The best validation among those reflectance-based Chls models was observed for the model with the predictor $\mathrm{RI}_{\mathrm{RDIb}/\rho\mathrm{REP}}$. Accordingly, we inferred that the reflectance variables ρ_{645}, ρ_{455}, and ρ_{REP} are able to capture the key spectral features of foliar chlorophyll status and hence are effective predictors of foliar Chls.

In addition, the models with only one reflectance feature of the prototype variables, i.e., ρ_{663}, ρ_{425}, ρ_{645}, or ρ_{455}, will not be able to successfully predict foliar Chls. Finally, we found a conflict of agreement between model adequacy and validation for the models with the predictor λ_{REP} and ρ_{REP}. Though λ_{REP} is fitted very well with high adequacy $R^2 = 0.90$ ($P < 0.01$), its model is validated with MAEa = 52 % and RMSEa = 0.95 mg g^{-1}; while ρ_{REP} is not fitted very well, its $R^2 = 0.30$ is still significant at the 0.05 probability, and this model is validated with MAEa = 20 % and RMSE = 0.48 mg g^{-1}. The reflectance at green peak wavelength (ρ_{Gmax}) could be partially useful in the Chls prediction of fresh leaves.

Oki (2010) showed that ratio of reflectivity is able to give a good estimation of chlorophyll a in lake water. Our results demonstrated that leaf chlorophyll concentration in cases of various water contents (fresh and/or water stressed) could be accurately predicted using spectral ratio indices such as RI (ratio index), SI (slope index), and NDI (normalized difference index) due to those indices effectively integrating the spectral features of chlorophyll b and additionally the red edge characteristics. The model adequacy and the prediction accuracy validation of the empirical models have the same agreement. This leads to the answers of the hypotheses H01 and H02. First, the reflectance of ChlsPn is linearly and negatively related to foliar Chls, while the reflectance of red edge and green peak is linearly and positively related to foliar Chls. Second, the ChlsPn, red edge, and green peak cannot achieve an acceptable accuracy in the estimation of foliar

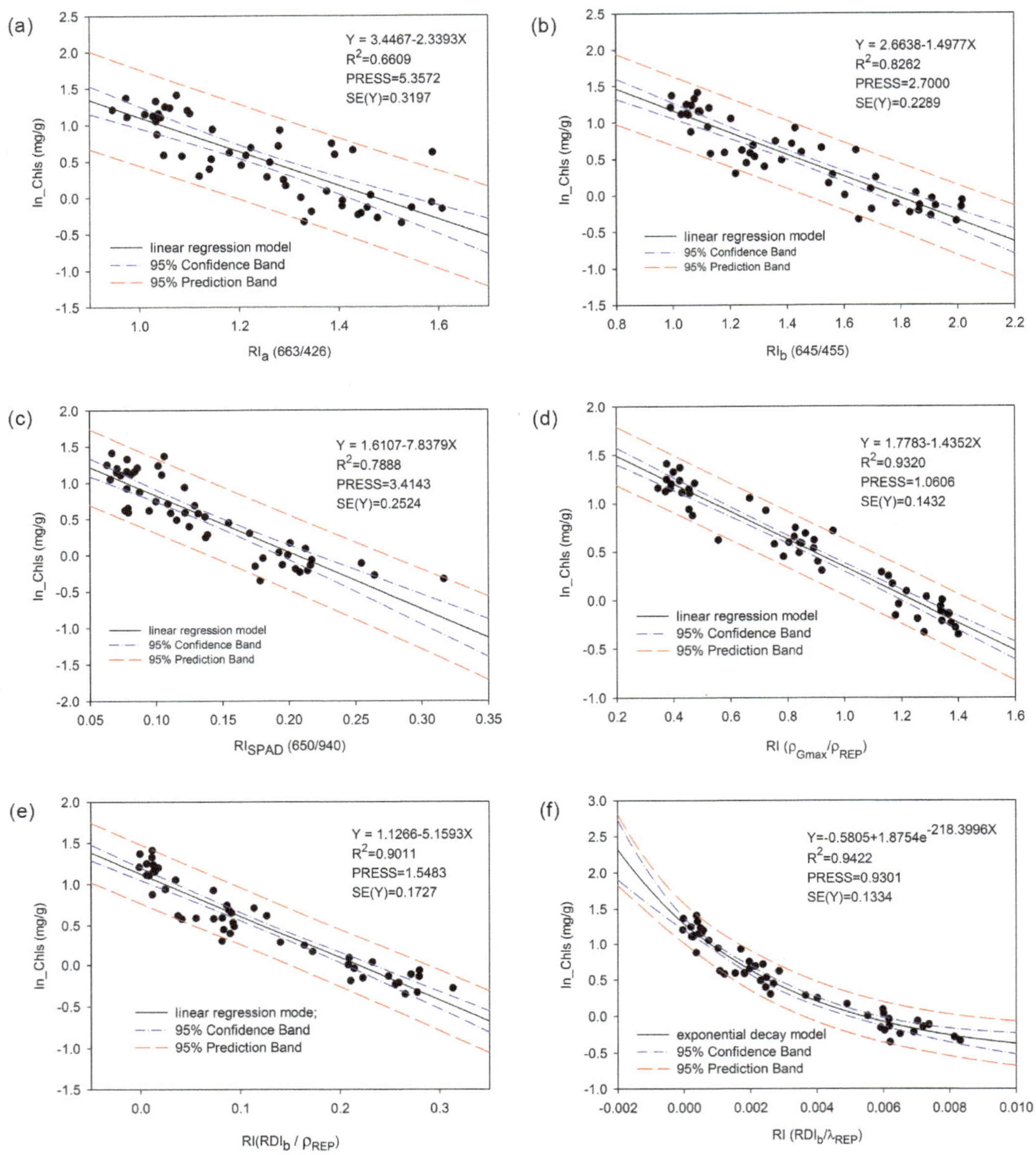

Figure 6. Ratio-index-based empirical models for foliar Chls concentration estimation.

Chls (for example MAE < 20 %) when they are used alone as predictors. Third, the ChlsPn variables can be integrated to produce a spectral difference index ($RDI_b = \rho_{645} - \rho_{455}$) or a spectral slope index $SI_b = (\rho_{645} - \rho_{455})/(\lambda_{645} - \lambda_{455})$ to achieve an acceptable accuracy. Finally, ChlsPn and red edge characteristics can also be integrated as new spectral indices by the combination of reflectance difference and simple ratio.

Specifically, foliar Chls is significantly more related to the reflectance of ρ_{645} and ρ_{63} than ρ_{455} and ρ_{426}. But the spectral difference index (RDI_b) and the slope index (SI_b) work much better than each of the four variables. Moreover, the prediction accuracy of the spectral difference index can be further improved by 17 % if it is synergized with the reflectance at red edge position. That is an appropriate predictor and can be derived by $(\rho_{645} - \rho_{455})/\rho_{REP}$, denoted as $RI_{RDI_b/\rho REP}$ in Table 2, for better prediction accuracy of foliar Chls in respect to various foliar water contents.

4.1.4 A comparison with previously developed spectral indices of chlorophyll indicators

Relationships between lnChls and the previous 18 spectral indices in Table 1 were developed using the training samples. The lnChls (y in Table 4) is mostly linearly and positively or negatively related to the indices, while an exponential decay relationship is observed between the lnChls and the indices TCI and SR_{775}. R^2 values for those models are mostly greater than 0.90, only the model with $NDVI_{680}$

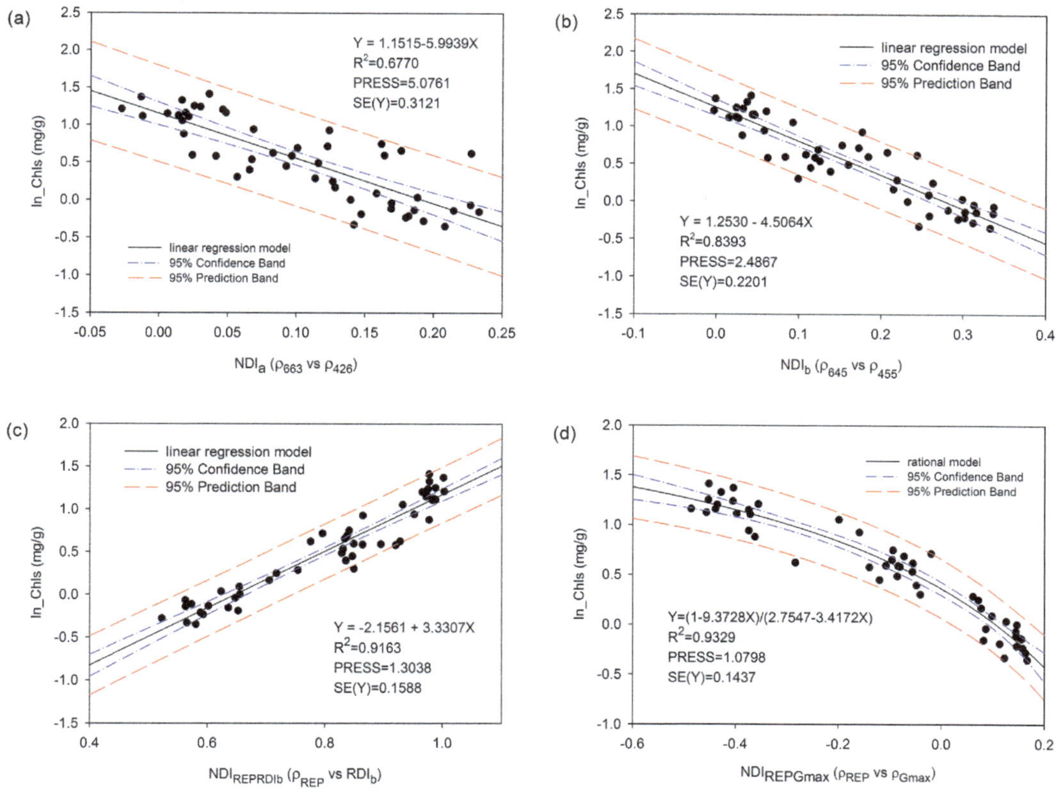

Figure 7. Normalized-difference-index-based empirical models for foliar Chls concentration estimation.

has an R^2 of 0.60, showing a relatively poor model adequacy. Chls prediction accuracy of those models is between RMSE 0.87–4.57 mg g^{-1} or MAE 39–177 % for both the fresh and water-stressed leaves (Table 5). The best accuracy was achieved by the indicator $CI_{red\,edge}$ with an RMSE of 0.72 and 1.00 mg g^{-1} and an MAE of 27 and 53 % for the fresh and water-stressed validation samples. The foliar Chls is linearly and positively related to $CI_{red\,edge}$, as shown in Gitelson et al. (2009).

Although Ollinger (2011) suggested that the near-infrared region is the most important to vegetation remote sensing, a spectral index that combines the reflectance of near-infrared and red edge wavelengths was not able to achieve the same accuracy level of the predictor $RI_{RDI_b/\lambda_{REP}}$. Specifically, taking the average of the prediction accuracy of both fresh and water-stressed samples, the indicator $CI_{red\,edge}$ could achieve an accuracy of RMSE $=0.87$ mg g^{-1} and MAE $=39$ %. That is almost 2.5 times the average accuracy (RMSE $=0.36$ mg g^{-1} and MAE $=14$ %) indicated by $RI_{RDI_b/\rho_{REP}}$. This is probably due to the structure of the foliar mesophyll which has changed when foliar water deficit happened. For example, Wuyts (2012) found that leaf thickness is conserved in response to water deficit under both high and low cumulative light regimes while mesophyll cells change in volume and shape. The change of the near-infrared reflectance is more complicated. Additional studies might be needed to explain the behavior of infrared reflectance in the future.

4.2 Evaluation of the absorptance-based Chls model (the SPAD-502 method)

4.2.1 Nonlinearity relationship between ln_transformed chlorophyll and SPAD readings

SPAD readings and total chlorophyll concentration relationship was explored from an independent experiment of 45 fresh leaf samples of *Cinnamomum camphora*. A three-parameter rational function, $Y=(1+aX)/(b+cX)$ was most appropriate for presenting the relationship of SPAD reading (X) and the lnChls (Y) based on the ANOVA F test of the fitted model and the t test of the model's parameters. Fig. 8 showed that SPAD readings are nonlinearly dependent on the natural-log-transformed Chls. Specifically, the coefficients a, b, and c of this fitted model were further t tested to be significant at the 0.01 level. In total, 95.77 percent of the variance of lnChls could be explained by the SPAD chlorophyll rational model. Compared with the measured value determined by the acetone method, the fitted rational model has an average accuracy of 0.22 mg g^{-1} RMSE and 15 % MAE that differ from the measured chlorophyll content for the training data set.

Table 4. Narrow-band-based spectral indices developed as chlorophyll indicators.

Indicator	Chls estimation models (y is lnChls)	R^2	PRESS	SE(Y)
Vog1	$y = -4.0302 + 3.4801\,\mathrm{Vog1}$	0.94	0.95	0.13
Vog2	$y = -0.4049 - 19.4238\,\mathrm{Vog2}$	0.94	0.95	0.13
Vog3	$y = -0.3766 - 17.3277\,\mathrm{Vog3}$	0.94	0.94	0.13
NDVI_{705}	$y = -0.8710 + 3.6844\,\mathrm{NDVI}_{705}$	0.92	1.21	0.15
NDVI_{700}	$y = -1.0487 + 3.1896\,\mathrm{NDVI}_{700}$	0.91	1.46	0.17
SR_{700}	$y = -0.8609 + 0.4175\,\mathrm{SR}_{700}$	0.94	1.00	0.14
wSR	$y = -0.3406 + 5.5226\,\mathrm{wSR}$	0.91	1.47	0.17
NDVI_{680}	$y = -5.5313 + 7.5436\,\mathrm{NDVI}_{680}$	0.60	6.29	0.35
MCARI	$y = 1.3432 - 0.0192\,\mathrm{MCARI}$	0.82	2.86	0.24
mSR	$y = -0.9494 + 0.5288\,\mathrm{mSR}$	0.94	0.91	0.13
mNDVI	$y = -0.8980 + 3.2961\,\mathrm{mNDVI}$	0.92	1.31	0.16
MTCI	$y = -0.4718 + 0.9146\,\mathrm{MTCI}$	0.94	1.01	0.14
rSR_{705}	$y = -0.4432 + 0.6887\,\mathrm{rSR}_{705}$	0.94	0.91	0.13
TCI	$y = 3.7126 \cdot \exp(-0.0743\,\mathrm{TCI})$	0.88	1.94	0.19
SR_{775}	$y = 22.5416 \cdot \exp(-8.1018\,\mathrm{SR}_{775})$	0.90	1.56	0.17
NDVI_{712}	$y = -0.6772 + 4.7009\,\mathrm{NDVI}_{712}$	0.93	1.15	0.15
TGI	$y = 1.7411 - 0.0011\,\mathrm{TGI}$	0.93	1.11	0.15
SR_{760}	$y = -0.3563 + 3.4783 \cdot \exp(-6.6687\,\mathrm{SR}_{760})$	0.91	1.50	0.17
SR_{420}	$y = 1.8611 - 0.4905\,\mathrm{SR}_{420}$	0.66	5.24	0.32
$\mathrm{CI}_{\mathrm{red\ edge}}$	$y = -0.5205 + 0.4737\,\mathrm{CI}_{\mathrm{red\ edge}}$	0.93	1.05	0.14

4.2.2 Limitation of the SPAD reading-based rational model

According to the experiment, we found that the SPAD reading increases positively relative to the chlorophyll concentration in fresh leaves. The SPAD reading remained at a high value even when leaves were under serious water stress, that is, the SPAD reading rises to a high level even the chlorophyll content is decreasing. It revealed that the previously mentioned nonlinear rational model is not suitable for those wilting leaves. Figure 9 showed a deficiency of the SPAD reading-based model in the estimation of foliar chlorophyll content. The Chls estimates of fresh leaves (presented as solid black circles) are mostly located under the real Chls line (i.e., under estimation), and the Chls estimates of water-stressed leaves (presented as open circles) are mainly located above the real Chls line (i.e., overestimation). The biases RMSE and MAE of Chls estimation were evaluated to be 0.29 mg g^{-1} and 16 % for the fresh leaves sample and 0.92 mg g^{-1} and 60 % for the water-stressed leaf sample. The fresh leaves sample has biases very close to the values (RMSE 0.22 mg g^{-1} and MAE 15 %) of the original modeling data set, but the water-stressed leaves sample has biases almost 4 times that of the original modeling data set. This finding indicates that the SPAD rational model can achieve a very good and acceptable Chls estimates in case of fresh leaves, while unfortunately it fails if the leaves are under water-stressed conditions. This result leads to the acceptance of the null hypothesis H03 and concludes that the determination of Chls using the chlorophyll meter (absorptance-based

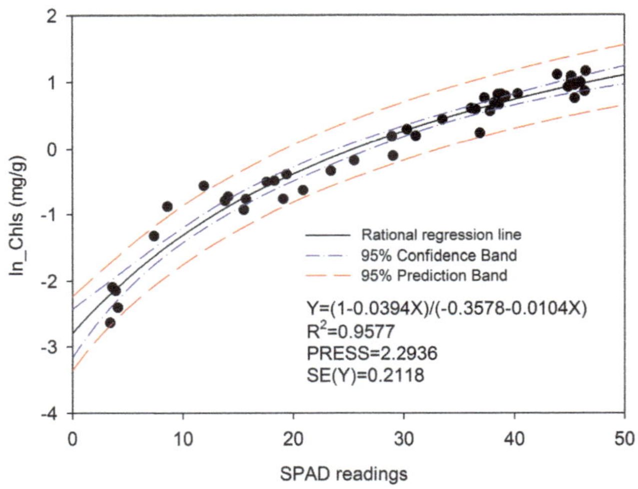

Figure 8. Rational regression model for foliar chlorophyll contents estimation. Number of data points is 45. In the rational model, the independent variable SPAD readings could be used to explain 95.77 % (R^2) variation of the dependent variable, lnChls, the natural-logarithm-transformed foliar chlorophyll content.

model) has a significant bias or uncertainty due to its failure in responding to the influence of water stress.

5 Conclusions

Typical vegetation reflectance is significantly related to foliar biochemistry and biophysical characteristics. A stronger negative relationship exists between the

Table 5. Foliar Chls prediction accuracy of the previously developed spectral indices.

Error index	RMSE ($mg\,g^{-1}$)			MAE (%)		
Chls indicator	Fresh sample	Wilted sample	All	Fresh sample	Wilted sample	All
Vog1	1.91	0.33	1.37	78.35	13.96	46.15
Vog2	3.01	0.41	2.15	111.33	27.99	69.66
Vog3	3.27	0.39	2.33	115.85	25.78	70.81
$NDVI_{705}$	1.43	0.54	1.08	78.32	39.56	58.94
$NDVI_{700}$	1.36	0.86	1.14	82.51	63.74	73.12
SR_{700}	4.28	0.48	3.04	155.11	34.37	94.74
wSR	3.65	2.49	3.12	93.48	170.68	132.08
$NDVI_{680}$	1.27	0.47	0.95	78.15	32.27	55.21
MCARI	1.06	1.55	1.33	71.17	119.34	95.26
mSR	6.44	0.41	4.57	173.85	28.68	101.26
mNDVI	1.46	0.57	1.11	78.08	42.21	60.14
MTCI	2.80	0.34	1.99	95.57	22.05	58.81
rSR_{705}	3.62	0.54	2.59	129.23	39.78	84.50
TCI	2.67	3.67	3.21	85.89	268.02	176.95
SR_{775}	5.98	0.38	4.24	171.87	26.43	99.15
$NDVI_{712}$	1.49	0.54	1.12	72.86	40.55	56.71
TGI	0.57	2.33	1.70	30.13	178.95	104.54
SR_{760}	1.93	0.54	1.41	102.96	38.92	70.94
SR_{420}	0.99	0.83	0.92	63.28	61.48	62.38
$CI_{broad\ band}$	0.71	1.00	0.87	27.10	51.86	39.48

chlorophyll concentration and the visible reflectance while the relationship of the chlorophyll concentration and the infrared reflectance is positive. As water stress occurred, the reflectance over the visible and infrared area will increase. The estimation of chlorophyll concentration using only the remotely sensed reflectance will be seriously affected by the reflectance changes caused by a departure of water content from the non-water-stressed conditions. The significant uncertainty for the estimation of chlorophyll concentration is caused by the reflectance changes induced by variations of the foliar water content. Red edge characteristics, such as position (λ_{REP}) and reflectance (ρ_{REP}), are also sensitive to water stress. Each of them, as a predictor of foliar Chls, has a significant positive linear relationship with foliar chlorophyll concentration. This is similar to the results of Matson et al. (1994) and Belanger et al. (1995).

The ChlsPn variables, such as the reflectance at the wavelengths 663, 645, 455, and 426 nm, are in particular directly related to the light absorption by chlorophyll *a* and chlorophyll *b* and therefore can characterize the foliar chlorophyll concentration. The relationship between Chls and each of the variables ρ_{426}, ρ_{455}, ρ_{645}, and ρ_{663} is statistically significant, but is still not good enough to be used alone for Chls estimation. The best adequacy (R^2) of the four reflectance-based ChlsPn models using one of the ChlsPn variables as the predictor is 0.77, meanwhile the best average accuracy achieved is MAE = 54 % and RMSE = 0.78 $mg\,g^{-1}$ for both fresh and water-stressed leaves. Spectral indices derived from ChlsPn variables by the methods of normalized difference, simple difference, slope transformation, and simple ratio can effectively improve the estimation accuracy of the reflectance-based Chls spectral index models. The better accuracy is the model using the slope index SI_b with MAE = 17 % and RMSE = 0.44 $mg\,g^{-1}$ or the difference index RDI_b with MAE = 18 % and RMSE=0.46 $mg\,g^{-1}$. By integrating the reflectance at the red edge position, the

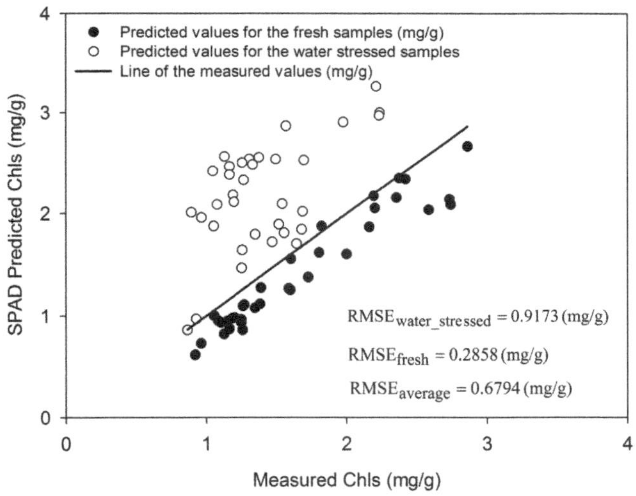

Figure 9. Evaluation of the prediction accuracy of the SPAD chlorophyll regression model. Number of data points is 70. For testing fresh samples, the estimation error is 0.2858 $mg\,g^{-1}$; for testing samples which are water stressed, the error is 0.9173 $mg\,g^{-1}$. Average error is 0.6794 $mg\,g^{-1}$ for all of the testing data points.

difference-based simple ratio index $(\rho_{645} - \rho_{455})/\rho_{\mathrm{REP}}$ can achieve the best accuracy of the Chls of fresh and water-stressed leaves. The MAE and RMSE are further decreased down to 14 % and 0.36 mg g^{-1}, respectively.

Plant growth and productivity are mostly affected by water shortage. This stress condition induces plant cell dehydration and then causes the decreased chlorophyll in older leaves. Since the influence of water stress on foliar spectral reflectance could be effectively reduced by the reflectance at the red edge and the wavelengths of 645 and 455 nm, we recommend the following three spectral indices as an effective chlorophyll indicator (ECI) for dealing with the potential influence of foliar water deficit for applications. The first predictor is the difference-based red edge reflectance ratio index ECI1 $= (\rho_{645} - \rho_{455})/\rho_{\mathrm{REP}}$, then the slope index ECI2 $= (\rho_{645} - \rho_{455})/(\lambda_{645} - \lambda_{455})$, and finally the difference-based red edge position ratio index ECI3 $= (\rho_{645} - \rho_{455})/\lambda_{\mathrm{REP}}$. ECI1 is negatively and linearly related to chlorophyll concentration, while ECI2 and ECI3 are exponential and negatively related to the natural-log-transformed foliar chlorophyll concentration. A temporal and spatial estimation of the chlorophyll content for the terrestrial ecosystems could be retrieved more feasibly and accurately using these effective chlorophyll indicators.

Acknowledgements. The authors would like to acknowledge the support provided by project NSC 102-2119-M-415-001 funded by the National Science Council, Taiwan.

Edited by: F. Chai

References

Arnon, D. I.: Copper enzymes in isolated chloroplasts. Polyphenoloxidase in *Beta vulgaris*, Plant Physiol., 24, 1–15, 1949.

Blackburn, G. A.: Quantifying chlorophyll and caroteniods at leaf and canopy scales: an evaluation of some hyperspectral approaches, Remote Sens. Environ., 66, 273–285, 1998.

Blackburn, G. A.: Hyperspectral remote sensing of plant pigments, J. Exp. Bot., 58, 855–867, 2007.

Blackburn, G. A. and Ferwerda, J. G.: Retrieval of chlorophyll concentration from leaf reflectance spectra using wavelet analysis, Remote Sens. Environ., 112, 1614–1632, 2008.

Blackmer, T. M. and Schepers, J. S.: Use of a chlorophyll meter to monitor nitrogen status and schedule fertigation for corn, J. Prod. Agric., 8, 56–60, 1995.

Belanger, M. J., Miller, J. R., and Boyer, M. G.: Comparative relationships between some red edge parameters and seasonal leaf chlorophyll concentrations, Can. J. Remote Sens., 21, 16–21, 1995.

Boegh, E., Houborg, R., Bienkowski, J., Braban, C. F., Dalgaard, T., van Dijk, N., Dragosits, U., Holmes, E., Magliulo, V., Schelde, K., Di Tommasi, P., Vitale, L., Theobald, M. R., Cellier, P., and Sutton, M. A.: Remote sensing of LAI, chlorophyll and leaf nitrogen pools of crop- and grasslands in five European landscapes, Biogeosciences, 10, 6279–6307, doi:10.5194/bg-10-6279-2013, 2013.

Carter, G. A.: Primary and secondary effects of water content on the spectral reflectance of leaves, Am. J. Bot., 78, 916–924, 1991.

Carter, G. A.: Responses of leaf spectral reflectance to plant stress, Am. J. Bot., 80, 239–243, 1993.

Carter, G. A.: Ratios of leaf reflectances in narrow wavebands as indicators of plant stress, Int. J. Remote Sens., 15, 697–703, 1994.

Carter, G. A. and Knapp, A. K.: Leaf optical properties in higher plants, Linking spectral characteristics to stress and chlorophyll concentration, Am. J. Bot., 88, 677–684, 2001.

Cate, T. M. and Perkins, T. D.: Chlorophyll content monitoring in sugar maple (*Acer saccharum*), Tree Physiol., 23, 1077–1079, 2003.

Ceccato, P., Flasse, S., Tarantola, S., Jacquemoud, S., and Gregoire, J. M.: Detecting vegetation leaf water content using reflectance in the optical domain, Remote Sens. Environ., 77, 22–33, 2001.

Cheng, T., Rivard, B., and Sanchez-Azofeifa, A.: Spectroscopic determination of leaf water content using continuous wavelet analysis, Remote Sens. Environ., 115, 659–670, 2011.

Curran, P. J., Dungan, J. L., and Gholz, H. L.: Exploring the relationship between reflectance red edge and chlorophyll content in slash pine, Tree Physiol., 7, 33–48, 1990.

Curran, P. J., Dungan, J. L., Macler, B. A., and Plummer, S. E.: The effect of a red leaf pigment on the relationship between red edge and chlorophyll concentration, Remote Sens. Environ., 35, 69–76, 1991.

Curtiss, B. and Goetz, A. F. H.: Field Spectrometry: Techniques and Instrumentation, Proceedings of ISSSR 1994, 10–15 July, San Diego, California, 1, 195–203, 1994.

Dash, J. and Curran, P J.: The MERIS terrestrial chlorophyll index, Int. J. Remote Sens., 25, 5403–5413, 2004.

Datt, B.: Remote sensing of chlorophyll a, chlorophyll b, chlorophyll $a+b$, and total carotenoid content in eucalyptus leaves, Remote Sens. Environ., 66, 111–121, 1998.

Daughtry, C. S. T., Walthall, C. L., Kim, M. S., de Colstoun, E. B., and McMurtrey III, J. E.: Estimating corn leaf chlorophyll concentration from leaf and canopy reflectance, Remote Sens. Environ., 74, 229–239, 2000.

Dawson, T. P., Curran, J. P., and Plummer, S. E.: LIBERTY – Modeling the effects of leaf biochemical concentration on reflectance spectra, Remote Sens. Environ., 65, 50–60, 1998.

Desotgiu, R., Pollastrini, M., Cascio, C., Gerosa, G., Marzuoli, R., and Bussotti, F.: Chlorophyll a fluorescence analysis along a vertical gradient of the crown in a poplar (Oxford clone) subjected to ozone and water stress, Tree Physiol., 32, 976–986, 2012.

Elvidge, C. D., and Chen, Z.: Comparison of Broad-band and Narrow-band Red and Near-Infrared Vegetation Indices, Remote Sens. Environ., 54, 38–48, 1995.

Emerson, R. and Lewis, C. M.: The dependence of the quantum yield of Chlorella photosynthesis on wavelength of light, Am. J. Bot., 30, 165–178, 1943.

Estep, L. and Carter, G. A.: Derivative Analysis of AVIRIS Data for Crop Stress Detection, Photogram. Eng. Rem. S., 71, 1417–1421, 2005.

Féret, J. B., Francois, C., Gitelson, A., Asner, G. P., Barry, K. M., Panigada, C., Richardson, A. D., and Jacquemoud, S.: Optimizing spectral indices and chemometric analysis of leaf chemical properties using radiative transfer modeling, Remote Sens. Environ., 115, 2742–2750, 2011.

Filella, I. and Peñuelas, J.: The red edge position and shape as indicative plant chlorophyll content, biomass and hydric states, Int. J. Remote Sens., 15, 1459–1470, 1994.

Gausman, H. W. and Allen, W. A.: Optical parameters of leaves of 30 plant species, Plant Physiol., 52, 57–62, 1973.

Gausman, H. W., Escobar, D. E., and Rodriguez, R. R.: Reflectance measurements of cotton leaf senescence altered by mepiquat chloride, USDA/Agricultural Research Service, Remote Sensing Research Unit, Weslaco, Texas, 1982.

Ghorbanli, M., Gafarabad, M., Amirkian, T., and Allahverdi Mamaghani, B.: Investigation of proline, total protein, chlorophyll, ascorbate and dehydroascorbate changes under drought stress in Akria and Mobil tomato cultivars, Iran. J. Plant Physiol., 3, 651–658, 2013.

Gitelson, A. A. and Merzlyak, M. N.: Spectral reflectance changes associated with autumn senescence of *Aesculus hippocastanum* L. and *Acer platanoides* L. leaves. Spectral features and relation to chlorophyll estimation, J. Plant Physiol., 143, 286–292, 1994a.

Gitelson, A. A. and Merzlyak, M. N.: Quantitative estimation of chlorophyll-a using reflectance spectra-experiments with autumn chestnut and maple leaves, J. Photochem. Photobiol., B, 22, 247–252, 1994b.

Gitelson, A. A. and Merzlyak, M. N.: Signature analysis of leaf reflectance spectra: algorithm development for remote sensing of chlorophyll, J. Plant Physiol., 148, 494–500, 1996.

Gitelson, A. A., and Merzlyak, M. N.: Remote estimation of chlorophyll content in higher plant leaves, Int. J. Remote Sens., 18, 2691-2697, 1997.

Gitelson, A. A., Gritz, U., and Merzlyak, M. N.: Relationships between leaf chlorophyll content and spectral reflectance and algorithms for non-destructive chlorophyll assessment in higher plant leaves, J. Plant Physiol., 160, 271–282, 2003.

Gitelson, A. A., Keydan, G. P., and Merzlyak, M. N.: Three-band model for noninvasive estimation of chlorophyll, carotenoids, and anthocyanin contents in higher plant leaves, Geophys. Res. Lett, 33, L11402, doi:10.1029/2006GL026457, 2006.

Gitelson, A. A., Chivkunova, O. B., and Nerzlyak, M. N.: Nondestructive estimation of anthocyanins and chlorophylls in anthocyanic leaves, Am. J. Bot., 96, 1861–1868, 2009.

Gond, V., de Pury, D. G. G., Veroustraete, F., and Ceulemans, R.: Seasonal variations in leaf area index, leaf chlorophyll, and water content; scaling-up to estimate fAPAR and carbon balance in a multilayer, multispecies temperate forest, Tree Physiol., 19, 673–679, 1999.

Haboudane, D., Tremblay, N., Miller, J. R., and Vigneault, P.: Remote estimation of crop chlorophyll content using spectral indices derived from hyperspectral data, IEEE Trans. Geosci. Remote Sens., 46, 423–437, 2008. Hatchell, D. C.: Analytical Spectral Devices. 4th edition, Technical Guide, ASD Inc, 1999.

Hawkins, T. S., Gardiner, E. S., and Comer, G. S.: Modeling the relationship between extractable chlorophyll and SPAD-502 readings for endangered plant species research, J. Nat. Conserv., 17, 125–129, 2009.

Hopkins, W. G. and Hüner, N. P. A.: Introduction to Plant Physiology, 3rd ed, John Wiley & Sons, NJ, 2004.

Horler, D. N. H., Dockray, M., and Barber, J.: The red edge of plant leaf reflectance, Int. J. Remote Sens., 4, 273–288, 1983.

Hunt, E. R., Daughtry, C. S. T., Eitel, J. U. H., and Long, D. S.: Remote sensing leaf chlorophyll content using a visible band index, Agron. J., 103, 1090–1099, 2011.

Hunt, E. R., Doraiswamy, P. C., McMurtrey, J. E., Daughtry, C. S. T., Perry, E. M., and Akhmedov, B.: A visible band index for remote sensing leaf chlorophyll content at the canopy scale, Int. J. Appl. Earth. Obs., 103, 1090–1099, 2013.

Hunt, E. R. and Rock, B. N.: Detection of changes in leaf water content using near- and middle-infrared reflectances, Remote Sens. Environ., 30, 43–54, 1989.

Jongschaap, R. E. E. and Booij, R.: Spectral measurements at different spatial scales in potato: relating leaf, plant and canopy nitrogen status, Int. J. Appl. Earth Obs., 5, 204–218, 2004.

Kaoau, M. E., Hsissou, D., and Belhadri, A.: Water Deprivation Effect on Pigments and Proline Content, and Growth in Cultivated Wheat Varieties in Morocco, Afr. Crop Sci. J., 15, 139–147, 2007.

Keenan, T., García, R., Friend, A. D., Zaehle, S., Gracia, C., and Sabate, S.: Improved understanding of drought controls on seasonal variation in Mediterranean forest canopy CO_2 and water fluxes through combined in situ measurements and ecosystem modelling, Biogeosciences, 6, 1423–1444, doi:10.5194/bg-6-1423-2009, 2009.

Kirnak, H., Kaya, C., Tas, I., and Higgs, D.: The influence of water deficit on vegetative growth, physiology, fruit yield and quality in eggplants, Bulg. J. Plant Physiol., 27, 34–46, 2001.

Knipling, E. B.: Physical and physiological basis for the reflectance of visible and near-infrared radiation from vegetation, Remote Sens. Environ., 1, 155–159, 1970.

Lavoir, A.-V., Staudt, M., Schnitzler, J. P., Landais, D., Massol, F., Rocheteau, A., Rodriguez, R., Zimmer, I., and Rambal, S.: Drought reduced monoterpene emissions from the evergreen Mediterranean oak Quercus ilex: results from a throughfall displacement experiment, Biogeosciences, 6, 1167–1180, doi:10.5194/bg-6-1167-2009, 2009.

Lin, C., Tsogt, K., and Chang, C. I.: An empirical model-based method for signal restoration of SWIR in ASD field spectroradiometry, Photogramm. Eng. Rem. S., 78, 119–127, 2012.

Ma, B. L., Morrison, M. J., and Voldeng, H. D.: Leaf greenness and photosynthetic rates in soybean, Crop Sci., 35, 1411–1414, 1995.

Matson, P., Johnson, L., Billow, C., Miller, J., and Pu, R.: Seasonal patterns and remote spectral estimation of canopy chemistry across the Oregon transect, Ecol. Appl., 4, 280–298, 1994.

Montagu, K. D. and Woo, K. C.: Recovery of tree photosynthetic capacity from seasonal drought in the wet-dry tropicsw: The role of phyllode and canopy processes in *Acacia auriculiformis*, Aust. J. plant Physiol., 26, 135–145, 1999.

Mutanga, O. and Skidmore, A. K.: Red edge shift and biochemical content in grass canopies, ISPRS J Photogramm. Rem. S., 62, 34–42, 2007.

Nave, L. E., Gough, C. M., Maurer, K. D., Bohrer, G., Hardiman, B. S., Le Moine, J., Munoz, A. B., Nadelhoffer, K. J., Sparks, J. P., Strahm, B. D., Vogel, C. S., and Curtis, P. S.: Disturbance and the resilience of coupled carbon and nitrogen cycling in a north temperate forest, J. Geophys. Res., 116, G04016, doi:10.1029/2011JG001758, 2011.

Nilsen, E. T. and Orcutt, D. M.: Physiology of plants under stress, Abiotic Factors, 2nd Edn. John Wiley & Sons Inc., NY, 1996.

Nobel, P. S.: Physicochemical and environmental plant physiology, 3rd ed, Academic Elsevier, Burlingtom, MA, 2005.

Novo, E., Gastil, M., and Melack, J.: An algorithm for chlorophyll using first difference transformations of AVIRIS reflectance spectra, Summaries of the Fifth Annual JPL Airborne Earth Science Workshop, NASA Jet Propulsion Laboratory publication 95-1, Pasadena, CA, 121–124, 1995.

Oki, K.: Why is the Ratio of Reflectivity Effective for Chlorophyll Estimation in the Lake Water?, Remote Sens., 2, 1722–1730, 2010.

Ollinger, S. V.: Sources of variability in canopy reflectance and the convergent properties of plants, New Phytol, 189, 375–394, 2011.

Pinar, A. and Curran, P. J.:, Grass chlorophyll and reflectance red edge, Int. J. Remote Sens., 17, 351–357, 1996.

Pirzad, A., Shakiba, M. R., Zehtab-Salmasi, S., Mohammadi, S. A., Darvishzadeh, R., and Samadi, A.: Effect of water stress on leaf relative water content, chlorophyll, praline and soluble carbohydrates in *Matricaria chamomilla* L., J. Med. Plants Res., 5, 2483–2488, 2011.

Pu, R., Ge, S., Kelly, N. M., and Gong, P.: Spectral absorption features as indicators of water status in coast live oak (*Quercus agrifolia*) leaves, Int. J. Remote Sens., 24, 1799–1810, 2003.

Rascher, U., Agati, G., Alonso, L., Cecchi, G., Champagne, S., Colombo, R., Damm, A., Daumard, F., de Miguel, E., Fernandez, G., Franch, B., Franke, J., Gerbig, C., Gioli, B., Gomeź, J. A., Goulas, Y., Guanter, L., Gutiérrez-de-la-Cámará, O., Hamdi, K., Hostert, P., Jimeneź, M., Kosvancova, M., Lognoli, D., Meroni, M., Miglietta, F., Moersch, A., Moreno, J., Moya, I., Neininger, B., Okujeni, A., Ounis, A., Palombi, L., Raimondi, V., Schickling, A., Sobrino, J. A., Stellmes, M., Toci, G., Toscano, P., Udelhoven, T., van der Linden, S., and Zaldei, A.: CEFLES2: the remote sensing component to quantify photosynthetic efficiency from the leaf to the region by measuring sun-induced fluorescence in the oxygen absorption bands, Biogeosciences, 6, 1181–1198, doi:10.5194/bg-6-1181-2009, 2009.

Read, J. J., Whaley, E. L., Tarpley, L., and Reddy, R.: Evaluation of a Hand-held Radiometer for Field Determination of Nitrogen Status in Cotton, in: Digital Imaging and Spectral Techniques: Applications to Precision Agriculture and Crop Physiology, edited by: Schepers J. and T. von Toai, ASA, Madison, Wis, 171-189, 2003.

Reddy, K. R. and Matcha, S. K.: Remote sensing algorithms for castor bean nitrogen and pigment assessment for fertility management, Ind. Crop. Prod., 32, 411–419, 2010.

Rossini, M., Cogliati, S., Meroni, M., Migliavacca, M., Galvagno, M., Busetto, L., Cremonese, E., Julitta, T., Siniscalco, C., Morra di Cella, U., and Colombo, R.: Remote sensing-based estimation of gross primary production in a subalpine grassland, Biogeosciences, 9, 2565–2584, doi:10.5194/bg-9-2565-2012, 2012.

Rowland, D., Blankenship, P., Puppala, N., Beasley, J., Burow, M., Gorbet, D., Jordan, D., Melouk, H., Simpson, C., and Bostick, J.: Variation in water-use efficiency of peanut varieties across peanut production regions, http://www.cropscience.org.au/icsc2004/poster/1/3/2/288_rowlanddl.htm (Last access: November 10, 2012), 2004.

Schelbert, S., Aubry, S., Burla, B., Agne, B., Kessler, F., Krupinska, K., and Hörtensteiner, S.: Pheophytin pheophorbide hydrolase (pheophytinase) is involved in chlorophyll breakdown during leaf senescence in Arabidopsis, Plant Cell, 21, 767–785, 2009.

Sims, D. A. and Gamon, J. A.: Relationships between leaf pigment content and spectral reflectance across a wide range of species, leaf structures and development stages, Remote Sens. Environ., 81, 337–354, 2002.

Takebe, M., Yoneyama, T., Inada, K., and Murakami, T.: Spectral reflectance ratio of rice canopy for estimating crop nitrogen status, Plant Soil, 122, 295–297, 1990.

Tan, S. C., Shi, G. Y., Shi, J. H., Gao, H. W., and Yao, X.: Correlation of Asian dust with chlorophyll and primary productivity in the coastal seas of China during the period from 1998 to 2008, J. Geophys. Res., 116, G02029, doi:10.1029/2010JG001456, 2011.

Ustin, S. L., Gitelson, A. A., Jacquemoud, S., Schaepman, M., Asner, G. P., Gamon, J. A., and Zarco-Tejada, P.: Retrieval of foliar information about plant pigment systems from high resolution spectroscopy, Remote Sens. Environ., 113, S67–S77, 2009.

Vogelmann, J. E., Rock, B. N., and Moss, D. M.: Red edge spectral measurements from sugar maple leaves. Int. J. Remote sens., 14, 1563–1575, 1993.

Weber, U., Jung, M., Reichstein, M., Beer, C., Braakhekke, M. C., Lehsten, V., Ghent, D., Kaduk, J., Viovy, N., Ciais, P., Gobron, N., and Rödenbeck C.: The interannual variability of Africa's ecosystem productivity: a multi-model analysis, Biogeosciences, 6, 285–295, doi:10.5194/bg-6-285-2009, 2009.

Wellburn, A. R.: The spectral determination of chlorophylls a and b, as well as total carotenoids, using various solvents with spectrophotometers of different resolution, J. Plant Physiol., 144, 307–313, 1994.

Whittaker, R. H. and Marks, P. L.: Methods of assessing terrestrial productivity, in Primary Productivity of the Biosphere, edited by: Lieth, H., and Whittaker, R. H., Springer-Verlag, NY, 55–118, 1975,

Wuyts, N., Massonnet, C., Dauzat, M., and Granier, C.: Structural assessment of the impact of environmental constraints on Arabidopsis thaliana leaf growth: a 3-D approach, Plant Cell Environ., 35, 1631–1646, 2012.

Zygielbaum, A. I., Gitelson, A. A., Arkebauer, T. J., and Rundquist, D. C.: Non-destructive detection of water stress and estimation of relative water content in maize, Geophys. Res. Lett., 36, L12403, doi:10.1029/2009GL038906, 2009.

2

Ocean acidification accelerates dissolution of experimental coral reef communities

S. Comeau, R. C. Carpenter, C. A. Lantz, and P. J. Edmunds

Department of Biology, California State University, 18111 Nordhoff Street, Northridge, CA 91330-8303, USA

Correspondence to: S. Comeau (steve.comeau@csun.edu)

Abstract. Ocean acidification (OA) poses a severe threat to tropical coral reefs, yet much of what is know about these effects comes from individual corals and algae incubated in isolation under high $p\mathrm{CO_2}$. Studies of similar effects on coral reef communities are scarce. To investigate the response of coral reef communities to OA, we used large outdoor flumes in which communities composed of calcified algae, corals, and sediment were combined to match the percentage cover of benthic communities in the shallow back reef of Moorea, French Polynesia. Reef communities in the flumes were exposed to ambient ($\sim 400\,\mu$atm) and high $p\mathrm{CO_2}$ ($\sim 1300\,\mu$atm) for 8 weeks, and calcification rates measured for the constructed communities including the sediments. Community calcification was reduced by 59 % under high $p\mathrm{CO_2}$, with sediment dissolution explaining ~ 50 % of this decrease; net calcification of corals and calcified algae remained positive but was reduced by 29 % under elevated $p\mathrm{CO_2}$. These results show that, despite the capacity of coral reef calcifiers to maintain positive net accretion of calcium carbonate under OA conditions, reef communities might transition to net dissolution as $p\mathrm{CO_2}$ increases, particularly at night, due to enhanced sediment dissolution.

1 Introduction

The calcium carbonate framework produced by coral reefs hosts the highest known marine biodiversity and protects tropical shores from wave erosion (Ferrario et al., 2014). However, in recent decades coral reefs have been impacted by a diversity of disturbances, and now they are threatened by an increase in seawater temperature and ocean acidification (OA) (Hoegh-Guldberg et al., 2007; Kleypas and Yates, 2009). OA is caused by the dissolution of atmospheric $\mathrm{CO_2}$ in seawater, which reduces pH, depresses carbonate ion concentration, and increases bicarbonate ion concentration with no change in total alkalinity (Feely et al., 2004). The net effects of OA on coral reefs remain unclear as most studies show a decrease in organismic calcification under OA conditions (Erez et al., 2011; Chan and Connolly, 2013), while recent laboratory work describes species-specific responses with some corals and calcifying algae resistant to decreasing pH (Comeau et al., 2013; Takahashi and Kurihara, 2013). Differential organismic sensitivities to OA potentially could lead to changes in coral community structure, and in turn this could affect habitat complexity (Fabricius et al., 2011, 2014).

Critically, most of the studies on coral reef organisms have been performed on individuals maintained in isolation in laboratory conditions, and studies performed at the scale of whole communities are scarce (Leclercq et al., 2002; Jokiel et al., 2008; Andersson et al., 2009; Dove et al., 2013). Generally there are three complementary approaches for studying the responses of coral reef communities to OA: firstly, in situ observations of communities living in naturally acidified water (Fabricius et al., 2011) due to volcanic activities or local conditions (Shamberger et al., 2014); secondly, carbonate chemistry can be manipulated directly in situ (Kline et al., 2012), although this approach is challenging technically and has not yet been used to study intact communities; and thirdly, reef communities can be constructed ex situ (Andersson et al., 2009; Dove et al., 2013) to allow precise control of the physical parameters predicted under future OA conditions. For our experiment, we chose to construct ex situ communities and used, for the first time, large outdoor flumes (after Atkinson and Bilger, 1992) to investigate the effects of OA on coral reef communities.

In addition to corals and macroalgae, it is important to incorporate sediments in OA experiments, as this component of reef ecosystems may be sensitive to decreasing pH (Cyronak et al., 2013a, b; Andersson et al., 2009). Dissolution occurs on coral reefs in sediment pore-waters, or in particular microenvironments where pCO_2 is elevated due to biological activity (Andersson and Gledhill, 2013). Observations in Bermuda have shown that the dissolution of Mg-calcite sediments occurs in a location with seawater pCO_2 naturally elevated to values expected by the end of the century (Andersson et al., 2007). Further, in situ manipulations show that elevated pCO_2 ($\sim 800\,\mu$atm) can transition the calcification budget of coral reef sediments from net precipitation to net dissolution (Cyronak et al., 2013a). Increasing pCO_2 likely will lead to increasing dissolution and decreased precipitation of calcium carbonate, resulting in coral reef community calcification changing from net precipitation to net dissolution (Yates and Halley, 2006; Silvermann et al., 2009; Andersson et al., 2009). Given the aforementioned results that highlight the importance of sediments in the community calcification of entire coral reefs, we included reef carbonate sediments into the constructed communities.

We investigated the response of constructed reef communities in flumes to OA filled with seawater maintained either at ambient pCO_2 (i.e., $\sim 400\,\mu$atm) or elevated pCO_2. Net calcification rates were measured at three levels of biological function: whole community, sediments, and macrocalcifiers to determine the sensitivity to OA of each compartment of the community.

2 Materials and methods

2.1 Collection and sample preparation

This study was carried out in August–October 2013 in Moorea, French Polynesia, using organisms collected from the back reef of the north shore at $\sim$ 1–2 m depth. The organisms used to construct communities in outdoor flumes were assembled to match the contemporary (in 2013) mean cover of a back reef in Moorea (Carpenter, 2014; Edmunds, 2014). Coral communities were built from the four dominant coral taxa found on the back reefs of Moorea: massive *Porites* spp. (11 % cover), *Porites rus* (6 %), *Montipora* spp. (3 %), and *Pocillopora* spp. (2 %), which together accounted for 98 % of the coral cover in this habitat. In addition to corals, 6 % of the planar floor surface of the flumes was covered by crustose coralline algae (66 % *Porolithon onkodes* and 33 % *Lithophyllum flavescens*), and 5 % by rubble (dead coral skeletons). After collection of corals and algae (all $\sim 10 \times 10$ cm), they were returned to the Richard B. Gump South Pacific Research Station and attached to plastic supports using epoxy glue. Following preparation, samples were left to recover in a seawater table for 3 days.

Sediments were collected from the lagoon on the north shore, $\sim$200 m from the reef crest, at 2 m depth using 24 custom-made boxes ($0.4 \times 0.3 \times 0.3$ m). Sediment boxes were inserted into the sediment and left in situ for 4 days to allow chemical stratification in the sediment to re-establish (note that chemical stratification was not monitored) before transferring the boxes to the flumes.

The four outdoor flumes consisted of a working section measuring $5.0 \times 0.3 \times 0.3$ m. Water was re-circulated using water pumps (W. Lim Wave II 373 J s^{-1}) to obtain a 10 cm s^{-1} flow. Flow was measured across the working section of the flume using a Nortek Vectrino Acoustic Doppler Velocimeter. At each end of the flume, seawater passed through an 88 cm long transition section (circular to rectangular) that housed 20 cm (length) flow straighteners made of stacked, 3 cm (inner diameter) PVC pipe, and then into a 12.5 cm (inner diameter) return section. Fresh sand-filtered seawater, pumped from Cook's Bay at 12 m depth, was dispensed continuously into the flume at 5 L min^{-1}. Flumes experienced natural sunlight that was attenuated using fiberglass screens to maintain irradiances similar to ambient irradiances in the back reefs of Moorea (daily maximum of $\sim 1500\,\mu$mol photons m^{-2} cm^{-1} over the incubation period determined with a 4π quantum sensor LI-193 and a LI-COR LI-1400 meter). Temperature in the flumes was maintained at ~ 27 °C to match the ambient temperature in the back reef of Moorea in September–October.

2.2 Carbonate chemistry control and measurements

As the pCO_2 in seawater flowing over the back reef of Moorea is close to open-ocean and current atmospheric values (e.g., Comeau et al., 2014a), pCO_2 levels for the incubations were chosen to match ambient pCO_2 ($\sim 400\,\mu$atm) and the pCO_2 expected in the atmosphere by the end of the present century under a pessimistic scenario of further anthropogenic activity (Representative Concentration Pathway 8.5, $\sim 1300\,\mu$atm; Moss et al., 2010). pCO_2 in the flumes was controlled using a pH-stat (AquaController, Neptune systems, USA) that actuated the bubbling of either pure CO_2 or CO_2-free air into the seawater. To match the natural diel variation in pH in the back reef of Moorea (Hofmann et al., 2011; Comeau et al., 2014a), pH was maintained 0.1 unit lower at night (from 18:00 to 06:00) than during the day.

pH was measured daily using a portable pH meter (Orion 3-stars, Thermo-Scientific, USA) fitted with a DG 115-SC pH probe (Mettler Toledo, Switzerland) calibrated every other day with Tris/HCl buffers (Dickson et al., 2007). pH also was measured spectrophotometrically using m-Cresol dye (Dickson et al., 2007) at regular intervals. pH measured spectrophotometrically or using a pH electrode provided similar results with means differing <0.01 pH unit. Measurement of total alkalinity (A_T) was made using open-cell potentiometric titrations (Dickson et al., 2007) using 50 mL samples of seawater collected every 2–3 days. Titrations

Table 1. Mean carbonate chemistry in the four flumes (F1–4) during the 8-week incubation. The partial pressure of CO_2 (pCO_2), the aragonite saturation state (Ω_{arag}), and the calcite saturation state (Ω_{calc}) were calculated from pH_T, total alkalinity (A_T), temperature, and salinity. The values presented are mean ±SE ($n = 56$). SE for salinity was <0.1.

Flume	Treatment	pH_T	A_T (μmol kg^{-1})	pCO_2 (μatm)	Ω_{arag}	Ω_{calc}	Temperature (°C)	Salinity
F1	High pCO_2	7.603 ± 0.008	2343 ± 1	1329 ± 28	1.60 ± 0.03	2.41 ± 0.04	27.0 ± 0.1	35.9
F2	Ambient	8.010 ± 0.012	2339 ± 1	456 ± 19	3.49 ± 0.07	5.26 ± 0.11	26.8 ± 0.1	35.9
F3	High pCO_2	7.617 ± 0.014	2345 ± 1	1306 ± 42	1.68 ± 0.05	2.53 ± 0.08	27.1 ± 0.1	35.9
F4	Ambient	8.015 ± 0.013	2339 ± 1	451 ± 18	3.53 ± 0.07	5.32 ± 0.11	26.9 ± 0.1	35.9

Figure 1. Photographs of the outdoor flumes. **(a)** The flumes consisted of a 5.00 × 0.30 × 0.30 m working section and a lower sediment chamber (2.50 × 0.30 × 0.25 m) in which sediments were maintained; together they contained ~600 L of seawater. **(b)** Communities matching the average composition (in 2013) of the back reef in Moorea were constructed in the flumes.

of certified reference materials provided by A. G. Dickson (batch 122) yielded A_T values within 3.5 μmol kg^{-1} of the nominal value (SE = 3.1 μmol kg^{-1}; $n = 14$). Parameters of the carbonate system in seawater were calculated using the R package seacarb (Lavigne and Gattuso, 2013).

2.3 Calcification measurements and sediment analysis

Net calcification rates were measured using the total alkalinity anomaly method (Chisholm and Gattuso, 1991), which is based on the stoichiometric relationship of two moles of A_T being removed/added for each mole of $CaCO_3$ precipitated/dissolved. Calcification measurements were made every 7 days on the constructed community, and in the analysis of sediments alone after 7, 30, and 56 days of treatments. During incubations, the addition of seawater was stopped so that each flume operated in a closed loop; seawater samples for A_T then were taken every 3 h during the day and every 6 h at night. To maintain A_T and nutrients close to ambient levels, water in the flumes was refreshed every 3–6 h for 30 min. Regular refreshing limited changes in alkalinity during incubations to <50–100 μmol kg^{-1}, which corresponded to variations in aragonite saturation state (Ω) of <0.1–0.2. Nutrient changes in the flumes were monitored during four incubations, and the changes in nitrate and ammonium during incubations were <2 μmol L^{-1}. To conduct incubations with sediments alone, corals and coralline algae were removed from the flumes for 24 h and held in a separate tank where conditions were identical to those in the flumes. Corals and coralline algal calcification was calculated by subtracting the mean light and dark net calcification of the sediments from the community calcification. For both corals and algae, buoyant weight (Davies, 1989) was recorded before and after the 8-week treatments and converted to dry weight to quantify the contribution of each functional group to the calcification budget. Sediment grain size of each flume was analyzed in triplicate using sediment sieves. Three vertical cross sections of sand (~600 g) were collected from each flume sediment chamber and dried at 60 °C to remove moisture. Sand then was sieved through five separate sediment sieves (149, 420, 840, 3360 μm) yielding six size class fractions for each flume ($n = 3$).

2.4 Statistical analysis

All analyses were performed using R software (R Foundation for Statistical Computing), and assumptions of normality and equality of variance were evaluated through graphical analyses of residuals. Calcification rates were analyzed using a repeated-measures ANOVA in which the within subject factor was time (week), pCO_2 was a fixed effect, and duplicate flumes was a nested effect.

3 Results

3.1 Carbonate chemistry and organism condition

Mean pCO_2 in the four flumes during the 8-week incubation was $456 \pm 21\,\mu$atm and $451 \pm 21\,\mu$atm in the ambient treatments, and $1329 \pm 28\,\mu$atm and $1306 \pm 41\,\mu$atm in the high-pCO_2 treatments ($\pm$SE, $n = 42$). pCO_2 differed between treatments (repeated-measures ANOVA, $F_{1,232} = 734.38$, $p<0.001$), but there was no difference within treatments ($F_{2,232} = 0.16$, $p = 0.852$). Communities were maintained in conditions within the flumes that were supersaturated with respect to aragonite, as $\Omega_{arag} \sim 3.5$ under ambient conditions and ~ 1.6 in the high pCO_2 treatment.

No *Pocillopora* spp. and *Montipora* spp. colonies died during the 8-week treatments, but 10 % of the *Porites* pooled across flumes died by the end of the experiment, regardless of treatment, because of an outbreak of corallivorous nudibranchs (*Phestilla* spp.), which consumed tissue of *Porites* spp. Most coralline algae ($\sim$ 70 %) had died by the end of the incubation, which was likely due to sediment abrasion. No difference in mortality or bleaching was observed between treatments for corals and calcified algae.

3.2 Community

Net calcification was higher at ambient versus high pCO_2 (Fig. 2a), both during the day and night (repeated-measures ANOVA, $F_{1,2} = 84.9$, $p = 0.012$ and $F_{1,2} = 44.9$, $p = 0.022$, respectively); there were no differences between flumes within each treatment, so the nested factor was removed from the final analysis. At night, treatment effects were more striking than during the day, as calcium carbonate dissolution exceeded precipitation at high pCO_2 ($-1.6 \pm 0.9\,gCaCO_3\,m^{-2}\,d^{-1}$), whereas net calcification remained positive at ambient pCO_2 ($2.6 \pm 0.6\,gCaCO_3\,m^{-2}\,d^{-1}$) (both means $\pm$ SE, $n = 16$). Calcification integrated over 24 h highlighted the difference between treatments ($F_{1,2} = 869.2$, $p = 0.001$), with calcification 59 % lower at high pCO_2 than at ambient pCO_2.

3.3 Sediments

Sediment grain sizes in the flumes were similar between flumes and fractionated (by weight) to $5.3 \pm 0.5\,\% < 149\,\mu m$ grain size, $56.5 \pm 1.4\,\% > 149\,\mu m$ and $< 420\,\mu m$, $25.9 \pm 0.4\,\% > 420\,\mu m$ and $< 840\,\mu m$, $10.1 \pm 0.5\,\% > 840\,\mu m$ and $< 3360\,\mu m$, and $2.2 \pm 0.9\,\% > 3360\,\mu m$. Net calcification of the sediments alone differed between treatments, during the day and night ($F_{1,2} = 344.2$, $p = 0.003$ and $F_{1,2} = 282.6$, $p = 0.003$, respectively) (Fig. 2b), but there were no differences between flumes within each treatment; hence the nested factor was removed from the final analysis. Net calcification pooled among treatments was negative during the day ($-0.7 \pm 0.5\,gCaCO_3\,m^{-2}\,d^{-1}$) and night ($-2.5 \pm 0.4\,gCaCO_3\,m^{-2}\,d^{-1}$) at high pCO_2, whereas net calcification was positive during the day ($0.9 \pm 0.7\,gCaCO_3\,m^{-2}\,d^{-1}$) and negative at night ($-0.6 \pm 0.8\,gCaCO_3\,m^{-2}\,d^{-1}$) in the ambient treatment. When calcification was integrated over 24 h, pCO_2 effects were significant ($F_{1,2} = 886.5$, $p = 0.001$), with dissolution exceeding precipitation at high pCO_2 ($-1.6 \pm 0.8\,gCaCO_3\,m^{-2}\,d^{-1}$) and a nearly balanced calcification budget under ambient pCO_2 ($0.1 \pm 0.6\,gCaCO_3\,m^{-2}\,d^{-1}$).

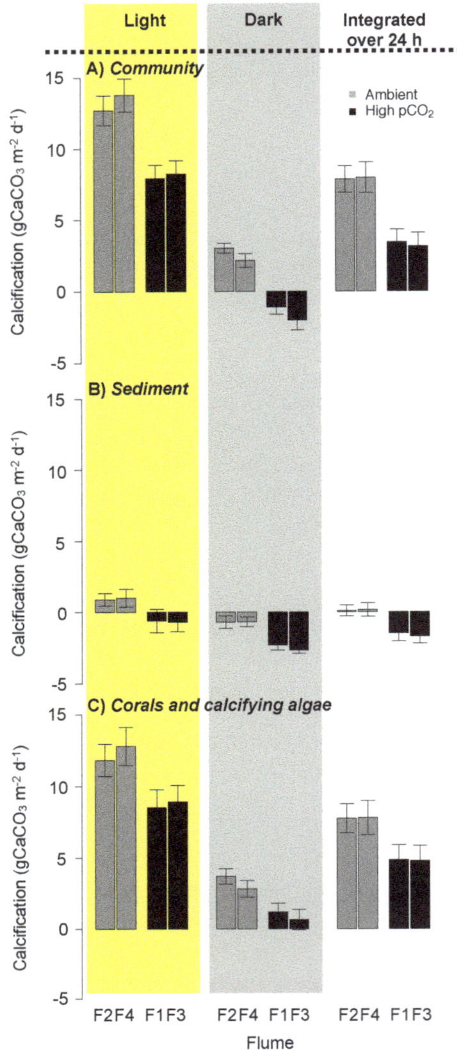

Figure 2. Calcification in the light, dark, and integrated over 24 h for intact communities **(a)**, sediment **(b)**, and corals and coralline algae **(c)** maintained under ambient and high pCO_2 ($\sim 1300\,\mu$atm). The grey bars represent the calcification measured in the ambient conditions, and the black bars are calcification in the elevated pCO_2 treatment. F1, F2, F3, and F4 indicate the different flumes.

3.4 Corals and calcifying algae

The total net calcification of corals and calcifying algae was estimated by subtracting the mean sediment calcification rates from the total community calcification in each flume. Net calcification of the corals and calcifying algae differed between treatments during the day ($F_{1,2} = 32.3$, $p = 0.030$) and night ($F_{1,2} = 22.9$, $p = 0.041$) (Fig. 2c). In contrast to the whole community and the sediments alone, net calcification of corals and calcifying algae was positive at night at high pCO_2 (0.9 ± 0.5 $gCaCO_3\,m^{-2}$ in 12 h), but it was 24 % and 44 % lower at high pCO_2 compared to ambient pCO_2 during the day and night, respectively. Net calcification integrated over 24 h also differed between treatments ($F_{1,2} = 2569$, $p < 0.001$), with calcification at ambient pCO_2 29 % higher than at high pCO_2.

Calcification of the constructed reef communities was driven principally by corals, since their contribution to the calcification budget, based on dry weight calculated from the changes in buoyant weight, was ~ 98 % of the total (Fig. 3). Massive *Porites* spp. were the main contributor among the corals, with an increased contribution to the calcification budget at high pCO_2 (40 % at ambient pCO_2, and 48.5 % at high pCO_2, Fig. 3). In contrast, the importance of *P. rus*, *Montipora* spp., and *Pocillopora* spp. was reduced at high pCO_2. The small contribution of coralline algae to the calcification budget was due to high mortality perhaps leading to potential dissolution during the last weeks of the incubation. Furthermore, while the ratio of planar area to surface area for crustose coralline algae is close to 1, corals have a disproportionately large surface area : planar area ratio due to their three-dimensional structure. With such a large actual surface area, the corals made a large contribution to the calcification budget of the communities assembled in the flumes.

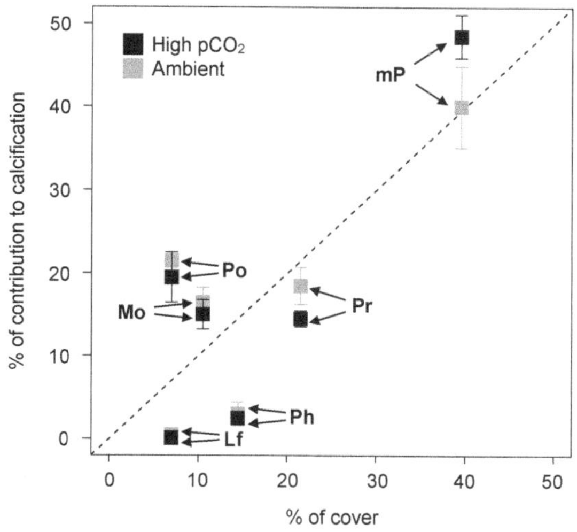

Figure 3. Relative contribution of each functional group of corals and calcifying algae to the calcification budget of communities as a function of their contribution to the planar surface area of calcifiers in the flumes. Contribution to the calcification budget was derived from the buoyant weight measurements made on each individual at the beginning and end of the 8-week incubation. The grey (ambient condition) and black (high pCO_2) squares correspond to the mean ± SD specific contributions of massive *Porites* (mP), *Porites rus* (Pr), *Pocillopora* spp. (Po), *Montipora* spp. (Mo), *Porolithon onkodes* (Ph), and *Lithophyllum flavescens* (Lf). The dashed line corresponds to a contribution to the calcification budget equivalent to the planar surface areas of calcifier in the flumes.

4 Discussion

Using outdoor flumes, we show that the effects of OA on coral reef communities are greater than estimates obtained by summing results obtained by incubating organisms in isolation under similar conditions and assuming their contribution to community calcification is proportional to their planar cover. Indeed, at the community level, the reduction in net calcification attributed to high pCO_2 was greater than the mean reduction of 26 % calculated in a recent meta-analysis of the effects of future conditions (~ 1300 μatm pCO_2) based on the consequences of high pCO_2 on organismic calcification (Chan and Connolly, 2013). This discrepancy likely is not caused by experimental bias, as rates of net community calcification in the flumes in the ambient treatment were similar to rates measured for back-reef communities on the north shore of Moorea. For instance, in 2012 and 2013 we measured calcification rates during the day that ranged from 5 to 25 $gCaCO_3\,m^{-2}\,d^{-1}$ (R. C. Carpenter, unpublished data), which spans the rates measured in flumes during the present study (i.e., 13.9 $gCaCO_3\,m^{-2}\,d^{-1}$ in the light, Fig. 2a). Net community calcification for the back reef of Moorea in 1991 (~ 19–25 $gCaCO_3\,m^{-2}\,d^{-1}$; Gattuso et al., 1996) was also similar to the rates measured in the flumes (this study) and in the field as described above. Rates of calcification in the present study under ambient conditions are also similar to the 7.9 $gCaCO_3\,m^{-2}\,d^{-1}$ reported by Andersson et al. (2009) for a reef community from Kaneohe Bay (Hawaii) that was assembled and incubated in mesocosms. However, while community calcification was still positive under high pCO_2 in the present study, Andersson et al. (2009) measured negative calcification (i.e., net dissolution) in their coral reef communities incubated at a pCO_2 twice that of current ambient values. The differences between the present study and that of Andersson et al. (2009) may be due to methodological effects. Andersson et al. (2009) manipulated pH through acid additions (we used CO_2 bubbling) and also used a different assemblage of species and sediments in dissimilar proportions compared to the present study.

The discrepancy in the evaluation of the effects of high pCO_2 at the community level (the present study) versus organismic level (previous studies) was the result of dissolution of sediments that represented up to 50 % of the decrease in calcification at high pCO_2. Increased dissolution of sediments at high pCO_2 likely was caused by the reduction of the seawater saturation state in the flumes, as we did not detect any difference in respiration and photosynthesis under elevated pCO_2 (results not shown) that could also affect sediment dissolution (Andersson and Gledhill, 2013). Our results reveal the sensitivity of carbonate sediments to dissolution at elevated pCO_2, and they are in agreement with a recent manipulative experiment conducted on Heron Island (Australia), where dissolution of in situ areas of sand (1.7 m depth) exceeded precipitation at $pCO_2 > 500\,\mu$atm (Cyronak et al., 2013a). During a mesocosm experiment, Dove et al. (2013) also demonstrated that a pH of 7.7 caused a change in sediment granularity to favor small-grained (i.e., ≤ 1 mm) sediments as a result of dissolution or increased bioerosion of larger grains. In this case, bioerosion was more likely than dissolution, as dissolution would favor a loss of the smallest grains as a result of their higher surface-area-to-volume ratio. Size-frequency distribution of sediment grain was not different between treatments at the end of our incubations and therefore is unlikely to have affected the treatment effects we detected. Sensitivity of coral reef communities to dissolution has been shown previously for communities constructed in mesocosms in Hawaii, where dissolution (-3.6 mmol $CaCO_3$ m^{-2} h^{-1}) was detected at night when CO_2 levels in the mesocosm were equivalent to 2-fold pCO_2 in ambient air (Andersson et al., 2009). In this case, dissolution was attributed to the thin layer of sediment that accumulated at the bottom of the mesocosms (Andersson et al., 2009).

In addition to chemical dissolution occurring in the communities constructed in the present study, we cannot exclude the possibility that at least some of the apparent community dissolution was caused by enhanced bioerosion, which previously has been shown to occur when blocks of *Porites lobata* are incubated under $750\,\mu$atm pCO_2 for 3 months (Tribollet et al., 2009). In future work it will be important to census the fragments of coral and rock to quantify the presence of bioeroders and their relative contribution to dissolution under ambient and OA conditions.

When the effect of sediment dissolution was subtracted from the overall net calcification rate for the communities assembled in our flumes, corals and coralline algae alone accounted for a decrease in net calcification of 29 % over 24 h at elevated pCO_2 versus ambient pCO_2. Such a decrease falls within the range of values we have previously reported for organismic effects of high pCO_2, in which the calcification rates of 16 calcifiers in Moorea declined 0–40 % at 1300 μatm pCO_2 compared to ambient pCO_2 (Comeau et al., 2013, 2014b). It is also within the range of the predicted changes for calcification of corals under a tripling of pCO_2 (relative to present values) estimated by meta-analysis (i.e., a ~ 26 % reduction; Chan and Connolly, 2013). The proportional decrease (i.e., ~ 29 %) in calcification rate for corals and coralline algae recorded in the present study under a tripling of present pCO_2 alone supports the validity of our experimental approach, which assumes that calcification of macrocalcifiers is equal to the difference between net sediment calcification and net community calcification. This "subtraction method" for calculating the calcification rate of corals and coralline algae included in community experiments has some limitations, as it assumes that the calcification of the sediments and the macrocalcifiers are independent. This assumption might be violated if, for example, sediment dissolution locally enhances total alkalinity that could favor calcification of nearby macrocalcifiers. Testing for such feedback mechanisms among the different compartments of the communities we built was beyond the scope of the present study, but it will be important to consider such effects in future experiments.

Our results demonstrate the suitability of large outdoor flumes for investigating the responses of coral reef communities to OA. Similar rates of calcification in the field and in the flumes suggest that the communities assembled in the flumes effectively mimicked both the biological communities and the physical and chemical conditions characterizing the back reef of Moorea. The ability to create ecologically relevant flow conditions in the flumes is likely to be especially important for establishing ecological relevance, as flow is critical in modulating mass transfer and metabolism of coral reef organisms (Atkinson and Gilmer, 1992; Carpenter and Williams, 2007; Comeau et al., 2014c). In the case of stony corals, for example, high flow speeds are suspected to enhance coral calcification by favoring proton export from coral tissue through boundary layers (Jokiel, 2011; Jokiel et al., 2014); for coralline algae, high flow speeds might increase sensitivity to OA by reducing the capacity to maintain high pH in the diffusion boundary layer adjacent to the algal thallus (Cornwall et al., 2013, 2014).

5 Conclusions

The present results suggest that, despite a reduction in calcification, calcifying reef organisms may maintain net positive calcification under pCO_2 as high as $1300\,\mu$atm. However, at the scale of coral reef communities in back-reef habitats, community net calcification will be affected strongly and negatively, at least for reefs similar in community structure to those in Moorea in 2013. The present experiments demonstrate the importance of living organisms on benthic surfaces in maintaining a positive balance between precipitation and dissolution of calcium carbonate. Whereas several reefs around the world are already at the threshold between precipitation and dissolution of calcium carbonate (Silverman et al., 2009, 2014), the susceptibility of coral reefs to net dissolution in the future likely will be linked directly to the proportion of the reef covered by macrocalcifiers and sediments. In addition to dissolution, it also is possible that coral reefs will be exposed to increased bioerosion at high pCO_2 (Wisshak et al., 2012; Crook et al., 2013) that will decrease the integrity of the carbonate framework. In addition to the direct effects of OA on reef builders, the associated loss of three-dimensional framework might impact a large variety of marine organisms by reducing habitat complexity and the availability of refuges (Fabricius et al., 2014). Our results suggest that, under OA conditions anticipated by the end of the current century, at least some tropical corals and calcifying algae will persist, but the function of the coral reef community as a net precipitator of calcium carbonate and as a physical structure to protect coasts against erosion (Ferrario et al., 2014) will be challenged.

Author contributions. S. Comeau designed and performed experiments, analyzed data, and wrote the paper; C. Lantz performed experiments and wrote the paper; B. Carpenter and P. Edmunds designed experiments, analyzed data, and wrote the paper.

Acknowledgements. We dedicate this paper to Marlin Atkinson, who pioneered the use of large outdoor flumes for the analysis of coral community metabolism, and whose work inspired our science and the present experiments. This study was funded by the National Science Foundation (OCE 10-41270) and the Moorea Coral Reef LTER (OCE 04-17413 and 10-26852). This is contribution number 225 of the CSUN Marine Biology Program.

Edited by: H. Niemann

References

Atkinson, M. J. and Bilger, R. W.: Effects of water velocity on phosphate uptake in coral reef-flat communities, Limnol. Oceanogr., 37, 273–279, 1992.

Andersson, A. J. and Gledhill, D.: Ocean acidification and coral reefs: effects on breakdown, dissolution, and net ecosystem calcification, Ann. Rev. Mar. Sci., 5, 321–348, 2013.

Andersson, A. J., Bates, N. R., and Mackenzie, F. T.: Dissolution of carbonate sediments under rising $p\mathrm{CO}_2$ and ocean acidification: observations from Devil's Hole, Bermuda, Aquatic. Geochem., 13, 237–264, 2007.

Andersson, A. J., Kuffner, I. B., Mackenzie, F. T., Jokiel, P. L., Rodgers, K. S., and Tan, A.: Net Loss of $CaCO_3$ from a subtropical calcifying community due to seawater acidification: mesocosm-scale experimental evidence, Biogeosciences, 6, 1811–1823, doi:10.5194/bg-6-1811-2009, 2009.

Atkinson, M. J. and Bilger, R. W.: Effects of water velocity on phosphate uptake in coral reef-flat communities, Limnol. Oceanogr., 37, 273–279, 1992.

Carpenter, R. C. of Moorea Coral Reef LTER: MCR LTER: Coral Reef: Long-term population and community dynamics: benthic algae and other community components, ongoing since 2005, knb-lter-mcr.8.27, available at: http://metacat.lternet.edu/knb/metacat/knb-lter-mcr.8.27/lter, 2014.

Carpenter, R. C. and Williams, S. L.: Mass transfer limitation of photosynthesis of coral reef algal turfs, Mar. Biol., 151, 435–450, 2007.

Chan, N. C. S. and Connolly, S. R.: Sensitivity of coral calcification to ocean acidification: a meta-analysis, Glob. Change Biol., 19, 282–290, 2013.

Chisholm, J. R. M. and Gattuso, J.-P.: Validation of the alkalinity anomaly technique for investigating calcification and photosynthesis in coral reef communities, Limnol. Oceanogr., 36, 1232–1239, 1991.

Comeau, S., Edmunds, P. J., Spindel, N. B., and Carpenter, R. C.: The responses of eight coral reef calcifiers to increasing partial pressure of CO_2 do not exhibit a tipping point, Limnol. Oceanogr., 58, 388–398, 2013.

Comeau, S., Edmunds, P. J., Spindel, N. B., and Carpenter, R. C.: Diel $p\mathrm{CO}_2$ oscillations modulate the response of the coral *Acropora hyacinthus* to ocean acidification, Mar. Ecol. Prog. Ser., 501, 99–111, 2014a.

Comeau, S., Edmunds, P. J., Spindel, N. B., and Carpenter, R. C.: Fast coral reef calcifiers are more sensitive to ocean acidification in short-term laboratory incubations, Limnol. Oceanogr., 59, 1081–1091, 2014b.

Comeau, S., Edmunds, P. J., Lantz, C. A., and Carpenter, R. C.: Water flow modulates the response of coral reef communities to ocean acidification, Scientific Reports, 4, 6681, doi:10.1038/srep06681, 2014c.

Cornwall, C. E., Hepburn, C. D., Pilditch, C. A., and Hurd, C. L.: Concentration boundary layers around complex assemblages of macroalgae: Implications for the effects of ocean acidification on understory coralline algae, Limnol. Oceanogr., 58, 121–130, 2013.

Cornwall, C. E., Boyd, P. W., McGraw, C. M., Hepburn, C. D., Pilditch, C. A., Morris, J. N., Smith, A. M., and Hurd, C. L.: Diffusion boundary layers ameliorate the negative effects of ocean acidification on the temperate coralline macroalga *Arthrocardia corymbosa*, PLoS One, 9, e97235, doi:10.1371/journal.pone.0097235, 2014.

Crook, E. D., Cohen, A. L., Rebolledo-Vieyra, M., Hernandez, L., and Paytan, A.: Reduced calcification and lack of acclimatization by coral colonies growing in areas of persistent natural acidification, P. Natl. Acad. Sci., 110, 11044–11049, 2013.

Cyronak, T., Santos, I. R., and Eyre, B. D.: Permeable coral reef sediment dissolution driven by elevated $p\mathrm{CO}_2$ and pore water advection, Geophys. Res. Lett., 40, 4876–4881, 2013a.

Cyronak, T., Santos, I. R., McMahon, A., and Eyre, B. D.: Carbon cycling hysteresis in permeable carbonate sands over a diurnal cycle: implications for ocean acidification, Limnol. Oceanogr., 58, 131–143, 2013b.

Davies, P. S.: Short-term growth measurements of corals using an accurate buoyant weighing technique, Mar. Biol., 101, 389–395, 1989.

Dickson, A. G., Sabine, C. L., and Christian, J. R. (Eds.): Guide to best practices for CO_2 measurements, PICES Special Publication, 3, 191 pp., 2007.

Dove, S. G., Kline, D. I., Pantos, S., Angly, F. E., Tyson, G. W., and Hoegh-Guldberg, O.: Future reef decalcification under a business-as-usual CO_2 emission scenario, P. Natl. Acac. Sci., 110, 15342–15347, 2013.

Edmunds, P. J. (of Moorea Coral Reef LTER): MCR LTER: Coral Reef: Long-term population and community dynamics: corals, knb-lter-mcr.4.31, available at: http://metacat.lternet.edu/knb/metacat/knb-lter-mcr.4.31/lter, 2014.

Erez, J., Reynaud, S., Silverman, J., Schneider, K., and Allemand, D.: Coral calcification under ocean acidification and global change, in: Coral reefs: An ecosystem in transition, edited by:

Dubinsky, Z. and Stambler, N., Springer, Germany, 151–176, 2011.

Fabricius, K. E., Langdon, C., Uthicke, S., Humphrey, C., Noonan, S., De'ath, G., Okazaki, R., Muehllehner, N., Glas, M. S., and Lough, J. M.: Losers and winners in coral reefs acclimatized to elevated carbon dioxide concentrations, Nature Clim. Change, 1, 165–169, 2011.

Fabricius, K. E., De'ath, G., Noonan, S., and Uthicke, S.: Ecological effects of ocean acidification and habitat complexity on reef-associated macroinvertebrate communities, P. Roy. Soc. B-Bio. Sci., 281, 20132479, doi:10.1098/rspb.2013.2479, 2014.

Feely, R. A., Sabine, C. L., Lee, K., Berelson, W., Kleypas, J., Fabry, V. J., and Millero, F. J.: Impact of anthropogenic CO_2 on the $CaCO_3$ system in the oceans, Science, 305, 362–366, 2004.

Ferrario, F., Beck, M. W., Storlazzi, C. D., Micheli, F., Shepard, C. C., and Airoldi, L.: The effectiveness of coral reefs for coastal hazard risk reduction and adaptation, Nature Comm., 5, 3794, 2014.

Gattuso, J.-P., Pichon, M., Delesalle, B., Canon, C., and Frankignoulle, M.: Carbon fluxes in coral reefs. I. Lagrangian measurement of community metabolism and resulting air-sea CO_2 disequilibrium, Mar. Ecol. Prog. Ser., 145, 109–121, 1996.

Hoegh-Guldberg, O., Mumby, P. J., Hooten, A. J., Steneck, R. S., Greenfield, P., Gomez, E., Harvell, C. D., Sale, P. F., Edwards, A. J., Caldeira, K., Knowlton, N., Eakin, C. M., Iglesias-Prieto, R., Muthiga, N., Bradbury, R. H., Dubi, A., and Hatziolos M. E.: Coral reefs under rapid climate change and ocean acidification, Science, 318, 1737–1742, 2007.

Hofmann, G. E., Smith, J. E., Johnson, K. S., Send, U., Levin, L. A., Micheli, F., Paytan, A., Price, N. N., Peterson, B., Takeshita, Y., Matson, P. G., Derse Crook, E., Kroeker, K. J., Gambi, M. C., Rivest, E. B., Frieder, C. A., Yu, P. C., and Martz, T. R.: High-frequency dynamics of ocean pH: a multi-ecosystem comparison, PLoS ONE, 6, e28983. doi:10.1371/journal.pone.0028983, 2011.

Jokiel, P. L.: The reef coral two compartment proton flux model: A new approach relating tissue-level physiological processes to gross corallum morphology, J. Exp. Mar. Biol. Ecol., 409, 1–12, 2011.

Jokiel, P. L., Rodgers, K. S., Kuffner, I. B., Andersson, A. J., Cox, E. F., and Mackenzie, F. T.: Ocean acidification and calcifying reef organisms: a mesocosm investigation, Coral Reefs, 27, 473–483, 2008.

Jokiel, P. L., Jury, C. P., and Rodgers, K. S.: Coral-algae metabolism and diurnal changes in the CO_2 -carbonate system of bulk sea water, PeerJ, 2, e378, doi:10.7717/peerj.378, 2014.

Kleypas, J. and Yates, K.: Coral reefs and ocean acidification, Oceanogr., 22, 108–117, 2009.

Kline, D. I., Teneva, L., Schneider, K., Miard, T., Chai, A., Marker, M., Headley, K., Opdyke, B., Nash, M., Valetich, M., Caves, J. K., Russell, B. D., Connell, S. D., Kirkwood, B. J., Brewer, P., Peltzer, E., Silverman, J., Caldeira, K., Dunbar, R. B., Koseff, J. R., Monismith, S. G., Mitchell, B. G., Dove, S., and Hoegh-Guldberg, O.: A short-term in situ CO_2 enrichment experiment on Heron Island (GBR), Scientific Reports, 2, 413, doi:10.1038/srep00413, 2012.

Lavigne, H. and Gattuso, J.-P.: seacarb, seawater carbonate chemistry with R. R package version 2.4.10, available at: http://CRAN.R-project.org/package=seacarb, 2013.

Leclercq, N., Gattuso, J.-P., and Jaubert, J.: Primary production, respiration, and calcification of a coral reef mesocosm under increased CO_2 partial pressure, Limnol. Oceanogr., 47, 558–564, 2002.

Moss, R. H., Edmonds, J. A., Hibbard, K. A., Manning, M. R., Rose, S. K., VanVuuren, D. P., Carter, T. R., Emori, S., Kainuma, M., Kram, T., Meehl, G. A., Mitchell, J. F., Nakicenovic, N., Riahi, K., Smith, S. J., Stouffer, R. J., Thomson, A. M., Weyant, J. P., and Wilbanks, T. J.: The next generation of scenarios for climate change research and assessment, Nature, 463, 747–756, 2010.

Shamberger, K. E. F., Cohen, A. L., Golbuu, Y., McCorkle, D. C., Lentz, S. J., and Barkley, H. C.: Diverse coral communities in naturally acidified waters of a Western Pacific reef, Geophys. Res. Lett., 41, 499–504, 2014.

Silverman, J., Lazar, B., Cao, L., Caldeira, K., and Erez, J.: Coral reefs may start dissolving when atmospheric CO_2 doubles, Geophys. Res. Lett., 36, L05606, doi:10.1029/2008GL036282, 2009.

Silverman, J., Schneider, K., Kline, D. I., Rivlin, T., Rivlin, A., Hamylton, S., Lazar, B., Erez, J., and Caldeira, K.: Community calcification in Lizard Island, Great Barrier Reef: A 33 year perspective, Geochim. Cosmochim. Acta, 144, 72–81, 2014.

Takahashi, A. and Kurihara, H.: Ocean acidification does not affect the physiology of the tropical coral *Acropora digitifera* during a 5-week experiment, Coral Reefs, 32, 305–314, 2013.

Tribollet, A., Godinot, C., Atkinson, M., and Langdon, C.: Effects of elevated pCO_2 on dissolution of coral carbonates by microbial euendoliths, Global Biogeochem. Cy., 23, GB3008, doi:10.1029/2008GB003286, 2009.

Wisshak, M., Schönberg, C. H. L., Form, A., and Freiwald, A.: Ocean acidification accelerates reef bioerosion, PLoS One, 7, e45124, doi:10.1371/journal.pone.0045124, 2012.

Yates, K. K. and Halley, R. B.: CO_3^{2-} concentration and pCO_2 thresholds for calcification and dissolution on the Molokai reef flat, Hawaii, Biogeosciences, 3, 357–369, doi:10.5194/bg-3-357-2006, 2006.

3

Intercomparison of fast response commercial gas analysers for nitrous oxide flux measurements under field conditions

Ü. Rannik[1], S. Haapanala[1], N. J. Shurpali[2], I. Mammarella[1], S. Lind[2], N. Hyvönen[2], O. Peltola[1], M. Zahniser[3], P. J. Martikainen[2], and T. Vesala[1]

[1]Department of Physics, P.O. Box 48, 00014 University of Helsinki, Helsinki, Finland
[2]Department of Environmental Science, University of Eastern Finland, Kuopio, Finland
[3]Center for Atmospheric and Environmental Chemistry, Aerodyne Research Inc., Billerica, MA, USA

Correspondence to: Ü. Rannik (ullar.rannik@heuristica.ee)

Abstract. Four gas analysers capable of measuring nitrous oxide (N_2O) concentration at a response time necessary for eddy covariance flux measurements were operated from spring until winter 2011 over a field cultivated with reed canary grass (RCG, *Phalaris arundinacea*, L.), a perennial bioenergy crop in eastern Finland. The instruments were TGA100A (Campbell Scientific Inc.), CW-TILDAS-CS (Aerodyne Research Inc.), N_2O / CO-23d (Los Gatos Research Inc.) and QC-TILDAS-76-CS (Aerodyne Research Inc.). The period with high emissions, lasting for about 2 weeks after fertilization in late May, was characterized by an up to 2 orders of magnitude higher emission, whereas during the rest of the campaign the N_2O fluxes were small, from 0.01 to 1 nmol m^{-2} s^{-1}. Two instruments, CW-TILDAS-CS and N_2O / CO-23d, determined the N_2O exchange with minor systematic difference throughout the campaign, when operated simultaneously. TGA100A produced the cumulatively highest N_2O estimates (with 29 % higher values during the period when all instruments were operational). QC-TILDAS-76-CS obtained 36 % lower fluxes than CW-TILDAS-CS during the first period, including the emission episode, whereas the correspondence with other instruments during the rest of the campaign was good. The reasons for systematic differences were not identified, suggesting further need for detailed evaluation of instrument performance under field conditions with emphasis on stability, calibration and any other factors that can systematically affect the accuracy of flux measurements. The instrument CW-TILDAS-CS was characterized by the lowest noise level (with a standard deviation of around 0.12 ppb at 10 Hz sampling rate) as compared to N_2O / CO-23d and QC-TILDAS-76-CS (around 0.50 ppb) and TGA100A (around 2 ppb). We identified that for all instruments except CW-TILDAS-CS the random error due to instrumental noise was an important source of uncertainty at the 30 min averaging level and the total stochastic error was frequently of the same magnitude as the fluxes when N_2O exchange was small at the measurement site. Both instruments based on continuous-wave quantum cascade laser, CW-TILDAS-CS and N_2O / CO-23d, were able to determine the same sample of low N_2O fluxes with a high mutual coefficient of determination at the 30 min averaging level and with minor systematic difference over the observation period of several months. This enables us to conclude that the new-generation instrumentation is capable of measuring small N_2O exchange with high precision and accuracy at sites with low fluxes.

1 Introduction

During the last few years there has been a rapid development in the application of laser spectroscopy for greenhouse gas measurements. In particular, the development of fast response N_2O analysers based on spectroscopic techniques (e.g. tunable diode laser (TDL) and quantum cascade laser (QCL) spectrometers) has facilitated the eddy covariance (EC) measurements of N_2O exchange in different ecosystems. Such measurements have been reported in literature and they have been carried out in different ecosystems such as agricultural (Smith et al., 1994; Wienhold et al., 1994;

Table 1. Instrumental characteristics. Experimental precision values are based on flux measurements during the period DOY 206–271 (period II). TDL – tunable diode laser; CW-QCL – continuous-wave quantum cascade laser; P-QCL – pulsed QCL. $P_{10}/P_{50}/P_{90}$ represent the lower percentile, median and upper percentile values.

Instrument model	TGA100A	CW-TILDAS-CS	N_2O / CO-23d	QC-TILDAS-76-CS
Manufacturer	Campbell Scientific Inc.	Aerodyne Research Inc.	Los Gatos Research Inc.	Aerodyne Research Inc.
Abbrev. used in current study	CS-TDL	AR-CW-QCL	LGR-CW-QCL	AR-P-QCL
Measured species	N_2O	N_2O, H_2O, CO	N_2O, H_2O, CO	N_2O, CO_2, H_2O
Sample cell volume (mL)	480	500		500 (76 m path length)
Sample cell pressure (hPa)	50	53	117	53
Spectroscopic correction coefficient b	0.00 (drier used in sampling line)	0.39	0.00 (built-in correction by the instrument)	0.0235
Precision, 10 Hz noise SD, $P_{10}/P_{50}/P_{90}$ this study (ppb)	1.89/1.98/2.1	0.12/0.12/0.14	0.46/0.60/0.78	0.43/0.46/0.51

Christensen et al., 1996; Laville et al., 1997; Scanlon and Kiely, 2003; Neftel et al., 2007; Kroon et al., 2007) and forest (Pihlatie et al., 2005; Eugster et al., 2007), and over urban canopies (Famulari et al., 2010; Järvi et al., 2014).

The observed N_2O emissions are episodic in nature, showing high spatial and temporal variability. Emission bursts of short duration, typically occurring after fertilizer application, or associated with thawing and rain events (Kroon et al., 2007; Pihlatie et al., 2010), are followed by long periods of small fluxes, when uptake of N_2O has also been observed (Flechard et al., 2005). Overall, N_2O fluxes reported by previous studies are characterized by large uncertainties and temporal variability, which are related to biogeochemical soil processes and several systematic and random error sources of the EC measurements. One of the sources of uncertainty for the N_2O fluxes measured by the EC technique is the performance and stability of fast response gas analysers. Some studies performed under field conditions (Eugster et al., 2007; Kroon et al., 2007; Neftel et al., 2009) have reported that laser drift can cause occasional over- or underestimation of EC flux. Instrumental drift is typically characteristic of TDL as well as QCL spectrometers (Werle et al., 1993; Nelson et al., 2002). Mammarella et al. (2010) thoroughly investigated the performance of TDL instruments in measurements of N_2O fluxes by the EC technique. They suggested that high-pass filtering could be used to remove the low-frequency signal drifting, which could otherwise contaminate the detected concentration time series and significantly increase the flux uncertainty.

Apart from episodic emissions, N_2O fluxes are typically small in magnitude (of the order of 1 to 100 μg N m^{-2} h^{-1}, which corresponds to N_2O flux range from 10^{-2} to 1 nmol m^{-2} s^{-1} as presented in the units used in the current study), being on the detection limit of the EC systems (e.g. Pihlatie et al., 2005; Wang et al., 2013). Small fluxes imply small turbulent fluctuations of concentration, requiring high instrument precision to resolve those fluctuations. In other words, the ratio of signal (turbulent fluctuations) to instrumental noise has to be high enough to achieve sufficiently low flux error arising from the noise present in measured signals (Lenschow and Kristensen, 1985).

The goals of this study are to compare the available equipment for N_2O flux measurements employing the EC technique and to evaluate their performance, ability to detect small fluxes and long-term stability in determining the N_2O exchange. The instruments used were TGA100A (Campbell Scientific Inc.), CW-TILDAS-CS (Aerodyne Research Inc.), N_2O / CO-23d (Los Gatos Research Inc.) and QC-TILDAS-76-CS (Aerodyne Research Inc.), which shall be further referred to as CS-TDL, AR-CW-QCL, LGR-CW-QCL and AR-P-QCL, respectively, throughout this study by using the combinations of acronyms for manufacturer and the laser type (see Table 1). In addition, the methods for flux calculation using the laser spectrometer data are evaluated and the magnitude and dynamics of N_2O fluxes during the reed canary grass (RCG) growing season are determined.

2 Materials and methods

2.1 Site

The measurement site was a 6.9 ha field cultivated with RCG, a perennial bioenergy crop. The site was located in the rural area of Maaninka (merged with the city of Kuopio 1 January 2015), eastern Finland (63°9′48.69″ N, 27°14′3.29″ E). Long-term (reference period 1981–2010; Pirinen et al., 2012) annual air temperature in the region is 3.2 °C, the coldest month of the year is February and the warmest is July, with monthly mean air temperature being −9.4 and 17.0 °C, respectively. The annual precipitation in the region is 612 mm. Part of this precipitation amount falls as snow. Snow cover season starts in October and lasts until the end of April with a maximum snow cover of approximately 50 cm. The RCG crop at the Maaninka site was fertilized in the beginning of the growing season (late May), resulting in a large emission pulse of N_2O. The site was applied with an N–P–K–S fertilizer containing 76 kg N ha^{-1}, based on ammonium nitrate (NO_3–N / NH_4–N = 47 : 53). The canopy height developed throughout the growing season from about 10 cm in mid-May to 1.7 m by late June. The increase in plant height was almost linear in the period between these two times, and from July onwards plant height grew slowly up to 1.9 m.

The soil at the study site is classified as fine sand to coarse silt (particle size 0.03–0.06 mm). According to the World Reference Base for Soil Resources (WRB) system (FAO, 2006), the soil is classified as Regosol. The soil pH varies from 5.4 to 6.1 within the ploughing depth from the surface to about 30 cm, electrical conductivity between 960 and 3060μs cm^{-1} and soil organic matter content between 3 and 11 %. The average C / N ratio in the ploughing depth is 14.9 (ranging from 14.1 to 15.7). The soil particle density is about 2.65 g cm^{-3} within the soil depth from the surface to about 20 cm.

2.2 Measurements

Measurements were conducted by the University of Helsinki (UH) and by the University of Eastern Finland (UEF), operating separate EC systems based on two different sonic anemometers. The UH measurement set-up included a 3-D ultrasonic anemometer (USA-1, METEK GmbH, Elmshorn, Germany) to acquire the wind components. The anemometer was installed on top of a pole, with a measurement height of 2.2 m. The measurement height was raised to 2.4 m on 30 June 2011 due to the RCG growth. The gas analysers were situated in an air conditioned cabin located about 15 m east from the anemometer pole. This wind direction (50–110° sector) was therefore discarded from further analysis due to possible disturbances to flux measurements. Sample inlets for gas analysers were located 10 cm below the anemometer. The N_2O instruments operated by the UH were the instrument based on tunable diode laser CS-TDL (model TGA100A, Campbell Scientific Inc.) and two instruments based on continuous-wave quantum cascade lasers, AR-CW-QCL (models CW-TILDAS-CS, Aerodyne Research Inc., see e.g. Zahniser et al., 2009; Lee et al., 2011) and LGR-CW-QCL (model N_2O / CO-23d, Los Gatos Research Inc., see e.g. Provencal et al., 2005). Sampling lines of AR-CW-QCL and LGR-CW-QCL were heated slightly above ambient temperature in order to prevent water condensing on the lines. CS-TDL used a dryer just before the instrument and no sampling line heating was used. The flow rates and tube dimensions were chosen to correspond to a turbulent flow regime except that the larger diameter of the sampling line of the LGR-CW-QCL analyser resulted in a laminar tube flow for that instrument (see Sect. 3.1 below). Further details of the instruments used are given in Table 1 and details of the different set-ups are given in Table 2.

The maintenance of CS-TDL was the most demanding of the compared instruments. It uses liquid nitrogen to keep the laser source at the operating temperature, and the Dewar was filled up twice a week. The instrument CS-TDL was calibrated in the beginning of the campaign. Further, the operating parameters of the analyser, such as laser current and laser, housing and detector temperatures, were checked once a week and after power failures. In addition, the shape and intensity of the absorption line were checked at the same time. These checks were assumed to guarantee calibration stability of the instrument to a reasonable degree. In addition, the inlet filter of CS-TDL was changed once a month.

The AR-CW-QCL was calibrated and its operating parameters were fine-tuned at the site after instrument installation. The instrument manufacturer provided a software upgrade during the campaign to conduct the real-time water vapour correction to the trace gas concentration data analysed by the instrument. In addition, the operating parameters were fine-tuned a few times on-line by the instrument manufacturer during the campaign.

LGR-CW-QCL was used in the campaign later (see Sect. 2.6 for details). The factory calibration of LGR-CW-QCL was checked but no deviation was observed within the uncertainty range of the calibration gases. After about 2 weeks of operation, the laser drifted out of the tuning range and the laser offset current was tuned manually to enable correct operation again. No calibration of the instruments AR-CW-QCL and LGR-CW-QCL was performed during the campaign as these analysers were expected to be very stable according to manufacturers' information.

The UEF set-up included a pulsed quantum cascade laser spectrometer AR-P-QCL (model QC-TILDAS-76-CS, Aerodyne Research Inc., Billerica, MS, USA, see McManus et al., 2005), an infrared gas analyser (IRGA, model Li-6262) and a 3-D sonic anemometer (model R3-50, Gill Instruments, Ltd., Hampshire, UK) for fast response gas concentration and

Table 2. Eddy covariance measurement set-up, flux calculation and quality screening parameters.

Instrument	CS-TDL	AR-CW-QCL	LGR-CW-QCL	AR-P-QCL
Sampling height (m)	2.2/2.4	2.2/2.4	2.4	2.0/2.5
Horizontal separation[1] (m)	0.05	0.05	0.07	0.1
Tube inner diameter (mm)	4	4	8	4
Tube length (m)	17.8	16	16	8.5
Flow rate (LPM)	17	13.2	11.6	13.5
Lag time from tube flow (s)	0.79	0.91	4.2	0.48
Lag time window used in flux calculation (s)	1.0 ± 0.0	1.0 ± 0.0	1.0 ± 0.0[2]	1.0 ± 0.8[3]
Time constant used in spectral corrections (s)	0.12	0.07	0.26	0.15

[1] Refers to separation of the sampling inlet from the centre position of the sonic anemometer. Vertical separation was 0.1 m for all instruments. [2] Prior to flux calculation the concentration records of LGR-CW-QCL were synchronized with AR-CW-QCL outputs. [3] The lag time window was used to determine the lag time for CO_2, which was assigned as the lag time for N_2O.

wind component measurements (Tables 1 and 2). The heated intake tubes for the laser spectrometer and IRGA were installed on either side of the sonic anemometer, all mounted on a boom on an adjustable instrument mast. The mast height was set at 2.0 m above the soil surface in the beginning of the campaign. To adjust to the increasing plant height, the mast was raised to 2.5 m during mid-June. AR-P-QCL was set up to measure the N_2O, CO_2 and water vapour mixing ratios simultaneously, while the IRGA was used to monitor the CO_2 and water vapour mixing ratios. Both trace gas analysers were calibrated against standard gases a minimum of once a month during the campaign; in particular, AR-P-QCL was calibrated every 2–3 weeks with two standard gases of 299 and 342 ppb. The calibration slope of AR-P-QCL did not change by more than 7.6 % throughout the campaign and maximum 6.1 % between consecutive calibrations. Thus, 6.1 % can be considered as the maximum flux systematic error arising from calibration accuracy of this instrument.

A weather station set up on another mast close to the EC mast monitored the supporting meteorological variables. The weather station mast height was also adjusted according to the changes in the EC mast height. Supporting measurements included air temperature and relative humidity (model: HMP45C, Vaisala Inc.) using radiation shield, atmospheric pressure (model CS106 Vaisala PTB110 Barometer), wind speed and direction (model 03002-5, R.M. Young Company) and several other variables not used in the current study. Data were collected using a datalogger (model CR3000, Campbell Scientific Inc.). Except air pressure (stored as hourly averages), meteorological data were stored as 30 min averages. Short gaps in the data were filled using linear interpolation, but when air temperature, relative humidity, pressure or rainfall data were missing for longer periods, data from Maaninka weather station operated by the Finnish Meteorological Institute located about 6 km south-east from the site were used.

2.3 Flux processing

Measurements were sampled at 10 Hz frequency. In order to eliminate spikes, filtering was performed according to the standard approach (Vickers and Mahrt, 1997), where the high-frequency EC data were de-spiked by comparing two adjacent measurements. If the difference between two adjacent concentration measurements of N_2O was greater than 20 ppb, the following point was replaced with the same value as the previous point.

The spectroscopic correction due to water vapour impact on the absorption line shape was applied along with the Webb–Pearman–Leuning (WPL) dilution correction due to water vapour on high-frequency raw concentration output X_C (mixing ratio with respect to moist air, uncorrected for spectroscopic effect) according to $\chi_C = \frac{X_C}{1-(1+b)\chi_V}$. Here χ_C and χ_V are the instantaneous mixing ratios of N_2O and water vapour with respect to dry air. The spectroscopic correction coefficient b was determined experimentally for each instrument (Table 1) by measuring the response of the instrument (output X_C) to sample air of standard gas (constant χ_C) with varying water content χ_V. The correction was not necessary for CS-TDL as a dryer installed after the air intake point on the sampling line dried the air sample before the optical cell. LGR-CW-QCL corrected for the water vapour effect using a built-in module in the LGR data acquisition software;

the same applied to AR-CW-QCL after a software update in July 2011.

Prior to calculating the turbulent fluxes, a 2-D rotation (mean lateral and vertical wind equal to zero) of sonic anemometer wind components was done according to Kaimal and Finnigan (1994) and all variables were linearly de-trended. The EC fluxes were calculated as 30 min covariances between the scalars and vertical wind velocity following commonly accepted procedures (e.g. Aubinet et al., 2000). Time lag between the concentration and wind measurements induced by the sampling lines was determined by maximizing the covariance. For CS-TDL, the lag was determined by maximizing the covariance for the high flux period only (day of year (DOY) 144–146) because in other periods the lag was not well defined by using this method. The final processing (instruments CS-TDL, AR-CW-QCL and LGR-CW-QCL) was done by fixing the time lag to avoid unphysical variations of lag due to random flux errors. For the AR-P-QCL system, the lag was determined by maximising the covariance for CO_2, and the same lag was assigned to N_2O. This was to use the instrument's ability to also measure CO_2, therefore enabling the use of a much better signal-to-noise ratio for determinating lag time. Spectral corrections were applied to account for the low- and high-frequency attenuation of the covariances (Sect. 2.4). Then, the humidity effect on temperature flux was accounted for after Schotanus et al. (1983). All data processing was performed with post-processing software EddyUH (http://www.atm.helsinki.fi/Eddy_Covariance/EddyUHsoftware.php).

2.4 Spectral corrections

Low- and high-frequency variations in the measured signal are attenuated due to data acquisition and processing, and by a non-ideal measurement system (e.g. Moore, 1986; Moncrieff et al., 1997; Rannik and Vesala, 1999; Massman, 2000). Block averaging and de-trending of data acts as a high-pass filter, thus damping low-frequency changes (Rannik and Vesala, 1999; Finnigan et al., 2003). Turbulent fluctuations occurring at high frequencies are attenuated due to the measurement system's limitations. Gas analyser's finite frequency response, attenuation of fluctuations in the sampling line, spatial separation between the anemometer measurement head and sampling line inlet affect the attenuation of high-frequency fluctuations in the signal.

The observed flux (F_m) can be formally presented as the integral over the convolution of the true co-spectrum (Co, unaffected by frequency attenuation) with the co-spectral transfer function as

$$F_m = \int_0^\infty T(f)\mathrm{Co}(f)\mathrm{d}f, \tag{1}$$

where the co-spectral transfer function $T(f)$ can be presented as the convolution of respective low-frequency $T_L(f)$ and high-frequency $T_H(f)$ transfer functions. For details on the low-frequency transfer function due to high-pass filtering and/or finite averaging period, see Rannik and Vesala (1999).

For evaluation of the instrument frequency performance and subsequent high-frequency flux corrections during post-processing, the high-frequency transfer function of the EC system was estimated (Aubinet et al., 2000) as the ratio of the observed and un-attenuated flux (Horst, 1997). The co-spectral transfer function $T_H(f)$ for a system behaving as a first-order response sensor can be described by

$$T_H(f) = \frac{1}{1+(2\pi f\tau)^2}, \tag{2}$$

where f is the natural frequency and τ the (first-order) response time of the attenuator (sensor or the system in total) (Horst, 1997). The effective transfer function of the EC system for different instruments was estimated as the ratio of co-spectral density of scalar flux relative to co-spectrum of sensible heat flux (Aubinet et al., 2000). Such a procedure assumed that temperature measurements were not affected by attenuation (true for the sonic anemometer) and includes normalization with integral over frequencies not affected by attenuation.

2.5 Estimation of random errors

Turbulent fluxes averaged over a limited time period have random errors because of the stochastic nature of turbulence (Lenschow et al., 1994; Rannik et al., 2006) as well as due to noise presented in measured signals (Lenschow and Kristensen, 1985).

The random error of the flux was evaluated as 1 standard deviation of the covariance error, hereafter in the manuscript denoted by δ_F. It was defined through the variance of the distribution of the individual flux realization around the ensemble mean (e.g. Lenschow et al., 1994). Theoretically, there are several approaches to approximate the same error estimate; see e.g. Rannik et al. (2009). Currently, the flux random error was calculated according to the method implemented in EddyUH, the method proposed by Finkelstein and Sims (2001). The method evaluates the error in the time domain through integration of the auto-covariance and cross-covariance functions of the vertical wind speed and the scalar concentration according to

$$\delta_F \approx \sqrt{\frac{1}{n}\left[\sum_{p=-m}^{m}\overline{w'w'}(p)\overline{c'c'}(p) + \sum_{p=-m}^{m}\overline{w'c'}(p)\overline{c'w'}(p)\right]}, \tag{3}$$

where $\overline{w'w'}(p) = \frac{1}{n}\sum_{i=1}^{n-p}(w(t_i)-\overline{w})\left(w(t_{i+p})-\overline{w}\right)$. In calculations, we used $m = 200$ (corresponding to 20 s) to ensure that integration of the covariance functions was performed over times exceeding the integral timescale of turbulence. This mathematically rigorous method provides estimates for the random uncertainty of the flux measurements for every flux averaging period.

Random uncertainty of the observed covariance due to presence of noise in instruments signal, giving essentially the lowest limit of the flux that the system is able to measure, was expressed in its simplest form as

$$\delta_{\mathrm{F,\,noise}} = \frac{\sigma_{\mathrm{w}}\sigma_{\mathrm{noise}}}{\sqrt{fT}}, \tag{4}$$

where σ_{w} and σ_{noise} denote the standard deviation of the turbulent record of vertical wind speed and the standard deviation of instrumental noise as observed at frequency f, and T denotes the flux averaging period. The expression above assumes that the noise component of the vertical wind speed measurement is negligible. In this study, we use the method developed by Lenschow et al. (2000) and applied to EC fluxes by Mauder et al. (2013) to estimate the flux error due to instrumental noise. Lenschow et al. (2000) derived the method to estimate the instrumental random noise variance $(\sigma_{\mathrm{noise}})^2$ from the auto-covariance function of the measured turbulent record close to zero-shift, enabling one to determine the error for each half-hour flux averaging period.

The random flux error δ_{F} is the result of limited sampling in time and/or in space of a stochastic turbulence realization. Its expression includes the covariance and cross-covariance functions of turbulent records; it therefore, in addition to variances and covariances, accounts for the respective integral timescales of turbulent records. The error δ_{F} also incorporates the contribution due to instrumental noise and is therefore larger than $\delta_{\mathrm{F,noise}}$.

The error $\delta_{\mathrm{F,noise}}$ instead does not depend on the integral timescale of turbulence; it is therefore mainly determined by the instrumental noise characteristics and less by the observation conditions (only via σ_{w}). Assuming no true turbulent variation of concentration and thus zero flux, the calculated flux will be generally non-zero due to noise in the instrumental signal. Evidently the system will not be able to detect the fluxes smaller than the ones obtained from the expression for $\delta_{\mathrm{F,noise}}$. Therefore, this is the minimum flux that the EC system can detect and $\delta_{\mathrm{F,noise}}$ proves useful in characterising the instrumental limitation to detect small fluxes.

If an average over fluxes F_i $(i = 1 \dots N)$ is calculated, each of these representing a flux value observed over averaging period T and being characterized by an error $\delta_{F,i}$, then the error of the average flux $\langle F \rangle = \frac{1}{N}\sum_{i=1}^{N} F_i$ is expressed as

$$\Delta_{<F>} = \sqrt{\frac{\sum_{i=1}^{N}(\delta_{F,i})^2}{N^2}}. \tag{5}$$

This expression will be used to estimate the random errors of the average fluxes in Sect. 3.4.

2.6 Periods of analysis and quality screening

The intercomparison measurements were performed from the beginning of the growing season in April until November 2011. According to instrumental data coverage, the period was divided into three sub-periods for the instrument evaluation and flux analysis purposes. During period I, DOY 110–181 (20 April–30 June 2011), the measurements of CS-TDL, AR-CW-QCL and AR-P-QCL were available; during period II, DOY 206–271 (25 July–28 September 2011), all instruments were measuring; and during period III, DOY 272–324 (29 September–20 November 2011), all other except CS-TDL were operational. Prior to analysis data quality screening was performed. The measurements corresponding to wind direction interval 50–110° were excluded as possibly affected by the instrumental cabin. In addition, quality screening was performed according to Vickers and Mahrt (1997) by applying the following statistics and selection thresholds: data with N_2O concentration skewness outside $(-2, 2)$, kurtosis outside $(1, 8)$ or Haar mean and Haar variance exceeding 3 were rejected. Applying the same statistics and thresholds as for N_2O, additional quality screening of N_2O fluxes was performed according to H_2O concentration statistics for AR-CW-QCL and AR-P-QCL due to the impact of the spectroscopic and dilution corrections on fluxes and according to CO_2 concentration statistics for AR-P-QCL because the lag obtained for CO_2 was assigned to N_2O in the case of this instrument.

The applied quality criteria were used to ensure exclusion of the system malfunctioning as well as unphysical and/or unusual occasions. No quality screening for stationarity was performed as the focus of the study was the instrumental intercomparison, which was not affected by occasional non-stationary conditions included in the analysed data set.

3 Results

The fluxes obtained for three periods are presented in Fig. 1, being averaged over daily period for the clarity of presentation. No gap-filling was used and for each day only the existing measurements, after applying the data quality screening described above, were averaged. In May, the fluxes increased significantly after fertilization and then decreased back to a low, although clearly positive, level after a few weeks. This was the only occasion of high N_2O emission followed by continuous decrease of fluxes towards the autumn. The soil temperature had an increasing trend until about DOY 205 (24 July 2011) and since August declining seasonal trend (Fig. 2). Soil water content (SWC) increased with occasional rain events. During the high emissions, starting on DOY 144 (24 July 2011) and lasting until approximately DOY 155 (4 June 2011), the SWC was relatively high at approximately $0.3\,\mathrm{m}^3\,\mathrm{m}^{-3}$.

The high fluxes observed during that period enable us to evaluate the frequency performance of three systems including CS-TDL, AR-CW-QCL and AR-P-QCL. The LGR-CW-QCL instrument was not operational then and the frequency

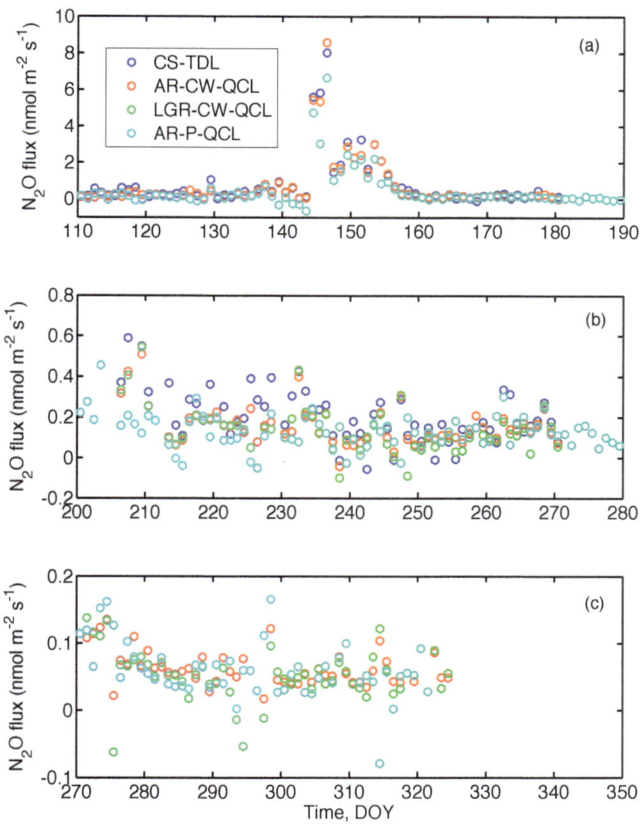

Figure 1. Daily average fluxes for four instruments containing period I DOY 110–181 (a), period II DOY 206–271 (b) and period III DOY 272–324 (c). No gap-filling was used in the calculation of daily average fluxes.

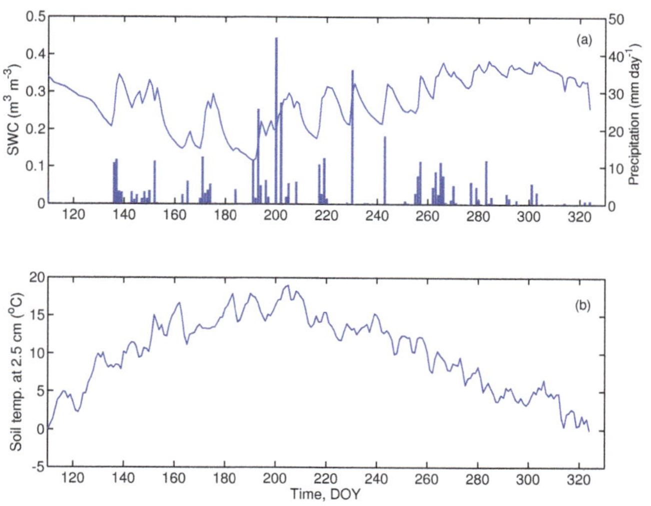

Figure 2. Soil water content (SWC) at 2.5 cm depth and precipitation **(a)** and soil temperature at 2.5 cm depth **(b)** during the measurement campaign.

response analysis for this instrument was performed based on the concurrently measured H_2O and CO signal analysis.

3.1 Spectral characteristics of the instruments

Spectral analysis was performed to study the frequency performance of the instruments. In general, averaging over long periods should lead to better spectral statistics. However, aggregating over different periods might lead to biased results as the spectra do not necessary follow the idealized normalizations in frequency scale, considering also that spectral scaling depends on stability. Therefore, we aimed to use optimal averaging period over several hours for similar conditions in terms of wind speed and stability. For the period of 26 May from 7:00 to 13:00 EET (eastern European time) when the conditions were moderately unstable (average wind speed of the period 3.2 m s^{-1} and sensible heat flux 50 W m^{-2}), the calculated spectra exhibited very clear and systematic patterns for temperature as well as N_2O concentration records measured by the three instruments (Fig. 3). In spite of high fluxes registered by the instruments during this period, the CS-TDL N_2O signal was dominated by noise almost over the whole frequency range presented. For AR-CW-QCL, almost no evidence of noise could be observed in the power spectral plot (multiplied with frequency). The older Aerodyne instrument, the AR-P-QCL, revealed an increase of the spectral density only at the high-frequency end of the power spectrum, characteristic of noise contribution. The co-spectra of all three instruments showed smooth patterns, the shape being consistent with the co-spectral model by Kaimal et al. (1972) but slightly shifted in frequency scale. At the high-frequency ends of the presented co-spectra the N_2O signal curves deviate from the theoretical as well as from temperature co-spectra, indicating attenuation of signals at high frequencies by the measurement systems.

The same time period was used to estimate the frequency response of the N_2O eddy covariance systems according to the method described in Sect. 2.4 (Fig. 4). The time constants estimated by making use of the co-spectra presented in Fig. 3 and Eq. (2) for CS-TDL, AR-CW-QCL and AR-P-QCL were 0.12, 0.07 and 0.08 s, respectively. Note that these time constants characterise the frequency response of the systems in total.

Although the response time obtained for the AR-P-QCL system from high flux period was 0.08 s, the analysis of the response time from measured CO_2 signal for several other periods yielded the average response time 0.15 s. The N_2O signal was synchronized with CO_2 by using the lag determined for CO_2 and theoretically the N_2O response time does not differ from that of CO_2 under turbulent tube flow regime; hence we choose the constant value 0.15 s for co-spectral corrections throughout the campaign for this instrument.

Spectral analysis was also performed for the period when LGR-CW-QCL measurements were available. For comparison purposes, the results of the time period of 4 August from 00:30 to 4:00 EET are presented for AR-CW-QCL and LGR-CW-QCL instruments (Fig. 5). The period was chosen with

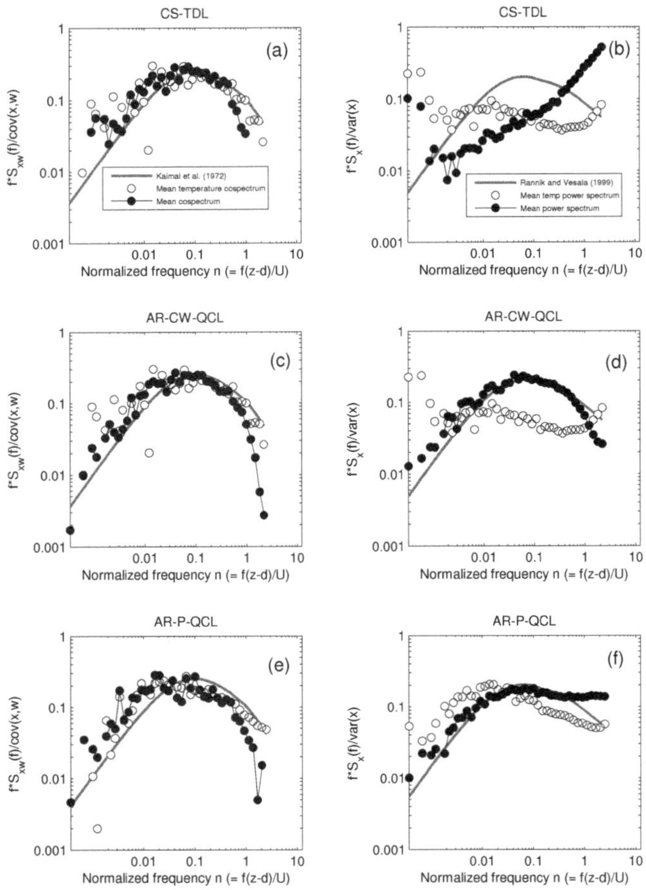

Figure 3. Normalized co-spectra (left panels) and spectra (right panels) of N_2O measurements by instruments CS-TDL **(a, b)**, AR-CW-QCL **(c, d)** and AR-P-QCL **(e, f)** during the high flux period, DOY 146 (26 May 2011) 7:00 to 13:00 EET. The RCG crop was about 0.4 m tall during the given period.

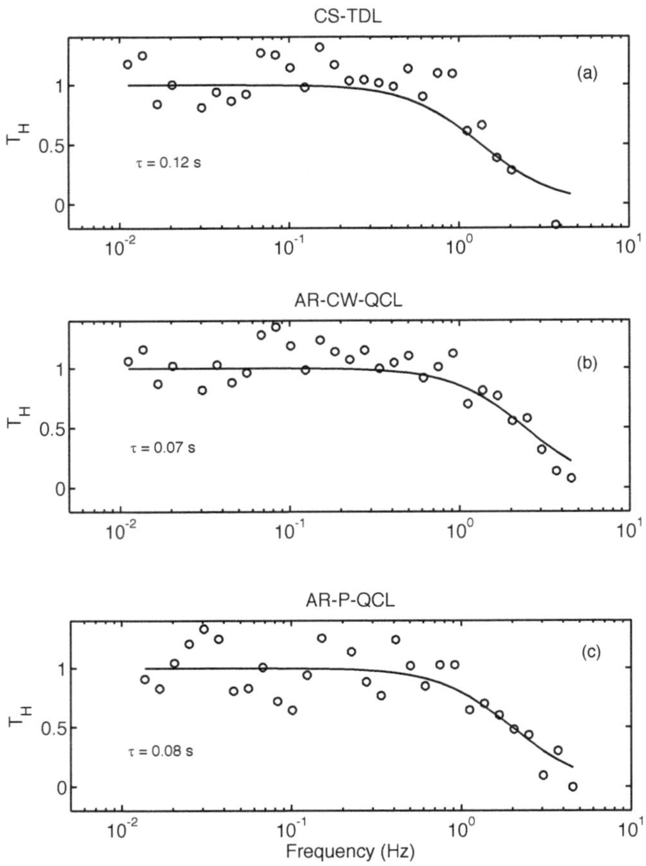

Figure 4. Co-spectral transfer functions derived for CS-TDL **(a)**, AR-CW-QCL **(b)** and AR-P-QCL **(c)** from the temperature and N_2O co-spectra presented in Fig. 2.

relatively high fluxes (with LGR-CW-QCL measurements available) and similar stability and wind conditions (average wind speed of the period of $0.94\ m\ s^{-1}$ and sensible heat flux of $-37.5\ W\ m^{-2}$). The power spectra of both instruments revealed a contribution of noise at the high-frequency end of the spectra, which was more pronounced for LGR-CW-QCL. The co-spectra were more scattered when compared to high flux period (Fig. 3). Estimation of the frequency response of the systems based on this period was uncertain due to scatter and could not be used as the basis for co-spectral corrections for LGR-CW-QCL.

The main difference in the flow set-ups of the systems concerned LGR-CW-QCL. With a larger tube diameter and slightly lower flow rate, the flow regime was likely laminar ($Re \approx 2000$), whereas for other instruments it was clearly turbulent ($Re \geq 4600$). It is well established that under laminar flow regime tube flow attenuates turbulent fluctuations of concentration much more than under turbulent flow. According to the expression for tube attenuation in laminar flow regime (Foken et al., 2012) the first-order response time for LGR-CW-QCL flow set-up would be 0.37 s (estimated for N_2O). For turbulent flow (ARI-CW-QCL set-up) the theoretical response time for tube damping is much smaller (0.01 s) than the response time obtained from the co-spectra (0.07 s), suggesting that the system's response was dominated by the instrumental response.

The frequency response of the LGR-CW-QCL system was further determined from the co-spectral analysis of the CO signal, and we obtained the value of 0.26 s. We also determined the experimental response time for water vapour from several periods corresponding to low-humidity conditions (RH < 40 %) and we consistently found the value around 0.35 s (for LGR-CW-QCL system). For comparison, the response time for H_2O measured by the ARI-CW-QCL system was determined to be 0.10 s. Damping of water fluctuations in sampling lines is stronger than for other scalars as evidenced by experimental studies (e.g. Mammarella et al., 2009). This is due to adsorption/desorption of water molecules on tube walls. This explains the difference between the response times obtained from CO and H_2O. Thus, we believe that a value of 0.26 s characterises well

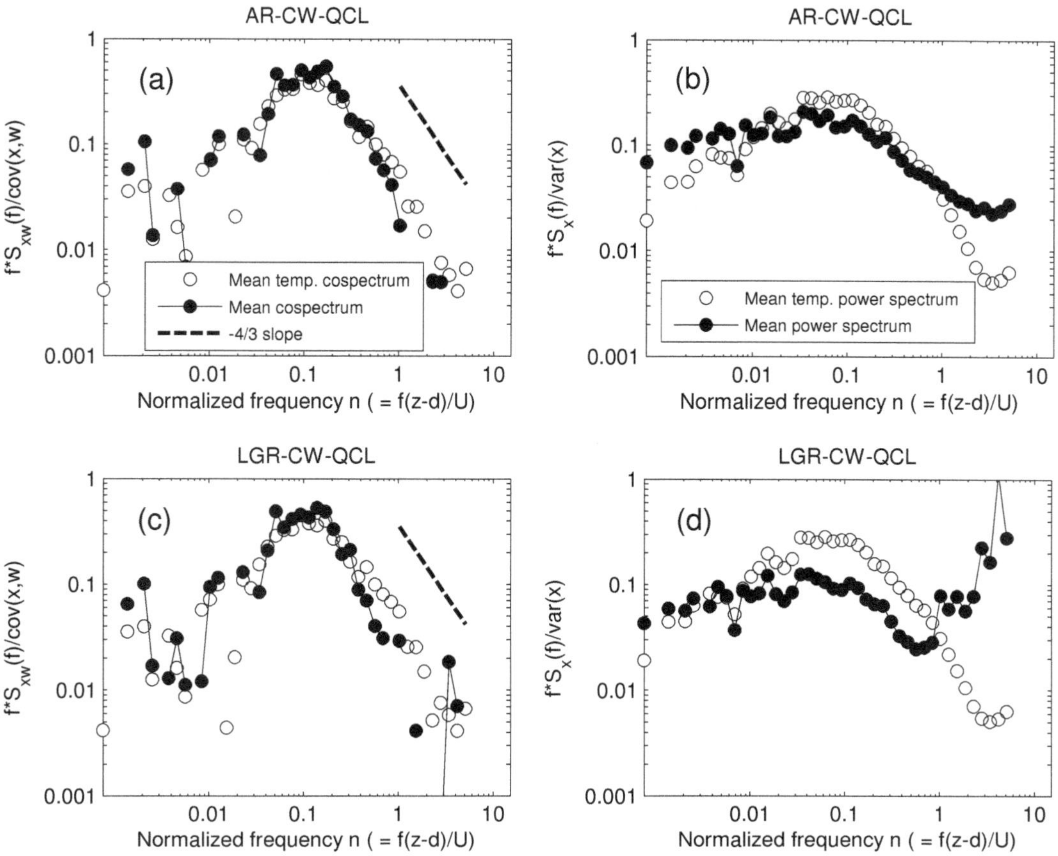

Figure 5. Normalized co-spectra (left panels) and spectra (right panels) of N_2O measurements by instruments AR-CW-QCL **(a, b)** and LGR-CW-QCL **(c, d)** during the period DOY 216 (4 August 2011) 00:30 to 4:00 EET. The RCG crop was about 1.8 m tall during the given period.

the first-order response time of the LGR-CW-QCL set-up for N_2O, and we use this value in co-spectral corrections. Note, however, that a higher response time of the LGR-CW-QCL system does not mean a slower instrument performance because the system has more damping primarily in the sampling line due to a lower flow rate and larger tube diameter (Table 2).

The frequency response times determined in this section were used in performing the co-spectral corrections (Table 2) as described in Sect. 2.4; the typical magnitudes of these corrections are presented in Table 3.

3.2 Random uncertainty of fluxes and instrumental noise

The method by Lenschow et al. (2000) described in Sect. 2.5 enables the calculation of the instrumental noise for each 30 min period and the resulting flux uncertainty due to instrumental noise. Figure 6a shows the estimated signal's noise statistics with upper and lower percentiles and quantiles (boxes), with a median value in the middle. For all instruments except LGR-CW-QCL, the distributions are very narrow and different percentiles cannot be separated from

Table 3. Statistics of spectral corrections of fluxes as percentage of raw uncorrected fluxes: lower percentile/median/upper percentile. Based on flux measurements during the period DOY 206–271 (period II) and data classified as qualified (Table 4). Daytime was defined by the elevation of sun higher than zero and night-time lower than zero, respectively. Statistics were derived for data when measurements were available for all four instruments.

	CS-TDL	AR-CW-QCL	LGR-CW-QCL	AR-P-QCL
All data	4.0/6.2/10.2	2.4/3.6/6.0	6.9/12.3/20.0	4.5/7.3/14.8
Daytime data	4.0/6.1/9.8	2.6/3.6/5.8	6.9/12.0/18.5	4.5/6.9/10.5
Night data	3.6/6.3/11.3	2.2/3.6/6.4	6.7/12.9/22.3	4.5/7.7/20.2

the plot (for values see Table 1). This tells us that the noise levels of the three instruments are very stable, but the noise level of LGR-CW-QCL somewhat varies. In a comparison of the instruments, AR-CW-QCL has by far the lowest noise level of around 0.12 ppb (standard deviation of the signal's noise at 10 Hz frequency). The two instruments, LGR-CW-QCL and AR-P-QCL, are characterized by a similar noise level (around 0.5 ppb), while CS-TDL signals show the highest noise level (2 ppb). Consequently, these instrumental noise levels are reflected in the random errors of

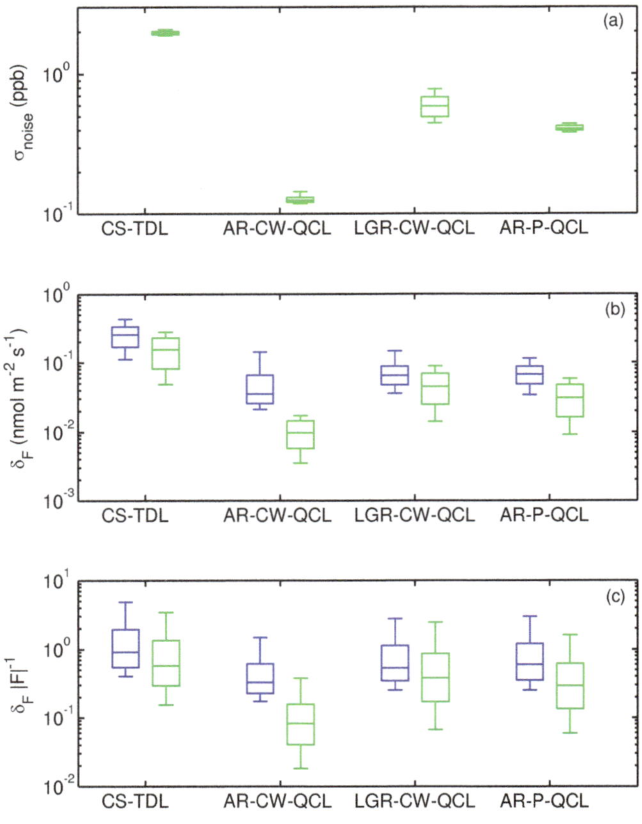

Figure 6. **(a)** Instrumental noise, presented as 1 standard deviation of the noise at 10 Hz frequency, **(b)** N_2O flux random error (blue) and flux random error due to instrumental noise (green) statistics; **(c)** the same as **(b)** but for relative fluxes. The boxplots present the lower and upper percentiles, quartiles and median values of the distributions. Based on flux measurements during the period DOY 206–271 (period II).

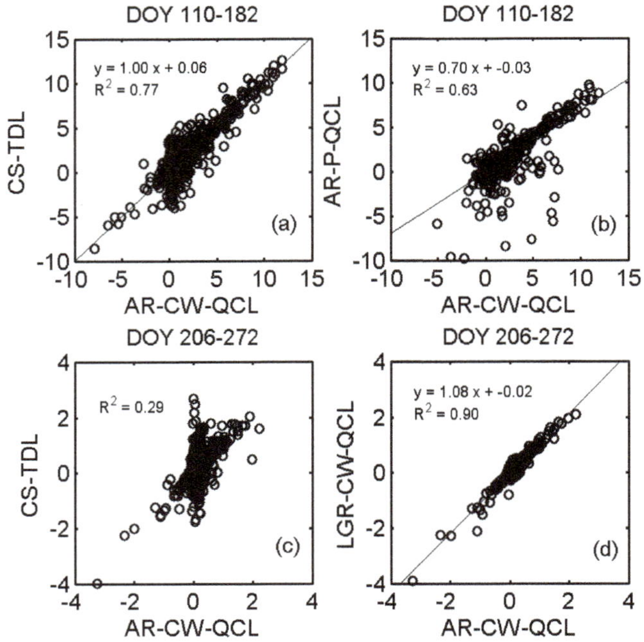

Figure 7. Correlation scatter plots of 30 min average N_2O fluxes (in nmol m^{-2} s^{-1}), as measured by CS-TDL and AR-P-QCL vs. AR-CW-QCL during period I DOY 110–181 **(a, b)**, and CS-TDL and LGR-CW-QCL vs. AR-CW-QCL during period II DOY 206–271 **(c, d)**. The lines present the linear fit with coefficients presented on the plots.

fluxes, determining essentially the minimum flux level that each instrument is able to measure at a given flux averaging interval (30 min period). For AR-CW-QCL, the respective lowest flux is around 10^{-2} nmol m^{-2} s^{-1} (as given by median in Fig. 6b), for LGR-CW-QCL and AR-P-QCL around 4×10^{-2} nmol m^{-2} s^{-1} and for CS-TDL 0.15 nmol m^{-2} s^{-1}.

The frequency distributions of the total flux random errors, calculated according to Eq. (3), are naturally higher than the flux error due to instrumental noise only. It can be observed that in the case of full flux random error, the difference between different instruments is reduced (Fig. 6b) because in addition to instrumental noise impact, this error statistic also incorporates the flux uncertainty due to the stochastic nature of turbulence. The relative random errors (Fig. 6c) are the largest for CS-TDL (being of the order of 100 % and in most cases less than ±300 %) and the smallest for AR-CW-QCL instruments (median around 30 % and the error mostly less than 100 %). It is the signal's noise of the instrument that contributes to the random error of the flux, determining which instrument is able to detect the lowest fluxes. In the case of CS-TDL the low-frequency signal drifting can also enlarge the total random error of the calculated flux.

3.3 Intercomparison of fluxes averaged over turbulent spectrum

It was observed that the fluxes calculated from CS-TDL measurements during the low flux period were dominated by stochastic uncertainty, being frequently of the order of the random uncertainties of the fluxes (Sect. 3.2). Therefore, the fluxes averaged over the 30 min period were compared for this instrument with AR-CW-QCL results over the period DOY 110–182, which included the high emissions episode starting on DOY 144 and exhibiting elevated fluxes until approximately DOY 155. In general, the fluxes with high magnitudes obtained by CS-TDL compared well with those of obtained by AR-CW-QCL (Fig. 7a). The AR-P-QCL system, as compared with AR-CW-QCL, showed systematically lower fluxes during the given period of high fluxes (slope 0.70). In spite of the lower noise level of this instrument, the coefficient of determination for this instrument (0.63) was lower than that for CS-TDL (0.77) in comparison to the fluxes as measured by AR-CW-QCL.

During the second observation period, when fluxes were much lower, CS-TDL was not able to determine fluxes with sufficiently small error and the correlation with AR-CW-QCL at the 30 min averaging level was very low (Fig. 7c).

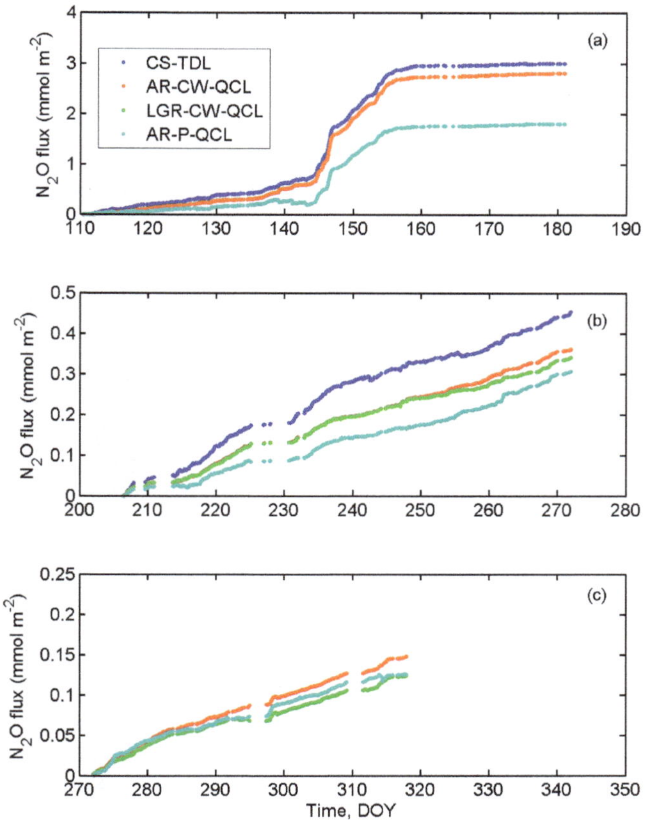

Figure 8. Cumulative sums of available flux data for three periods: **(a)** period I DOY 110–181 (20 April–30 June 2011), **(b)** period II DOY 206–271 (25 July–28 September 2011) and **(c)** period III DOY 272–324 (29 September–20 November 2011). Accumulation of fluxes for each instrument was performed only for data if measurements were available for all instruments used in respective period. No gap filling was used.

At around zero fluxes as measured by AR-CW-QCL, the results of CS-TDL showed scattered values visually between $\pm 2\,\mathrm{nmol\,m^{-2}\,s^{-1}}$. The noise level of CS-TDL around 2 ppb translates into a flux uncertainty due to instrumental noise of about 0.05 to $0.3\,\mathrm{nmol\,m^{-2}\,s^{-1}}$. The total flux error δ_F was within the range of 0.1 to $0.45\,\mathrm{nmol\,m^{-2}\,s^{-1}}$ (upper and lower quantiles of the distribution in Fig. 6b). We analysed the range of variation of CS-TDL fluxes during the given period DOY 206–272, conditionally selecting the observations when the observed fluxes by AR-CW-QCL were absolutely smaller than $0.15\,\mathrm{nmol\,m^{-2}\,s^{-1}}$ (90 % of N_2O flux random errors for AR-CW-QCL less than this value during the given period). The respective N_2O fluxes as determined by CS-TDL were characterized by the upper and lower quantiles of -0.27 and $0.52\,\mathrm{nmol\,m^{-2}\,s^{-1}}$. This is consistent with the upper quantile of the flux error distribution for CS-TDL. Therefore, the fluxes of CS-TDL, corresponding to close-to-zero fluxes as determined by AR-CW-QCL, were consistent with the flux error estimates.

The comparison of the 30 min average fluxes calculated from two instruments, AR-CW-QCL and LGR-CW-QCL, revealed very good correspondence and high correlation ($R^2 = 0.90$) even though those measurements corresponded to very low N_2O fluxes. The slope close to unity and a negligible intercept indicates that there is no systematic bias between the measurements of these systems (Fig. 7d).

3.4 Long-term averages and systematic differences

In order to evaluate the possible systematic differences, cumulative curves of the flux observations were calculated. No gap-filling of missing data was done, but instead only the half-hour periods were used when the results for all instruments were available. Thus, the cumulative sums are not assumed to represent the total emissions over the given periods, although rough estimates could be calculated by using the data coverage percentages presented in Table 4 to account for missing flux data. The summation of fluxes over the first and second periods reveals that CS-TDL gives the highest flux sums and AR-P-QCL the lowest, in particular during the first period (Fig. 8). The cumulative sums for fluxes obtained from AR-CW-QCL and LGR-CW-QCL measurements converge over periods II and III and show only small differences. Also, the cumulative fluxes measured by AR-P-QCL during these periods are very close to fluxes measured by the two other instruments. In order to assess the magnitude of the random errors in these differences, the random errors of the fluxes averaged over the three periods were calculated according to Eq. (5). The analysis revealed that the average fluxes for period II, obtained from the measurements of AR-CW-QCL and LGR-CW-QCL instruments, did not differ within calculated error limits, and were very close during period III to the result for AR-P-QCL (Table 4).

However, CS-TDL produced a 7 % higher total sum for the period of high fluxes (DOY 110–181 with an average flux of $0.87\,\mathrm{nmol\,m^{-2}\,s^{-1}}$ as determined by AR-CW-QCL) and a 29 % higher sum for the second period (DOY 206–271) compared to an average flux $0.142\,\mathrm{nmol\,m^{-2}\,s^{-1}}$ (average of AR-CW-QCL and LGR-CW-QCL results). The AR-P-QCL instrument determined for these two periods 36 and 13 % lower average fluxes, respectively. The possible reasons for this will be discussed in the next section. For the third period, the results of AR-P-QCL did not differ much from the results of the other two instruments.

4 Discussion

Performance of four instruments (see Tables 1 and 2) capable of fast response measurement of N_2O was studied throughout the 2011 growing season over a field cultivated with RCG in eastern Finland. The N_2O fluxes were small in the beginning of the season, increased significantly after fertilization (late May) and then decreased back to low, positive values after a

Table 4. Average fluxes (nmol $m^{-2} s^{-1}$) ±random error of the average. Period I DOY 110–181 (20 April–30 June 2011), period II DOY 206–271 (25 July–28 September 2011), period III DOY 272–324 (29 September–20 November 2011). Percent data available represents the fraction of half-hour periods when data from all instruments (three in periods I and III and all four in period II) was available (data from wind direction interval 50–110° excluded), relative to full time period length. Averaging of fluxes for each instrument was performed only for data if measurements were available for all instruments used in respective period. No gap filling was used.

	% data available	% data qualified (out of available)	no. of 30 min periods averaged	CS-TDL	AR-CW-QCL	LGR-CW-QCL	AR-P-QCL
Period I	69.2	75.2	1797	0.931 ± 0.018	0.870 ± 0.009	–	0.560 ± 0.011
Period II	55.0	79.4	1383	0.183 ± 0.010	0.146 ± 0.006	0.138 ± 0.007	0.124 ± 0.003
Period III	61.4	78.2	1220	–	0.067 ± 0.002	0.057 ± 0.002	0.058 ± 0.003

few weeks. Three instruments, CS-TDL, AR-CW-QCL and AR-P-QCL were operational during this high emissions period. During this period, all instruments detected the same flux dynamics, whereas the fluxes obtained by AR-P-QCL, the previous instrument version by Aerodyne, were lower compared to the other two instruments.

For many applications, the systematic errors of micrometeorological flux measurements of atmospheric trace gases are more important than the random errors. For example, for determination of annual balances (e.g. Kroon et al., 2010b) or for the comparison of exchange of different ecosystems (e.g. Nicolini et al., 2013) the systematic errors become very important. The two CW-QCL instruments compared very well on a half-hourly basis and produced statistically close cumulative fluxes over the period when the two instruments were simultaneously operational (25 July 2011–20 November 2011). The cumulative emission estimate obtained by CS-TDL for period II (25 July–28 September 2011) was 29 % higher than the average result for instruments based on the continuous-wave quantum cascade lasers, AR-CW-QCL and LGR-CW-QCL. AR-P-QCL obtained 36 % lower fluxes than AR-CW-QCL during the first period including the emission episode, whereas the correspondence with other instruments during the rest of the campaign was relatively good.

The systematic differences in fluxes could be the result of calibration and/or limited stability of the system over time. The impact of the instruments calibration (sensitivity shift) impact on flux systematic differences can be assessed by using calibration information (Sect. 2.2) as well as comparison of average concentrations measured by different instruments. The two analysers based on CW-QCL-s are expected to be very stable, which was confirmed by the measurements: the concentrations measured by these two instruments were very consistent and the slope (characterising sensitivity) of the 30 min average concentration comparison did not deviate from unity by more than 5 % (with the coefficient of determination of linear regression $R^2 = 0.86$).

The sensitivity of AR-P-QCL did not change more than 6.1 % between consecutive calibrations, and this can be considered as the maximum flux error arising from the calibration accuracy of this instrument (Sec. 2.2). Nevertheless, the correlation of the 30 min average concentration measured by this instrument as compared to AR-CW-QCL was not as good (for the period DOY 206–272, a slope of 1.05 was determined with $R^2 = 0.63$). The concentration comparison presented here does not imply that the calibration bias was the reason for the observed flux systematic difference for the instrument AR-P-QCL.

The analyser CS-TDL is known for its signal drifting as illustrated and discussed by Mammarella et al. (2010), and the absolute concentrations were not well determined during our campaign. Therefore, accurate measurement of absolute concentration by this instrument over a long period of time cannot be expected, and the concentration comparison was not used as the method for evaluation of the instrument's calibration impact on flux systematic bias. Note that signal drifting makes the time series produced by the instrument essentially non-stationary and therefore enhances the random variability of the flux estimate around the true value. However, such enhanced random uncertainty does not systematically affect the cumulative sums over longer periods.

In the case of low fluxes the water vapour dilution and spectral line broadening effects are the primary suspects for the reasons in systematic differences in fluxes (e.g. Peltola et al., 2014). Close correspondence of the concentrations and fluxes as measured by AR-CW-QCL and LGR-CW-QCL let us conclude that the spectroscopic and water vapour dilution corrections for these instruments were adequate. Note that those corrections were done by built-in functionality in the case of LGR-CW-QCL. For AR-CW-QCL, the respective corrections were done in post-processing phase for period I and by built-in software for the rest of the campaign.

The only evident systematic flux error source that could affect performance of CS-TDL would be incomplete drying of sample air. If that was the case, then the calculated fluxes would have suffered from missing partial density and spectroscopic corrections. Since the water fluxes are dominantly upward, a respective correction would tend to increase the flux values, therefore increasing even more the systematic difference relative to other instruments.

The instrument ARI-P-QCL is based on the pulsed quantum cascade laser. For this instrument, the experimentally

Table 5. Average micrometeorological conditions during the experimental periods. Period I DOY 110–181 (20 April–30 June 2011), period II DOY 206–271 (25 July–28 September 2011), period III DOY 272–324 (29 September–20 November 2011). Daytime was defined by the elevation of sun higher than zero and night-time lower than zero, respectively. Average latent heat fluxes were determined from IRGA measurements.

	Temperature	Air rel. humidity, %	Wind speed, $m\,s^{-1}$	Friction velocity, $m\,s^{-1}$	Sensible heat flux, $W\,m^{-2}$	Latent heat flux, $W\,m^{-2}$
Day, I	11.6	62.9	2.21	0.28	27.5	78.9
Night, I	6.5	78.3	1.34	0.14	−20.2	8.1
Day, II	15.3	75.2	1.35	0.26	9.7	109.3
Night, II	11.2	90.3	1.06	0.17	−18.6	10.1
Day, III	6.1	85.0	1.46	0.29	−10.8	41.5
Night, III	4.8	90.6	1.21	0.23	−23.5	11.5

determined spectroscopic correction coefficient was much lower than the coefficient for AR-CW-QCL (Table 1). The reason for systematically lower values of fluxes determined by AR-P-QCL from the beginning of the experiment in April until June 2011, but subsequent relatively good comparison with other instruments towards the end of the experiment in November 2011, is not known. Two types of corrections were applied to N_2O fluxes: the spectroscopic correction to account for the impact of water vapour on the absorption line shape, and the co-spectral correction. The latter correction was comparable to all instruments (Table 3) and does not introduce a significant difference between instruments. The spectroscopic correction was applied together with the water vapour dilution correction (Sect. 2.3) and can constitute a major correction depending on the value of the coefficient b. The correction is related to the water vapour flux, which was during the daytime on the average around $100\,Wm^{-2}$ (periods I and II, Table 5), with mid-day averages around 150 to $200\,Wm^{-2}$. Considering an average concentration of N_2O around 330 ppb and spectroscopic correction value $b = 0.39$ (the value for AR-CW-QCL), the spectroscopic correction can be a few tenths of $nmol\,m^{-2}\,s^{-1}$ during mid-day, which is of the order of the flux magnitude. We used all auxiliary data available to investigate the possible reasons for the systematic differences, but found no explaining variable or reason. In particular, no systematic variation of the residual between AR-P-QCL and AR-CW-QCL fluxes was found over wide range of latent heat fluxes from −20 to $250\,W\,m^{-2}$. This proves that the dilution and spectroscopic corrections were properly accounted for. In addition, larger spectroscopic correction would not explain the systematic difference observed only during the first period.

Thus the reasons for flux underestimation by AR-P-QCL during period I are not known, and we suggest that extreme care should be exercised during long-term measurement campaigns both with N_2O and H_2O calibrations due to the strong impact of the water vapour on the N_2O flux through spectroscopic and dilution corrections.

A comment should be made regarding the observation level used in the study. When RCG was grown high, the measurement level was only about 0.5 m above the canopy top. The measurements within the roughness sublayer can be disturbed in terms of several statistics, but the impact can be expected to reveal more in spectral shapes than in integral statistics. The spectra obtained for N_2O (Figs. 3 and 5) were dominated by white noise over wider (CS-TDL) or narrow (AR-CW-QCL) frequency ranges depending on the instrument in question. The temperature spectra were similarly affected by the noise but only at the high-frequency end of spectra and we believe do not show evidence of a canopy impact on spectral shapes. We checked also the spectra for vertical wind speed (not shown). The spectra exhibited smooth and consistent shapes, without the particular impact of the canopy foliage on spectral forms usually observed inside canopies. Launiainen et al. (2007) studied the turbulence statistics and spectral shapes within pine forest canopy. They did not observe deviation of spectral shapes above canopy at height $z/h = 1.47$ (h being the canopy height) from the atmospheric surface layer forms; within the crown space ($z/h = 0.78$), the spectra deviated only slightly from the above-canopy forms. Within the trunk space ($z/h = 0.4$), the spectra were distorted due to the drag imposed by the canopy elements. This supports that the spectra measured close to but above canopy are weakly affected by the canopy presence. Thus, we do not expect that the relatively low observation level biases the overall N_2O flux magnitude and that the comparison of instrumentation is affected. Also, the effect on the instrumental noise and flux random uncertainty analysis is expected to be very limited through the influence on the covariance functions. The positive impact of the close positioning of the system could be its higher sensitivity in detecting the low fluxes through higher concentration fluctuations expected (more) close to the source level.

Important characteristics of the instruments for performing the EC measurements are the response time and the noise level. The response times for CS-TDL, AR-CW-QCL and AR-P-QCL flux measurements systems were determined to be 0.12 and 0.07 and 0.08 s, respectively. The main factors affecting the response time of the closed-path EC system are the damping of fluctuations in the sampling line and the instrumental response. Since the flow rate of the CS-TDL system was higher, it can be concluded that the response characteristics of other two instruments are superior. The response time of the EC system including LGR-CW-QCL was larger due to the laminar tube flow regime, but the instrumental response was not determined based on the current field measurements.

In order to understand drivers of exchange and infer the broad average fluxes such as seasonal or annual sums by using gap-filling methodologies, it is important that the exchange at a shorter timescale is distinguishable from random variation. Therefore, an understanding of random errors is important when working with low fluxes as is frequently the case with N_2O. At the half-hour averaging timescale, the flux estimates for AR-CW-QCL and LGR-CW-QCL instruments were very well correlated and showed good correspondence. Apart from high N_2O fluxes exceeding a few nmol m^{-2} s^{-1} during the high emissions period, CS-TDL was not able to resolve the emission fluxes at half-hourly timescale. Therefore, one can conclude that CS-TDL is not suitable for measuring such low fluxes if the aim is to resolve fluxes at hourly timescale and not the daily or longer averages.

Aerodyne AR-CW-QCL had the lowest noise level (around 0.12 ppb at 10 Hz sampling rate) compared to Los Gatos LGR-CW-QCL instrument (SD of noise 0.60 ppb) and has therefore an advantage in resolving low fluxes over short averaging periods. The noise level of AR-P-QCL was comparable to LGR-CW-QCL, but the old-generation instrument Campbell CS-TDL suffered clearly from higher noise level (around 2 ppb). Huang et al. (2014) reported for the instrument similar to AR-CW-QCL the precision 0.066 ppb for 10 Hz. The value obtained by us was higher roughly by a factor of 2. According to the manufacturer, the precision of LGR-CW-QCL is 0.1 ppb at 1 Hz averaging; at 10 Hz this would correspond to 0.32 ppb. We have determined again a median value roughly twice higher than this. Kroon et al. (2007) reported for the instrument similar to AR-P-QCL the precision value of 0.5 ppb $Hz^{-1/2}$ (equivalent to 1.6 ppb at 10 Hz), whereas Neftel et al. (2007) and Eugster et al. (2007) report 0.3 ppb $Hz^{-1/2}$(equivalent to 0.95 ppb at 10 Hz). Pihlatie et al. (2005) and Wang et al. (2013) report noise for CS-TDL as 1 ppb and 1.5 ppb (at 10 Hz), respectively. Under field conditions the instrumental noise can be somewhat higher compared to laboratory conditions where the instrumental characteristics are typically studied. Also, the estimation method from the field records where the turbulent variation is superimposed by the instrumental noise can introduce some uncertainty. In summary, the observed instrumental noise characteristics for instruments compare well with the results reported by others and are useful in characterising instrumental performance.

The flux errors due to instrumental noise for the observation conditions prevailing at the site were determined to be around 10^{-2} nmol m^{-2} s^{-1} for AR-CW-QCL, 4×10^{-2} nmol m^{-2} s^{-1} for LGR-CW-QCL and AR-P-QCL and 0.15 nmol m^{-2} s^{-1} for CS-TDL. Based on a half-hour and long-term flux comparison, the best correspondence was observed between the systems with the new-generation instruments AR-CW-QCL and LGR-CW-QCL, of which the former has the advantage of detecting lower fluxes at a half-hourly averaging basis (lower noise level).

The signal's noise of the anemometer used by the UH (USA1 by METEK) was determined to be 0.037 m s^{-1} at 10 Hz sampling frequency for the vertical wind speed component. The noise level of the anemometer employed by the UEF was similar. The flux error due to anemometer's noise for the observation conditions prevailing at the site during the period DOY 206–271 (the period for the statistics presented in Fig. 6) was determined to be around 2×10^{-3} nmol m^{-2} s^{-1} (the median value). This was much less than the respective flux error around 10^{-2} nmol m^{-2} s^{-1} for the instrument AR-CW-QCL, which had the lowest noise level 0.012 ppb (median value) of all instruments compared. Therefore, the assumption that the anemometer's noise affects flux detection much less than the gas analysers was well justified.

The chamber techniques are widely used to measure the soil N_2O exchange. The traditional way to perform chamber measurements is to determine the gas concentration at several moments during the chamber operation (called deployment time, DT). In such data collection the sources of uncertainty are the imprecision related to gas sampling (either manual or automatic) as well as instrumental uncertainty (e.g. Venterea et al., 2009), leading to a measurement precision which is called the detection limit of chamber-based flux measurement system. Neftel et al. (2007) report a flux detection limit of about 0.23 nmol m^{-2} s^{-1} for their chamber system with DT of 10 min and the concentration sampling interval of 1 min. The measurement cycle of the system was however 2 h. Wang et al. (2013) found for their automatic and manual chamber systems detection limits of about 5 μg m^{-2} h^{-1} (0.05 nmol m^{-2} s^{-1}) for hourly DT. Their instrument precision was high, around 0.4 % relative to ambient N_2O concentration. By using the methodology and scaled results presented by Parkin et al. (2012), we estimated the flux detection limit of a chamber system with an assumed chamber height of 0.5 m, area of 0.25 m^2, deployment time 30 min and instrumental precision as high as 0.1 % to be 0.03 nmol m^{-2} s^{-1}. It has to be noted that the flux detection limit of the chamber systems depends on several factors such as the type of the chamber and respective sampling method, the precision of the instrument, chamber dimensions and operation time (DT). Nevertheless, the obtained result is well comparable with the EC systems. The random error of N_2O fluxes for 30 min averaging time for the instrument with lowest noise, the AR-CW-QCL instrument, was found to be

0.036 nmol m^{-2} s^{-1} (the median value). Note that here we compare the flux detection limit of the chamber-based systems (which accounts for all possible sources of uncertainty) with the total stochastic error of the EC fluxes. The results are of the same magnitude.

In this study we followed the methodology proposed by Mauder et al. (2013) in quantification of the random errors in EC fluxes, i.e. the stochastic error and the error due to instrumental noise in flux. The relative random errors obtained in our study were much larger than the respective errors reported by Mauder et al. (2013) for CO_2 measurements, evidencing that the importance of random errors depends on the trace gas of interest via instrumental precision and the flux magnitude ratio. Kroon et al. (2010a) focus on the evaluation of the EC flux measurements of CH_4 and N_2O specifically. They observed over a dairy farm site the fluxes in the range of 15 to 110 ng N m^{-2} s^{-1} (0.5 to 4 nmol m^{-2} s^{-1}), which they classified in low to high flux classes. They performed calibration of the instrument similar to our AR-P-QCL weekly and considered the respective uncertainty random over longer periods of time. Kroon et al. (2010a) reported the average daily and monthly flux relative uncertainties of 31 and 7 %, respectively. In our study the N_2O fluxes were typically much smaller (excluding the fertilization episode), around 0.1 to 0.3 nmol m^{-2} s^{-1}. We measured with the similar instrument 36 % lower fluxes than obtained by AR-CW-QCL over the period DOY 110–181 and 13 % lower fluxes than obtained by two new-generation instruments over the period DOY 206–271. Evidently our measurements performance was affected by an unidentified error source, systematic in nature. In evaluation of the annual balances of CH_4 and N_2O fluxes over a managed fen meadow, Kroon et al. (2010b) made an assumption that the uncertainty in EC fluxes was random and was neglected in the evaluation of long-term averages. In our results, this assumption was contradicted, and we suggest that all possible systematic error sources should be considered very carefully in planning, implementing and evaluating the flux measurements of trace gases.

In analysing the random errors of the fluxes Kroon et al. (2010a) assumed that the flux error due to instrumental precision in concentration measurement was negligible. We observed that this was not necessarily the case for N_2O when low flux levels were measured and demonstrated that the method originally proposed by Lenchow et al. (2000) to determine instrumental noise variance worked well in field conditions over a long period of time.

5 Conclusions

The new instruments based on continuous-wave quantum cascade lasers, AR-CW-QCL and LGR-CW-QCL, were stable throughout of the campaign in terms of determining the absolute concentrations and obtaining close fluxes.

The older instruments CS-TDL and AR-P-QCL measured systematically different fluxes over subperiods of the campaign up to +29 and −36 %, respectively, compared to the new instruments based on CW-QCL-s, whereas the systematic differences did not prevail throughout the campaign. The reasons for the systematic differences were not identified. We suggest that special emphasis should be on the instrumental stability and correcting procedures that can systematically affect the accuracy of measured fluxes when conducting long-term measurements of prevailingly low fluxes.

The lowest noise level was determined for AR-CW-QCL (0.12 ppb at 10 Hz sampling rate) and the highest for the old-generation instrument CS-TDL (precision 2 ppb at 10 Hz sampling rate). During the period DOY 206–272, when all instruments were operational, the lower-quantile/median/upper-quantile statistics of the fluxes measured by AR-CW-QCL instrument were 0.008/0.11/0.31 nmol m^{-2} s^{-1}, respectively.

The random errors of fluxes originate from the stochastic nature of turbulence (one-point sampling over limited time interval). Additionally, the instrumental noise contributes to the random flux error. The median values for flux errors during the period DOY 206–272 (error due to instrumental noise/total error) were detected for the instruments as follows: for CS-TDL 0.155/0.255, AR-CW-QCL 0.010/0.036, LGR-CW-QCL 0.046/0.065 and AR-P-QCL 0.031/0.068 nmol m^{-2} s^{-1}. These error statistics indicate that (i) the major component of the flux random error source is the instrumental noise and (ii) the flux errors for CS-TDL are dominantly larger than the flux magnitude, and only in the case of AR-CW-QCL can the flux error due to instrumental noise be said to be much smaller than the typical flux value.

The following fractions of fluxes were smaller than the stochastic flux error: in the case of CS-TDL, 47 %; AR-CW-QCL, 15 %; LGR-CW-QCL, 28 %; and AR-P-QCL, 30 %. We conclude that apart from AR-CW-QCL, a large fraction of the fluxes were within the error magnitude of single half-hour observations.

With the new-generation analysers based on continuous-wave QCL-s, N_2O fluxes can be measured with the EC at locations where the fluxes are small, well below the detection limit of older instruments (CS-TDL for instance). According to our analysis, the new instruments enable one to attain the flux precision as good as the precision of the modern chamber systems. Thus, the new instruments open up the possibility of studying N_2O exchange in new ecosystems, broadening scientific perspectives.

Acknowledgements. This work was supported by the Academy of Finland (project nos. 118780 and 127456). ICOS (271878), ICOS-Finland (281255) and ICOS-ERIC (281250), DEFROST Nordic Centre of Excellence and InGOS EU are gratefully acknowledged for funding this work. This work was also supported by institutional research funding (IUT20-11) of the Estonian Ministry of Education and Research. The UEF part of the research work was supported by the funding from the UEF infrastructure funding, Academy of Finland FidiPro programme (PIs – Profs Pertti Martikainen and Seppo Kellomäki) and the Ministry of Agriculture and Forestry, Finland.

Edited by: A. Neftel

References

Aubinet, M., Grelle, A., Ibrom, A., Rannik, Ü., Moncrieff, J., Foken, T., Kowalski, A.S., Martin, P.H., Berbigier, P., Bernhofer, Ch., Clement, R., Elbers, J., Granier, A., Grünwald, T., Morgenstern K., Pilegaard K., Rebmann C., Snijders W., Valentini R. and Vesala, T.: Estimates of the annual net carbon and water exchange of European forests: the EUROFLUX methodology, Advances Ecol, Res., 30, 113-175, 2000.

Christensen, S., Ambus, P., Arah, J. R., Clayton, H., Galle, B., Griffith, D. W. T., Hargreaves, K. J., Klemedtsson, L., Lind, A. M., Maag, M., Scott, A., Skiba, U., Smith, K. A., Welling, M., and Wienhold, F. G.: Nitrous oxide emissions from an agricultural field: comparison between measurements by flux chamber and micrometeorological techniques, Atmos. Environ., 30, 4183–4190, 1996.

Eugster, W., Zeyer, K., Zeeman, M., Michna, P., Zingg, A., Buchmann, N., and Emmenegger, L.: Methodical study of nitrous oxide eddy covariance measurements using quantum cascade laser spectrometery over a Swiss forest, Biogeosciences, 4, 927–939, doi:10.5194/bg-4-927-2007, 2007.

Famulari, D., Nemitz, E., Di Marco, C., Phillips, G.J., Thomas, R., House, E. and Fowler, D.: Eddy-Covariance measurements of nitrous oxide fluxes above a city, Agric. For. Meteorol. 150, 786–793, 2010.

FAO, World reference base for soil resources 2006, World soil resources reports 103, Rome, Italy, 2006.

Finkelstein, P. L. and Sims, P. F.: Sampling error in eddy correlation flux measurements, J. Geophys. Res., 106, 3503–3509, 2001.

Finnigan, J. J., Clement, R., Malhi, Y., Leuning, R., and Cleugh, H. A.: A re-evaluation of longterm flux measurement techniques – Part I: Averaging and coordinate rotation, Bound.-Layer Meteorol., 107, 1–48, 2003.

Flechard, C., Neftel, A., Jocher, M., and Amman, C.: Bi-directional soil-atmosphere N2O exchange over two mown grassland systems with contrasting management practices, Glob. Change Biol., 11, 2114–2127, 2005.

Foken, T., Leuning, R., Oncley, S. R., Mauder, M., and Aubinet, M.: Corrections and Data Quality Control, in Eddy Covariance. A Practical Guide to Measurement and Data Analysis, Springer Science+Business Media B.V., 85–131, 2012.

Foken, T. and Wichura, B.: Tools for quality assessment of surface-based flux measurements, Agr. Forest Meteorol., 78, 83–105, 1996.

Horst, T. W.: A simple formula for attenuation of eddy fluxes measured with first-order-response scalar sensors, Bound.-Layer Meteorol., 82, 219–233, 1997.

Huang, H., Wang, J., Hui, D., Miller, D. R., Bhattarai, S., Dennis, S., Smart, D., Sammis, T., and Reddy, K. C.: Nitrous oxide emissions from a commercial cornfield (Zea mays) measured using the eddy-covariance technique, Atmos. Chem. Phys. Discuss., 14, 20417–20460, doi:10.5194/acpd-14-20417-2014, 2014.

Järvi, L., Nordbo, A., Rannik, Ü., Haapanala, S., Riikonen, A., Mammarella, I., Pihlatie, M., and Vesala, T.: Urban nitrous oxide fluxes measured using the eddy-covariance technique in Helsinki, Finland. Boreal Env. Res., 19, 108–121, 2014.

Kaimal, J. C. and Finnigan, J. J.: Atmospheric Boundary Layer Flows. Their Structure and Measurement, Oxford University Press, New York, 1994.

Kaimal, J. C., Wyngaard, J. C., Izumi, Y., and Cotè, O. R.: Spectral characteristics of surface layer turbulence, Q. J. Roy. Meteor. Soc., 98, 563–589, 1972.

Korhonen, J. F. J., Pihlatie, M., Pumpanen, J., Aaltonen, H., Hari, P., Levula, J., Kieloaho, A.-J., Nikinmaa, E., Vesala, T., and Ilvesniemi, H.: Nitrogen balance of a boreal Scots pine forest, Biogeosciences, 10, 1083–1095, doi:10.5194/bg-10-1083-2013, 2013.

Kroon, P. S., Hensen, A., Jonker, H. J. J., Zahniser, M. S., van 't Veen, W. H., and Vermeulen, A. T.: Suitability of quantum cascade laser spectroscopy for CH_4 and N_2O eddy covariance flux measurements, Biogeosciences, 4, 715–728, doi:10.5194/bg-4-715-2007, 2007.

Kroon, P. S., Hensen, A., Jonker, H. J. J., Ouwersloot, H. G., Vermeulen, A. T., and Bosveld, F. C.: Uncertainties in eddy covariance flux measurements assessed from CH_4 and N_2O observations, Agr. Forest Meteorol., 150, 806–816, 2010a.

Kroon, P. S., Schrier-Uijl, A. P., Hensen, A. , Veenendaal, E. M., and Jonker, H. J. J.: Annual balances of CH_4 and N_2O from a managed fen meadow using eddy covariance flux measurements, Europ. J. Soil Sci., 61, 773–784, 2010b.

Launiainen, S., Vesala, T., Mölder, M., Mammarella, I., Smolander, S., Rannik, Ü., Kolari, P., Hari, P., Lindroth, A., and Katul, G.G.: Vertical variability and effect of stability on turbulence characteristics down to the floor of a pine forest, Tellus B, 59, 919–936, 2007.

Laville, P., Henault, C., Renault, P., Cellier, P., Oriol, A., Devis, X., Flura, D., and Germon, J. C.: Field comparison of nitrous oxide emission measurements using micrometeorological and chamber methods, Agronomie, 17, 375–388, 1997.

Lee, B. H., Wood, E. C., Zahniser, M. S., McManus, J. B., Nelson, D. D., Herndon, S. C., Santoni, G. W., Wofsy, S. C., and Munger, J. W.: Simultaneous measurements of atmospheric HONO and NO_2 via absorption spectroscopy using tunable mid-infrared continuous-wave quantum cascade lasers, Appl. Phys. B, 102, 417–423, 2011.

Lenschow, D. H. and Kristensen, L.: Uncorrelated noise in turbulence measurements, J. Atm. Oceanic Technol., 2, 68-81, 1985.

Lenschow, D. H., Mann, J., and Kristensen, L.: How long is long enough when measuring fluxes and other turbulence statistics?, J. Atm. Oceanic Technol., 18, 661–673, 1994.

Lenschow, D., Wulfmeyer, V., and Senff, C.: Measuring second-through fourth-order moments in noisy data, J. Atmos. Ocean. Technol., 17, 1330–1347, 2000. Mammarella, I., Launiainen, S., Gronholm, T., Keronen, P., Pumpanen, J., Rannik, Ü., and Vesala, T.: Relative humidity effect on the high frequency attenuation of water vapour flux measured by a closed-path eddy covariance system, J. Atmos. Ocean. Tech., 26, 1856–1866, 2009.

Mammarella, I., Werle, P., Pihlatie, M., Eugster, W., Haapanala, S., Kiese, R., Markkanen, T., Rannik, Ü., and Vesala, T.: A case study of eddy covariance flux of N_2O measured within forest ecosystems: quality control and flux error analysis, Biogeosciences, 7, 427–440, doi:10.5194/bg-7-427-2010, 2010.

Massman, W.: A simple method for estimating frequency response corrections for eddy covariance systems, Agr. Forest Meteorol., 104, 185–198, 2000.

Mauder, M., Cuntz, M., Druee, C., Graf, A., Rebmann, C., Schmid, H. P., Schmidt, M., and Steinbrecher, R.: A strategy for quality and uncertainty assessment of long-term eddy covariance measurements, Agric. For. Meteorol., 169, 122–135, 2013.

McManus, J. B., Nelson, D. D., Shorter, J. H., Jiménez, R., Herndon, S., Saleska, S., and Zahniser, M. S.: A high precision pulsed QCL spectrometer for measurements of stable isotopes of carbon dioxide, J. Modern Optics, 52, 2309–2321, 2005.

Moncrieff, J. B., Massheder, J. M., de Bruin, H., Elbers, J., Friborg, T., Heusinkveld, B., Kabat, P., Scott, S., Soegaard, H., and Verhoef, A., : A system to measure surface fluxes of momentum, sensible heat, water vapour and carbon dioxide, J. Hydrol., 188/189, 589–611, 1997.

Moore, C. J.: Frequency response corrections for eddy correlation systems, Boundary-Layer Meteorol., 37, 17–35, 1986.

Neftel, A., Flechard, C., Ammann, C., Conen, F., Emmenegger, L. and Zeyer, K.: Experimental assessment of N_2O background fluxes in grassland systems, Tellus B, 59, 470–482, 2007.

Neftel, A., Ammann, C., Fischer, C., Spirig, C., Conen, F., Emmenegger, L., Tuzson, B., and Wahlen, S.: N_2O exchange over managed grassland: application of a quantum cascade laser spectrometer for micrometeorological flux measurements, Agr. Forest Meteorol., 150, 775–785, 2010.

Nelson, D. D., Shorter, J. H., McManus, J. B., and Zahniser, M. S.: Sub-part-per-billion detection of nitric oxide in air using a thermoelectrically cooled mid-infrared quantum cascade laser spectrometer, Appl. Phys. B, 75, 343–350, 2002.

Nicolini, G., Castaldi, S., Fratini, G., and Valentini, R.: A literature overview of micrometeorological CH_4 and N_2O flux measurements in terrestrial ecosystems, Atmos. Environ., 81, 311–319, 2013.

Parkin, T. B., Venterea, R. T., and Hargreaves, S. K.: Calculating the Detection Limits of Chamber-based Soil Greenhouse Gas Flux Measurements, J. Environ. Qual., 41, 705–715, 2012.

Peltola, O., Hensen, A., Helfter, C., Belelli Marchesini, L., Bosveld, F. C., van den Bulk, W. C. M., Elbers, J. A., Haapanala, S., Holst, J., Laurila, T., Lindroth, A., Nemitz, E., Röckmann, T., Vermeulen, A. T., and Mammarella, I.: Evaluating the performance of commonly used gas analysers for methane eddy covariance flux measurements: the InGOS inter-comparison field experiment, Biogeosciences, 11, 3163–3186, doi:10.5194/bg-11-3163-2014, 2014.

Pihlatie, M., Rinne, J., Ambus, P., Pilegaard, K., Dorsey, J. R., Rannik, Ü., Markkanen, T., Launiainen, S., and Vesala, T.: Nitrous oxide emissions from a beech forest floor measured by eddy covariance and soil enclosure techniques, Biogeosciences, 2, 377–387, doi:10.5194/bg-2-377-2005, 2005.

Pihlatie, M., Kiese, R., Brüggemann, N., Butterbach-Bahl, K., Kieloaho, A.-J., Laurila, T., Lohila, A., Mammarella, I., Minkkinen, K., Penttilä, T., Schönborn, J., and Vesala, T.: Greenhouse gas fluxes in a drained peatland forest, Biogeosciences, 7, 1715–1727, doi:10.5194/bg-7-1715-2010, 2010.

Pirinen, P., Simola, H., Aalto, J., Kaukoranta, J., Karlsson, P., and Ruuhela, R.: Tilastoja Suomen ilmastosta 1981–2010, Finnish Meteorological Institute, Helsinki, 2012.

Provencal, R., Gupta, M., Owano, T. G., Baer, D. S., Ricci, K. N., O'Keefe, A., and Podolske, J. R.: Cavity-enhanced quantum-cascade laser-based instrument for carbon monoxide measurements, Appl. Optics, 44, 6712–6717, 2005.

Rannik, Ü., and Vesala, T.: Autoregressive filtering versus linear detrending in estimation of fluxes by the eddy covariance method, Bound.-Layer Meteorol., 91, 259–280, 1999.

Rannik, Ü., Kolari, P., Vesala, T., and Hari, P.: Uncertainties in measurement and modelling of net ecosystem exchange of a forest ecosystem at different time scales, Agric. Forest. Meteorol., 138, 244–257, 2006.

Rannik, Ü., Mammarella, I., Aalto, P., Keronen, P., Vesala, T., and Kulmala, M.: Long-term particle flux observations Part I: Uncertainties and time-average statistics, Atmos. Environ., 43, 3431–3439, 2009.

Scanlon, T. M. and Kiely, G.: Ecosystem-scale measurements of nitrous oxide fluxes for an intensely grazed, fertilized grassland, Geophys. Res. Letter, 30, 1852, doi:10.1029/2003GL017454, 2003.

Schotanus, P., Nieuwstadt, F. T. M., and Debruin, H. A. R.: Temperature-measurement with a sonic anemometer and its application to heat and moisture fluxes, Bound.-Layer Meteorol., 26, 81–93, 1983.

Smith, K. A., Clayton, H., Arah, J. R. M., Christensen, S., Ambus, P., Fowler, D., Hargreaves, K. J., Skiba, U., Harris, G. W., Wienhold, F. G., Klemedtsson, L., and Galle, B.: Micrometeorological and chamber methods for measurement of nitrous oxide fluxes between soils and the atmosphere: Overview and conclusions, J. Geophys. Res., 99, 16541–16548, 1994.

Venterea, R. T., Spokas, K. A., and Baker, J. M.: Accuracy and Precision Analysis of Chamber-Based Nitrous Oxide Gas Flux Estimates, Soil Sci. Soc. Am. J., 73, 1087–1093, 2009.

Vickers, D. and Mahrt, L.: Quality control and flux sampling problems for tower and aircraft data, J. Atmos. Ocean. Tech., 14, 512–526, 1997.

Wang, K, Zheng, X., Pihlatie, M., Vesala, T., Liu, C., Haapanala, S., Mammarella, I., Rannik, Ü., and Liu, H.: Comparison between static chamber and tunable diode laser-based eddy covariance techniques for measuring nitrous oxide fluxes from a cotton field, Agric. Forest Meteorol., 171/172, 9–19, 2013.

Werle, P., Muecke, R., and Slemr, F.: The limits of signal averaging in atmospheric trace gas monitoring by tunable diode-laser absorption spectroscopy, Appl. Phys. B 57, 131–139, 1993.

Wienhold, F. G., Frahm, H., and Harris, G. W.: Measurements of N_2O fluxes from fertilized grassland using a fast response tunable diode laser spectrometer, J. Geophys. Res., 99, 16557–16567, 1994.

Zahniser, M. S., Nelson, D. D., McManus, J. B., Hern-don, S. C., Wood, E. C., Shorter, J. H., Lee, B. H., Santoni, G. W., Jimenez, R., Daube, B. C., Park, S., Kort, E. A., and Wofsy, S. C.: Infrared QC laser applications to field measurements of atmospheric trace gas sources and sinks in environmental research: enhanced capabilities using continuous wave QCLs, Proc. SPIE, 7222, doi:10.1117/12.815172, 2009.

4

CH_4 and N_2O dynamics in the boreal forest–mire ecotone

B. Ťupek[1], K. Minkkinen[1], J. Pumpanen[1], T. Vesala[2], and E. Nikinmaa[1]

[1]Department of Forest Sciences, P.O. Box 27, 00014 University of Helsinki, Finland
[2]Department of Physics, P.O. Box 48, 00014 University of Helsinki, Finland

Correspondence to: B. Ťupek (boris.tupek@helsinki.fi)

Abstract. In spite of advances in greenhouse gas research, the spatiotemporal CH_4 and N_2O dynamics of boreal landscapes remain challenging, e.g., we need clarification of whether forest–mire transitions are occasional hotspots of landscape CH_4 and N_2O emissions during exceptionally high and low ground water level events.

In our study, we tested the differences and drivers of CH_4 and N_2O dynamics of forest/mire types in field conditions along the soil moisture gradient of the forest–mire ecotone. Soils changed from Podzols to Histosols and ground water rose downslope from a depth of 10 m in upland sites to 0.1 m in mires. Yearly meteorological conditions changed from being exceptionally wet to typical and exceptionally dry for the local climate. The median fluxes measured with a static chamber technique varied from -51 to $586\,\mu g\,m^{-2}\,h^{-1}$ for CH_4 and from 0 to $6\,\mu g\,m^{-2}\,h^{-1}$ for N_2O between forest and mire types throughout the entire wet–dry period.

In spite of the highly dynamic soil water fluctuations in carbon rich soils in forest–mire transitions, there were no large peak emissions in CH_4 and N_2O fluxes and the flux rates changed minimally between years. Methane uptake was significantly lower in poorly drained transitions than in the well-drained uplands. Water-saturated mires showed large CH_4 emissions, which were reduced entirely during the exceptional summer drought period. Near-zero N_2O fluxes did not differ significantly between the forest and mire types probably due to their low nitrification potential. When upscaling boreal landscapes, pristine forest–mire transitions should be regarded as CH_4 sinks and minor N_2O sources instead of CH_4 and N_2O emission hotspots.

1 Introduction

Soil fertility, soil water content, and soil carbon storage of boreal forests varies between well-drained mineral soils mainly found in uplands and poorly drained organic soils mainly found in peatlands (Seibert et al., 2007; Weishampel et al., 2009). The CH_4 and N_2O fluxes from mineral and organic soils are impacted by varying soil moisture conditions (Solondz et al., 2008; Pihlatie et al., 2004). Typical mineral soil forests are small sinks of CH_4 and small sources or sinks of N_2O (Moosavi and Crill, 1997; Pihlatie et al., 2007). Sparsely forested peatlands are typically large or small sources of CH_4 and small sources or sinks of N_2O (Martikainen et al., 1995; Nykänen et al., 1995; D'Angelo and Reddy, 1998). Field CH_4 and N_2O studies of natural boreal forest–mire ecotones are rare (e.g., Ullah et al., 2009; Ullah and Moore, 2011) in comparison to those of typical forests or mires. However, the area of forest–mire transitions is relatively large, e.g., in Finland, forested mires with an organic horizon < 30 cm cover 1.5 million hectare or approximately 7 % of the total forest area (Finnish statistical yearbook of forestry, 2013), and at the present time it is not clear whether the terrestrial–aquatic interfaces, such as the forest–mire transition, represents a biogeochemical hotspot of CH_4 and N_2O emissions (McClain et al., 2003).

The lagg transitional zone in the forest–mire ecotone receives nutrients from the adjacent mineral soil runoff, and is thus more minerotrophic, biologically diverse, and productive than open mires or bogs (Howie and Meerveld, 2011). Furthermore, ecotones between forests and mires are ecological switches (Agnew et al., 1993), where the vegetation of forests and mires coincide and soils frequently undergo fluctuations in water level position and chemistry (Hartshorn et al., 2003; Howie and Meerveld, 2011), and where the CH_4

and N_2O dynamics of forest–mire transitions may be expected to differ generally and on a year-to-year basis from those of typical forests and mires.

The CH_4 uptake of forest soils is a result of CH_4 oxidizing aerobic methanotrophs sensitive to water saturation, soil porosity, moisture, temperature, pH, and ammonium (Moosavi and Crill, 1997; Saari et al., 2004; Jaatinen et al., 2004). Unsaturated upland forest soils oxidize CH_4 at higher rates than more water-saturated, acidic, and ammonium rich forested peat soils (Saari et al., 2004). In contrast to the CH_4 sinks of upland forest soils, and drained peatlands, natural mires emit CH_4 to the atmosphere (Bubier et al., 1995; Nykänen et al., 1998; Kettunen et al., 1999). CH_4 production in peat soil is a result of methanogenic and methanotrophic active bacteria, whose activity depends on anoxic and oxic conditions below and above the water level, temperature, and availability of carbon substrate (Kettunen et al., 1999). Increasing soil wetness increases anoxic conditions necessary for increased methanogenesis (Juottonen et al., 2005), and as a result CH_4 emissions increase (Saarnio et al., 1997; Ojanen et al., 2010; Yrjälä et al., 2011). Methane production potential in peat soils generally increases positively with pH (Juottonen et al., 2005; Ye et al., 2012), whereas CH_4 oxidation of forested peatlands has a narrow pH optimum around 5.5 (Saari et al., 2004). Increased pH levels, e.g., through the inflow of less acidic mineral soil water, typically containing greater calcium and bicarbonate concentrations than peat water (Howie and Meerveld, 2011), could increase CH_4 emissions from transitions.

N_2O emissions in well-drained boreal forest soils are controlled by soil moisture, pH, available nitrate, ammonium, oxygen, and carbon concentrations (Regina et al., 1996; Ullah et al., 2008). N_2O production is limited by the amount of nitrogen and is subject to denitrification and nitrification processes (Ambus et al., 2006). In well-drained soils NO_3 limitation, anoxic microsites, and larger soil porosity may also promote N_2O consumption (Frasier et al., 2010). N_2O consumption of soils correlates with dehydrogenase activity, which is affected by oxidation-reduction status and possibly controlled by soil moisture (Wlodarczyk et al., 2005). The N_2O consumption by soils is attributed to respiratory reduction (Conrad, 1996) caused by denitrifiers and nitrifiers (Rosenkranz et al., 2006). N_2O emissions increase during drier periods through increased ammonification and nitrification (Regina et al., 1996; Nykänen et al., 1995; Von Arnold et al., 2005). In water-saturated minerotrophic peatlands nitrification supplies nitrate (Wrage et al., 2001) for denitrification, which is the main but small N_2O source (Wray et al., 2007; Frasier et al., 2010). In nutrient rich mires, N_2O emissions increase during drier periods through increased ammonification and nitrification (Regina et al., 1996; Nykänen et al., 1995; Von Arnold et al., 2005). Nitrification and the supply of nitrate for denitrification increases with higher pH (Regina et al., 1996). However, if nitrate is available, low pH increases N_2O emissions (Weslien et al., 2009). Therefore, if nitrate were present during water level drawdown, the forest–mire transitions could become sources of N_2O.

Our aims were (1) to test whether forest floor CH_4 and N_2O fluxes of the forest–mire transition differ from the typical upland forests and lowland mires of natural boreal landscapes and (2) how meteorologically different years, i.e., exceptionally wet (2004), typical (2005), and exceptionally dry (2006), affect the fluxes.

We addressed the question of whether increasing wetness in forest–mire transitions promotes CH_4 production, and whether dry conditions reduce CH_4 production and increase N_2O emissions. We hypothesized that forest/mire types exhibit distinct levels of CH_4 and N_2O fluxes due to the changing soil structure from Podzols to Histosols and due to increasing soil water content from xeric to saturated. We expected that the occasionally saturated organo-mineral soils of forest–mire transitions are variable sources of CH_4 and N_2O fluxes. In order to evaluate the underlying factors behind CH_4 and N_2O forest floor fluxes, we measured the fluxes and environmental variables, such as soil temperature, soil moisture, water table depth, and soil water pH, in nine sites along the forest–mire ecotone during exceptionally different meteorological conditions. In order to detect statistically significant differences between CH_4 and N_2O fluxes of nine sites we used two-way analysis of variance, and for better understanding of flux responses to environmental factors we used linear and nonlinear regression models, and residual sensitivity analysis.

2 Material and methods

2.1 Study site characteristics

The Vatiharju–Lakkasuo ecotone of nine forest and mire study sites forms a gradient in vegetation communities, soil moisture and nutrient conditions in central Finland (61°47′, 24°19′) (Ťupek et al., 2008). Forest/mire types were classified using the Finnish classification systems (Cajander, 1949; Laine et al., 2004) based on soil fertility reflected by the composition and abundance of forest floor vegetation, and by the site location on the slope. The ecotone study sites are situated along a 450 m transect on a hillslope with a relative relief of 15 m and a 3.3 % slope facing NE (Fig. 1a). The fertility of the forest/mire sites increase from the poorly fertile sites at the xeric and saturated edges of the ecotone towards the most fertile *Oxalis-Myrtillus* type forest (OMT) in the middle of the hillslope (Fig. 1b).

Dominant vegetation composition changes with increasing soil moisture down the slope. Xeric Scots pine forest (CT – *Calluna* type) on the summit of glacial sandy esker gives way to subxeric Scots pine Norway spruce forest (VT – *Vaccinium vitis-idaea* type) on the shoulder, and mesic and herb rich Norway spruce dominated types on the back slope and footslope (MT – *Vaccinium myrtillus*

Table 1. Site soil water solution pH and soil properties.

	CT		VT		MT		OMT		OMT+		KgK		KR		VSR1		VSR2	
	mean	SE	mean	SE	mean	SE	mean	SE	mean	SE	mean	SE	mean	SE	mean	SE	mean	SE
pH 10 cm	5.57	0.36	5.14	0.42	5.24	0.08	4.68	0.39	4.58	0.30	4.46	0.14	4.37	0.22	5.06	0.39	4.80	0.44
pH 30 cm	6.20	0.06	6.18	0.02	5.91	0.13	5.30	0.11	5.53	0.04	4.91	0.10	4.55	0.08	5.32	0.15	4.79	0.19
Bulk density 0–10 cm	0.37	0.09	0.28	0.04	0.48	0.03	0.27	0.09	0.31	0.13	0.33	0.05	0.24	0.02	0.40	0.12	0.40	0.12
Bulk density 10–30 cm									0.92	0.07	0.31	0.12	0.85	0.03	0.90	0.07	0.90	0.07
Tot C (%) 0–10 cm	43.17		24.22		49.63		47.09		45.36		48.68		50.30		45.76		48.20	
Tot C (%) 10–30 cm									21.76		53.31		48.33		47.70		49.97	
Tot N (%) 0–10 cm	1.02		0.61		1.18		1.59		2.19		1.47		1.12		1.29		0.96	
Tot N (%) 10–30 cm									0.96		1.95		1.45		1.87		1.81	
C / N 0–10 cm	42.32		39.70		42.06		29.62		20.71		33.12		44.91		35.47		50.21	
C / N 10–30 cm									22.67		27.34		33.33		25.51		27.61	

Figure 1. (a) Airborne infrared photograph shows a 450 m long boreal forest–mire ecotone located on the NE slope of the glacial Vatiharju–Lakkasuo esker in Finland (61°47′, 24°19′). **(b)** The fish-eye photographs show tree stands of xeric (1), subxeric (2), mesic (3), herb rich (4), paludified (5–7), and saturated (8–9) forest/mire types. **(c)** Photographs show ground vegetation and **(d)** soil profiles of nine forest/mire types. Upland forests: 1 CT – *Calluna*, 2 VT – *Vaccinium vitis-idaea*, 3 MT – *Vaccinium myrtillus*, 4 OMT – *Oxalis-Myrtillus*; forest–mire transition types: 5 OMT+ – *Oxalis-Myrtillus* paludified, 6 KgK – *Myrtillus* spruce forest paludified, 7 KR – spruce pine swamp; sparsely forested wet mire types: 8 VSR1 and 9 VSR2 – tall sedge pine fen.

type, OMT – *Oxalis-Myrtillus* type). The toe slope contains forest–mire transitions of paludified mixed spruce–pine–birch forests (OMT+ – *Oxalis-Myrtillus* paludified, KgK – *Myrtillus* spruce forest paludified). There is a permanently wet mixed spruce–pine–birch swamp (KR – spruce pine swamp) at the mire edge of the forest–mire transitions. On the level of the hillslope there are birch–pine fen mires with open tree canopies (VSR1 and VSR2 – tall sedge pine fen) (Fig. 1b). The forest floor vegetation is composed of site-specific mosses and vascular plants (Fig. 1c).

Soils are formed by well-drained Haplic Podzols on the hillslope, intermediately drained Histic and Gleyic-Histic Podzols in the forest–mire transitions on the toe of the slope, and permanently wet Hemic Histosols downslope (Fig. 1d).

We measured pH during summer campaign 2005 from soil water data collected on all sites by suction cup lysimeters. Three lysimeters were installed in 10 cm and one in 30 cm depth below the soil surface in each site. Detailed description of the lysimeters and sampling procedure can be found in Starr (1985). The pH was measured on the day of water sampling in the laboratory by pH meter equipped with a glass electrode. The mean acidity level of the sites of forest–mire ecotone was gradually increasing from pH 5.6 in uplands (CT) to 4.4 in transitions (KR), whereas mires were less acid than transitions with pH 5.1 and 4.8 (VSR1 and VSR2, respectively) (Table 1). Collected soil water from 30 cm depth showed generally higher pH than soil water pH at 10 cm depth. Three soil cores for each plot were taken in July 2006 from the top soil (0–10 cm) in upland forests and from the two profile depths (0–10, 10–30 cm) in forest–mire transitions and in peatlands. The volume of samples was measured before the oven drying at 70 °C to determine the bulk density. The bulk density of the upper organic layer ranged from 0.24 $g\,cm^{-3}$ (KR) to 0.48 $g\,cm^{-3}$ (MT) and was approximately half of the bulk density of the organic layer from 10 to 30 cm depth (mean of transitions and mires 0.77 $g\,cm^{-3}$) (Table 1). The C / N ratio was determined once for each plot from the soil organic matter analyzed by dry combustion with Leco CNS-1000 (Leco Corp., USA). The C / N ratio was wider in the 0–10 cm profile (mean 37) than in the 10–30 cm profile (mean 27). The highest N content as well as the lowest C / N ratio along the ecotone was found in forest–mire transitions OMT+ and KgK (Table 1). A more detailed forest/mire type characterization is given by Ťupek et al. (2008).

2.2 Micrometeorological conditions

The micrometeorological measurements along the Vatiharju–Lakkasuo forest–mire ecotone were taken weekly during the summers of 2004 (July–November), 2005 (May–November), 2006 (May–September), and monthly during the winters (December–April). The forest floor soil temperatures (°C) at depths of 5, 15, and 30 cm (T_5, T_{15}, and T_{30}) were measured using a portable thermometer connected to thermocouples installed permanently in the soil. The volumetric soil moisture (%) at depths of 5, 10, and 30 cm (soil water content – SWC_5, SWC_{10}, and SWC_{30}) was measured by a portable ThetaProbe (Delta-T Devices Ltd.) in diagonally installed perforated PVC tubes, to ensure the same compactness of the soil. The depth of water table was measured inside PVC tubes (⌀ 30 mm) installed at each site. Precipitation was measured by an automated bucket system at a station for monitoring forest – atmosphere relations, SMEARII (Hari and Kulmala, 2005), located 6 km north – west from the forest–mire ecotone. Missing soil temperature and moisture data of ecotone were gap filled by linear regression between continuous measurements of soil temperature and moisture at SMEARII.

2.3 CH_4 and N_2O fluxes

The field gas sampling was conducted weekly in the 2004 and 2005 seasons, bi-weekly during the 2006 season, and monthly during the winters. The gas sampling was done within 3-days interval of the micrometeorological measurements. If there was packed snow on the ground, the gas samples would be taken from the top and bottom layers; and the CH_4 ($\mu g\, m^{-2}\, h^{-1}$) and N_2O ($\mu g\, m^{-2}\, h^{-1}$) fluxes were calculated by the snowpack diffusion method using each gas concentration difference, snow depth, porosity and temperature, and gas diffusion coefficients as in Sommerfeld et al. (1993). Otherwise, if there was no snowpack, the samples would be taken from three opaque, vented, closed, static chambers (⌀ 315 mm, h 295 mm) placed air tightly on pre-installed collars. On each measuring occasion a sample of ambient gas and four 15 ml samples from each of the three chambers were drawn in syringes at intervals of 5, 10, 15, and 20 min from chamber closure, totaling 13 samples for each site. Chamber temperature was monitored during the sampling. After the sampling event, the gas samples were stored in coolers at +4 °C and analyzed within 36 h in a laboratory with a gas chromatograph. The gas chromatograph (Hewlett-Packard, USA) model number HP-5890A was fitted with a flame ionization detector (FID) for CH_4 and an electron capture detector (ECD) for N_2O detection. The gas chromatograph was also equipped with a moisture trap. Prior to analysis of field samples and after each set of 13 samples a reference gas sample of known CH_4 and N_2O concentration was analyzed. The CH_4 ($\mu g\, m^{-2}\, h^{-1}$) and N_2O ($\mu g\, m^{-2}\, h^{-1}$) fluxes were calculated from the slope of linear regression between the set of four gas concentrations and sampling time, time elapsed after the chamber closure, and by applying temperature correction. For the flux calculation we used a MATLAB (The Mathworks Inc.) script developed at the Dept. of Physics, University of Helsinki.

The method quantification limit (MQL) of the gas chromatograph was based on 100 subsequently analyzed samples of reference gas of known CH_4 and N_2O concentrations (mean ± two SD: 1.837 ± 0.055 and 0.295 ± 0.023 ppm, respectively) and reference gas samples analyzed before the set of field samples for each site. The MQL was a gas-specific standard deviation of the random fluxes derived from 1000 random sets of four CH_4 or N_2O concentrations of reference gas samples (22 $\mu g\, m^{-2}\, h^{-1}$ for CH_4 and 18 $\mu g\, m^{-2}\, h^{-1}$ for N_2O). In order to minimize the random error related to gas sampling in the field, fluxes were verified using the ambient field air sample analyzed before each sequence of chamber samples adopting similar criteria as used in Alm et al. (2007). Due to gas sampling disturbances in the field and poor gas chromatograph accuracy 17 % of CH_4 and 49 % of N_2O fluxes were discarded.

2.4 Statistical analysis

Two-way analysis of variance (ANOVA) was used to test whether CH_4 and N_2O fluxes of forest/mire types have common means in wet, typical, and dry years. Post hoc Tukey HSD (honest significant difference) tests were used to test the pairwise differences between the forest and mire types and years changing from wet to dry. For CH_4 fluxes we ran ANOVA tests twice, first on the whole data set including nine forest/mire types and then on a subset of data including upland forests and forest–mire transitions, and excluding mires. For testing significant differences between the two groups of data we performed Welch's two sample t test, e.g., between the N_2O fluxes from the snow on the ground season (January–April in 2006) and the N_2O fluxes from the snowless seasons (May–November in 2005 and May–September in 2006).

In addition to ANOVA, we tested the dependence between the measured CH_4 ($\mu g\, m^{-2}\, h^{-1}$) and the gap filled half-hourly environmental variables in separate models for: (a) the upland forests on mineral soils (CT, VT, MT, OMT), and (b) forest–mire transitions on organo-mineral soils and (OMT+, KgK, and KR) (c) mires (VSR1, VSR2).

CH_4 fluxes ($\mu g\, m^{-2}\, h^{-1}$) of uplands and transitions were fitted by two linear mixed-effects regression models with a random effect for forest types (Pinheiro et al., 2013). For both groups of forest types, we evaluated the effect of all our environmental variables on CH_4 together and their combinations iteratively by selecting the model combination of variables that were significant.

The CH_4 fluxes for upland forests and transitions included soil moisture at 10 cm (%) (SWC_{10}) and soil temperature at 5 cm (°C) (T_5) as predictors in separate models (Eqs. 1 and 2):

$$yu_{ij} = \beta_{CT}SWC_{10} + \beta_{VT}SWC_{10} + \beta_{MT}SWC_{10} + \beta_{OMT}SWC_{10} + \beta_{CT}T_5 + \beta_{VT}T_5 + \beta_{MT}T_5 + \beta_{OMT}T_5 + b_{CT} + b_{VT} + b_{MT} + b_{OMT} + \varepsilon_{ij}, \quad (1)$$

$$yt_{ij} = \beta_{OMT+}SWC_{10} + \beta_{KgK}SWC_{10} + \beta_{KR}SWC_{10} + \beta_{OMT+}T_5 + \beta_{KgK}T_5 + \beta_{KR}T_5 + b_{OMT+} + b_{KgK} + b_{KR} + \varepsilon_{ij}, \quad (2)$$

where yu_{ij} and yt_{ij} are the CH_4 flux ($\mu g\, m^{-2}\, h^{-1}$) for upland forests or transitions and for a particular ith forest type and the jth observation, β_{CT} through β_{KR} are the fixed effect coefficients for a particular ith forest type (CT, VT, MT, OMT Eq. 1, or OMT+, KgK, and KR Eq. 2), SWC_{10}, and T_5 are the fixed effect variables (predictors) for observation j in forest type i where each forest type's predictor is assumed to be multivariate normally distributed, b_{CT} through b_{KR} are intercepts for the random effect for a particular ith forest type, and ε_{ij} is the error for case j in forest type i where each forest type's error is assumed to be multivariate normally distributed (Table 2).

The CH_4 fluxes ($\mu g\, m^{-2}\, h^{-1}$) of mires were fitted by using a multiplicative nonlinear regression model with a combined response to water table depth and soil temperature at 5 cm Eq. (1):

$$y_{ij} = a_0 e^{\left(-0.5\left(\frac{\text{WT-WTopt}}{\text{WTtol}}\right)^2\right)} e^{\left(-0.5\left(\frac{T5\text{-Topt}}{\text{Ttol}}\right)^2\right)} + \varepsilon_{ij}, \quad (3)$$

where y_{ij} is the CH_4 flux ($\mu g\, m^{-2}\, h^{-1}$) for the ith mire (VSR1,VSR2) and for the jth case, WT (cm) is water table depth, $T5$ (°C) is soil temperature at 5 cm, and a_0, WTopt, WTtol, Topt, and Ttol are parameters (Table 3).

The N_2O fluxes ($\mu g\, m^{-2}\, h^{-1}$) of all forest/mire types were fitted by using one multiplicative nonlinear regression model with a combined response to soil moisture and soil temperature at 5 cm Eq. (4):

$$z_{ij} = a_0 SWC_5 e^{\left(-0.5\left(\frac{T5\text{-Topt}}{\text{Ttol}}\right)^2\right)} + \varepsilon_{ij}, \quad (4)$$

where z_{ij} is the N_2O flux ($\mu g\, m^{-2}\, h^{-1}$) for the ith mire (VSR1,VSR2) and for the jth case, SWC_5 (%) is soil moisture at 5 cm, and T5 (°C) is soil temperature at 5 cm, and a_0, Topt, and Ttol are parameters (Table 4).

To illustrate the sensitivity of CH_4 and N_2O flux response to environmental factors we performed a residual analysis by simulating a value for each data point with only one factor allowed to vary and the other set to its mean level. To examine correlations between CH_4 and N_2O fluxes and pH, and soil properties we preformed the Pearson's correlation tests. The statistical analyses were performed in MATLAB R2012a (The Mathworks Inc.) and in R (R Core Team 2013) software environments.

3 Results

3.1 Micrometeorological conditions

The largest differences between years 2004, 2005, and 2006 were seen in changing summer precipitation patterns (measured nearby the SMEARII station). The average June–August monthly precipitation was reduced from 94 to 44 mm from a wet 2004 to a dry 2006, while ambient temperature increased from 14 to 17 °C. In the coldest summer (2004) the average precipitation in June and July was over 117 mm, and dropped to 47 mm in August. In the typically warm summer of 2005 the monthly precipitation gradually increased up to 123 mm in August, and dropped to 58 mm in September. However, in the warmest summer (2006) the monthly precipitation never reached more than 48 mm. In July 2006, two rainless weeks induced a drought. By drought we mean that the soil water content in the upper soil layer (in mineral soils) was so low that mosses wilted and dried (all along the ecotone). The drought conditions lessened in mid-August and ended in September with increasing rains towards autumn. Late autumn was exceptionally warm and snowless.

Monthly median soil temperatures at 5 cm (T_5) ranged from around 5 °C in May, culminated to around 15–16 °C in July and August, and subsided again to around 5 °C in October. The non-vegetative season T_5 minimum was close to 0 °C. The warmest T_5 was in upland forest CT and the coldest was in upper forest–mire transition OMT+. Soil temperature slightly increased from forest–mire transitions towards mires. In spite of the ambient air temperature difference throughout all the months in the 3 years, we detected differences mainly during early and late season in 2004, 2005, and 2006 T_5 (Fig. 2a).

The median water table (WT) showed the obvious rise from 10 m at the summit of the hill, to around 1 m in the mid-slope, between 0.5 and 0.1 m at the toe slope, and close to 0.01 m on the level (Fig. 2b). The seasonal WT rise in 2005 was observed between the July and August medians. During the drought of 2006, the WT values dropped less than 0.1 m for the uppermost forest sites, but dropped heavily by ~ 1 m in the forest–mire transitions, and more than 0.5 m in the lowermost peatland sites.

Volumetric SWC in 10 cm depth ranged from a dry value of around 10 % in the mineral soils to a water-saturated value of around 80 % in swamp and mires (Fig. 2c). The largest drought reduction of SWC was in August 2006 on the well-drained sandy Podzols at the summit of the hill, and also on the poorly drained Histic Podzols on the toe slope.

3.2 CH_4 fluxes

The median fluxes from the forest floor varied from −51 to 586 $\mu g\, m^{-2}\, h^{-1}$ for CH_4 among individual sites during the entire period (Fig. 3a). The small negative CH_4 fluxes associated with prevailing oxidation were mostly observed

Table 2. Parameter estimates and their standard errors for trend coefficients of CH_4 fluxes ($\mu g\,m^{-2}\,h^{-1}$) of the upland forest types (CT, VT, MT, and OMT, Eq. 1), and for the forest–mire transitions (OMT+, KgK, and KR, Eq. 2). Both equations are functions of volumetric soil moisture at 10 cm (%) and soil temperature at a depth of 5 cm (°C).

Eq. (1)	bi	Group bi	Group bi SE	$\beta i1$	$\beta i1$ SE	$\beta i2$	$\beta i2$ SE	N	RMSE
CT	−39.345	−43.632	9.102	0.762[a]	0.299	−1.249	0.223	137	35.2
VT	−26.213							143	25.1
MT	−50.984							139	25.2
OMT	−57.985							144	32.1
Eq. (2)									
OMT+	−49.898	−50.248	7.507	0.638	0.105	−0.109[b]	0.226	139	22.3
KgK	−48.216							146	17.9
KR	−52.630							149	31.5
Eq. (2) soil temperature excluded from fitting									
OMT+	−51.799	−52.466	6.341	0.660	0.099			139	22.3
KgK	−50.404							146	17.9
KR	−55.196							149	31.5

$p < 0.001$ for all parameters, except [a] $p = 0.011$, [b] $p = 0.629$. $\beta i1$ – soil moisture at 10 cm, $\beta i2$ – soil temperature at 5 cm.

Table 3. Parameter estimates and their standard errors for trend coefficients of CH_4 fluxes ($\mu g\,m^{-2}\,h^{-1}$) of the mires (VSR1, VSR2, Eq. 3). Equation (3) is a function of water table depth (cm) and soil temperature at a depth of 5 cm (°C).

Eq. (3)	$a0$	$a0$ SE	T_{opt}	T_{opt} SE	T_{tol}	T_{tol} SE	WT_{opt}	WT_{opt} SE	WT_{tol}	WT_{tol} SE	N	RMSE
mires	1207.1	126.7	13.9	1.4	6.4	1.3	−18.0	2.2	16.6	2.8	324	656
VSR1	1570.3	155.1	13.0	0.8	5.8	0.8	−18.6	1.6	15.5	1.7	162	424
VSR2	801.3	190.8	16.6[a]	6.8	8.7[b]	4.5	−17.3[c]	5.3	20.7[d]	9.7	162	558

p values < 0.001, except [a] $p = 0.016$, [b] $p = 0.053$, [c] $p = 0.002$, [d] $p = 0.035$.

in uplands and in transitions, while mires typically showed large positive CH_4 fluxes associated with prevailing production. The CH_4 flux dynamics changed exponentially with increasing levels of the ground water table from small uptake to large emissions (Figs. 2, 3). The median CH_4 fluxes of uplands (CT, VT, MT, OMT), transitions (OMT+, KgK, KR), and mires (VSR1, VSR2) varied from −38, −48, and 392 $\mu g\,m^{-2}\,h^{-1}$, respectively (Fig. 3b). Momentary CH_4 fluxes of uplands and transitions ranged from −342 to 143 $\mu g\,m^{-2}\,h^{-1}$, whereas in mires the fluxes ranged from −12 to 6808 $\mu g\,m^{-2}\,h^{-1}$ (Fig. 3b). The median CH_4 fluxes for one upland (VT) and all the transitions (OMT+, KgK, KR) were found inside the range of the gas chromatograph detection limits ($MQL_{CH_4} = 22\,\mu g\,m^{-2}\,h^{-1}$). In forest–mire transitions the ground water level in August 2005 increased towards the surface and approached the levels typically found in mires (Fig. 2b), but the soil water saturation in transitions was not followed by CH_4 emissions such as those found in mires.

ANOVA showed that forest floor CH_4 fluxes differed significantly for the nine forest/mire types of the ecotone $F(8, 1252) = 108$, $p < 0.001$ and for the wet, typical, and dry years $F(2, 1252) = 10$, $p < 0.001$. There was a significant interaction between CH_4 fluxes of forest/mire types and wet, typical, and dry years $F(16, 1252) = 5$, $p < 0.001$. The post hoc Tukey comparison of the nine forest/mire types indicated that the mires had significantly higher CH_4 fluxes than the forests. Differences in means (M) and 95 % confidence limits (CI) ranged from minimum VSR2–KgK ($M = 481$, 95 % CI [352, 610]) to maximum VSR1–OMT ($M = 793$, 95 % CI [668, 918]) at $p < 0.001$. Also the CH_4 fluxes of the mires were significantly different from each other VSR2–VSR1 ($M = -260$, 95 % CI [−384, -137]), $p < 0.001$. Differences between the years were significant at $p < 0.001$ for dry–typical ($M = -96$, 95 % CI [−149, −43]) when CH_4 fluxes of mires were highly reduced. The comparison of mean CH_4 fluxes of typical–wet ($M = 51$, 95 % CI [−6, 108]), $p = 0.089$, and dry–wet years did not show a significant difference ($M = -45$, 95 % CI [−111, 20]), $p = 0.237$.

Differences between the forest types (transitions, uplands) were not significant when analyzed together with the CH_4 fluxes of mires, but became significantly different $F(6, 976) = 71$, $p < 0.001$, when ANOVA was run without mires. Though unlike the nine forest/mire type data set, for the

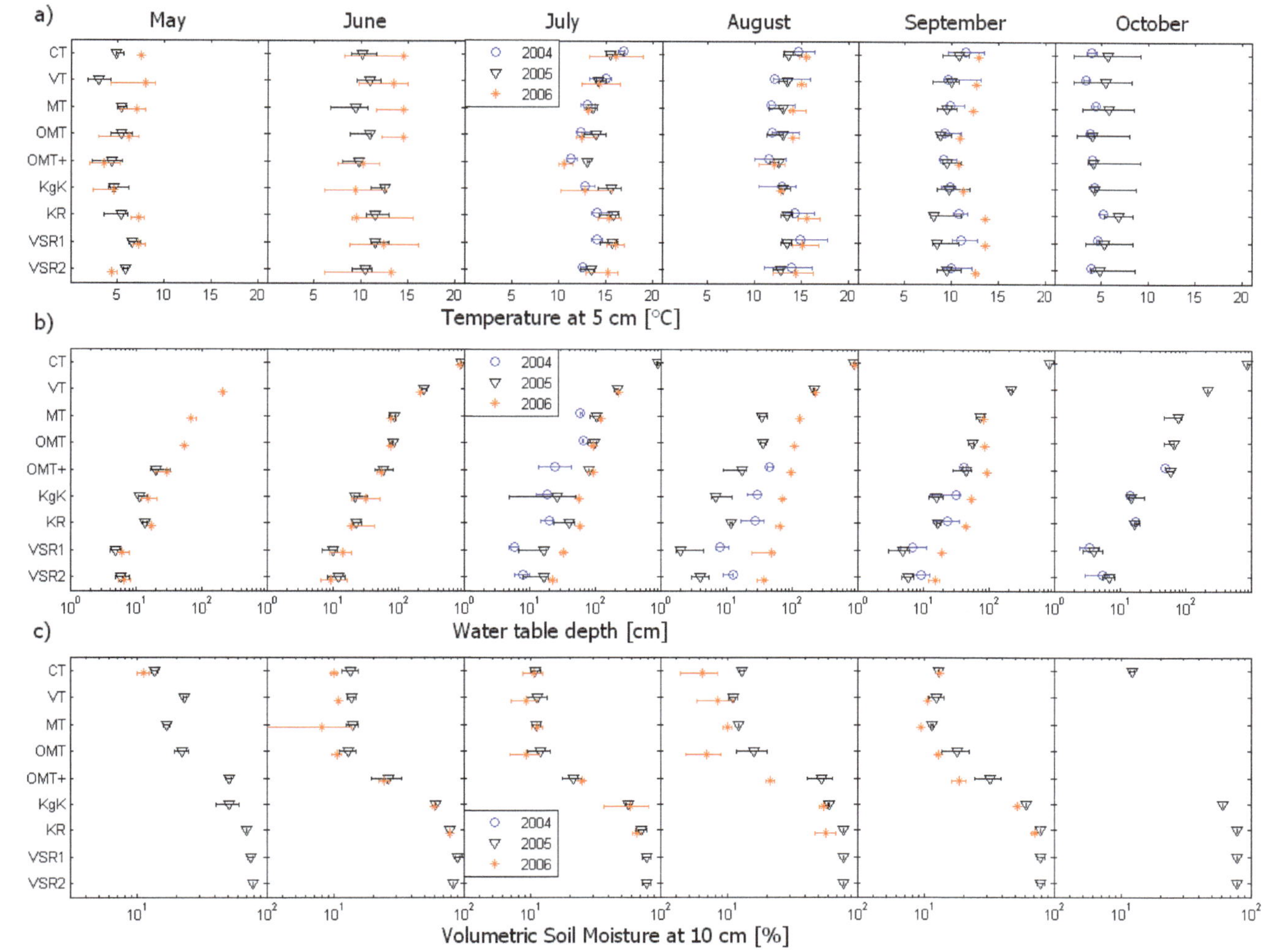

Figure 2. The panels **(a–c)** show the monthly medians of environmental variables: **(a)** soil temperature at a depth of 5 cm, **(b)** ground water level, and **(c)** volumetric soil moisture at 10 cm depth observed along the forest–mire ecotone during wet (2004), intermediate (2005), and dry years (2006). The top–down arrangement of sites mimics the locations on the slope (see Fig. 1). The error bars represent the 25th and 75th percentiles.

Table 4. Parameter estimates and their standard errors for forest floor N_2O fluxes ($\mu g\,m^{-2}\,h^{-1}$) of all forest/mire types (CT–VSR2) in one group Eq. (4). Eq. (4) is function of volumetric soil moisture at 5 cm (%) and soil temperature at a depth of 5 cm (°C).

Eq. (4)	$a0$	$a0$ SE	T_{opt}	T_{opt} SE	T_{tol}	T_{tol} SE	N	RMSE
forests/mires	4.034	0.635	11.268	0.183	1.414	0.181	400	36.2

$p < 0.001$ for all parameters.

group of uplands with transitions there was no difference between wet, typical, and dry years $F(2, 976) = 1$, $p = 0.292$, or their interactions $F(12, 976) = 1$, $p = 0.135$. The mean CH_4 uptake of the upland forests ($-42.9\,\mu g\,m^{-2}\,h^{-1}$) was for the whole period significantly larger than the mean CH_4 uptake of the forest–mire transitions ($-12.8\,\mu g\,m^{-2}\,h^{-1}$) according to Welch's two sample t test $t(994) = 15.56$, $p < 0.001$. The post hoc Tukey comparison of the differences in the mean CH_4 fluxes for 21 pairs of seven upland and transitional forest types was significant for 17 pairs at $p < 0.001$ and ranged from OMT–VT ($M = -35$, 95 % CI [−45, −25]) to KR–OMT ($M = 51$, 95 % CI [41, 61]). The post hoc Tukey comparisons showed non-significant p values for 4 of the 21 pairs of CH_4 fluxes of transitional and upland forest types (MT–CT 0.056, OMT+–VT 0.965, OMT–MT 0.431, and KR–KgK 0.999).

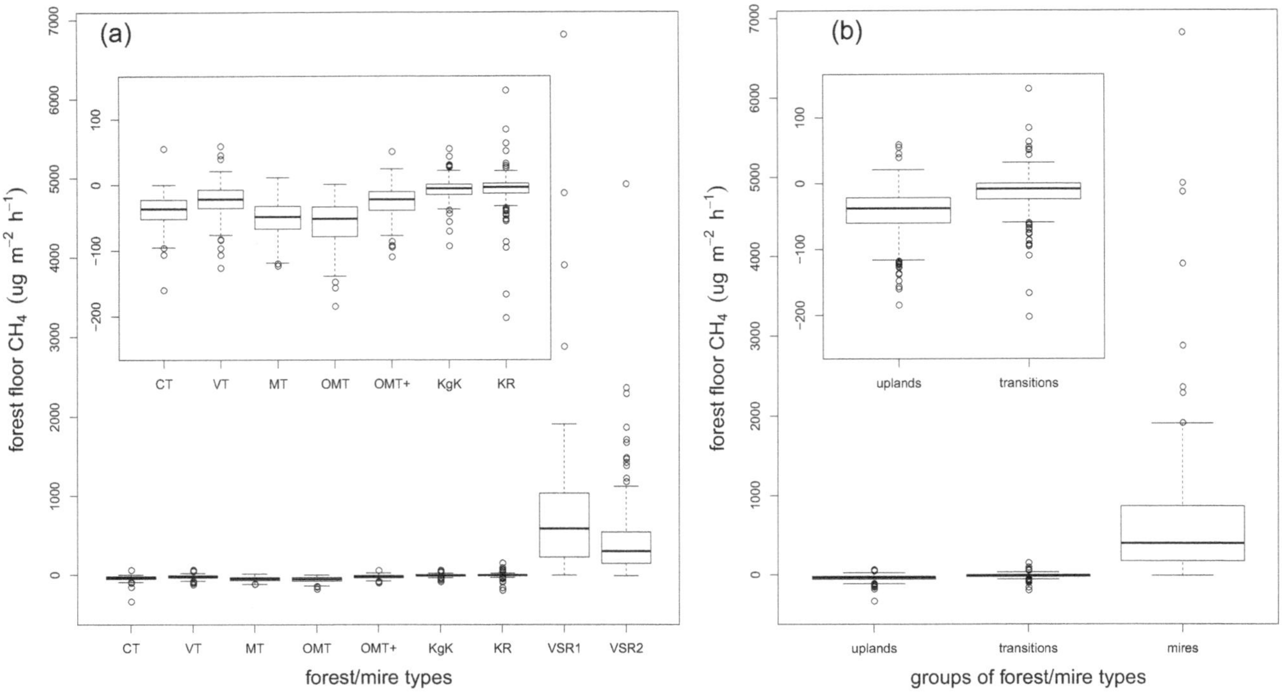

Figure 3. The box plots of forest floor CH_4 fluxes ($\mu g\,m^{-2}\,h^{-1}$) for each forest/mire type (a), and (b) for uplands (CT, VT, MT, OMT), transitions (OMT+, KgK, KR), and mires (VSR1, VSR2) during the whole period. The left–right arrangement of sites mimics the locations on the slope (see Fig. 1).

3.3 Factors controlling CH_4 fluxes

The mean level of CH_4 fluxes of upland and transitional forests differed (Table 2, parameter group bi), though the sensitivity response to environmental factors was similar (Fig. 4). The largest part of the CH_4 fluxes remained unexplained with our models, as the proportion of explained variance was relatively low for uplands (10 %) and transitions (15 %) and slightly higher for mires (22 %). The modeled CH_4 flux response for the upland and transitional forest sites to soil moisture at 10 cm was nearly flat, although the soil moisture parameter was significant ($p = 0.011$, Table 2). In the transitional *Oxalis-Myrtillus* paludified forest type OMT+, where the soil moisture at 10 cm ranged from 20 % (in the uplands) to over 70 % (in the mires), the modeled CH_4 flux response between dry and water-saturated soil differed by 50 $\mu g\,m^{-2}\,h^{-1}$. A stronger gradient than that in the soil moisture was detected by modeling stronger temperature responses of CH_4 fluxes for the uplands and the nearly flat response for the transitions (Fig. 4). The model parameter to soil temperature at 5 cm in the uplands was highly significant at $p < 0.001$, in contrast to transitions where the temperature parameter was insignificant $p = 0.629$ (Table 2). In the mires the observed range of water level during wet, typical, and dry years spanned from the surface to a depth of 54 cm and showed a sigmoidal response with lower CH_4 fluxes towards the extreme ends. The optimum water level for CH_4 emissions was 18 cm below the surface with 16.6 cm tolerance which is deviation of water level up to 60 % of CH_4 flux maximum (Fig. 4; $p < 0.001$, WT_{opt} and WT_{tol} in Table 3). Optimum near-surface peat temperature for the CH_4 emissions was found at 13.9 °C with 6.4 °C tolerance (Fig. 4; $p < 0.001$, T_{opt} and T_{tol} in Table 3).

3.4 N_2O fluxes

During the typical and dry years the momentary forest floor N_2O fluxes of forest/mire types ranged from −107 to 248 $\mu g\,m^{-2}\,h^{-1}$. The median N_2O fluxes were similar for the forest/mire types and ranged only from 0 to 6 $\mu g\,m^2\,h^{-1}$ (Fig. 5). The median N_2O fluxes of all forest/mire types were found inside the range of the method quantification limits ($MQL_{N_2O} = 18\,\mu g\,m^{-2}\,h^{-1}$). The N_2O fluxes of the snow on the ground period were significantly lower than the N_2O fluxes of the snowless period according to Welch's two sample t test $t(297) = 5.094$, $p < 0.001$. Forest floor N_2O fluxes did not differ significantly for the nine forest/mire types of the ecotone for the snowless periods $F(8, 284) = 0.708$, $p = 0.684$. Though, the momentary N_2O fluxes were significantly different in typical and dry snowless seasons $F(1, 284) = 6.157$, $p < 0.014$. N_2O fluxes were lower during dry snowless seasons and a small increase was observed only in one forest–mire transition (KR – spruce pine swamp) and in one mire (VSR2 – tall sedge pine fen) (Fig. 6).

In general N_2O fluxes were low and did not show clear spatial differences in relation to increasing soil moisture from

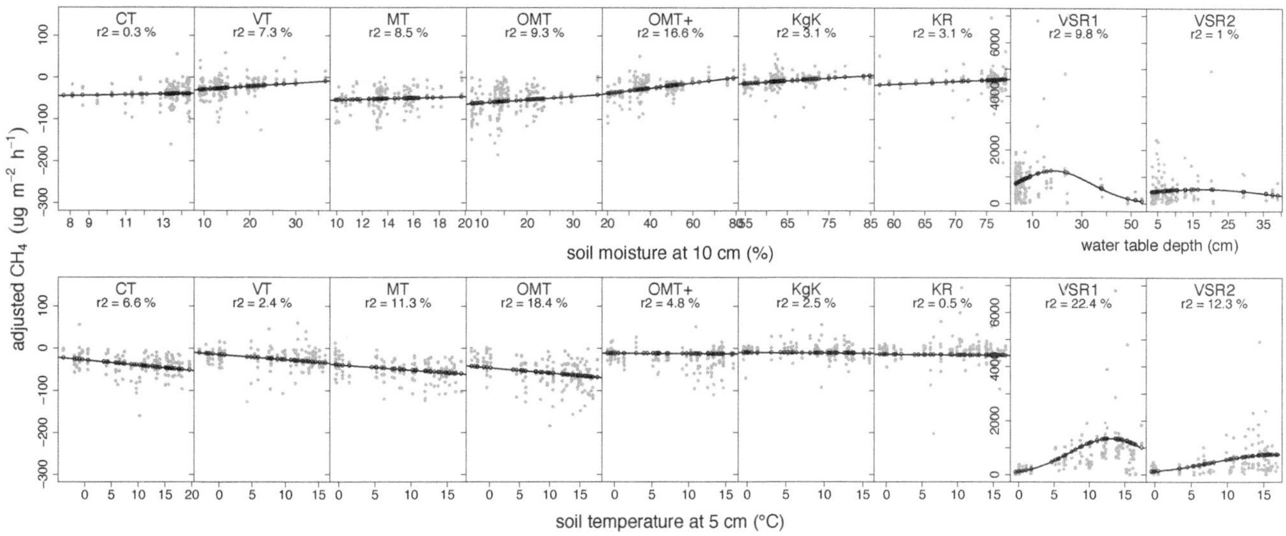

Figure 4. Comparison of sensitivity of forest floor CH_4 fluxes ($\mu g\,m^{-2}\,h^{-1}$) to environmental factors for nine forest/mire types. Modeled in the upper panels is CH_4 flux response to soil moisture at 10 cm (uplands and transitions) or to water table depth (cm) (mires) for uplands (CT, VT, MT, OMT) Eq. (1), for transitions (OMT+, KgK, KR) Eq. (2), and for mires (VSR1, VSR2) Eq. (3). Water table depth is indicated as negative when it is above the soil surface. In the lower panels, CH_4 flux response (Eqs. 1–3) is modeled to soil temperature at 5 cm of the same forest/mires types and during the same period as in the upper panel. The CH_4 flux response for each individual environmental factor is illustrated so that the simulated value for each data point was recalculated by allowing only one factor at a time to vary while the other was set to its mean level. To the adjusted CH_4 flux responses (black points) the corresponding residual of each data point was added in order to describe the unexplained model variation (gray points). The r^2 (%) is the proportion of explained variance. The left–right arrangement of sites mimics the locations on the slope (see Fig. 1).

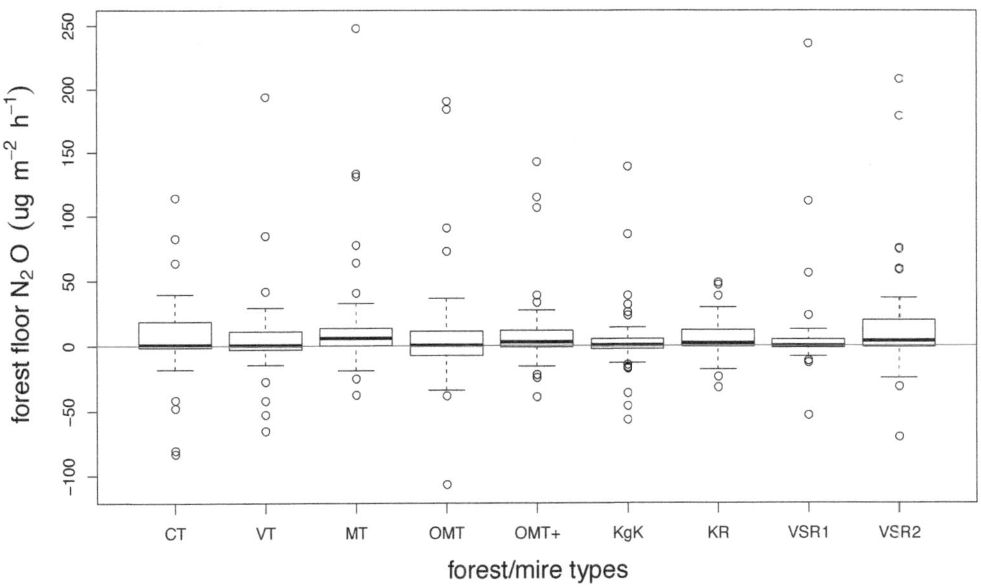

Figure 5. The box plot of forest floor N_2O fluxes ($\mu g\,m^{-2}\,h^{-1}$) for each forest/mire type (uplands – CT, VT, MT, OMT; transitions – OMT+, KgK, KR; and mires – VSR1, VSR2) during the period including typical and dry years. The left–right arrangement of sites mimics the locations on the slope (see Fig. 1).

xeric uplands to water-saturated mires, but the N_2O fluxes were lower in the dry than in the typical year. The post hoc Tukey tests of means and 95 % confidence limits of N_2O fluxes for all pairs (except one) showed insignificant forest/mire type pairwise differences during the whole period and also during the snowless periods of wet or dry years (Fig. 6). The significant N_2O flux difference for VSR2–OMT

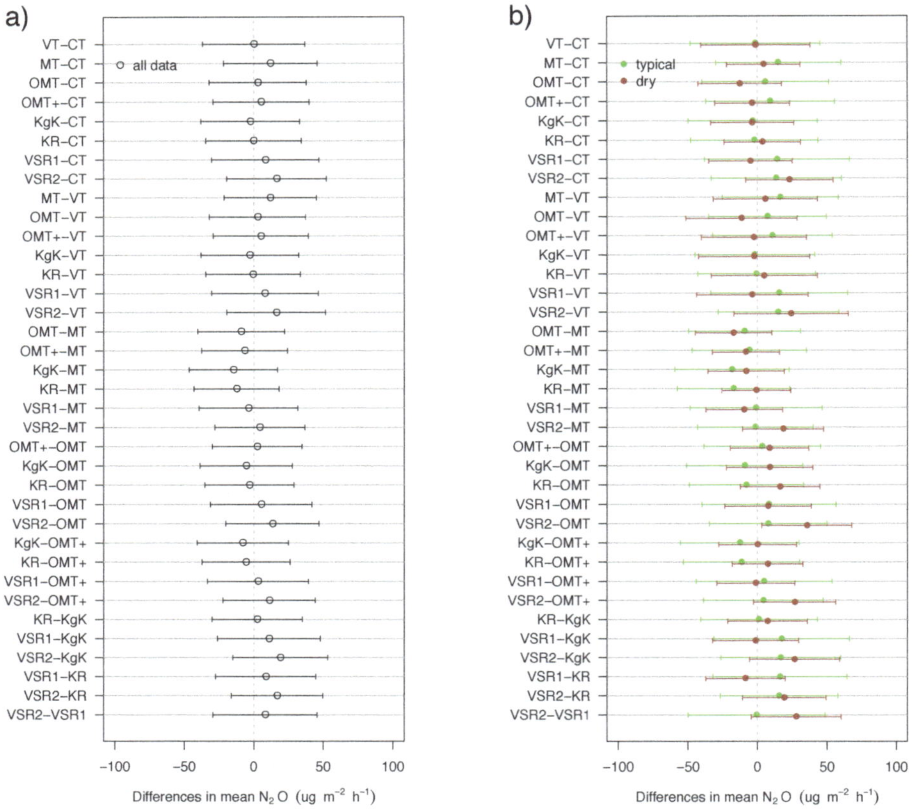

Figure 6. The post hoc Tukey differences (error bars for 95 % confidence intervals) of mean N_2O ($\mu g\, m^{-2}\, h^{-1}$) fluxes from the forest floor for the pairwise comparisons of forest/mire types (uplands – CT, VT, MT, OMT; transitions – OMT+, KgK, KR; and mires – VSR1, VSR2): **(a)** the N_2O flux differences over the whole period for a typical and dry year, **(b)** the N_2O flux differences only for snowless seasons and separately for typical and dry years.

in a dry year ($M = 35$, 95 % CI [3, 68], $p = 0.02$) was caused by a small decrease in OMT and increase in VSR2 fluxes.

3.5 Factors controlling N_2O fluxes

The sensitivity response of fluxes was weak in relation to soil moisture at 5 cm and had a somewhat clearer and significant relation with soil temperature at 5 cm ($p < 0.001$, Table 4, Fig. 7). The modeled Gaussian type response showed optimum N_2O production at 11.3 (°C) soil temperature at a depth of 5 cm with a very narrow temperature range increasing from 7 °C and subsiding at 14 °C.

3.6 Effects of pH and soil properties on CH_4 and N_2O flux

The site-specific momentary CH_4 and N_2O fluxes did not show significant correlation with varying soil water pH (except for one correlation coefficient $r = -0.45$, $p = 0.02$ on MT for N_2O and pH at 10 cm). Neither was any correlation found between pH and momentary CH_4 on the ecotone level. However, when mires were excluded, Pearson correlation between momentary CH_4 fluxes and soil water pH was significant ($r = -0.32$, $p < 0.001$). Mean values of summer 2005 CH_4 of upland forests and forest–mire transition were negatively correlated with mean pH ($CH_4 = 129.35–33.36 \times pH$, $r^2 = 0.49$; Fig. 8a). The ecotone N_2O fluxes were significantly correlated with pH ($r = 0.174$, $p = 0.004$). The mean N_2O values of sites increased with mean pH ($N_2O = -117.07 + 27.33 \times pH$, $r^2 = 0.32$; Fig. 8b). However, the post hoc Tukey differences of mean N_2O fluxes from the forest floor for the pairwise comparisons of forest/mire types were not significant for 31 pairs and mean N_2O flux differences were significant only for 5 pairs (KgK–CT, VSR1–KgK, VSR1–KR, VSR1–MT, VSR1–OMT, Fig. 9). We did not find significant correlation between site-specific mean CH_4 and N_2O flux and bulk density and/or C / N ratio.

4 Discussion

4.1 CH_4 dynamics

The forest/mire types significantly differed in forest floor CH_4 fluxes and between wet, typical, and dry years. As expected, the largest difference was found between emissions of mires and the small uptake of other forest types. However, CH_4 uptake also showed significant differences between the

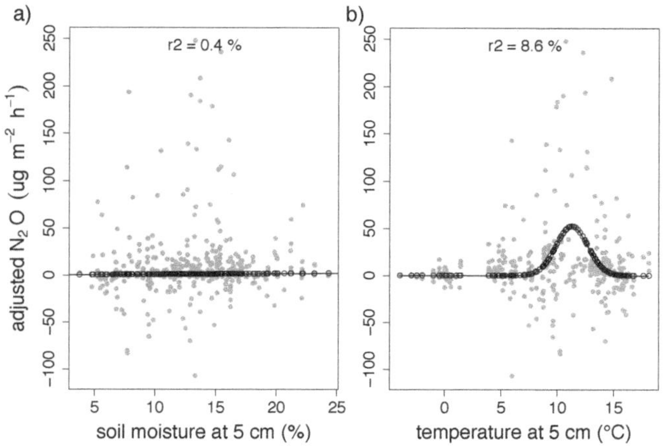

Figure 7. Sensitivity of forest floor N_2O fluxes ($\mu g\,m^{-2}\,h^{-1}$) of forest/mire types together with environmental factors: **(a)** N_2O flux response to soil moisture at 5 cm, and **(b)** N_2O flux response to soil temperature at 5 cm during the period including wet, typical, and dry years. The N_2O flux response form to each individual environmental factor is illustrated so that the simulated value by Eq. (4) for each data point was recalculated by allowing only one factor at a time to vary, while the other was set to its mean level. To the adjusted N_2O flux responses (black points) the corresponding residual of each data point was added in order to describe the unexplained model variation (gray points). The r^2 (%) is the proportion of explained variance.

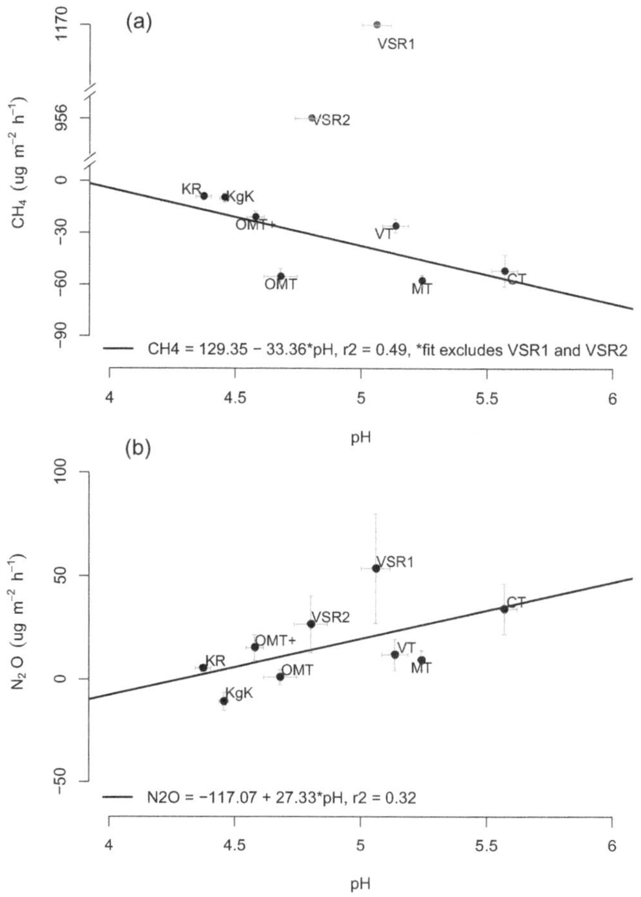

Figure 8. Scatterplot between site-specific mean pH and mean flux ($\mu gm^{-2}\,h^{-1}$) of **(a)** CH_4 or **(b)** N_2O of summer with intermediate moisture over the period of soil water sampling campaign (July–September 2005). Error bars show standard error. The CH_4 error bars for VSR1 and VSR2 are not shown.

forest types on mineral soil (uplands) and organo-mineral soil (transitions). Our study demonstrated that the CH_4 flux response to soil moisture changes with the relatively small mesoscale levels of a forest–mire ecotone (450 m long transect) (Fig. 4). The CH_4 flux sensitivity to soil moisture showed a positive linear response to CH_4 oxidation for the drier soils of transitions and uplands. Alternatively, CH_4 emission in mires showed a Gaussian form response to water level depth with a reduction of the optimum under saturated or dry peat conditions (Fig. 4). We have complemented the few studies on forest–mire gradients (e.g., Moosavi and Crill, 1997; Ullah et al., 2009; Ullah and Moore, 2011) and have lowered the likelihood of forest–mire transitions being biogeochemical hotspots of CH_4 emissions during short-term water level fluctuations.

The lack of an increase in CH_4 emissions during increased ground water levels in the transitions in our study could be attributed more to the relatively slow response of CH_4 producing bacteria than to the effectiveness of CH_4 oxidation which was reduced by a reduction in the aerated soil layer. Mäkiranta et al. (2009) showed that in forested peatlands the highest abundance of respiratory microbes could be found in the zone around the average water level. It is also known that the depth of maximum CH_4 production and oxidation is strongly related to 30-day average water level depth with time lag differences between the drier and wetter microsites (Kettunen et al., 1999). The duration of exceptionally increased high water levels was probably too short for CH_4 producing bacteria to relocate and/or adapt to water-saturated conditions. The methane production potential of mire varies in relation to methanogen communities, substrate availability, pH, and temperature (Juottonen et al., 2005, Juottonen et al., 2008). Unlike open mires, in drier conditions (similar to our forest–mire margin) decreases in the methanogen community are associated with low CH_4 production potential and with low emissions (Yrjälä et al., 2011). In the forest–mire margin, a relatively small population of methanotrophic microbes coupled with *Sphagnum* mosses and low CH_4 oxidation potential, related to low CH_4 concentrations in moss layer, could indicate low production potential (Larmola et al., 2010). It is known that water level depth is a major control of CH_4 oxidation, and that *Sphagnum* species originally not oxidising CH_4 need from several days to a month to activate methanotrophs through a water phase (Larmola et al., 2010; Putkinen et al., 2012).

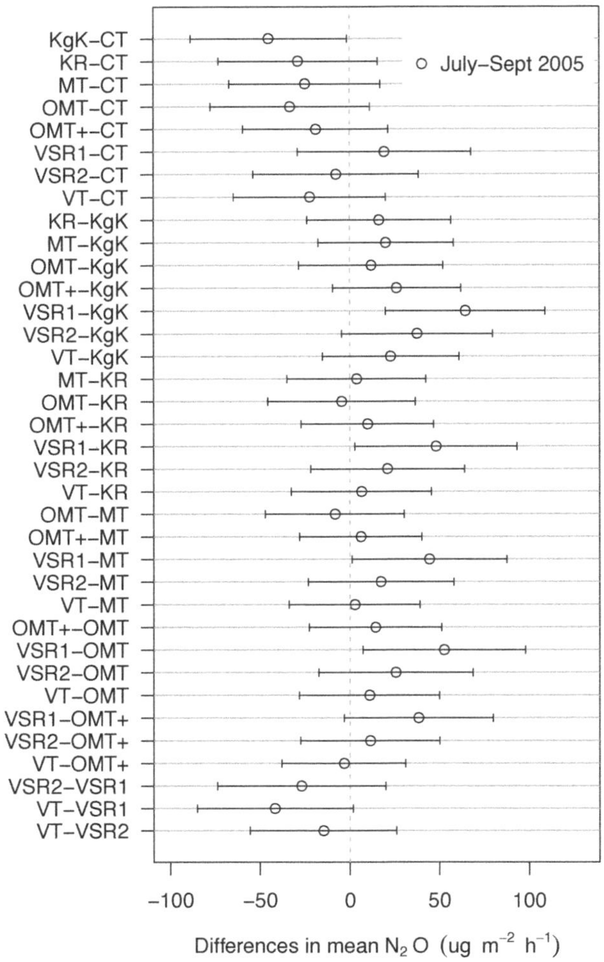

Figure 9. The post hoc Tukey differences (error bars for 95 % confidence intervals) of mean N_2O ($\mu gm^{-2}\,h^{-1}$) fluxes from the forest floor for the pairwise comparisons of forest/mire types (uplands – CT, VT, MT, OMT; transitions – OMT+, KgK, KR; and mires – VSR1, VSR2) over the period of the soil water sampling campaign (July–September 2005).

Temporally water-saturated soil layers of pristine forest–mire transitions had low CH_4 emissions possibly due to low pH, imposing physiological restrictions on soil microbial communities. Methanogenic activity in water-saturated organic soils can be reduced by high acidity (e.g., Ye et al., 2012). Activity of methanotrophic microbes is also pH dependent with optimum above 5 (Danilova and Dedysh, 2013; Saari et al., 2004). Our forest–mire transitions had mean pH below 5 and demonstrated lower net CH_4 uptake rates in comparison to upland forests on mineral soils (Fig. 8), which is in line with Saari et al. (2004). In spite of positive pH and CH_4 correlation found for the group of transitions and uplands together, the net CH_4 sink of upland well-drained mineral soil sites was primarily determined by high oxygen content. Small momentary CH_4 emissions (Supplement Fig. S3a) observed in forest–mire transitions also indicated potential for occasionally higher production than oxidation.

Beside differences in microsite soil water saturation, pH, and microbial communities, also plant communities (e.g., Saarnio et al., 1997; Strom et al., 2003; Riutta et al., 2007; Dorodnikov et al., 2011) play an important role in explaining net CH_4 emissions. In the forest–mire margin sites (KR and KgK) vascular plants (Fig. 1c) contributed to net forest floor CH_4 emissions (Fig. S3), if methane production occasionally increased. It is known that transport of recently photosynthesized carbon downwards to plant roots feeds microbial methane production (Alm et al., 1997; Strom et al., 2003; Dorodnikov et al., 2011). Aerenchyma of vascular plants transports most of produced CH_4 from peat to the atmosphere without oxidation in the acrotelm, and increases net CH_4 emissions (Hornibrook et al., 2009, Dorodnikov et al., 2011). A smaller amount of produced methane that is transported by pore water diffusion is efficiently oxidized by methanotrophs in the aerobic layer of peat and *Sphagnum* mosses (Hornibrook et al., 2009; Larmola et al., 2010).

Small CH_4 emissions as observed in relatively dry Scots pine dominated forests (VT – *Vaccinium vitis-idaea* type) (Fig. 3) with sandy Podzols soil and ground water depths around 2 m, have been occasionally found in mineral soil forests in other studies. The occasional mineral soil CH_4 effluxes suggested that plants' deepest roots transport CH_4 via the transpiration stream (Megonigal and Guenther 2008). Ullah et al., (2009) found that Spruce forest soils produced CH_4 only during the spring thaw season but later under drier summer conditions soils switched to CH_4 consumption. In our study the rare occurrence of small CH_4 emissions from forest soils differed between forest types and cannot only be attributed to increased soil moisture levels of microsites or transport from deep ground water sources. Small CH_4 emissions could be also partly attributed to the random noise in measurements. However, all the data showed a significant reduction of CH_4 uptake with increasing soil moisture at 10 cm, this may be associated with oxidation processes.

The form of dependence of CH_4 flux on soil moisture is better known from soil incubation studies (Pihlatie et al., 2004; Ullah et al., 2007) than from field studies, as field soil moisture ranges may be narrow (e.g., Nakamo et al., 2004). In order to describe the sensitivity of CH_4 uptake to moisture in the field we need a large amount of data covering a wide range of soil conditions (e.g., Hashimoto et al., 2011). In our study soil moisture varied between xeric and saturated conditions both spatially along the ecotone and temporally between years. Temporal soil water saturation in transitional forest–mire sites rather reduced CH_4 oxidations than promoted such CH_4 emissions as found in nearby permanently saturated mires. Beside the sensitivity of CH_4 fluxes to moisture we also observed sensitivity to soil temperature (Fig. 4) possibly also reflecting the role of soil physiochemical properties and/or the activity of methanogens. The linearly increasing CH_4 oxidation rates with temperature in upland forest types could reflect the importance of soil physiochemical properties, whereas the Gaussian form may also reflect a biological driven response in mires.

In our upland forests the role of soil physiochemical and microbiological drivers may have contributed to the fact that the temperature and moisture explained just 10 % of the variation. Although our mean CH_4 data did not show significant correlations with bulk density, the porous organic horizon is known to enable larger diffusion and CH_4 oxidation (Nakamo et al., 2004; Ullah and Moore 2011). It was difficult to assess the differences in sensitivity of CH_4 oxidation because of poor MQL and low fluxes of CH_4 oxidation. The absolute levels of the temperature effect on CH_4 fluxes in forest–mire transitions caused part of the signal to be mixed with variable sources of sampling errors and gas chromatograph precision errors. Though, in transitions both soil physiochemical and microbiological drivers may be important for CH_4 oxidations, as our forest–mire transitions showed a significant relation to soil moisture but not to temperature. The weak response of CH_4 oxidation to temperature was in contrast to the strong response to moisture and bulk density found in forests growing on mineral soils (Hashimoto et al., 2011). However, Nakamo et al. (2004) reported a clear relation with temperature but not with moisture for boreal birch forest (similar to our KR – spruce pine swamp).

In mires, the form of CH_4 sensitivity to temperature and water table depth may be also determined by differences in pH, and the composition of microbial and plant functional communities (Bubier et al., 1995; Jaatinen et al., 2004; Juottonen et al., 2005, 2008; Larmola et al., 2010; Riutta et al., 2007; Saarnio et al., 1997; Saari et al., 2004; Yrjälä et al., 2011). The CH_4 emissions in VSR1 were larger than in VSR2 (Fig. 4). Differences in pH could favor methanogen activity in less acid fen (Jouttonen et al., 2005; Yrjälä et al., 2011; Ye et al., 2012). Different coverage of vascular aerenchymous plants and *Sphagnum* mosses between VSR1 and VSR2 could affect site-specific CH_4 production and oxidation potentials. For example in VSR1 the water level was closer to the surface, and the lawn microsites had abundance of *Menyanthes* species (Fig. 1c), which are known to mediate higher CH_4 transport (Bubier et al., 1995; Macdonald et al., 1998), whereas in VSR2 *Menyanthes* species was absent. Shallower form of CH_4 sensitivity to water table in a hummock type fen VSR2 than in lawn type of fen VSR1 could resulted from differences in plant mediated CH_4 emissions (e.g., Riutta et al., 2007; Hornibrook et al., 2009; Dorodnikov et al., 2011) or CH_4 oxidation potential between *Sphagnum* species (Larmola et al., 2010). For example in the study by Saarnio et al. (1997) the CH_4 flux response to water level would be exponential if it accounted only for emissions from hummock and *Carex* lawn microsites, but the response was Gaussian for flark, hummock, *Eriophorum* lawn and *Carex* lawn microsites taken together.

4.2 N_2O dynamics

The momentary N_2O fluxes in the range from -107 to 248 ($\mu g\, m^{-2}\, h^{-1}$) and median emissions close to 0 ($\mu g\, m^{-2}\, h^{-1}$) for forest/mire types (Fig. 5) were in the proximity of values for soils in similar climates (Von Arnold et al., 2005a, b, Pihlatie et al., 2007; Matson et al., 2009; Ullah et al., 2009; Ojanen et al., 2010). Forest floor N_2O fluxes did not differ significantly for the nine forest/mire types of the ecotone $p = 0.637$ for the whole period from May 2005 to September 2006, probably due to the low nitrification potential of boreal forests in natural conditions (Regina et al., 1996). In contrast to our results, Ullah and Moore (2009, 2011) found that soil drainage and dominant tree species strongly control net nitrification rates, and that N_2O emissions from poorly drained soils can be three times larger than those from well-drained soils due to slower denitrification than nitrification activity. Statistically significant differences were also found between drained and undrained forests growing on organic soils and between evergreens and deciduous plants (Von Arnold et al., 2005a, b).

Soil incubation studies under various moisture and temperature regimes (Pihlatie et al., 2004; Szukics et al., 2010) imply that our higher forest floor N_2O emissions during typical summer 2005 than during dry summer 2006 (Supplement Fig. 3b) were probably induced by stimulated N turnover through the soil wetting and drying cycle under favorable temperature. During conditions with intermediate moisture (July–September 2005) we also observed mean N_2O flux of dry pine forest significantly larger than that of paludified spruce forest (larger CT than KgK), whereas mean N_2O flux of water-saturated mire was larger than four sites (VSR1–KgK, VSR1–KR, VSR1–MT, VSR1–OMT) (Figs. 8, 9). Therefore, during fluctuating soil moisture, we could expect increased N_2O fluxes of a normally xeric (CT) and water-saturated (VSR1) site due to stimulated nitrification (CT in rewetting phase, and VSR1 in drying phase). During July–September 2005, CT and VSR1 sites were also least acid along the ecotone which could favor nitrification and consequently N_2O emissions through denitrification (Regina et al., 1996; Ste-Marie and Pareé, 1999; Paavolainen et al., 2000). These studies reported that the increase of pH by rewetting could initiate nitrification. In contrast to less acid CT and VSR1, the more acid forest–mire transitions with the widest range of water level fluctuations ranked into a group of sites with lower N_2O fluxes. Highly acid conditions prevent development of nitrifiers, substrate affinity, and nitrification, even if ammonium is available (Ste-Marie and Pareé, 1999; Paavolainen et al., 2000). The fact that net nitrification of acid sensitive nitrifiers positively increases with forest floor pH, whereas acidification reduces it, suggests that nitrifiers in our sites were acid sensitive and not acid tolerant. The lack of nitrate renders denitrification potential to be negligible. Although, if nitrate were present, low pH would enhance N_2O emissions due to inhibiting di-nitrogenoxide reductase and increasing N_2O / N_2 ratio of denitrification (e.g., Weslien et al., 2009).

In pristine mires nitrification positively depended on pH and negatively on water level (Regina et al., 1996) in supply of nitrate for denitrification, as the main source of N_2O emissions (Regina et al., 1996; Nykänen et al., 1995; Wray et al., 2007). Thus, during drying–rewetting periods as in July–September 2005, our sites could initiate short-term significant differences, but for the whole measurement period the lack of a statistically significant difference in N_2O fluxes was probably due to low nitrification potential. Generally, low pH and high C / N ratios of our forest floors suggest conditions of low nitrification potential. Thus, the lack of a statistically significant difference in N_2O fluxes was probably due to low nitrification potential. Other reasons could be the low field sampling frequency and relatively high noise in the data (MQL compared to low fluxes). Measuring three microsites per site could lead to missing some peak N_2O emission events due to a large microscale spatial variation (Von Arnold et al., 2005a). With our weekly or bi-weekly sampling frequency we could not identify larger microsite specific peak events possibly occurring after N was mobilized from, e.g., fast decomposition of deciduous foliage during the drought related early peak in litterfall or during sudden soil freeze–thaw cycles (Pihlatie et al., 2007). However, during the active growing season these events might be rare in typical boreal conditions where plants are adapted to a rapid uptake of limited rates of soil N mineralization (Hikosaka, 2003; Korhonen et al., 2013; Lupi et al., 2013).

Several studies (Martikainen et al., 1995; Regina et al., 1996) reported that peatlands in a pristine state showed small N_2O emissions, but when drained, nitrification rates were enhanced depending on nutrient status (a large increase for rich sites and no increase for poor sites). The limited increase in N_2O emissions during the summer drought in our mires may be therefore attributed to low nutrient levels, a low supply of nitrate and/or low nitrification potential. Relatively low fertility may also be expected to limit the N_2O emissions during the dry season of our forests and forest–mire transitions as the N_2O emissions are also known to correlate with site fertility, e.g., expressed as C / N ratio (Klemedtsson et al., 2005; Ojanen et al., 2010; Hashimoto et al., 2011).

The N_2O fluxes of forest/mire types fitted by nonlinear regression models showed positive linear response to soil moisture at a depth of 5 cm and significant Gaussian type response to temperature at depths of 5 cm (Table 4, Fig. 7). However, the residuals of the moisture and temperature model were large (Fig. 7) and r^2 was only 10 %. Luo et al. (2012) demonstrated for temperate forests that N_2O emissions depended nonlinearly on the soil moisture and positively on soil temperature. In our study, the weak linear response of soil moisture to N_2O fluxes could be an artifact of fitting several N_2O processes of different sensitivity to different forest/mire types. For example in well-drained uplands the N_2O fluxes may be mainly due to processes of ammonification and nitrification, while in mires nitrification in the drier surface layer may be coupled with denitrification in deeper water-saturated layers (Ambus et al., 2006; Regina et al., 1996). The soil moisture and temperature from deeper layers did not significantly explain the N_2O fluxes (results not shown). An active depth of 5 cm corresponding to the top of the organic layer is in agreement with Pihlatie et al. (2007) who demonstrated that N turnover in poor boreal forest soil takes place in the litter layer and that N_2O emissions originate mainly from the top soil. The N_2O production in our study increased with rising soil temperature of the humus layer from 7 °C typically found after the soil thawed during spring warming and in autumn during soil cooling. These could be the periods when the nitrification potential increased; in spring probably due to mobilization of nitrogen during freeze–thaw cycles and in autumn probably due to mobilization of nitrogen from the quickly decomposing foliar litterfall (Pihlatie et al., 2007, 2010; Luo et al., 2012).

5 Conclusions

The CH_4 fluxes of forest–mire ecotone were significantly different not only between sources or sink type forests but also between sinks (upland and transitional types) and between sources (mires). The forest–mire transitions showed CH_4 oxidation rather than emission with very small sensitivity to wet and dry events. The N_2O fluxes of forest mire types were generally low. Despite small N_2O peaks in spring and autumn, the N_2O fluxes showed low sensitivity to soil moisture probably due to poor soil nitrogen content and the low nitrification potential of the forest/mire types in pristine conditions. In spite of the potential of pristine forest–mire transitions to represent biogeochemical hotspots in the landscape, the CH_4 and N_2O flux levels in the transitions changed minimally during extremely large range of weather conditions. Our pristine forest–mire transitions did not act as biogeochemical hotspots for CH_4 and N_2O emissions. Therefore, when making attempts to upscale boreal landscape carbon and nitrogen cycles, the organo-mineral soils of pristine forest–mire transitions should be regarded as CH_4 sinks and minor N_2O sources rather than having the peak emissions on the landscape level.

Acknowledgements. This work was supported by Academy of Finland projects ICOS 271878, ICOS-Finland 281255 and ICOS-ERIC 281250; EU projects; NordForsk, through the Nordic Centre of Excellence (project DEFROST); the Finnish Centre of Excellence in Physics, Chemistry, Biology and Meteorology of

Atmospheric Composition and Climate Change (FCoE); and the Academy of Finland Center of Excellence program (project number 1118615). We also thank Jukka Laine, Jukka Alm, Mike Starr and Frank Berninger for valuable discussions; Mike Starr for providing suction cup lysimeters; Hannu Ilvesniemi for the soil moisture probe; Mari Pihlatie for providing us with the flux calculation script; Ilkka Korpela for providing the aerial image, courtesy of Finnish National Land Survey 2004; and Donald Smart for English language revision. We appreciate the useful comments of the editor Donatella Zona and three anonymous reviewers who improved the manuscript.

Edited by: D. Zona

References

Agnew, A. D. Q., Wilson, J. B., and Sykes, M. T.: A vegetation switch as the cause of a forest/mire ecotone in New Zealand, J. Veget. Sci., 4, 273–278, 1993.

Alm, J., Talanov, A., Saarnio, S., Silvola, J., Ilkkonen, E., Aaltonen, H., Nykänen, H., and Martikainen, P. J.: Reconstruction of carbon balance for microsites in a boreal oligotrophic pine fen, Finland, Oecologia, 110, 423–431, 1997.

Alm, J., Shurpali, N. J., Tuittila, E.-S., Laurila, T., Maljanen, M., Saarnio, S., and Minkkinen, K.: Methods for determining emission factors for the use of peat and peatlands – flux measurements and modelling, Boreal Env. Res., 12, 85–100, 2007.

Ambus, P., Zechmeister-Boltenstern, S., and Butterbach-Bahl, K.: Sources of nitrous oxide emitted from European forest soils, Biogeosciences, 3, 135–145, doi:10.5194/bg-3-135-2006, 2006.

Bubier, J., Moore, T., and Juggins, S.: Predicting methane emissions from bryophyte distribution in northern Canadian peatlands, Ecology, 76, 677–693, 1995.

Cajander, A. K.: Forest types and their significance, Acta Forestalia Fennica, 56, 1–69, 1949.

D'Angelo, E. and Reddy, K.: Regulators of heterotrophic microbial potentials in wetland soils, Soil Biol. Biochem., 31, 815–830, 1999.

Danilova, O. V. and Dedysh, S. N.: Abundance and diversity of methanotrophic *Gammaproteobacteria* in northern wetlands, Microbiology, 83, 67–76, 2014.

Dorodnikov, M., Knorr, K.-H., Kuzyakov, Y., and Wilmking, M.: Plant-mediated CH4 transport and contribution of photosynthates to methanogenesis at a boreal mire: a ^{14}C pulse-labeling study, Biogeosciences, 8, 2365–2375, 2011, http://www.biogeosciences.net/8/2365/2011/.

Frasier, R., Ullah, S., and Moore, T. R.: Nitrous oxide consumption potentials of well-drained forest soils in southern Quebec, Canada, Geomicrobiol. J., 27, 53–60, 2010.

Finnish statistical yearbook of forestry 2013. Finnish Forest Research Institute, Metsäntutkimuslaitos, Finland, http://www.metla.fi/julkaisut/metsatilastollinenvsk/index-en.htm, 2014.

Hari, P. and Kulmala, M.: Station for Measuring Ecosystem–Atmosphere Relations (SMEAR II), Boreal Env. Res., 10, 315–322, 2005.

Hartshorn, A. S., Southard, R. J., and Bledsoe, C. S.: Structure and function of peatland-forest ecotones in southeastern Alaska, Soil Sci. Soc. Am. J., 67, 1572–1581, 2003.

Hashimoto, S., Morishita, T., Sakata, T., Ishizuka, S., Kaneko, S., and Takahashi, M.: Simple models for soil CO_2, CH_4, and N_2O fluxes calibrated using a Bayesian approach and multi-site data, Ecol. Modell., 222, 1283–1292, 2011.

Hikosaka K.: A model of dynamics of leaves and nitrogen in a plant canopy: an integration of canopy photosynthesis, leaf life span, and nitrogen use efficiency, The American Naturalist , 162, 149–164, 2003.

Hornibrook, E. R. C., Bowes, H. L., Culbert, A., and Gallego-Sala, A. V.: Methanotrophy potential versus methane supply by pore water diffusion in peatlands, Biogeosciences, 6, 1491–1504, doi:10.5194/bg-6-1491-2009, 2009.

Howie, S. A. and Tromp-van Meerveld, I.: The essential role of the lagg in raised bog function and restoration: a review, Wetlands, 31, 613–622, 2011.

Huttunen, J., Nykänen, H., Martikainen, P., and Nieminen, M.: Fluxes of nitrous oxide and methane from drained peatlands following forest clear-felling in Southern Finland, Plant Soil, 255, 457–462, 2003.

Jaatinen, K., Knief, C., Dunfield, P. F., Yrjälä, K., and Fritze, H.: Methanotrophic bacteria in boreal forest soil after fire, FEMS Microbiol. Ecol., 50, 195–202, 2004.

Juottonen, H., Galand, P. E., Tuittila, E., Laine, J., Fritze, H., and Yrjälä, K.: Methanogen communities and bacteria along an ecohydrological gradient in a northern raised bog complex, Environ. Microbiol., 7, 1547–1557, 2005.

Juottonen, H., Tuittila, E., Juutinen, S., Fritze, H., and Yrjälä, K.: Seasonality of rDNA- and rRNA-derived archaeal communities and methanogenic potential in a boreal mire, The ISME J., 2, 1157–1168, 2008.

Kettunen, A., Kaitala, V., Lehtinen, A., Lohila, A., Alm, J., Silvola, J. and Martikainen, P. J.: Methane production and oxidation potentials in relation to water table fluctuations in two boreal mires, Soil Biol. Biochem., 31, 1741–1749, 1999.

Klemedtsson, L., Von Arnold, K., Weslien, P., and Gundersen, P.: Soil CN ratio as a scalar parameter to predict nitrous oxide emissions, Global Change Biol., 11, 1142–1147, 2005.

Korhonen, J. F. J., Pihlatie, M., Pumpanen, J., Aaltonen, H., Hari, P., Levula, J., Kieloaho, A.-J., Nikinmaa, E., Vesala, T., and Ilvesniemi, H.: Nitrogen balance of a boreal Scots pine forest, Biogeosciences, 10, 1083–1095, doi:10.5194/bg-10-1083-2013, 2013.

Laine, J., Komulainen, V.-M., Laiho, R., Minkkinen, K., Rasinmäki, A., Sallantaus, T., Sarkkola, S., Silvan, N., Tolonen, K., Tuittila, E-S., Vasander, H., and Päivänen, J.: Lakkasuo, a guide to mire ecosystems, University of Helsinki Department of Forest Ecology Publications 31, Helsinki, Finland, 123 pp., 2004.

Larmola, T., Tuittila, E.-S., Tiirola, M., Nykäne, H., Martikainen, J.P., Yrjälä, K., Tuomivirta, T., and Fritze, H.: The role of *Sphagnum* mosses in the methane cycling of a boreal mire, Ecology, 91, 2356–2365, 2010.

Luo, G. J., Brüggemann, N., Wolf, B., Gasche, R., Grote, R., and Butterbach-Bahl, K.: Decadal variability of soil CO_2, NO, N_2O, and CH_4 fluxes at the Höglwald Forest, Germany, Biogeosciences, 9, 1741–1763, doi:10.5194/bg-9-1741-2012, 2012.

Lupi, C., Morin, H., Deslauriers, A., Rossi, S., Houle, D.: Role of soil nitrogen for the conifers of the boreal forest: a critical review, International Journal of Plant and Soil Science, 2, 155-189, 2012.

Macdonald, J. A., Fowler, D., Hargreaves, K. J., Skiba, U., Leith, I. D., and Murray, M. B.: Methane emission rates from a northern wetland; response to temperature, water table and transport, Atmos. Environ., 32, 3219–3227, 1998.

Mäkiranta, P., Laiho, R., Fritze, H., Hytönen, J., Laine, J., and Minkkinen, K.: Indirect regulation of heterotrophic peat soil respiration by water level via microbial community structure and temperature sensitivity, Soil Biol. Biochem., 41, 695–703, 2009.

Martikainen, P. J., Nykanen, H., Crill, P., and Silvola, J.: Effect of a lowered water table on nitrous oxide fluxes from northern peatlands, Nature, 366, 51–53, 1993.

Martikainen, P. J., Nykänen, H., Alm, J., and Silvola, J.: Change in fluxes of carbon dioxide, methane and nitrous oxide due to forest drainage of mire sites of different trophy, Plant Soil, 168/169, 571–577, 1995.

Matson, A., Pennock, D., and Bedard-Haughn A.: Methane and nitrous oxide emissions from mature forest stands in the boreal forest, Saskatchewan, Canada, For. Ecol. Manage., 258, 1073–1083, 2009.

McClain, M. E., Boyer, E. W.,Dent, C. L., Gergel, S. E., Grimm, N. B., Groffman, P. M., Hart, S. C., Harvey, J. W., Johnston, C. A., Mayorga, E., McDowell, W. H., and Pinay, G.: Biogeochemical hot spots and hot moments at the interface of terrestrial and aquatic ecosystems, Ecosystems 6, 301–312, 2003.

Megonigal, J. P and Guenther, A. B.: Methane emissions from upland forest soils and vegetation, Tree Physiol., 28, 491–498, 2008.

Moosavi, S. C. and Crill, P. M.: Controls on CH_4 and CO_2 emissions along two moisture gradients in the canadian boreal zone, J.Geophys.Res., 102, 29261–29277, 1997.

Nakano, T., Tnoue, G., and Fukuda, M.: Methane consumption and soil respiration by a birch forest soil in West Siberia, Tellus B, 56, 223–229, 2004.

Nykänen, H., Alm, J., Lang, K., Silvola, J., and Martikainen, P.: Emissions of CH_4, N_2O and CO_2 from a virgin fen and a fen drained for grassland in Finland, J. Biogeogr., 22, 351–357, 1995.

Nykänen, H., Alm, J., Silvola, J., Tolonen, K., and Martikainen, P. J.: Methane fluxes on boreal peatlands of different fertility and the effect of long-term experimental lowering of the water table on flux rates, Global Biogeochem. Cy., 12, 53–69, 1998.

Ojanen, P., Minkkinen, K., and Alm, J.: Soil–atmosphere CO_2, CH_4 and N_2O fluxes in boreal forestry-drained peatlands, Forest Ecol. Manage., 260, 411–421, 2010.

Paavolainen, L., Fox, M., and Smolander, A.: Nitrification and denitrification in forest soil subjected to sprinkling infiltration, Soil Biol. Biochem., 32, 669–678, 2000.

Pihlatie, M., Syväsalo, E., Simojoki, A., Esala, M., and Regina, K.: Contribution of nitrification and denitrification to N_2O production in peat, clay and loamy sand soils under different soil moisture conditions, Nutr. Cy. Agroecosyst., 70, 135–141, 2004.

Pihlatie, M., Pumpanen, J., Rinne, J., Ilvesniemi, H., Simojoki, A., Hari, P., and Vesala, T.: Gas concentration driven fluxes of nitrous oxide and carbon dioxide in boreal forest soil, Tellus B, 59, 458–469, 2007.

Pihlatie, M. K., Kiese, R., Brüggemann, N., Butterbach-Bahl, K., Kieloaho, A.-J., Laurila, T., Lohila, A., Mammarella, I., Minkkinen, K., Penttilä, T., Schönborn, J., and Vesala, T.: Greenhouse gas fluxes in a drained peatland forest during spring frost-thaw event, Biogeosciences, 7, 1715–1727, doi:10.5194/bg-7-1715-2010, 2010.

Pinheiro, J., Bates, D., DebRoy, S., Sarkar, D., and the R Development Core Team: nlme: Linear and Nonlinear Mixed Effects Models, R package version, 3.1, 113, http://cran.r-project.org/web/packages/nlme/nlme.pdf, 2013.

Putkinen, A., Larmola, T., Tuomivirta, T., Siljanen, H.M.P., Bodrossy, L., Tuittila, E.-S., and Fritze, H.: Water dispersal of methanotrophic bacteria maintains functional methane oxidation in Sphagnum mosses, Front. Microbio., 3, 15, doi:10.3389/fmicb.2012.00015, 2012.

R Core Team, R: A language and environment for statistical computing, R Foundation for Statistical Computing, Vienna, Austria, http://www.R-project.org/, 2013.

Regina, K., Nykänen, H., Silvola, J., and Martikainen, P.: Fluxes of nitrous oxide from boreal peatlands as affected by peatland type, water table level and nitrification capacity, Biogeochemistry, 35, 401–418, 1996.

Riutta, T., Laine, J., Aurela, M., Rinne, J., Vesala, T., Laurila, T., Haapanala, S., Pihlatie, M., and Tuittila, E.: Spatial variation in plant community functions regulates carbon gas dynamics in a boreal fen ecosystem, Tellus B, 59, 838–852, 2007.

Rosenkranz, P., Brűggemann, N., Papen, H., Xu, Z., Seufert, G., and Butterbach-Bahl, K.: N_2O, NO and CH_4 exchange and microbial N turnover over a Mediterranean pine forest soil, Biogeosciences, 3, 121–133, doi:10.5194/bg-3-121-2006, 2006.

Saari, A., Rinnan, R., and Martikainen, P. J.: Methane oxidation in boreal forest soils: Kinetics and sensitivity to pH and ammonium, Soil Biology and Biochemistry, 36, 1037–1046, 2004.

Saarnio, S., Alm, J., Silvola, J., Lohila, A., Nykanen, H., and Martikainen, P.: Seasonal variation in CH_4 emissions and production and oxidation potentials at microsites on an oligotrophic pine fen, Oecologia, 110, 414–422, 1997.

Seibert, J., Stendahl, J., and Sørensen, R.: Topographical influences on soil properties in boreal forests, Geoderma, 141, 139–148, 2007.

Solondz, D. S., Petrone, R. M., and Devito, K. J.: Forest floor carbon dioxide fluxes within an upland-peatland complex in the Western Boreal Plain, Canada, Ecohydrology, 1, 361–376, 2008.

Sommerfeld, R. A., Mosier, A. R., and Musselman, R. C.: CO_2, CH_4 and N_2O flux through a Wyoming snowpack and implications for global budgets, Nature, 361, 140–142, 1993.

Starr, M. R.: Variation in the quality of tension lysimeter soil water samples from a Finnish forest soil. Soil Sci., 140, 453–461, 1985.

Ste-Marie, C. and Pareé, D.: Soil, pH and N availability effects on net nitrification in the forest floors of a range of boreal forest stands, Soil Biol. Biochem., 31, 1579–1589, 1999.

Strom, L., Ekberg, A., Mastepanov, M., and Christensen, T. R.: The effect of vascular plants on carbon turnover and methane emissions from a tundra wetland, Global Change Biol., 9, 1185–1192, doi:10.1046/j.1365-2486.2003.00655.x., 2003.

Szukics, U., Abell, G.C., Hödl, V., Mitter, B., Sessitsch, A., Hackl, E., and Zechmeister-Boltenstern, S.: Nitrifiers and denitrifiers respond rapidly to changed moisture and increasing temperature in a pristine forest soil, FEMS Microbiol. Ecol., 72, 395–406, 2010.

Ťupek, B., Minkkinen, K., Kolari, P., Starr, M., Chan, T., Alm, J., Vesala, T., Laine, J., and Nikinmaa, E.: Forest floor versus ecosystem CO_2 exchange along boreal ecotone between upland forest and lowland mire, Tellus B, 60, 153–166, 2008.

Ullah, S. and Moore, T. R.: Soil drainage and vegetation controls of nitrogen transformation rates in forest soils, southern Quebec, J. Geophys. Res., 114, 01014, doi:10.1029/2008JG000824, 2009.

Ullah, S. and Moore, T. R.: Biogeochemical controls on methane, nitrous oxide, and carbon dioxide fluxes from deciduous forest soils in eastern Canada, J. Geophys. Res., Biogeosciences, 116, G03010, doi:10.1029/2010JG001525, 2011.

Ullah, S., Frasier, R., King, L., Picotte-Anderson, N., and Moore, T. R.: Potential fluxes of N_2O and CH_4 from three forests type soils in eastern Canada, Soil Biol. Biochem., 40, 986–994, 2008.

Ullah, S., Frasier, R., Pelletier, L., and Moore, T. R.: Greenhouse gas fluxes from boreal forest soils during the snow-free period in Quebec, Canada, Canadian Journal of Forest Research-Revue Canadienne De Recherche Forestiere, 39, 666-680, 2009.

Von Arnold, K., Ivarsson, M., Öqvist, M., Majdi, H., Björk, R.G., Weslien P., and Klemedtsson, L.: Can distribution of trees explain variation in nitrous oxide fluxes?, Scand. J. Forest. Res., 20, 481–489, 2005a.

Von Arnold, K., Weslien, P., Nilsson, M., Svensson, B., and Klemedtsson, L.: Fluxes of CO_2, CH_4 and N_2O from drained coniferous forests on organic soils, Forest Ecol. Manage., 210, 239–254, 2005b.

Weishampel, P., Kolka, R., and King, J. Y.: Carbon pools and productivity in a 1 km^2 heterogeneous forest and peatland mosaic in Minnesota, USA, Forest Ecol. Manage., 257, 747–754, 2009.

Weslien, P., Kasimir Klemedtsson, Å., Börjesson, G., and Klemedtsson, L.: Strong pH influence on N_2O and CH_4 fluxes from forested organic soils, Europ. J. Soil Sci., 60, 311–320, 2009.

Włodarczyk, T., Szarlip, P., and Brzezińska, M.: Nitrous oxide consumption and dehydrogenase activity in Calcaric Regosols, Polish J. Soil Sci., 2, 97–110, 2005.

Wrage, N., Velthof, G. L., van Beusichem, M. L., and Oenema, O.: Role of nitrifier denitrification in the production of nitrous oxide, Soil Biol. Biochem., 33, 1723–1732, 2001.

Wray, H. E. and Bayley, S. E.: Denitrification rates in marsh fringes and fens in two boreal peatlands in Alberta, Canada, Wetlands, 27, 1036–1045, 2007.

Ye, R. Z., Jin, Q. S., Bohannan, B., Keller, J. K., McAllister, S. A., and Bridgham, S. D.: pH controls over anaerobic carbon mineralization, the efficiency of methane production, and methanogenic pathways in peatlands across an ombrotrophic-minerotrophic gradient, Soil Biol. Biochem., 54, 36–47, 2012.

Yrjälä, K., Tuomivirta, T., Juottonen, H., Putkinen, A., Lappi, K., Tuittila, E., Penttilä, T., Minkkinen, K., Laine, J., Peltoniemi, K., and Fritze, H.: CH_4 production and oxidation processes in a boreal fen ecosystem after long-term water table drawdown, Glob. Change Biol., 17, 1311–1320, 2011.

Quantifying environmental stress-induced emissions of algal isoprene and monoterpenes using laboratory measurements

N. Meskhidze[1], **A. Sabolis**[1,*], **R. Reed**[2], **and D. Kamykowski**[1]

[1]Marine, Earth, and Atmospheric Science, North Carolina State University, Raleigh, NC, USA
[2]Center for Applied Aquatic Ecology, North Carolina State University, Raleigh, NC 27606, USA
[*]now at: Scott Safety, Monroe, NC 28110, USA

Correspondence to: N. Meskhidze (nmeskhidze@ncsu.edu)

Abstract. We report here production rates of isoprene and monoterpene compounds (α-pinene, β-pinene, camphene and d-limonene) from six phytoplankton monocultures as a function of irradiance and temperature. Irradiance experiments were carried out for diatom strains (*Thalassiosira weissflogii* and *Thalassiosira pseudonana*), prymnesiophyte strains (*Pleurochrysis carterae*), dinoflagellate strains (*Karenia brevis* and *Prorocentrum minimum*), and cryptophyte strains (*Rhodomonas salina*), while temperature experiments were carried out for diatom strains (*Thalassiosira weissflogii* and *Thalassiosira pseudonana*). Phytoplankton species, incubated in a climate-controlled room, were subject to variable light (90 to 900 μmol m^{-2} s^{-1}) and temperature (18 to 30 °C) regimes. Compared to isoprene, monoterpene emissions were an order of magnitude lower at all light and temperature levels. Emission rates are normalized by cell count and Chlorophyll *a* (Chl *a*) content. Diatom strains were the largest emitters, with $\sim 2 \times 10^{-17}$ g(cell)$^{-1}$ h^{-1} ($\sim$ 35 μg (g Chl *a*)$^{-1}$ h^{-1}) for isoprene and $\sim 5 \times 10^{-19}$ g (cell)$^{-1}$ h^{-1} ($\sim$ 1 μg (g Chl *a*)$^{-1}$) h^{-1}) for α-pinene. The contribution to the total monoterpene production was $\sim$ 70 % from α-pinene, $\sim$ 20 % for d-limonene, and < 10 % for camphene and β-pinene. Phytoplankton species showed a rapid increase in production rates at low irradiance (< 150 μmol m^{-2} s^{-1}) and a gradual increase at high (> 250 μmol m^{-2} s^{-1}) irradiance. Measurements revealed different patterns for time-averaged emissions rates over two successive days. On the first day, most of the species showed a distinct increase in production rates within the first 4 h while, on the second day, the emission rates were overall higher, but less variable. The data suggest that enhanced amounts of isoprene and monoterpenes are emitted from phytoplankton as a result of perturbations in environmental conditions that cause imbalance in chloroplasts and force primary producers to acclimate physiologically. This relationship could be a valuable tool for development of dynamic ecosystem modeling approaches for global marine isoprene and monoterpene emissions based on phytoplankton physiological responses to a changing environment.

1 Introduction

The age of global warming has brought about a heightened amount of attention to the interactions between life and the Earth's climate. Global climate models with advanced treatment of manmade aerosols and pollutants and their interactions with clouds have been widely used for calculations from climate sensitivity to greenhouse gas emissions. Over the last decade, a large number of studies have been devoted to reducing the uncertainty in anthropogenic aerosol radiative forcing of climate. However, a lesser-known uncertainty in climate predictions as a result of natural aerosols has only recently been explored (Meskhidze et al., 2011; Carslaw et al., 2013). As estimates of anthropogenic forcing are based on a pre-industrial atmosphere composed mainly of natural aerosols, the representation of natural aerosols in climate models strongly influences the predictions of current and future climate effects of anthropogenic aerosols (Hoose et al., 2009). Data analysis shows that up to 45 % of the variance of aerosol forcing arises from uncertainties in natural emissions of volcanic sulfur dioxide, marine dimethyl sulfide, biogenic volatile organic carbon (BVOC), biomass burning, and sea spray (Carslaw et al., 2013). Moreover, recent studies have revealed that several biotic and abiotic stresses associated with changing environmental factors increase the emission of BVOC from both terrestrial and aquatic plants (Rinnan et al., 2014). Therefore, the Earth's climate models should be able not only to characterize aerosol and trace gas emissions from the Earth's various ecospheres and quantify the uncertainties associated with these emissions, but also to predict the changes in the emission rates that are related to the human activity.

While climate is regulated by numerous factors, some of the largest effects of natural aerosols on climate radiative forcing occur over the oceans, where boundary layer clouds cover the vast expanse of the Earth's surface and have some of the highest sensitivity of cloud albedo to cloud condensation nuclei concentrations (Platnick and Twomey, 1994; Hoose et al., 2009). Thus, factors that regulate the concentration of aerosols over the oceans and the resulting reflectivity of low-level marine clouds can critically affect the climate system as a whole (e.g., Randall et al., 1984; Stevens et al., 2005). The existence of physical relationships between marine biota, gas emissions, aerosols, clouds, and radiative forcing has been hypothesized for more than several decades (Shaw, 1983). If such a relationship exists, the observed tight coupling between environmental change and plankton dynamics (such as community structure, timing of seasonal abundance, and geographical range) (Hays et al., 2005) may have a considerable impact on the current and future climates.

Marine aerosols associated with ocean biota can be derived from both primary (through bubble bursting) and secondary (through oxidation of phytoplankton-emitted trace gases) processes (Blanchard, 1964; Hoffman and Duce, 1976; Charlson et al., 1987; O'Dowd et al., 2004; Meskhidze and Nenes, 2006; O'Dowd and Leeuw, 2007). Recent reviews summarized the state of the art and remaining uncertainties in production mechanisms, number concentration, size distribution, chemical composition, and optical properties of sea spray aerosols (de Leeuw et al., 2011; Meskhidze et al., 2013). In addition to primary aerosols, many oceanic trace gases produced as a consequence of marine biological activity have also been shown to have global-scale impacts on biogeochemical cycling, tropospheric and stratospheric ozone depletion, photochemical processing, and secondary aerosol formation (Carpenter et al., 2012, and references therein). Although the global BVOC emission rate of phytoplankton is considerably small compared to terrestrial vegetation, the emissions occur over relatively pristine regions and therefore have the ability to influence oxidation capacity and natural aerosol formation in remote marine and coastal regions (Luo and Yu, 2010).

Following an initial finding for a marine source of isoprene (Bonsang et al., 1992; Broadgate et al., 1997) and monoterpenes (Yassaa et al., 2008), an increasing number of studies have focused recently on improved understanding of their potential effects on coastal air quality (Palmer and Shaw, 2005; Liakakou et al., 2007; Gantt et al., 2010a, b) and climate (Gantt et al., 2009; Roelofs, 2008; Meskhidze et al., 2011). Isoprene production rates have been determined for microalgae (Moore et al., 1994; Milne et al., 1995; Mckay et al., 1996; Shaw et al., 2003; Gantt et al., 2009; Bonsang et al., 2010; Exton et al., 2013), macroalgae (Broadgate et al., 2004), and microbial communities (Acuña Alvarez et al., 2009). Despite this, large discrepancies remain in bottom-up and top-down estimates of marine sources of isoprene, having a wide range from ~ 0.1 to $1.9\,\mathrm{Tg\,C\,yr^{-1}}$ and $\sim 11.6\,\mathrm{Tg\,C\,yr^{-1}}$, respectively (Milne et al., 1995; Palmer and Shaw, 2005; Gantt et al., 2009; Lou and Yu, 2010). An even greater level of uncertainty exists for sources of ocean-derived monoterpenes that were proposed to range from 0.013 to $29.5\,\mathrm{Tg\,C\,yr^{-1}}$ (Luo and Yu, 2010). Past studies also revealed that isoprene emissions from phytoplankton are strongly correlated with incoming radiation (Shaw et al., 2003; Gantt et al., 2009). Both studies showed a rapid increase in isoprene production at low irradiance levels ($<150\,\mu\mathrm{mol\,m^{-2}\,s^{-1}}$) and a gradual increase at higher irradiance levels ($>250\,\mu\mathrm{mol\,m^{-2}\,s^{-1}}$). Emission rates of monoterpenes from nine algal species have also been reported, but only at significantly lower (30 to $100\,\mu\mathrm{mol\,m^{-2}\,s^{-1}}$) irradiance levels (Yassaa et al., 2008). When studying emission of isoprene and monoterpenes from different algal species, most of the research has been focused on either long-term emission rates from normal growth conditions or short-term emissions as responses to different environmental stress factors (i.e., light and thermal stress). Little attention has been devoted to production of isoprene and monoterpenes from phytoplankton as a function of time when they are forced to adjust or acclimate physiologically.

During their lifetime, phytoplankton can also be exposed to temperature values considerably different from the ones at which cultures have been acclimated due to vertical mixing in the water column. Eppley (1972) pioneered the study of phytoplankton growth rates at acclimated temperature. Schofield et al. (1998) investigated temperature effects on dinoflagellate photosynthetic light and dark reactions, while Staehr and Birkeland (2006) discussed temperature effects on broader aspects of phytoplankton biochemistry and metabolism. Production of isoprene and monoterpenes from marine algae as a result of simultaneous changes in temperature and light regimes has not been previously investigated.

This study explores isoprene and monoterpene production rates that arise in response to variations in environmental factors for light-stressed and temperature-stressed regimes from variable phytoplankton species. Four broadly distributed and globally common phytoplankton classes (Cermeño et al., 2010) and more than one species of diatom and dinoflagellate were used to generalize the class coverage beyond just diatoms and dinoflagellates and only one species per class. The six algal species were selected based on a variety of factors including use in previous photosynthetic and physiological studies, production of secondary metabolites and toxins, rapid growth rate and high nutritional value. The term "stress" is debatable in the ecological sense; here, it denotes an external constraint limiting the rates of resource acquisition, growth or reproduction by marine algae (Grime, 1989). The 12 h light/temperature cycles, different from those at which the bulk cultures were incubated in the climate control room, were used to observe a possible stress-related response of BVOC production. The 12 h dark cycle at the standard temperature in the climate-controlled room was intended to give phytoplankton some time to repair the photo/temperature damage that they may have sustained. Although full acclimation to a severely altered environment may require more than 12 h to reach a new equilibrium, timescales (hours to days) were selected to explore BVOC production rates that could be a response of phytoplankton to variable light/temperature conditions potentially occurring as a result of turbulent motions in the surface mixed layer (e.g., Cullen and Lewis, 1988; Geider et al., 1996). The experimental irradiance conditions applied here are within the global range of natural environmental conditions. Boreal/austral summer noon surface photosynthetically active radiation (PAR) approaches 2000 μmol m^{-2} s^{-1} seasonally and regularly from 40° N to 40° S, and can do so between 60° N and 60° S (Bouvet et al., 2002). Therefore, the challenges of abrupt irradiance intensity increases used here are conceivable in the upper ocean due to displacement in the vertical light gradient; however, the challenges of abrupt temperature changes become less likely at the extremes of the applied range. For temperature, these experiments represent the natural condition under a limited temperature range ($\pm$4 °C around acclimation temperature), but become a test of physiological capability over a larger temperature range. The temperature values were based on the previous works documented in Eppley (1972), Schofield et al. (1998) and Staehr and Birkeland (2006).

2 Instrumentation and method development

2.1 Incubation and sampling methodology

Figure 1 shows a schematic diagram for the incubation and sampling methodology. Phytoplankton monocultures were grown in a climate control room with a constant temperature of 22 °C and a 12 h on/12 h off light cycle at 90 μmol m^{-2} s^{-1} and 0 μmol m^{-2} s^{-1}, respectively. In order to ensure similar growth conditions for each monoculture, larger bulk samples were grown for each batch of the cultures; then, smaller volumes were extracted and used for the analysis. The following species were grown in 9 L Pyrex bottles for analysis: diatom strains – *Thalassiosira weissflogii* (*T. weiss.*) (CCMP 1336) and *Thalassiosira pseudonana* (*T. pseud.*) (CCMP 1335); prymnesiophyte strains – *Pleurochrysis carterae* (*P. carter.*) (CCMP 645); dinoflagellate strains – *Karenia brevis* (*K. brevis*) (CCMP 718, CCMP 2229) and *Prorocentrum minimum* (*P. minim.*) (CCMP 1329); and cryptophyte strains – *Rhodomonas salina* (*R. salina*) (UTEX 2423). Larger bulk samples were prepared in autoclaved filtered seawater and with an L1-based medium (Sigma-Aldrich, St. Louis, MO). Samples were then sealed at the top with aluminum foil, to allow for air transfer to the culture, and incubated in the climate control room. Bulk samples were incubated in the climate control room for 2 weeks for the diatoms, cryptophyte, and prymnesiophyte species and 3 weeks for the slower growing dinoflagellate species. These were the approximate time frames in which each monoculture was able to reach a maximum/stable biomass, where growth rate was light-limited due to self-shading. Monocultures were not axenic, but multiple steps such as acid washing all glassware and autoclaving seawater and nutrients were used to reduce the likelihood of bacterial contamination within the cultures.

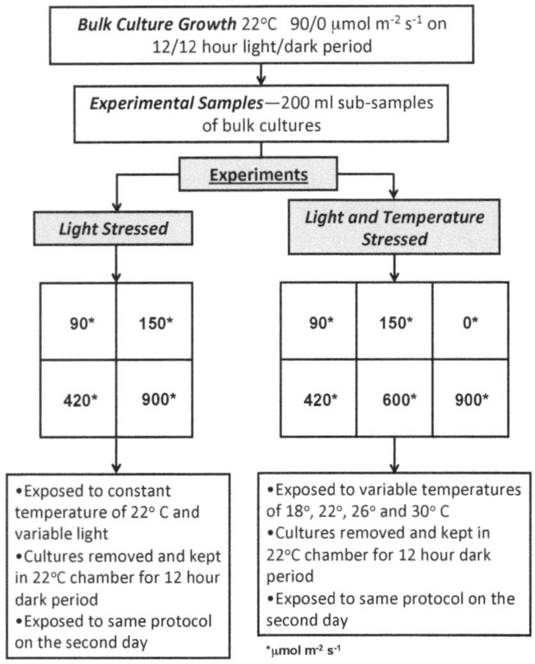

Figure 1. Schematic diagram of the experimental setup.

Smaller samples (200 mL) of each monoculture were extracted from the bulk containers and transferred to 250 mL borosilicate glass flasks for analysis purposes, leaving 50 mL of gas headspace in each purging vessel. The smaller volume monocultures were then subjected to different light and temperature regimes. These different regimes were used to assess differences between normal BVOC emissions and the emissions due to physiological stress-induced effects. An apparatus used for assessing the effect of changing incoming solar radiation for BVOC production of phytoplankton was composed of a tank suspended over six halogen lights (Philips 250 W projector lamp no. 13095) and a circulating water bath used to control the temperature within the tank. To vary the light intensities within the tank, the bottom of the water bath was lined with several different layers of semi-translucent fiberglass screens that attenuate a portion of the light reaching the samples. Irradiance inside the water bath was then measured by a QSL-100 laboratory quantum scalar irradiance meter (Biospherical Instruments, San Diego, CA). A Neslab CFT-33 refrigerated recirculator (Thermo Fisher Scientific, Waltham, MA) was used to control the water temperature. Emission analysis experiments were carried out to examine the effects of light and temperature stress on different phytoplankton species.

For the light-dependent production experiments, four different irradiance intensities were used: 90, 150, 420 and 900 μmol m^{-2} s^{-1}. At the end of the 12 h dark cycle, the bulk cultures from the climate-controlled room were transferred to four 250 mL purging vessels placed in the circulating water bath kept at 22 °C (the same as the climate-controlled room). The samples were initially purged (in the dark) to remove the excess BVOC that may have been generated during their growth in the climate-controlled room. At time zero, the lights were turned on and the individual samples were subjected to one of the respective irradiance intensities defined above. During the 12 h incubation time, samples were purged and analyzed every ~2 h over 12 h, leading to a total of six measurements for each sample. Since cultures were originally incubated at 12 h of dark and 12 h of light conditions, BVOC emission rates measured at irradiances higher than 90 μmol m^{-2} s^{-1} are defined as the light-stress-induced production. After the 12 h lights-on cycle, the phytoplankton samples were returned back to the climate-controlled room, where they were subjected to a 12 h dark cycle. After this dark cycle, species were placed in the circulating water (at 22 °C) bath again and the same sampling procedure was repeated.

For temperature- and light-dependent production experiments, four different water temperatures (18, 22, 26 and 30 °C) and six different irradiance intensities (0, 90, 150, 420, 600 and 900 μmol m^{-2} s^{-1}) were used. The sampling methodology was similar to the light-dependent production experiments described above. However, due to the large number of experiments, only two diatom strains (*T. weiss.* and *T. pseud.*) were used for the analysis. The sampling frequency was also reduced to one purging at the end of the 12 h cycle to obtain a broader range of data for variable temperature and light regimes. At the end of a 12 h cycle, the bulk cultures were transferred to a climate control room and kept in a full 12 h dark cycle at 22 °C, after which the same sampling procedure was repeated. Since cultures were originally incubated to an irradiance intensity of 0 and 90 μmolm^{-2} s^{-1} (on a 12 h cycle) at 22 °C, the BVOC emission rates measured for such a sampling setup were defined as the light-stressed/temperature-stressed production.

2.2 BVOC analysis

The experimental system consisted of a Varian 450-GC gas chromatograph (GC) attached to a 220-MS ion trap mass spectrometer (MS). The GC column was a 25 m × 0.32 mm I.D. CP-PoraBOND PLOT Q fused silica column (Varian, Inc.). Ultra high purity (UHP) helium was used as a carrier gas for the system. Data collection and analysis were performed with Varian MS Workstation (version 6.X) software equipped with the National Institute of Standards and Technology (NIST) library for chemical compound identification. A pre-concentrating system consisted of a purge and trap CDS 8000 sample concentrator unit (CDS Analytical, PA) equipped with a Vocarb 3000 (K Trap) attached in-line with the GC/MS system.

The following sampling procedure was used for all emission analyses. The purging vessel (a 250 mL borosilicate glass Erlenmeyer flask) was sealed at the top with a silicone stopper and borosilicate glass tubing was used as inlet and outlet ports. To avoid VOC contamination (e.g., emissions from plastics or rubber) in connection lines, purging vessels, or port connections, only borosilicate glass or 316 stainless steel or high purity silicone tubing (McMaster-Carr Inc., Atlanta, GA) was used. The inlet was attached to a mixture of 350 ppm CO_2 balanced with 78 % nitrogen and 21 % oxygen. This gas mixture (a "blanket gas") was bubbled through the liquid sample to liberate BVOC in the water into the headspace of the flask. The blanket gas was used for purging, as the lab air may have been contaminated by VOCs. The CO_2 level in the blanket gas was equivalent to that of current atmospheric levels. The gas stream was then passed into the wet trap of the CDS concentrator (set to 200 °C), followed by the sorbent trap (set to 40 °C), where the BVOCs in the gas stream are trapped. After 35 min of purging, the sample vessel at a flow rate of 0.40 slpm, a solenoid valve in the CDS concentrator was switched to inline the sorbent trap with the GC/MS through a transfer line. The trap was desorbed for 5 min at 250 °C and transferred to the GC inlet (set at 250 °C and a split ratio of 10 : 1) through the transfer line maintained at 250 °C.

The GC column temperature was held at 50 °C for 2 min followed by a ramp-up of temperature to 250 °C at a rate of 6 °C min^{-1}. The split ratio was initially set to 10 : 1 for 0.75 min, then 100 : 1 for 2.25 min and finally 20 : 1 till the end of the run. To avoid a "dead volume" signal in the chromatograph, the MS was set at a 5 min hold. The MS had a mass detection range of 35–300 m z^{-1} and ran for the full duration of the GC program (35.33 min). Each sample was screened for 38 different compounds (see Table S1 and Fig. S1 in the Supplement).

Custom-made calibration standards were obtained from Supelco Analytical at a 1000 μg L^{-1} concentration dissolved in methanol. A five-point calibration curve was made for each compound by injecting five decreasing amounts of the standards into the CDS concentrator. Error or uncertainty of the sampling system was quantified through the accuracy, precision and detection limit. The following statistics were calculated based on spiked seawater samples subjected to the same purging analysis as the phytoplankton monocultures. Seawater was spiked with a known concentration of analytes, and purged through the CDS concentrator and GC/MS. A total of eight spiked seawater analyses were conducted. The relative % error (RE) was computed as

$$\mathrm{RE} = \left| \frac{(x-u)}{u} \right| \times 100,$$

where x is the measured analyte concentration and u is the actual analyte concentration.

Accuracy was determined by averaging RE over the eight identical spiked seawater samples. Precision was determined by the relative standard deviation (RSD), which was computed as

$$\mathrm{RSD} = \frac{\sum(x-\overline{x})^2}{N-1} \times 100,$$

where $\overline{x}$ is the mean analyte concentration and N is the number of analyses. The detection limit of an analytical system is normally determined by the lowest detected signal above a blank background within a 99 % confidence interval, assuming a signal to noise (S/N) ratio of 5 (WDNR, 1996). A blank run through the analytical system showed no discernible detection of the analytes above a S/N of 5. An alternative detection limit calculation can be obtained by the method detection limit (MDL). The MDL is the minimum concentration of an analyte measured within a 99 % confidence interval by repetitive samples of the analyte in the sample matrix (WDNR, 1996). Due to the fact that BVOC analysis includes a three-step process (purge and trap by the CDS concentrator, compound separation by the GC, and detection by the MS), a comprehensive MDL was calculated for each compound:

$$\mathrm{MDL} = t_{(n-1,1-\infty=0.99)}(S),$$

where t is the student t value based on the number of analyses (8) at the 99 % confidence level and S is the standard deviation of the analyte concentration. For all analytes, the precision and accuracy are less than 20 %. The MDL ranges from 3.83 to 7.15 pptv. Table 1 summarizes compound attributes and error analyses for isoprene and monoterpene species.

2.3 Purging analysis

The water sample was bubbled inside the 250 mL purging vessels through a Mist Air® glass air stone. The gas inlet to the vessel was attached to the blanket gas, while the outlet was attached to a purge and trap CDS concentrator. Each sample was purged for 35 min at a flow rate of 0.40 slpm. Following the purge, the monocultures were sealed and placed back in the water tank. The analysis was based on the principle of liberating BVOC from the water samples into the gas stream. To quantify the amount of BVOC recovered by the purging analysis, purging efficiencies were calculated separately for each compound and are reported in Table 1. The purging efficiencies were calculated by spiking seawater with a known concentration of the standards. Successive purging steps from the same sample vial were performed until the compound concentrations were below the detection limit. The purging efficiency was calculated by taking the ratio of the initial purge BVOC concentrations divided by the sum of the BVOC concentrations over all the purging steps. The purging procedure was optimized for >90 % purging efficiency for isoprene and monoterpene species. Calculated purging efficiencies were comparable to values of >90 % and >95 % obtained by a similar analysis from Broadgate et al. (1997) and Shaw et al. (2003), respectively.

2.4 Biological parameters

At the end of each 12 h sampling period, 30 mL of each monoculture were filtered onto a Whatman GF/F filter. Chlorophyll *a* (Chl *a*) was extracted from the filters with 90 % acetone and the concentration was determined following the method of Holm-Hansen and Riemann (1978) using a Turner fluorometer model no. 450. For the light stress experiments, another 30 mL of each sample were analyzed for cell counts and size distributions of phytoplankton cultures. A Coulter® Sample Stand II connected to a COULTER COUNTER analyzer (Coulter Electronics, Inc., Hialeah, FL) was used to determine the number and size of particles that were suspended in seawater. Duplicates of both Chl *a* and cell counts were taken for each sample container.

Precision was calculated for both Chl *a* and cell counts from multiple runs of the same monoculture sample. Similarly to the BVOC calculation, eight replicate samples were conducted for both Chl *a* and cell counts. The RSD for Chl *a* and cell counts was 19.3 and 15.6 %, respectively.

2.5 Production rate calculation

The corrected analyte mass (μg) was calculated by dividing the measured mass (μg) by the purging efficiency using the method similar to that of Shaw et al. (2003). To account for the possible presence of the analyte in a blank seawater sample or in autoclaved seawater after inoculation with the nutrient medium, a correction was applied by subtracting the purging efficiency normalized blank analyte mass from the corrected analyte mass. The analyte mass was therefore corrected for both purging efficiency and the background mass in seawater samples. To account for the residual or carry-over due to the incomplete purging of the compounds from each

Table 1. Compound attributes and error analysis.

Compound	Formula	Ret. time (min)	Quan ion ($m\,z^{-1}$)	Relative error (%)	Relative standard deviation (%)	MDL (pptv)	Purging efficiencies (%)	RSD_{Total} (%)
Isoprene	C_5H_8	13.594	66.8	8.53	4.29	6.04	95.73	25.18
α-Pinene	$C_{10}H_{16}$	31.083	92.9	10.70	5.10	5.40	94.65	25.34
β-Pinene	$C_{10}H_{16}$	31.835	93.1	17.43	14.78	7.15	91.28	28.89
Camphene	$C_{10}H_{16}$	31.134	93.0	11.55	5.06	5.10	92.30	25.33
d-Limonene	$C_{10}H_{16}$	34.072	67.0	12.60	12.61	3.83	93.70	27.84

sample flask, the final analyte mass was calculated as

$$\text{Final} = \text{blank corrected analyte mass}_{P2} - \left[(1 - \text{purging efficiency}) \times \text{blank corrected analyte mass}_{P1}\right],$$

where P1 and P2 denote the purging step and (1 − purging efficiency) indicates the residual analyte that remained in the sample flask from the previous purging step.

The calculated final production rates for each trace gas were normalized by the incubation time (or the time between purging steps), the volume of the headspace and aqueous sample, and Chl *a* and cell counts. The total uncertainty (RSD_{Total}) of production rates for each BVOC was quantified using the following weighting calculation:

$$RSD_{Total} = \left[(RSD_i)^2 + \left(RSD_{Chl\,a/CC}\right)^2\right]^{0.5},$$

where RSD_{Total} is the total uncertainty for each BVOC, i stands for different BVOCs (i.e., isoprene and monoterpene compounds), RSD_i is the uncertainty for each compound (listed in Table 1), and $RSD_{Chl\,a/CC}$ is the 19.3 % uncertainty for Chl *a* measurements or the 15.6 % uncertainty for cell count measurements. These values were used to constrain the error bars for each production rate calculation per sample run.

3 Results

3.1 Light stress

3.1.1 Isoprene production rates

Figure 2 shows isoprene production rates as a function of time for different light intensities on two successive days. Since species response to stress can be related to changes in cellular chlorophyll content, we have chosen to use phytoplankton cell number as the base and Chl *a* concentration as a supplement to normalize isoprene and monoterpene emission rates for these experiments (see Figs. 2 and S2 in the Supplement). Figure 2 shows that isoprene production rates increase as a function of light intensity on the first day for all six phytoplankton species examined in this study. This result is consistent with most previous studies (e.g., Shaw et al., 2010), suggesting that isoprene production in algae species is linked to photosynthetic activities. Isoprene production rates in four of the phytoplankton species (*T. weiss.*, *T. pseud.*, *P. carter.*, and *R. salina*), excluding the two dinoflagelletes (i.e., *P. minim.* and *K. brevis*), also reveals a large increase within the first 4 h of the experiment, followed by a considerable fall (see Fig. 2). A particularly large increase by a factor of 4 was observed for the diatom species (Fig. 2a, b), followed by the prymnesiophyte and cryptophyte strains (Fig. 2c, d). The dinoflagellates (*P. minim.*, *K. brevis*) exhibited high isoprene production rates on initial exposure to high light without a lag period. After 4 h, the isoprene production rates decreased for all species and remained constant for the rest of the 12 h sampling period. It should be noted that the enhancements in isoprene production rates are particularly pronounced for higher irradiance intensities (i.e., 420 or 900 $\mu mol\,m^{-2}\,s^{-1}$). The highest isoprene production rates of 1.8×10^{-17} and 1.5×10^{-17} $g\,cell^{-1}\,h^{-1}$ (i.e., 31.5 and 34.7 $\mu g\,(g\,Chl\,a)^{-1}\,h^{-1}$) were recorded for *T. weiss.*, and *T. pseud.*, respectively, followed by *R. salina.*, *P. minim.*, *P. carter.*, and *K. brevis* of 9.0×10^{-18}, 8.9×10^{-18}, 6.0×10^{-18}, 5.3×10^{-18} $g\,cell^{-1}h^{-1}$ (i.e., 6.4, 15.3, 12.9 and 9.9 $\mu g\,(g\,Chl\,a)^{-1}\,h^{-1}$), respectively. Recently, Bonsang et al. (2010) also reported similar light-intensity-dependent isoprene production rates from some of the temperate diatom species, although the maximum intensity of 100 $\mu mol\,m^{-2}\,s^{-1}$ for their simulated diurnal cycle is considerably lower compared to the irradiances used in this study. Overall, when averaged over the entire 12 h cycle, all species show a rapid increase in isoprene production at low irradiance levels (< 150 $\mu mol\,m^{-2}\,s^{-1}$), and a gradual increase at intermediate irradiance levels (150 to 420 $\mu mol\,m^{-2}\,s^{-1}$) until the light saturated production rates were reached (see Fig. S3 in the Supplement).

Figure 2 reveals very different isoprene production rates for *T. weiss.*, *T. pseud.*, *P. carter.* and *R. salina* on the second day (after 12 h of the dark cycle). These species show generally higher isoprene production rates at the beginning of the experiment. Isoprene production rates for the diatoms are a factor of 5 higher compared to the first 2 h of day 1, but comparable to the emission levels reached at the end of

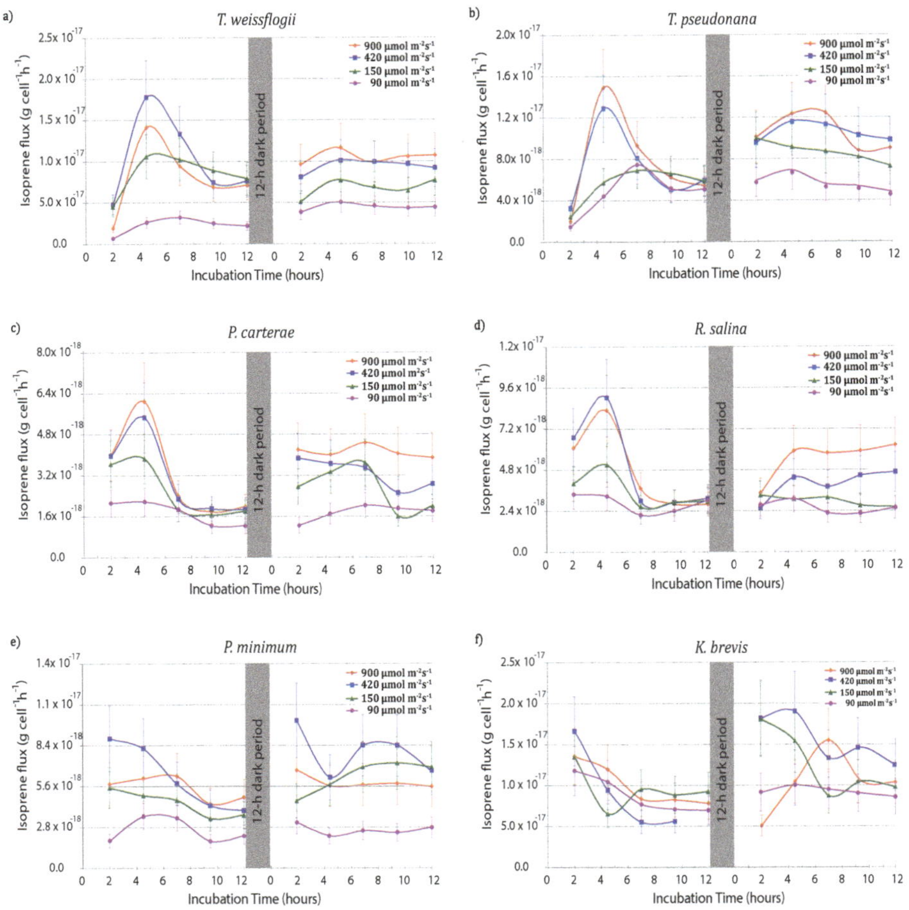

Figure 2. Light stress production rates for isoprene ($g\,cell^{-1}\,h^{-1}$) as a function of time during the first (0 to 12 h) and second (12 to 24 h) light cycles. Note that there was a 12 h dark period between the two five-point sample sets and that the first samples in each set were collected 2 h after lights on. Error bars denote the RSD_{Total} value listed in Table 1.

the 12 h light cycle on the first day. Diatoms also do not reveal the light-saturated maximum rate in isoprene production characteristic of the first day. The cumulative production rates of isoprene for these species are generally higher during the second day compared to the first one (see Fig. S3). For the dinoflagellates (*P. minim.*, and *K. brevis*), the second-day isoprene production rates exhibit patterns that more closely resemble those of the first 12 h. Table 2 summarizes isoprene production rates from all six phytoplankton species at different irradiance levels averaged over a 12 h period on the first day.

3.1.2 Monoterpene production rates

Figure 3 shows phytoplankton production rates for α-pinene, which was the highest of all monoterpene species examined in this study. Comparison of Figs. 2 and 3 shows that, for the first 12 h period, diatom production rates for isoprene and α-pinene exhibit a considerable increase within the first 4 h of the experiment, followed by a decline. According to Fig. 3a, b, the highest α-pinene production rates in diatoms are 5.1 and $4.4 \times 10^{-19}\,g\,cell^{-1}\,h^{-1}$ (i.e., 1.0 and 1.1 $\mu g\,(g\,Chl\,a)^{-1})\,h^{-1})$ for *T. weiss.*, and *T. pseud.*, respectively, and are more than an order of magnitude lower compared to isoprene. Other strains (i.e., prymnesiophyte, dinoflagellate and cryptophyte) do not exhibit marked changes in α-pinene production rates with time on day 1 (see Fig. 3c, d). During a second 12 h exposure on day 2, there was no increase in α-pinene production rates for *T. Weiss.*, while the rest of the algae species either showed considerable increases (i.e., *T. pseud.* and *P. carter.*) or exhibited few discernible differences. According to Fig. 3, α-pinene production rates for *K. brevis* exhibit a considerable increase within the first 4 h of the light exposure on day 2. Increases in monoterpene emissions on day 2 seem to be common for *K. brevis*, as it was detected for all the monoterpenes measured in this study. Overall, when averaged over the entire 12 h cycle, α-pinene production patterns are similar to those of isoprene, with a rapid increase at low irradiance levels ($< 150\,\mu mol\,m^{-2}\,s^{-1}$) and with a gradual increase for intermediate irradiance levels (150 to $420\,\mu mol\,m^{-2}\,s^{-1}$) until the light saturated production rates are reached (see Fig. S4 in the Supplement).

Table 2. Isoprene and monoterpene production rates from phytoplankton monocultures averaged over a 12 h period on the first day.

	T. weiss.				*T. pseud.*				*P. carter.*				*R. salina*				*P. minim.*				*K. brevis*			
	900	420	150	90	900	420	150	90	900	420	150	90	900	420	150	90	900	420	150	90	900	420	150	90
	μmol m^{-2} s^{-1}				μmol m^{-2} s^{-1}				μmol m^{-2} s^{-1}				μmol m^{-2} s^{-1}				μmol m^{-2} s^{-1}				μmol m^{-2} s^{-1}			
Isoprene μgram (g Chl *a*)$^{-1}$ h^{-1}	17.2	18.0	13.7	7.0	17.6	15.7	11.8	8.4	6.9	6.4	5.4	4.9	8.7	9.9	7.3	5.7	12.2	10.7	9.0	5.6	6.3	5.8	5.8	5.4
Isoprene $\times 10^{20}$ moles (cell)$^{-1}$ h^{-1}	11.5	14.9	12.3	3.18	10.9	10.2	7.96	6.75	4.67	4.46	3.70	2.46	6.90	7.23	5.16	4.14	8.01	9.10	6.49	3.75	4.08	4.75	3.91	3.60
α-Pinene $\times 10$ μgram (g Chl *a*)$^{-1}$ h^{-1}	4.65	5.57	3.84	1.32	5.51	5.68	3.52	2.19	2.91	2.84	2.33	1.64	2.69	4.19	4.64	2.99	3.99	3.98	2.93	2.03	5.99	5.37	4.01	2.97
α-Pinene $\times 10^{21}$ moles (cell)$^{-1}$ h^{-1}	1.55	2.31	1.73	0.30	1.71	1.84	1.21	0.88	1.00	0.99	0.81	0.41	1.06	1.52	1.74	1.09	1.31	1.70	1.05	0.60	1.93	2.22	1.35	0.99
Camphene $\times 100$ μgram (g Chl *a*)$^{-1}$ h^{-1}	0.66	1.66	0.28	0.75	1.07	0.41	1.86	0.15	0.74	0.72	0.45	0.23	0.52	0.76	0.41	0.29	1.20	0.44	0.82	1.07	0.82	0.81	0.60	0.52
Camphene $\times 10^{22}$ moles (cell)$^{-1}$ h^{-1}	2.20	2.23	1.27	1.35	3.30	1.33	3.12	0.59	1.97	1.51	0.75	0.58	1.71	2.76	1.55	0.90	3.10	1.54	2.94	3.55	2.18	3.33	1.65	0.76
β-Pinene $\times 100$ μgram (g Chl *a*)$^{-1}$ h^{-1}	2.54	4.36	3.34	7.44	7.98	12.2	4.15	6.02	5.66	4.65	0.57	1.82	1.14	2.27	4.09	2.02	1.25	1.82	1.59	2.65	0.91	1.25	1.02	0.68
β-Pinene $\times 10^{22}$ moles (cell)$^{-1}$ h^{-1}	0.85	1.81	1.51	1.69	2.47	3.95	2.01	2.43	1.93	1.62	0.21	0.57	0.47	0.82	1.54	0.73	0.41	0.77	0.56	0.90	0.28	0.53	0.36	0.24
d-Limonene $\times 10$ μgram (g Chl *a*)$^{-1}$ h^{-1}	3.61	4.86	1.67	0.41	1.66	5.04	2.13	0.42	0.77	1.16	1.94	0.36	1.17	1.37	0.76	1.18	3.46	5.20	2.45	1.99	1.96	1.61	1.86	0.88
d-Limonene $\times 10^{22}$ moles (cell)$^{-1}$ h^{-1}	12.1	20.0	7.50	0.93	5.15	13.1	8.71	1.70	2.62	4.02	6.72	0.91	4.63	5.02	2.84	4.32	11.4	17.7	8.77	6.60	6.34	6.66	6.27	2.90

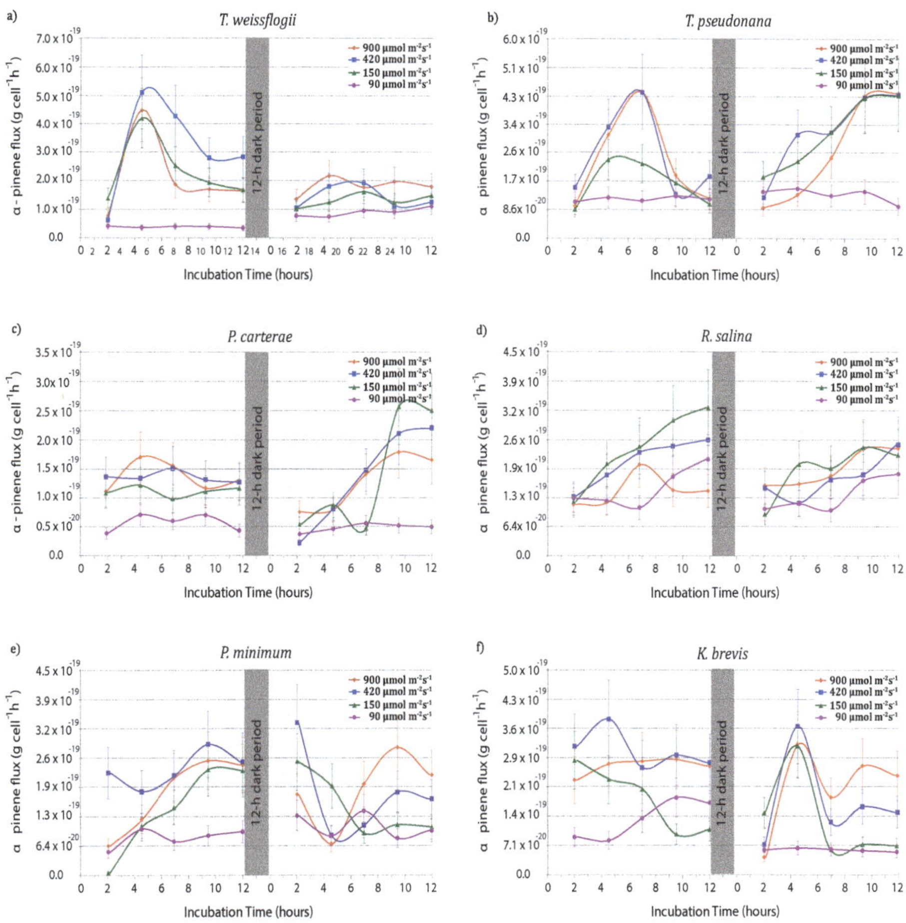

Figure 3. Light stress production rates for α-pinene (g cell^{-1} h^{-1}) as a function of time during the first (0 to 12 h) and second (12 to 24 h) light cycles. Note that there was a 12 h dark period between the two five-point sample sets and that the first samples in each set were collected 2 h after lights on. Error bars denote the $\mathrm{RSD}_{\mathrm{Total}}$ value listed in Table 1.

Emission patterns of d-limonene, camphene and β-pinene (not shown) were similar to that of α-pinene, but at considerably lower rates. Although monoterpene species were detected at all levels (with enhanced rates for higher levels of irradiance), very low emission rates made detection sporadic. For example, β-pinene was present at some purging steps, but not others. Table 2 summarizes 12 h averaged monoterpene production rates from six phytoplankton species at different irradiance levels. According to Table 2, the average contribution to the total monoterpene production for all phytoplankton species was $\sim 70\,\%$ from α-pinene, $\sim 20\,\%$ for d-limonene, and $< 10\,\%$ for camphene and β-pinene. Compared to isoprene, total terpene emission was an order of magnitude lower for all phytoplankton species at all light levels (see Table 2).

3.2 Temperature and light stress

3.2.1 Isoprene production rates

Temperature dependence of isoprene emissions has been noted for both terrestrial and marine photosynthetic organisms. Increasing temperature was shown to enhance isoprene emissions to a maximum production rate until a certain temperature is reached, beyond which further increases in temperature were shown to cause a decline in isoprene emissions (Guenther et al., 1993; Harley et al., 1996; Shaw et al., 2003). As mentioned above, bulk cultures in this study were acclimated to an irradiance intensity of $90\,\mu\text{mol m}^{-2}\,\text{s}^{-1}$ and 22 °C. So, any temperature and/or irradiance levels different from these values are expected to cause short- and long-term changes to the plankton photosystem intended to establish balanced growth. The effects of such changes on isoprene production by algae species are explored below.

Temperature- and light-dependent experiments were conducted for the two diatom species, *T. weiss.* and *T. pseud.*, and samples were averaged over two successive 12 h cycles. Such a long averaging time does not allow monitoring of short-term responses of phytoplankton to temperature and light stresses. Also, due to the absence of the cell count data, all flux values are normalized with respect to Chl *a*, which may change due to photoacclimation. Nevertheless, Fig. 4 shows some interesting features for isoprene production rates. According to Fig. 4, under the low irradiance regime (i.e., $\leq 90\,\mu\text{mol m}^{-2}\,\text{s}^{-1}$), temperature variability has a minor effect on isoprene production for two diatom species. Isoprene emission rates by *T. weiss.* and *T. pseud.* tend to be lower at 18 °C for all light levels. Emissions increase, with the temperature reaching its highest levels around 22 to 26 °C. Further increases in temperature cause a considerable decline in isoprene production rates. However, careful inspection of Fig. 4 shows that temperature can influence isoprene production rates in diatoms exposed to the high irradiance regime (i.e., $\geq 150\,\mu\text{mol m}^{-2}\,\text{s}^{-1}$). Figure 4 also shows differences in the 12 h averaged isoprene production rate between the two successive days. For *T. pseud.*, the maximum production rate on the first day is reached at 22 °C for all irradiance levels while, on the second day, the production rate peaked at 26 °C for lower irradiance levels (i.e., below $150\,\mu\text{mol m}^{-2}\,\text{s}^{-1}$) and at 22 °C for higher irradiance levels (i.e., above $150\,\mu\text{mol m}^{-2}\,\text{s}^{-1}$). Overall, Fig. 4 shows that, for all temperature and light levels, higher isoprene production rates are reached on the second day. A similar conclusion was previously reached for isoprene production as a function of light stress only (see Sect. 3.1.1). Table 3 summarizes 12 h averaged isoprene production rates for two diatom species at different temperature and irradiance levels.

3.2.2 Monoterpene production rates

Unlike isoprene, monoterpenes in terrestrial vegetation can be stored in leaves and be released relatively quickly on a timescale of minutes to hours. Monoterpene emitting terrestrial vegetation has been found to have a higher temperature than light dependencies (Koppmann and Wildt, 2007). If something similar was true for phytoplankton, one would expect enhanced monoterpene emission rates for temperatures higher than 22 °C (the temperature at which species were acclimated) with emission patterns distinct from those of isoprene.

Figure 5 shows that 12 h averaged α-pinene emission on two successive days is similar to that of isoprene (see Fig. 4), although at considerably lower rates. For *T. weiss.*, α-pinene production rates for both days are highest at 26 °C. For *T. pseud.*, α-pinene production rates peak at a lower temperature (22 °C) at irradiance levels $\geq 150\,\mu\text{mol m}^{-2}\,\text{s}^{-1}$. Table 3 summarizes 12 h averaged monoterpene production rates for two diatom species at different temperature and irradiance levels.

4 Discussion

A large number of recent studies have attempted to elucidate the mechanisms as well as the purpose behind the formation and emission of isoprene and monoterpenes from terrestrial plants. Despite this, the biological as well as ecological role and evolutionary aspect for the production of these BVOCs by vascular plants is not fully understood. Even less is known about their emissions from phytoplankton. However, as both isoprene and monoterpenes are formed in primary producers as a by-product of photosynthetic metabolism, to explain our findings, we will take a reductionist approach and presume similarities in isoprene and monoterpene biosynthesis between better known terrestrial vegetation and lesser known phytoplankton. Production of isoprene and monoterpenes in plants proceeds through a methylerythritol phosphate (MEP) metabolic pathway that is strongly tied to photosynthesis of primary producers through chlorophyll and carotenoid synthesis (Lichtenthaler, 2009). Lichtenthaler (1999) provided

Table 3. Isoprene and monoterpene production rates from phytoplankton monocultures averaged over a 12 h period on the first day*.

		T. weiss.						*T. pseud.*					
		900	600	420	150	90	0	900	600	420	150	90	0
		μmol m^{-2} s^{-1}						μmol m^{-2} s^{-1}					
Isoprene μgram (g Chl a)$^{-1}$ h^{-1}	18 °C	12.5	11.8	10.7	10.4	5.87	1.43	16.4	14.2	7.99	7.48	3.71	1.33
	22 °C	18.9	17.7	16.1	15.6	8.85	2.16	20.2	23.2	18.0	13.3	6.94	1.3
	26 °C	20.8	16.3	9.43	8.04	6.64	2.38	12.5	14.7	14.7	13.1	7.04	1.68
	30 °C	13.9	10.9	7.71	7.46	6.81	1.82	14.1	16.2	12.6	9.30	4.85	0.91
α-Pinene × 10 μgram (g Chl a)$^{-1}$ h^{-1}	18 °C	1.93	2.33	2.04	2.33	0.40	0.284	2.55	2.84	2.61	3.80	1.53	0.284
	22 °C	2.90	3.46	3.12	3.52	0.57	0.40	3.52	3.80	3.12	4.26	1.65	0.341
	26 °C	4.09	7.04	5.28	3.35	0.45	0.45	2.67	2.78	2.72	1.82	1.36	0.114
	30 °C	1.36	0.80	0.68	0.85	0.45	0.34	2.44	2.67	2.21	2.95	1.14	0.227
Camphene × 100 μgram (g Chl a)$^{-1}$ h^{-1}	18 °C	3.42	3.41	3.43	3.41	7.95	<DL	3.41	2.27	2.18	3.41	5.68	1.14
	22 °C	4.54	4.58	4.54	4.54	12.5	1.14	6.81	3.41	11.4	17.0	2.27	<DL
	26 °C	1.14	4.54	1.14	1.14	9.08	1.12	3.41	3.18	4.54	6.81	4.54	1.14
	30 °C	6.82	3.42	2.27	3.41	9.10	<DL	4.54	2.27	7.95	11.4	2.27	<DL
d-Limonene × 10 μgram (g Chl a)$^{-1}$ h^{-1}	18 °C	2.84	1.48	0.34	2.38	0.34	<DL	2.16	1.14	1.14	0.91	1.7	0.34
	22 °C	4.20	2.27	0.45	3.52	0.45	<DL	1.59	0.68	3.18	0.45	0.45	0.14
	26 °C	7.15	2.95	0.34	3.86	0.34	0.01	0.91	1.59	1.48	1.36	0.34	0.23
	30 °C	1.02	1.02	0.23	0.68	0.34	<DL	1.14	0.57	2.27	0.34	0.32	0.11

* DL – detection limit

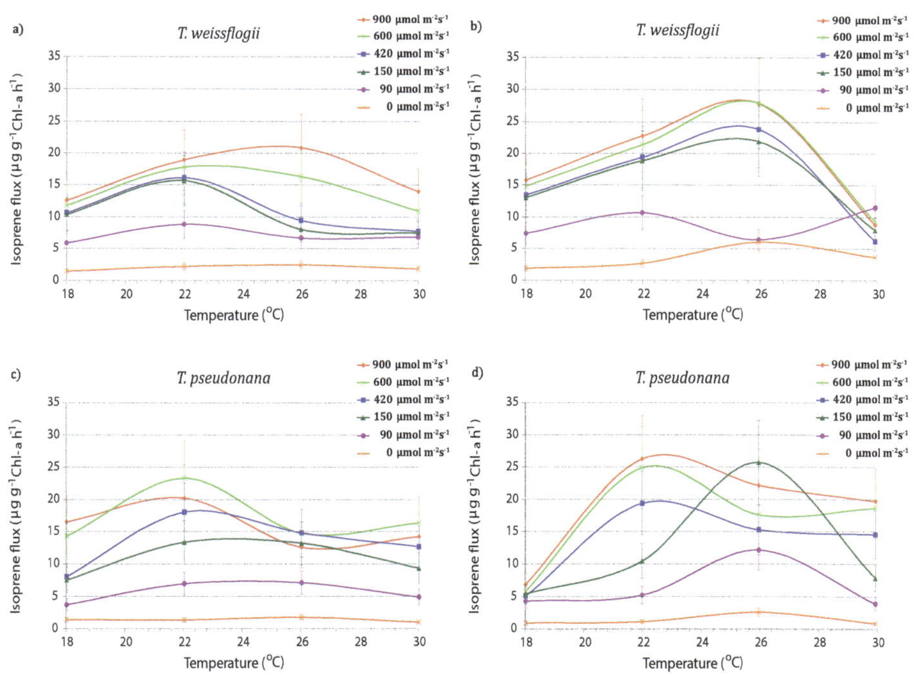

Figure 4. The 12 h averaged isoprene production rates (μg (gChl a)$^{-1}$ h^{-1}) for the first (left column) and second (right column) light cycles as a function of temperature. Error bars denote the RSD_{Total} value listed in Table 1.

a detailed description of the MEP pathway, where a chain of enzymatic processes leads to the formation of dimethylallyl diphosphate (DMAPP) and isopentenyl diphosphate (IPP). DMAPP then through isoprene synthase can lead to isoprene formation, while both DMAPP and IPP can form monoterpene species. Production and expulsion of isoprene and monoterpenes from primary producers is very energy costly (Sharkey and Yeh, 2001). The basic hypothesis is that

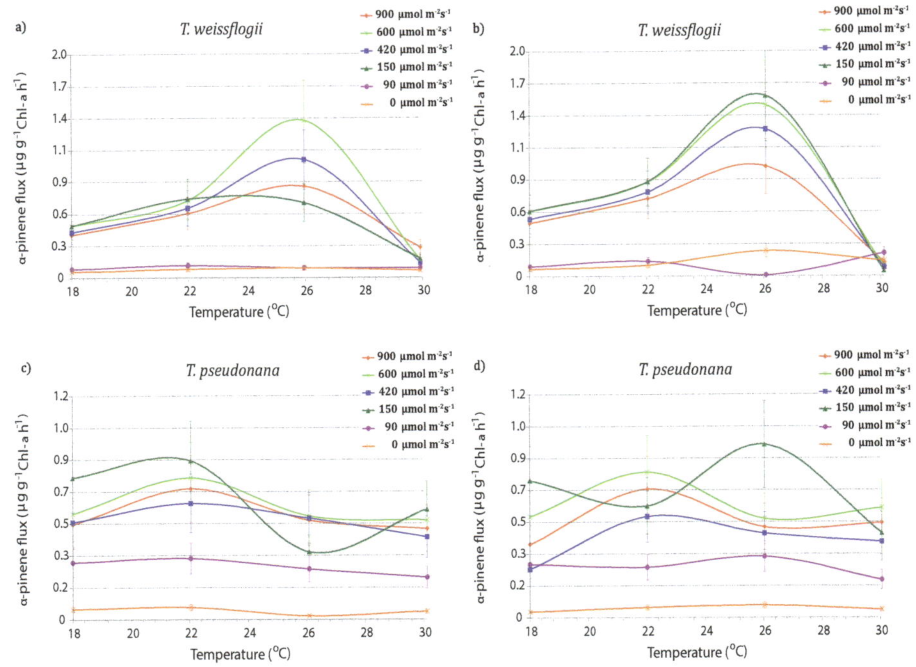

Figure 5. The 12 h averaged α-pinene production rates (μg (gChl $a)^{-1}$ h^{-1}) for the first (left column) and second (right column) light cycles as a function of temperature. Error bars denote the RSD_{Total} value listed in Table 1.

there is an advantage to gain by photosynthesizing organisms to emit isoprene. Some advantages in terrestrial plants include thermotolerance, use as an anti-oxidant from atmospheric oxidants like ozone, an energy "release valve" or a mechanism to get rid of unwanted metabolites or energy, flowering response, and insect defense (Sharkey et al., 2008). Unlike isoprene that is not stored within the leaf and is emitted as soon as it is formed (Delwiche and Sharkey, 1993), many terrestrial plants have special storage organelles either in leaf-internal structures (e.g., secretory cavities and secretory canals or ducts) or in structures located on the surface of the leaf (e.g., trichomes) where monoterpenes can be stored. Other parameters such as light, drought, and herbivore attack can also affect the monoterpene emissions from terrestrial vegetation, though the emission rate has been shown to increase exponentially with leaf temperature (Koppmann and Wildt, 2007).

Similar to vascular plants, phytoplankton attempt to find a balance between energy from light reactions in chloroplasts and the energy used for metabolic processes and carbon fixation (Geider et al., 1996). Perturbations under different environmental conditions (e.g., light and temperature) can disturb the balance in chloroplasts and force primary producers to adjust or acclimate physiologically (Cullen and Lewis, 1988; Huner et al., 1998). For a short-term modulation of energy flow to the photosynthetic reaction centers, accessory pigments in plants can help dissipate excessive photon flux (non-photochemical quenching) on the timescale of minutes (MacIntyre et al., 2002). When the short-term responses are insufficient for dealing with prolonged changes in light, longer-term responses (on the order of hours or days) to high irradiances can include the reduction in the quantum efficiency of photosystems by a decline in Chl a, the light harvesting antennae, and accessory pigments, as well as a decrease in the number of photosystem reaction centers and changes in electron transfer chain components and Calvin cycle enzymes (MacIntyre et al., 2002). Photoacclimation is complete when a condition of "balanced" growth has been achieved (Geider et al., 1996). Results from the first day of the experiment seem to be consistent with this picture. For the experimental irradiances above 150 μmol m^{-2} s^{-1} (that well exceeded the 90 μmol m^{-2} s^{-1} at which species were acclimated), two diatoms, primnesiophyte, and cryptophyte strains showed large increases in the production rates of isoprene and monoterpenes within the first 4 h of the experiment. This result indicates that initial pigment adjustment, related to the xanthophyll cycle or other non-photochemical quenching, was not enough to dissipate the excess energy, and the balance between the photosynthetic reaction centers was disrupted; i.e., species experienced initial photoinhibition. After several hours, the photoacclimation of species caused a sharp reduction in isoprene and monoterpene emission rates. It should be noted that the two dinoflagellates showed a different pattern in BVOC production that may be related either to the mechanical stress on phytoplankton from bubbling of air through the sample and/or to differences in photoacclimation response. Previous work has shown that mechanical stress, by sparging or shaking dinoflagellate cells, can cause an increase in BVOC emissions (Wolfe et al., 2002). Different methods for the partitioning of BVOCs into the headspace may be needed to avoid the shear stress on the dinoflagellates caused by the air bubbling. Past studies have also revealed interspecific variability in quantum efficiency of photosynthesis between different algae species acclimated under the same conditions. The lowest values of quantum efficiency were reported for dinoflagellates, while the highest values were found in diatoms (Langdon, 1988).

On the second day of the experiment, species showed higher and relatively steady BVOC emission rates compared to the first day. As production of isoprene and monoterpenes is directly related to the cell metabolism, elevated emission rates at higher light levels may simply indicate a delayed adjustment in dark reactions. The Chl *a* to cell number ratio for cultures was generally lower on the second day compared to the first one, suggesting the reduction in the Chl *a* synthesis rate as part of the species' photoacclimation process. Elevated emission rates on day 2 may also suggest that the photoresponse mechanism applied during the first day was associated with chlorophyll decline and that the second-day exposure required a continued adjustment in dark reactions to re-establish long-term photoacclimation. Overall, the results suggest that 2 days were not sufficient for the tested species to attain steady-state photoacclimation to the imposed light changes and that longer experiments may be needed to describe isoprene and monoterpene production fully over the full photoacclimation period (Geider et al., 1996). These results also suggest that future studies in this area should report emission rates normalized to cell number as well as Chl *a*.

Although temperature is perhaps one of the most important environmental factors that influences algal growth rate, cell size, biochemical composition and nutrient requirements, our measurements with simultaneous occurrences of elevated light and temperature levels show that temperature variability alone has a minor effect on isoprene and monoterpene production rates from the phytoplankton. Previous studies of marine cyanobacteria acclimated to a different temperature over the entire range for growth did not reveal a clear relationship between isoprene emission rates and temperature, leading to the suggestion that, unlike terrestrial plants, phytoplankton are exposed to more stable and lower temperatures on average, and would not need to emit isoprene for thermal protection (Shaw et al., 2003). However, our results suggest that, when combined with the light stress, temperature stress can influence isoprene and monoterpene emissions from phytoplankton. Based on our results, BVOC emission patterns for dual (temperature and light) stress conditions can be clustered in three major regions. For temperatures lower than the one at which phytoplankton was acclimated, production rates of isoprene and monoterpenes are consistently lower for all light levels examined in this study. Such behavior is in agreement with exponential reduction of specific growth rates of algae cultures suggested by Eppley (1972) related to a Q_{10} temperature relationship. Lower temperatures also cause an increase in the levels of unsaturated fatty acids inside the cell, making the membranes more susceptible to damage by free radicals (Juneja et al., 2013). Increasing temperatures beyond the species' tolerance induces the marked reduction in the growth rate (Boyd et al., 2013). Both diatoms examined in this study reveal consistently lower BVOC emission rates at 30 °C. Discrepancies between *T. weiss.* and *T. pseud.* are likely to be associated with species- and strain-specific levels of supraoptimal temperatures (Juneja et al., 2013). In between two temperature extremes, isoprene and monoterpene emission rates reveal complex patterns that cannot be easily interpreted, but that probably are related to variable species-specific imbalances between light and dark reactions (Geider et al., 1996). Since simultaneous occurrences of elevated ocean temperature and stratification are among the most likely phenomena associated with global change, additional measurements (including time-dependent fluxes of BVOC under temperature and light stress conditions) need to be conducted for both acclimated and instantaneous environmental responses. Real ocean phytoplankton populations will encounter and will respond to both short-term (vertical mixing) and long-term (interannual increases) exposures to altered global climate change environments.

Despite large discrepancies in emission rates, the great similarity between emission patterns of isoprene and monoterpenes revealed in this study suggests that monoterpenes – like isoprene – may be emitted from phytoplankton as soon as they are produced. Shipboard measurements during a field campaign in the southern Atlantic Ocean also revealed a positive correlation ($R^2 = 0.57$) between α-pinene and isoprene within the bloom region (Yassaa et al., 2008).

5 Conclusions

In this study, we quantified production of isoprene and four different monoterpene compounds (α-pinene, β-pinene, camphene and d-limonene) from six phytoplankton monocultures as a function of different irradiance and temperature regimes. Our measurements confirmed that both isoprene and monoterpene production from marine algae can be influenced by variability in light and temperature. Isoprene and monoterpene emission patterns were similar regardless of whether production rates were normalized by cell number or Chl *a* concentration. Emission rates were found to increase sharply when low-light acclimated phytoplankton were subject to higher irradiances of up to 900 μmol m^{-2} s^{-1}, reaching the maximum isoprene and sum of all monoterpene production rates of $\sim$25 and $\sim$1.5 μg (g Chl *a*)$^{-1}$ s^{-1}, respectively, for diatom species. On average, it was found that the contribution to the total monoterpene production for all phytoplankton examined here was $\sim$70 % from α-pinene, $\sim$20 % for d-limonene, and <10 % for camphene and β-pinene. Production rates were found to increase sharply starting at 150 μmol m^{-2} s^{-1}, and continued at a higher rate up to

900 μmol m^{-2} s^{-1}. This result is consistent with most previous studies, suggesting that isoprene and monoterpene production from phytoplankton is linked to photosynthetic activities. However, unlike previous studies that examined time-averaged BVOC emission rates from phytoplankton subject to temperature and light regimes different from the one at which bulk cultures were incubated, here we explore time-dependent emissions over two successive days. Moreover, this study explored phytoplankton response to the joint effect of variable light and temperature conditions.

To our knowledge, this is the first work that has explored isoprene and monoterpene production rates from microalgae species as a function of both step changes in light and temperature over 2 days, making it difficult to compare our results with other studies. Despite this, we believe that results presented in this study contribute to understanding the physiology behind the production of BVOCs by marine algae. For BVOCs examined in this study, air–sea exchange is the dominant loss term at all oceanic latitudes due to their low Henry law constants (e.g., Palmer and Shaw, 2005; Gantt et al., 2009). Therefore, experimental results presented here can lead to improved prediction of changes in natural emissions and help to understand variability in the abundance and spatial distribution of different algae species under changing environmental conditions. As changes in isoprene and monoterpene production were suggested to be associated with sequential responses to photoinhibition experienced over an extended time period, experimental length needs to be extended to more than 2 days to determine if prolonged exposure to altered light and temperature conditions would result in reduction of isoprene and monoterpene production once photoacclimation is complete. Future studies should also assess the effect of growth phase (Kameyama et al., 2011) on BVOC production rates from various phytoplankton species. The experiments should also be rerun using the same phytoplankton species and experimental conditions to check the repeatability of the results. Finally, in addition to light intensity and temperature, future studies should also examine the effects of CO_2 levels, seawater pH, nutrient availability, and the presence of other organisms on isoprene and monoterpene production from phytoplankton.

Acknowledgements. This research was supported by the Office of Science (BER), US Department of Energy, grant no. DE-FG02-08ER64508, and the National Science Foundation through grants ATM-0826117 and AGS-1249273. A. Sabolis also acknowledges support from the North Carolina Space Grant Graduate Research Fellowship.

Edited by: K. Suzuki

References

Acuña-Alvarez, L. I., Exton, D. A., Timmis, K. N., Suggett, D. J., and McGenity, T. J.: Characterization of marine isoprene-degrading communities, Env. Microbiol., 11, 3280–3291, doi:10.1111/j.1462-2920.2009.02069.x, 2009.

Blanchard, D. C.: Sea-to-air transport of surface active material, Science, 146, 396–397, doi:10.1126/science.146.3642.396, 1964.

Bonsang, B., Polle, C., and Lambert, G.: Evidence for marine production of isoprene, Geophys. Res. Lett., 19, 1129–1132, 1992.

Bonsang, B., Gros, V., Peeken, I., Yassaa, N., Bluhm, K., Zoellner, E., Sarda-Esteve, R., and Williams, J.: Isoprene emission from phytoplankton monocultures, relationship with chlorophyll, cell volume, and carbon content, Environ. Chem., 7, 554–563, doi:10.1071/EN09156, 2010.

Bouvet, M., Hoepffner, N., and Dowell, M. D.: Parameterization of a spectral solar irradiance model for the global ocean using multiple satellite sensors, J. Geophys. Res., 107, 3215, doi:10.1029/2001JC001126, 2002.

Boyd, P. W., Rynearson, T. A., Armstrong, E. A., Fu, F., Hayashi, K., Hu, Z., Hutchins, D. A., Kudela, R. M., Litchman, E., Mulholland, M. R., Passow, U., Strzepek, R. F., Whittaker, K. A., Yu, E., and Thomas, M. K.: Marine Phytoplankton Temperature versus Growth Responses from Polar to Tropical Waters – Outcome of a Scientific Community-Wide Study, PLoS ONE, 8, e6309, doi:10.1371/journal.pone.0063091, 2013.

Broadgate, W. G., Malin, G., Kupper, F. C., Thompson. A., and Liss, P. S.: Isoprene and other non-methane hydrocarbons from seaweeds: a source of reactive hydrocarbons to the atmosphere, Mar. Chem., 88, 61–73, doi:10.1016/j.marchem.2004.03.002, 2004.

Broadgate, W. J., Liss, P. S., and Penkett, S. A.: Seasonal emissions of isoprene and other reactive hydrocarbon gases from the ocean, Geophys. Res. Lett., 24, 2675–2678, 1997.

Carpenter, L. J., Archer, S. D., and Beale, R.: Ocean-atmosphere trace gas exchange, Chem. Soc. Rev., 41, 6473–6506, doi:10.1039/C2CS35121H, 2012.

Carslaw, K. S., Lee, L. A., Reddingtonet, C. L., Pringle, K. J., Rap, A., Forster, P. M., and Mann, G.: Large contribution of natural aerosols to uncertainty in indirect forcing, Nature, 503, 67–71, doi:10.1038/nature12674, 2013.

Cermeño, P., de Vargas, C., Abrantes, F., and Falkowski, P. G.: Phytoplankton Biogeography and Community Stability in the Ocean, PLoS ONE 5, e10037, doi:10.1371/journal.pone.0010037, 2010.

Charlson, R. J., Lovelock, J. E., Andreae, M. O., and Warren, S. G.: Oceanic phytoplankton, atmospheric sulphur, cloud albedo and climate, Nature, 326, 655–661, 1987.

Cullen, J. J. and Lewis, M. R.: The kinetics of algal photoadaptation in the content of vertical mixing, J. Plankton Res., 10, 1039–1063, 1988.

de Leeuw, G., Andreas, E. L., Anguelova, M. D., Fairall, C. W., Lewis, E. R., O'Dowd, C., Schulz, M., and Schwartz, S. E.: Production flux of sea spray aerosol, Rev. Geophys., 49, RG2001, doi:10.1029/2010RG000349, 2011.

Delwiche, C. F. and Sharkey, T. D.: Rapid appearance of ^{13}C in biogenic isoprene when $^{13}CO_2$ is fed to intact leaves, Plant Cell Environ., 16, 587–591, 1993.

Eppley, R. W.: Temperature and phytoplankton growth in the sea, Fish. B.-NOAA, 70, 1063–1085, 1972.

Exton, D. A., Suggett, D. J., McGenity, T. J., and Steinke, M.: Chlorophyll-normalized isoprene production in laboratory cultures of marine microalgae and implications for global models, Limnol. Oceanogr., 58, 1301–1311, doi:10.4319/lo.2013.58.4.1301, 2013.

Gantt, B., Meskhidze, N., and Kamykowski, D.: A new physically-based quantification of marine isoprene and primary organic aerosol emissions, Atmos. Chem. Phys., 9, 4915–4927, doi:10.5194/acp-9-4915-2009, 2009.

Gantt, B., Meskhidze, N., Zhang, Y., and Xu, J.: The effect of marine isoprene emissions on secondary organic aerosol and ozone formation in the coastal United States, Atmos. Environ., 44, 115–121, doi:10.1016/j.atmosenv.2009.08.027, 2010a.

Gantt, B., Meskhidze, N., and Carlton, A. G.: The contribution of marine organics to the air quality of the western United States, Atmos. Chem. Phys., 10, 7415–7423, doi:10.5194/acp-10-7415-2010, 2010b.

Geider, R. J., MacIntyre, H. L., and Kana, T.: A dynamic model of photoacclimation in phytoplankton, Limnol. Oceanogr., 41, 1–15, 1996.

Grime, J. P.: The stress debate: symptom of impending synthesis?, Biol. J. Linn. Soc., 37, 3–17, 1989.

Guenther, A. B., Zimmerman, P. R., and Harley, P. C.: Isoprene and monoterpene emission rate variability: model evaluations and sensitivity analysis, J. Geophys. Res., 98, 12609–12617, 1993.

Harley, P., Deem, G., Flint, S., and Caldwell, M.: Effects of growth under elevated UVB on photosynthesis and isoprene emission in Quercus gambelii and Mucuna pruriens, Glob. Change Biol., 2, 149–154, 1996.

Hays, G. C., Richardson, A. J., and Robinson, C.: Climate change and plankton, Trends Ecol. Evol., 20, 337–344, doi:10.1016/j.tree.2005.03.004, 2005.

Hoffman, E. J. and Duce, R. A.: Factors influencing the organic carbon content of marine aerosols: A laboratory study, J. Geophys. Res., 81, 3667–3670, doi:10.1029/JC081i021p03667, 1976.

Holm-Hansen, D. and Riemann, B.: Chlorophyll *a* determination: improvements in methods, Oikos, 30, 438–447, 1978.

Hoose, C., Kristjánsson, J. E., Iversen, T., Kirkevåg, A., Seland, Ø., and Gettelman, A.: Constraining cloud droplet number concentration in GCMs suppresses the aerosol indirect effect, Geophys. Res. Lett., 36, L12807, doi:10.1029/2009GL038568, 2009.

Huner, P. A., Oquist, G., and Sarhan, F.: Energy balance and acclimation to light and cold, Trends Plant Sci., 3, 224–230, 1998.

Juneja, A., Ceballos, R. M., and Murthy, G. S.: Effects of Environmental Factors and Nutrient Availability on the Biochemical Composition of Algae for Biofuels Production: A Review, Energies, 6, 4607–4638, doi:10.3390/en6094607, 2013.

Kameyama, S., Tanimoto, H., Inomata, S., Suzuki, K., Komatsu, D. D., Hirota, A., Konno, U., and Tsunogai, U.: Application of PTR-MS to incubation experiments of the marine diatom Thalassiosira pseudonana, Geochem. J., 45, 355–363, 2011.

Koppmann, R. and Wildt, J.: Oxygenated Volatile Organic Compounds, In: Volatile Organic Compounds in the Atmosphere, edited by: Koppmann, R., 129–172, Blackwell Publishing Ltd., Oxford, 2007.

Langdon, C.: On the causes of interspecific differences in the growth–irradiance relationship for phytoplankton, Part II, A general review, J. Plankton Res., 10, 1291–312, 1988.

Liakakou, E., Vrekoussis, M., Bonsang, B., Donousis, C., Kanakidou, M., and Mihalopoulos, N.: Isoprene above the Eastern Mediterranean: Seasonal variation and contribution to the oxidation capacity of the atmosphere, Atmos. Environ., 4, 1002–1010, doi:10.1016/j.atmosenv.2006.09.034, 2007.

Lichtenthaler, H. K.: The 1-deoxy-D-xylulose-5-phosphate pathway of isoprenoid biosynthesis in plants, Annu. Rev. Plant Physiol. Plant Mol. Biol., 50, 47–65, 1999.

Lichtenthaler, H. K.: Biosynthesis and Accumulation of Isoprenoid Carotenoids and Chlorophylls and Emission of Isoprene by Leaf Chloroplasts, Bull. Georg. Natl. Acad. Sci., 3, 81–94, 2009.

Luo, G. and Yu, F.: A numerical evaluation of global oceanic emissions of a-pinene and isoprene, Atmos. Chem. Phys., 10, 2007–2015, doi:10.5194/acp-10-2007-2010, 2010.

MacIntyre, H. L, Kana, T. M., Anning, J., and Geider, R.: Photoacclimation of photosynthesis irradiance response curves and photosynthetic pigments in microalgae and cyanobacteria, J. Phycol., 38, 17–38, doi:10.1046/j.1529-8817.2002.00094.x, 2002.

McKay, W. A., Turner, M. F., Jones, B. M. R., and Halliwell, C. M.: Emissions of hydrocarbons from marine phytoplankton-some results from controlled laboratory experiments, Atmos. Environ., 30, 2583–2593, 1996.

Meskhidze, N. and Nenes, A.: Phytoplankton and cloudiness in the Southern Ocean, Science, 314, 1419–1423, doi:10.1126/science.1131779, 2006.

Meskhidze, N., Xu, J., Gantt, B., Zhang, Y., Nenes, A., Ghan, S. J., Liu, X., Easter, R., and Zaveri, R.: Global distribution and climate forcing of marine organic aerosol: 1. Model improvements and evaluation, Atmos. Chem. Phys., 11, 11689–11705, doi:10.5194/acp-11-11689-2011, 2011.

Meskhidze, N., Petters, M. D., Tsigaridis, K., Bates, T., O'Dowd, C., Reid, J., Lewis, E. R., Gantt, B., Anguelova, M. D., Bhave, P. V., Bird, J., Callaghan, A. H., Ceburnis, D., Chang, R., Clarke, A., de Leeuw, G., Deane, G., DeMott, P. J., Elliot, S., Facchini, M. C., Fairall, C. W., Hawkins, L., Hu, Y., Hudson, J. G., Johnson, M. S., Kaku, K. C., Keene, W. C., Kieber, D. J., Long, M. S., Martensson, M., Modini, R. L., Osburn, C. L., Prather, K. A., Pszenny, A., Rinaldi, M., Russell, L. M., Salter, M., Sayer, A. M., Smirnov, A., Suda, S. R., Toth, T. D., Worsnop, D. R., Wozniak, A., and Zorn, S. R.: Production mechanisms, number concentration, size distribution, chemical composition, and optical properties of sea spray aerosols, Atmos. Sci. Lett., 14, 207–213, doi:10.1002/asl2.441, 2013.

Milne, P. J., Riemer, D. D., Zika, R. G., and Brand, L. E.: Measurement of vertical distribution of isoprene in surface seawater, it chemical fate, and its emission from several phytoplankton monocultures, Mar. Chem., 48, 237–244, 1995.

Moore, R. M., Oram, D. E., Penkett, S. A.: Production of isoprene by marine-phytoplankton cultures, Geophys. Res. Lett., 21, 2507–2510, doi:10.1029/94GL02363, 1994.

O'Dowd, C. D. and de Leeuw, G.: Marine aerosol production: a review of the current knowledge, Phil. Trans. R. Soc. A, 365, 1753–1774, 2007.

O'Dowd, C. D., Facchini, M. C., Cavalli, F., Ceburnis, D., Mircea, M., Decesari, S., Fuzzi, S., Yoon, Y. J., and Putaud, J. P.: Biogenically driven organic contribution to marine aerosol, Nature, 431, 676–680, doi:10.1038/nature02959, 2004.

Palmer, P. I. and Shaw, S. L.: Quantifying global marine isoprene fluxes using MODIS chlorophyll observations, Geophys. Res. Lett., 32, L09805, doi:10.1029/2005GL022592, 2005.
Platnick, S. and Twomey, S.: Determining the susceptibility of cloud albedo to changes in droplet concentration with the Advanced Very High Resolution Radiometer, J. Appl. Meteorol., 33, 334–347, 1994.
Randall, D. A., Coakley, Jr., J. A., Fairall, C. W., Kropfli, R. A., and Lenschow, D. H.: Outlook for research on subtropical marine stratocumulus clouds, B. Am. Meteorol. Soc., 65, 1290–1301, 1984.
Rinnan, R., Steinke, M., McGenity, T., and Loreto, F.: Plant volatiles in extreme terrestrial and marine environments, Plant Cell Environ., 37, 1776–1789, doi:10.1111/pce.12320, 2014.
Roelofs, G. J.: A GCM study of organic matter in marine aerosol and its potential contribution to cloud drop activation, Atmos. Chem. Phys., 8, 709–719, doi:10.5194/acp-8-709-2008, 2008.
Schofield, O., Grzymski, J., Moline, M. M. A., and Jovine, R. V. M.: Impact of temperature acclimation on photosynthesis in the toxic red-tide dinoflagellate *Alexandrium fundyense* (Ca28), J. Plankton Res., 20, 1241–1258, 1998.
Sharkey, T. D. and Yeh, S.: Isoprene emission from plants, Annu. Rev. Plant Phys., 52, 407–436, 2001.
Sharkey, T. D., Wiberley, A. E, and Donohue, A. R.: Isoprene emission from plants: why and how, Ann. Bot., 101, 5–18, doi:10.1093/aob/mcm240, 2008.
Shaw, G. E.: Bio-controlled thermostasis involving the sulfur cycle, Climatic Change, 5, 297–303, 1983.
Shaw, S. L., Chisholm, S. W., and Prinn, R.: Isoprene production by Prochlorococcus, a marine cyanobacterium, and other phytoplankton, Mar. Chem, 80, 227–245. 2003.
Shaw, S. L., Gantt, B., and Meskhidze, N.: Production and Emissions of Marine Isoprene and Monoterpenes: A Review, Adv. Meteorol., 2010, 408696, doi:10.1155/2010/408696, 2010.
Staehr, P. A. and Birkeland, M. J.: Temperature acclimation of growth, photosynthesis and respiration in two mesophilic phytoplankton species, Phycologia, 45, 648–656, doi:10.2216/06-04.1, 2006.
Stevens, B., Vali, G., Comstock, K., Wood, R., Van Zanten, M. C., Austin, P. H., Bretherton, C. S., and Lenschow, D. H.: Pockets of Open Cells (POCs) and Drizzle in Marine Stratocumulus, B. Am. Meteorol. Soc., 86, 51–57, doi:10.1175/BAMS-86-1-51, 2005.
WDNR: Analytical Detection Limit Guidance and Laboratory Guide for Determining Method Detection Limits, PUBL-TS-056-96, Wisconsin Department of Natural Resources, 24 pp., 1996.
Wolfe, G. V., Strom, S. L., Holmes, J. L, Radzio, T., and Olson, M. B.: Dimethylsulfoniopropionate cleavage by marine phytoplankton in response to mechanical, chemical, or dark stress, J. Phycol, 38, 948–960, doi:10.1046/j.1529-8817.2002.t01-1-01100.x, 2002.
Yassaa, N., Peeken, I., Zollner, E., Bluhm, K., Arnold, S., Spracklen, D., and Williams, J.: Evidence for marine production of monoterpenes, Environ. Chem. Lett., 5, 391–401, doi:10.1071/EN08047, 2008.

6

On the role of circulation and mixing in the ventilation of oxygen minimum zones with a focus on the eastern tropical North Atlantic

P. Brandt[1], **H. W. Bange**[1], **D. Banyte**[1], **M. Dengler**[1], **S.-H. Didwischus**[1], **T. Fischer**[1], **R. J. Greatbatch**[1], **J. Hahn**[1], **T. Kanzow**[1,*], **J. Karstensen**[1], **A. Körtzinger**[1], **G. Krahmann**[1], **S. Schmidtko**[1], **L. Stramma**[1], **T. Tanhua**[1], **and M. Visbeck**[1]

[1]GEOMAR Helmholtz-Zentrum für Ozeanforschung Kiel, Kiel, Germany
[*]now at: Alfred-Wegener-Institut Helmholtz-Zentrum für Polar- und Meeresforschung, Bremerhaven, Germany

Correspondence to: P. Brandt (pbrandt@geomar.de)

Abstract. Ocean observations are analysed in the framework of Collaborative Research Center 754 (SFB 754) "Climate-Biogeochemistry Interactions in the Tropical Ocean" to study (1) the structure of tropical oxygen minimum zones (OMZs), (2) the processes that contribute to the oxygen budget, and (3) long-term changes in the oxygen distribution. The OMZ of the eastern tropical North Atlantic (ETNA), located between the well-ventilated subtropical gyre and the equatorial oxygen maximum, is composed of a deep OMZ at about 400 m in depth with its core region centred at about 20° W, 10° N and a shallow OMZ at about 100 m in depth, with the lowest oxygen concentrations in proximity to the coastal upwelling region off Mauritania and Senegal. The oxygen budget of the deep OMZ is given by oxygen consumption mainly balanced by the oxygen supply due to meridional eddy fluxes (about 60 %) and vertical mixing (about 20 %, locally up to 30 %). Advection by zonal jets is crucial for the establishment of the equatorial oxygen maximum. In the latitude range of the deep OMZ, it dominates the oxygen supply in the upper 300 to 400 m and generates the intermediate oxygen maximum between deep and shallow OMZs. Water mass ages from transient tracers indicate substantially older water masses in the core of the deep OMZ (about 120–180 years) compared to regions north and south of it. The deoxygenation of the ETNA OMZ during recent decades suggests a substantial imbalance in the oxygen budget: about 10 % of the oxygen consumption during that period was not balanced by ventilation. Long-term oxygen observations show variability on interannual, decadal and multidecadal timescales that can partly be attributed to circulation changes. In comparison to the ETNA OMZ, the eastern tropical South Pacific OMZ shows a similar structure, including an equatorial oxygen maximum driven by zonal advection but overall much lower oxygen concentrations approaching zero in extended regions. As the shape of the OMZs is set by ocean circulation, the widespread misrepresentation of the intermediate circulation in ocean circulation models substantially contributes to their oxygen bias, which might have significant impacts on predictions of future oxygen levels.

1 Introduction

The oceanic oxygen distribution is generally characterised by slightly supersaturated oxygen levels in the surface layer, an intermediate oxygen minimum, and higher oxygen levels at depth. This vertical structure is a consequence of the delicate balance between the supply of oxygen through ventilation and circulation, oxygen production by photosynthesis, and oxygen consumption by remineralisation of sinking organic matter. The horizontal distribution of oxygen shows major large-scale open ocean subsurface oxygen minimum zones (OMZs) in the eastern parts of the tropical Atlantic and Pacific oceans as well as in the northern Indian Ocean. By analysing a combination of historical and modern observations, an expansion and intensification of OMZs in the tropical oceans has been detected (Stramma et al., 2008b). However, numerical simulations with global or regional models are not able consistently to reproduce such trends and thus up

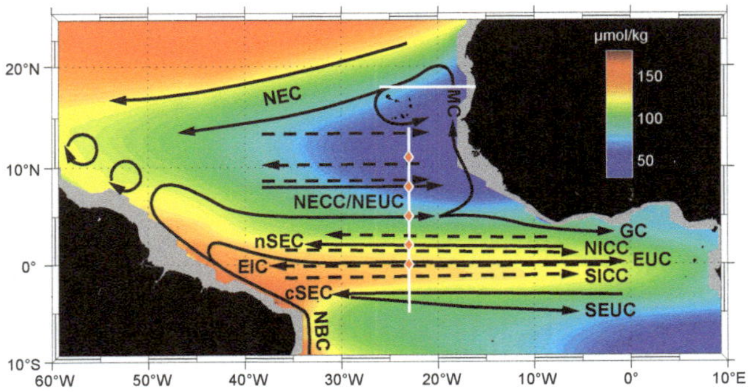

Figure 1. Oxygen concentration (μmol kg^{-1}) in the tropical Atlantic at $\sigma_\theta = 27.1\,\text{kg m}^{-3}$ (close to the deep oxygen minimum) as obtained from the MIMOC climatology (Schmidtko et al., 2013) with circulation schematic superimposed. Surface and thermocline current branches shown (black solid arrows) are the North Equatorial Current (NEC), the Mauritania Current (MC), the northern and central branches of the South Equatorial Current (nSEC and cSEC), the North Equatorial Countercurrent (NECC), the Guinea Current (GC), the North Brazil Current (NBC), the North and South Equatorial undercurrents (NEUC and SEUC), and the Equatorial Undercurrent (EUC). Intermediate current branches shown (black dashed arrows) are the North and South Intermediate countercurrents (NICC and SICC) or "flanking jets", and the Equatorial Intermediate Current (EIC). The 23° W and 18° N repeat sections are marked by white lines, mooring positions by red diamonds.

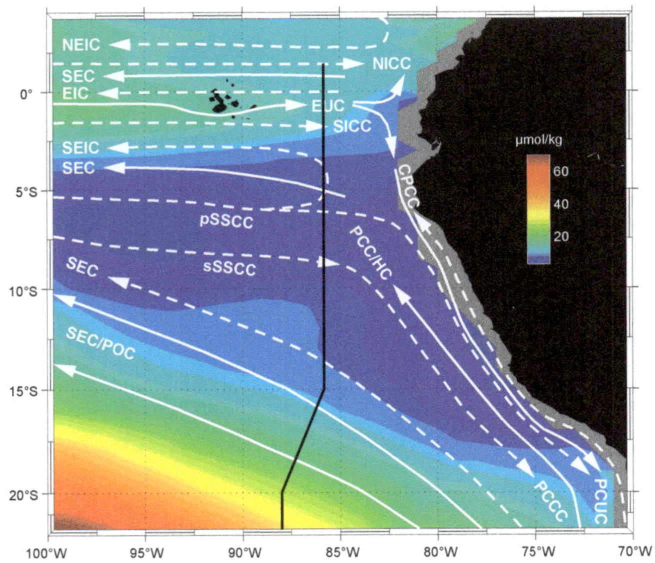

Figure 2. Oxygen concentration (μmol kg^{-1}) in the eastern tropical Pacific at $\sigma_\theta = 26.8\,\text{kg m}^{-3}$ (close to the deep oxygen minimum) as obtained from the MIMOC climatology (Schmidtko et al., 2013) with circulation schematic superimposed. Current bands displayed are, for the surface layer (white solid arrows), the South Equatorial Current (SEC), the Equatorial Undercurrent (EUC), the Peru–Chile or Humboldt Current (PCC/HC), the Peru Oceanic Current (POC), and for the thermocline layer (white dashed arrows), the North Equatorial Intermediate Current (NEIC), the North Intermediate Countercurrent (NICC), the Equatorial Intermediate Current (EIC), the South Intermediate Countercurrent (SICC), the primary and secondary Southern Subsurface countercurrents (pSSCC, sSSCC), the deeper layer of the SEC, the Chile–Peru Coastal Current (CPCC), the Peru–Chile Undercurrent (PCUC) and the Peru–Chile Countercurrent (PCCC). The location of the ~86° W section is marked by the black line .

to now have failed to provide an explanation for the observed oxygen trends in the tropical ocean (Stramma et al., 2012).

OMZs in the tropical Atlantic were first identified by analysing hydrographic data from the German *Meteor* expedition from 1925 to 1927 (Wattenberg, 1938). This data set revealed the existence of OMZs in both hemispheres of the eastern tropical Atlantic at depths between 300 and 700 m, situated equatorward of the subtropical gyres and separated by an equatorial oxygen maximum. Based on data, including those from the German *Meteor* expedition, and theoretical considerations, Wyrtki (1962) concluded that the boundaries of these OMZs are set by advection with the lowest oxygen levels occurring in almost stagnant water bodies. A plausible theory of thermocline ventilation was delivered by Luyten et al. (1983b). The basis of their theory is of an ocean forced by subtropical Ekman pumping and otherwise obeying circulation pathways that are governed by potential vorticity conservation. This theory explains the existence of non-ventilated, near-stagnant shadow zones in the eastern tropics. The remaining slow ventilation of such shadow zones, which under the assumption of steady state is required to balance oxygen consumption, is expected to be the consequence of lateral fluxes of oxygen from oxygen-rich water masses of the subtropics as well as due to diapycnal oxygen fluxes from oxygen-rich layers above and below the thermocline of the OMZs.

The near-surface layers (upper ~250 m) of the tropical oceans are characterised by the presence of energetic zonal current bands. In the Atlantic below that layer, substantial mean zonal currents are also found, particularly in the depth range of the OMZs (Fig. 1). Close to the Equator, the strongest intermediate currents are observed with eastward flow at 2° N and 2° S and westward flow in between. The eastward current bands have been found to ventilate the central and eastern equatorial regions with oxygen-rich waters from the western boundary (Tsuchiya et al., 1992; Schott et al., 1995, 1998). Together with time-varying equatorial jets they produce an equatorial oxygen maximum at intermediate depths (Brandt et al., 2012). Further poleward alternating zonal jets are present at intermediate depths including the latitude range of the OMZs. Their strengths have been quantified using subsurface drift trajectories from floats (Maximenko et al., 2005; Ollitrault et al., 2006) and repeated shipboard sections (Brandt et al., 2010). Such currents have been reproduced by idealised process modelling (Ménesguen et al., 2009; Ascani et al., 2010; Qiu et al., 2013), but are typically not found (or are unrealistically weak) in ocean circulation models. They contribute to the ventilation of the eastern tropical North Atlantic (ETNA) at intermediate depth, and decadal to multidecadal changes in the strengths of these jets might play a significant role in modulating long-term oxygen changes in the ETNA OMZ (Brandt et al., 2010).

The Atlantic and Pacific OMZs have many similarities, particularly regarding OMZ shape and circulation pattern. The ETNA and the eastern tropical South Pacific (ETSP) OMZs (Figs. 1, 2) are both located in the shadow zones of the ventilated thermocline and are ventilated by lateral and vertical mixing as well as by zonal advection in the equatorial band. However, the striking difference between both OMZs is that the ETNA OMZ is hypoxic (oxygen below ~ 60 to $120\,\mu\text{mol kg}^{-1}$) and the ETSP is suboxic (oxygen below about $10\,\mu\text{mol kg}^{-1}$). Karstensen et al. (2008) concluded that this difference is the result of reduced oxygen levels in the eastward current bands of the Pacific OMZs compared to the Atlantic OMZs, which they argue can be traced back to the larger ratio of the total volume of the OMZ layer to the renewal or subduction rate in the Pacific compared to the Atlantic.

As part of the Collaborative Research Center 754 (Sonderforschungsbereich, SFB 754) "Climate-Biogeochemistry Interactions in the Tropical Ocean" (first phase 2008–2011 and second phase 2012–2015), physical processes responsible for the ventilation of the ETNA OMZ have been studied using an extended observational programme including repeat hydrography by shipboard and glider measurements, an array of subsurface moorings, microstructure measurements and two tracer release experiments. The goals of the research programme are to deliver an improved understanding of the ventilation physics of the ETNA OMZ, to come up with a quantitative understanding of the functioning of the OMZs, to monitor regional oxygen variability and trends, and to analyse their causes. The ETSP OMZ has been studied as well using a reduced observational programme. However, the comparison between the hypoxic ETNA and the suboxic ETSP is of particular interest here, as the observed deoxygenation in the ETNA, or future climate change, might lead to a shift from hypoxic to suboxic conditions. The present paper provides an overview of the current status of the science regarding these topics. The paper is organised as follows: in Sect. 2, data and methods used in this study are described. In Sect. 3, the current system and the OMZ structure in the ETNA are characterised. Results for the quantification of the strength of different ventilation processes, i.e. vertical mixing, lateral mixing, and advection, are presented in Sect. 4. In Sect. 5, the current knowledge of oxygen consumption estimates is presented. The OMZ structure and processes at the continental margin are presented in Sect. 6. Long-term oxygen variability with a special focus on the period of enhanced data coverage is presented in Sect. 7. The results obtained for the ETNA OMZ are then compared with results obtained for the ETSP in Sect. 8 and, finally, in Sect. 9, the results are summarised and discussed.

2 Ocean observations

A major focus of the observational work presented here has been on circulation, ventilation physics, and water mass distribution. In the tropical North Atlantic, observations have been concentrated on the 23° W section with repeat hydrography, microstructure measurements, velocity measurements (Table 1), and moored observations (Table 2). The 23° W section cuts through the ETNA OMZ from south of the Cape Verde archipelago to slightly south of the Equator (Fig. 1). Along the 23° W section, moorings with instrumentation to observe temperature, salinity, oxygen and velocity continuously were deployed at 8 and 5° N, delivering multi-year time series. Additionally, oxygen sensors were installed at 300 and 500 m in depth at selected moorings (23° W, 4° and 11.5° N) of the Prediction and Research Moored Array in the Tropical Atlantic (PIRATA; Bourles et al., 2008) and at subsurface moorings at 23° W, 2° N and 0° (Fig. 1). For the analysis of hydrographic and velocity data acquired along 23° W, we used the measurements given in Table 1. Besides the 23° W section, here we shall also present data acquired along 18° N at the northern boundary of the ETNA OMZ (Fig. 1, Table 1). Moreover, two tracer release experiments (TREs) were carried out in the ETNA OMZ. During the first TRE, GUTRE (Guinea Upwelling Tracer Release Experiment), in April 2008, 92 kg of the halocarbon tracer trifluoromethyl sulfur pentafluoride (CF_3SF_5) were released at 23° W, 8° N on the potential density surface, $\sigma_\theta = 26.88\,\text{kg m}^{-3}$. The depth of release, of about 330 m, corresponds to the depth of the oxycline above the deep oxygen minimum. During the following 2.5 years, three tracer surveys were carried out to measure the vertical and horizontal spreading of the tracer (Banyte et al. (2012), Table 1). During the second TRE, OSTRE (Oxygen Supply Tracer Release Experiment), in November 2012, 88.5 kg of the same tracer were released at 21° W, 11° N on the potential density surface $\sigma_\theta = 27.03\,\text{kg m}^{-3}$, corresponding to about 500 m in depth, which is in the core region of the ETNA OMZ.

In the ETSP OMZ, a particular focus was on the $\sim 86°$ W section (section located at 85°50′ W north of 15° S with a westward shift to 88° W south of 20° S, called the $\sim 86°$ W section in the following) with hydrographic and current measurements from 2° N to about 22° S (Fig. 2). Two recent cruises covered that section repeating measurements taken during the RV *Knorr* cruise in March 1993 (Table 1). Additionally, four cruises were carried out along the continental margin of Peru (Table 1) to investigate the circulation along the continental slope and shelf off Peru as well as the physical processes contributing to the redistribution of oxygen, nutrients and other solutes.

Table 1. Research cruises to the tropical eastern Atlantic and Pacific oceans. Depending on the measurements carried out and the geographical area covered on the different cruises, up to 22 sections were used to determine the mean 23° W section, 7 sections for the mean 18° N section and 3 sections for the mean ~86° W section.

Vessel and cruise (date)	Main work	Region
	Tropical Atlantic, 5° S–14° N/~23° W and OMZ area	
Thalassa (Jul–Aug 1999)	23° W section	5° S–6° N
Seaward Johnson (Jan 2000)	23° W section	5° S–4° N
Meteor 47/1 (Apr 2000)	23° W section	5° S–4° N
Meteor 55 (Oct 2002)	24° W section	0–10° N
Polarstern Ant XXII/5 (Jun 2005)	23° W section	5° S–14° N
Meteor 68/1 (May 2006)	23° W section	2° S–0.5° N
Ron Brown (Jun 2006)	23° W section	5° S–14° N
Meteor 68/2 (Jun–Jul 2006)	23° W section, moorings	4° S–14° N
Ron Brown (May 2007)	23° W section	4° N–14° N
L'Atalante GEOMAR 4 (Feb–Mar 2008)	23° W section, moorings	2° S–14° N
Maria S. Merian 08/1 (Apr–May 2008)	23° W section, GUTRE tracer release	7.5–14° N, 23° W, 8° N at 330 m
Maria S. Merian 10/1 (Nov–Dec 2008)	GUTRE tracer survey	4–14° N / 27.5–17.5° W
Polarstern Ant XXV/5 (May 2009)	23° W section	5° S–14° N
Endeavor 463 (May–Jun 2009)	23° W section	5° S–3° N
Ron Brown (Jul–Aug 2009)	23° W section	0–14° N
Meteor 80/1 (Oct–Nov 2009)	23° W section, moorings	5° S–14° N
Polarstern Ant XXVI/1 (Nov 2009)	23° W section	5° S–14° N
Meteor 80/2 (Dec 2009)	GUTRE tracer survey	4–14° N/31–15° W
Meteor 81/1 (Feb–Mar 2010)	22° W section	5° S–13° N
Polarstern Ant XXVI/4 (May 2010)	23° W section	5° S–13.5° N
Meteor 83/1 (Oct–Nov 2010)	GUTRE tracer survey	2–15° N / 28–15° W
Maria S. Merian 18/2 (May–Jun 2011)	23° W section, moorings	5° S–14° N
Maria S. Merian 18/3 (Jun–Jul 2011)	23° W section	4–14° N
Ron Brown (Jul–Aug 2011)	23° W section	0–14° N
Maria S. Merian 22 (Oct–Nov 2012)	23° W section, moorings	5° S–14° N
Maria S. Merian 23 (Dec 2012)	23° W section, OSTRE tracer release	4° S–5° N, 21° W, 11° N at 500 m
Meteor 97 (May–Jun 2013)	OSTRE tracer survey	8–12° N/23–19° W
Meteor 106 (Apr–May 2014)	23° W section, moorings	5° S–14° N
	Tropical Atlantic, 26–16° W/18° N	
P320/1 (Mar–Apr 2005)	18° N section	19–16.4° W
Meteor M68/3 (Jul–Aug 2006)	18° N section	26–16.3° W
P347 (Jan-Feb 2007)	18° N section	17.5–16.3° W
P348 (Mar 2007)	18° N section	23.2–16.4° W
L'Atalante GEOMAR 3 (Feb 2008)	18° N section	24.3–16.3° W
P399/2 (Jun 2010)	18° N section	21–16.5° W
Maria S. Merian 22 (Nov 2012)	18° N section	26–20° W
	Tropical Pacific, 22° S–2° N/~86° W and continental slope	
Knorr (Mar–Apr 1993)	~86° W section	22° S–2° N
Meteor 77/3 (Jan 2009)	Continental slope	18–10° S
Meteor 77/4 (Feb 2009)	~86° W section	14° S–2° N
Meteor 90 (Nov 2012)	~86° W section	22° S–2° N
Meteor 91 (Dec 2012)	Continental slope	17–5° S
Meteor 92 (Jan 2013)	Continental slope	13–10° S
Meteor 93 (Feb 2013)	Continental slope	14–10° S

Table 2. Moored oxygen observations in the eastern tropical Atlantic along 23° W.

Position	Period	Mooring type	Depth (m)
0°/23° W	Mar 2008–Oct 2012	Subsurface	300, 500
2° N/23° W	Feb 2008–May 2011	Subsurface	300, 500
4° N/23° W	Jul 2009–Jan 2013	PIRATA	300, 500
5° N/23° W	Nov 2009–Oct 2012	Subsurface	100–800
8° N/23° W	Nov 2009–Oct 2012	Subsurface	100–800
11.5° N/23° W	Jul 2009–Jan 2013	PIRATA	300, 500

3 Structure of the ETNA OMZ

The subtropical gyre circulation of the Northern Hemisphere is, to first order, determined by the negative wind stress curl associated with mid-latitude westerlies and northeasterly trade winds. The resulting Ekman pumping drives subduction of oxygen-rich surface water masses in the subtropics. According to theory, equatorward and westward propagation of subducted water masses forms the northern boundary of the shadow zone of the ventilated thermocline (Luyten et al., 1983b). Within the shadow zone, which is characterised by a weak mean circulation, the ETNA OMZ with a core depth at about 400 m is found. Lowest oxygen concentrations at the core depth are found away from the continental margin at about 20° W, 10° N (Fig. 3). North of the ETNA OMZ, the North Equatorial Current (NEC) flows southwestward along the Cape Verde Frontal Zone. It transports oxygen-rich Central Water (CW) formed by subduction in the subtropics as well as intermediate water masses in the deeper layers that have their origin mainly in the Labrador Sea and the Mediterranean outflow. To the south, the ETNA OMZ is bounded by the energetic zonal flows near the Equator forming the equatorial oxygen maximum (Brandt et al., 2012). Above the main deep OMZ, at a depth of about 100 m, a shallow OMZ is situated, defined as the secondary oxygen minimum below the surface mixed layer and above 200 m (Fig. 4). It is characterised by generally higher oxygen levels compared to the deep OMZ, while occasionally, extremely low oxygen levels are possible, and are most pronounced in the northeastern part of the shadow zone close to the highly productive eastern boundary upwelling region off Mauritania and Senegal (Fischer et al., 2013). The mean 18° N section shows shallow mixed-layer depths over the continental margin typical of coastal upwelling regions as well as lower salinities in the CW layer that are a consequence of the northward transport of Southern Hemisphere water along the continental slope within the Poleward Undercurrent (Barton, 1989) and the surface flow associated with the Mauritania Current (Mittelstaedt, 1983) (Fig. 5).

The western boundary of the Atlantic Ocean is associated with relatively high oxygen levels at all latitudes (Fig. 1). At the density of the OMZ layer, the North Brazil Undercurrent

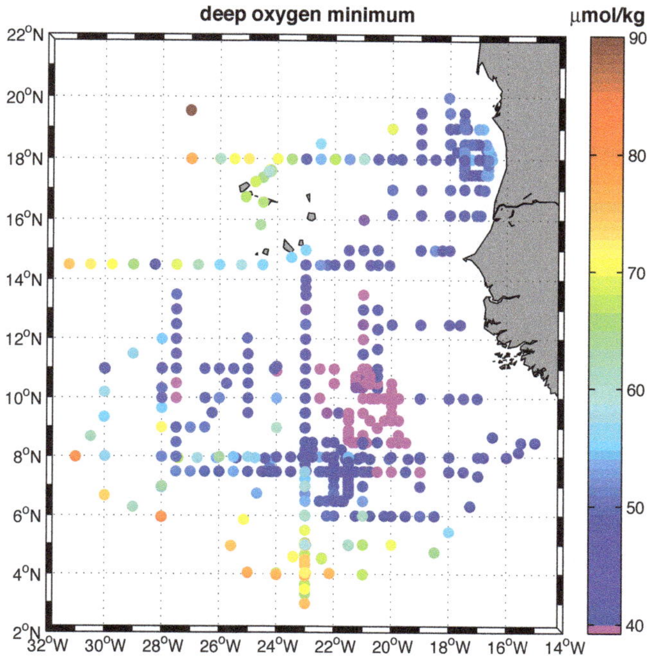

Figure 3. Minimum oxygen concentration below 200 m (representing the deep oxygen minimum) as obtained from CTD station data taken during the period 2006–2013. Oxygen concentration at the deep oxygen minimum below 40 μmol kg^{-1} is marked by purple dots.

(NBUC)/North Brazil Current (NBC) (Schott et al., 2005) transports central and intermediate water masses of Southern Hemisphere origin northward. The high oxygen concentrations in the CW layer of the NBUC can be traced back along the southern branches of the South Equatorial Current (SEC) to the subduction region in the eastern subtropical gyre (Tsuchiya, 1986; Stramma and England, 1999). The CW also includes water from the Indian Ocean that is brought into the Atlantic by eddy shedding from the Agulhas retroflection.

The Antarctic Intermediate Water (AAIW) below the CW originates mainly from the Drake Passage and is transported around the Southern Hemisphere subtropical gyre to feed into the NBUC (Suga and Talley, 1995). Of importance for the ventilation of the ETNA OMZ is the northward flow of CW and AAIW across the Equator. The northward penetration of Southern Hemisphere water masses at the western boundary changes with depth: AAIW dominates as far as 15° N, and the upper CW only as far as 10° N, because of the presence of water masses of Northern Hemisphere origin (Kirchner et al., 2009).

A substantial part of the water masses transported northward within the NBUC forms the upper branch of the Atlantic Meridional Overturning Circulation (AMOC), a circulation known since the German *Meteor* cruises in the 1920s, as documented by Wüst (1935). The presence of the AMOC under present climate conditions is identified as the main

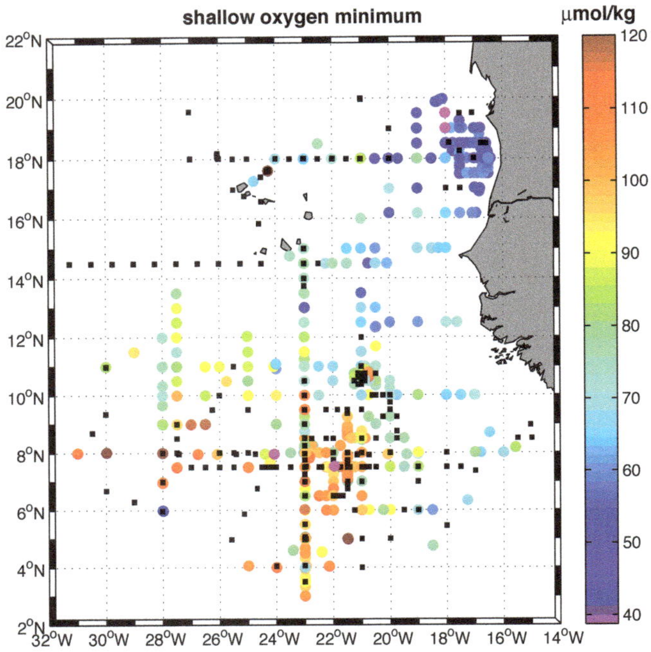

Figure 4. Minimum oxygen concentration above 200 m (representing the shallow oxygen minimum) as obtained from CTD station data taken during the period 2006–2013. Black squares indicate profiles without a shallow oxygen minimum (i.e. the minimum oxygen concentration was found at the lower boundary of the chosen depth range, that is 200 m). Oxygen concentration at the shallow oxygen minimum below 40 μmol kg^{-1} is marked by purple dots.

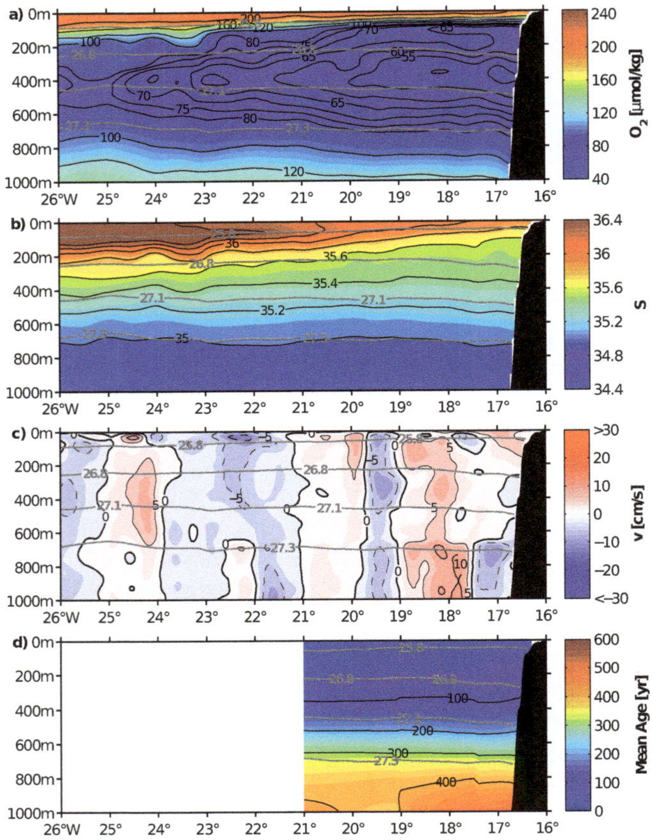

Figure 5. **(a)** Mean oxygen content, **(b)** salinity, **(c)** meridional velocity (positive northward), and **(d)** mean age as obtained from zonal ship sections taken along 18° N from 2005 to 2012. Grey contours mark potential density (kg m^{-3}). Besides the deep oxygen minimum at about 400 m in depth, there is a shallow oxygen minimum at about 100 m in proximity to the shelf **(a)**.

reason for the dominance of Southern Hemisphere water masses in the tropical North Atlantic discussed above. It contributes to the asymmetric shallow overturning circulations in both hemispheres as well, the subtropical cell (STC) of the Northern Hemisphere being much weaker than its counterpart in the Southern Hemisphere (Schott et al., 2004). The STC connects the subduction regions of the eastern subtropical gyres to the equatorial and coastal upwelling regions. In the Northern Hemisphere, the subducted water masses mostly do not reach the Equator. Instead, they contribute to the eastward flow within the North Equatorial Counter Current (NECC)/North Equatorial Undercurrent (NEUC) at about 5° N (Zhang et al., 2003). A particular feature in the ETNA is the presence of an open ocean upwelling regime within the cyclonic circulation of the Guinea Dome south of the Cape Verde archipelago. Associated with the presence of the Guinea Dome are changes in the potential vorticity distribution that further limit the flow of newly subducted water masses from the Northern Hemisphere subtropics toward the Equator within the STC (Malanotte-Rizzoli et al., 2000).

The 23° W section (Fig. 6) cuts through the ETNA OMZ, which can be identified by low oxygen levels as well as by the high age of the water masses. The gradual change in salinity on density surfaces along this section defines the transition between low- and high-saline water masses of southern and northern origin, respectively. Along this section, mean eastward and westward flow is typically identified by positive and negative oxygen anomalies relative to the background oxygen distribution (Brandt et al., 2010).

Ventilation timescales of the interior ocean can be quantified by analysing transient tracer distributions. A comprehensive set of CFC-12 and SF_6 measurements in the ETNA (these tracers were measured in parallel to the deliberately released CF_3SF_5 tracer) has been explored in detail by Schneider et al. (2012) using the concept of transit time distributions (TTDs) (e.g. Waugh et al., 2004). The mean age in the centre of the OMZ ($\sigma_\theta = 27.0$ kg m^{-3}) is in the range of 120 to 180 years (Fig. 7). The mean age refers to the average time it takes for a water parcel to reach a certain location in the interior ocean from the time it was last in contact with the surface ocean and hence the atmosphere (see Sect. 5 for more discussion on the TTD concept). In contrast to waters in the OMZ centre, water south of about 5° N is significantly better ventilated, with mean ages close to 100 years, reflecting the more energetic circulation in the equatorial region. Roughly

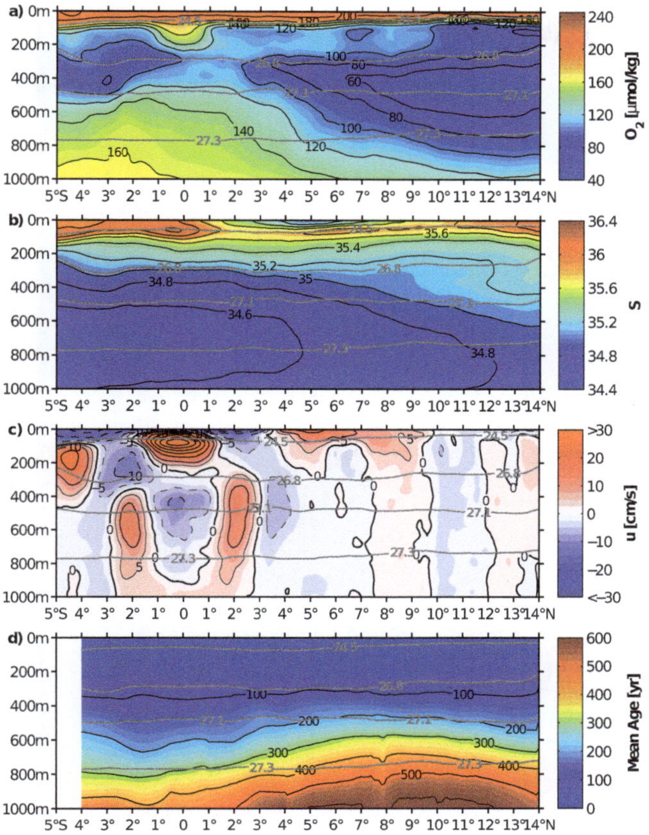

Figure 6. (a) Mean oxygen content, (b) salinity, (c) zonal velocity (positive eastward), and (d) mean age as obtained from meridional ship sections taken along 23° W during 1999–2012. Grey contours mark potential density ($\mathrm{kg\,m^{-3}}$). Eastward current bands, marked by reddish colours, are generally associated with elevated oxygen content.

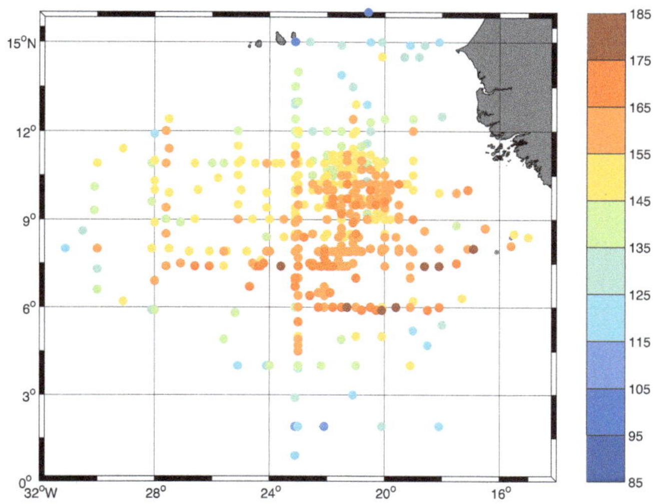

Figure 7. Mean age (yr) at $\sigma_\theta = 27.0\,\mathrm{kg\,m^{-3}}$, which corresponds approximately to the depth of the deep oxygen minimum.

the same age is found north of about 13° N close to the Cape Verde Islands despite lower oxygen values in the northern region compared to the southern region. Below the poorly ventilated OMZ, the even older AAIW ($\sigma_\theta = 27.3\,\mathrm{kg\,m^{-3}}$) with ventilation times in excess of 500 years is found (close to the detection limit of the CFCs, and thus difficult to quantify accurately), although this water mass has a high oxygen concentration. Again, at this density layer, the area south of 5° N is significantly better ventilated than north of 5° N (Schneider et al., 2012).

4 Ventilation processes

The oxygen budget of the OMZ takes account of consumption, advection, and diffusion of oxygen. Any imbalance of these terms results in decreasing or increasing oxygen concentrations. While consumption is an oxygen sink, advection and diffusion might be sources or sinks, depending on the background conditions. Mean advection of oxygen manifests itself in the mean oxygen and velocity distributions: along 23° W, mean eastward current bands are generally associated with an elevated oxygen content (Fig. 6) representing an advective ventilation pathway from the western boundary toward the OMZ (Brandt et al., 2010). Horizontal and vertical diffusion act on the mean horizontal and vertical oxygen gradients, respectively. The associated variance production by mesoscale eddy stirring and small-scale turbulence (Ferrari and Polzin, 2005) results in locally elevated oxygen variance. The Eulerian variance along 23° W, as obtained from ship sections, might additionally result from lateral meandering of zonal currents or from vertical movements of isopycnals associated with internal waves and eddies. Moored time series reflect this variability pattern. There is generally higher oxygen variance at 300 m in depth close to the oxycline above the deep OMZ core compared to at 500 m in depth (cf. Figs. 8, 9). Timescales of processes driving the variance in moored time series cover a wide range from those associated with internal waves and tides, inertial oscillations, and the mesoscale eddy field to seasonal and interannual variability, including planetary waves (Hahn et al., 2014). Using repeat ship sections, the effect of vertical motion of isopycnals can be removed by calculating oxygen variance on potential density surfaces and projecting back onto depth space (Fig. 10). The remaining oxygen variance in regions of weak mean flow surrounding the ETNA OMZ might be associated with processes responsible for the vertical and lateral mixing that is discussed in the following subsections.

4.1 Vertical mixing

Vertical mixing acts on the vertical oxygen gradients and leads to an oxygen supply to the OMZ via down-gradient oxygen fluxes. In order to estimate the vertical or diapycnal oxygen supply, the diapycnal diffusivity K_ρ as a measure for diapycnal mixing is required. From the diapycnal spread of

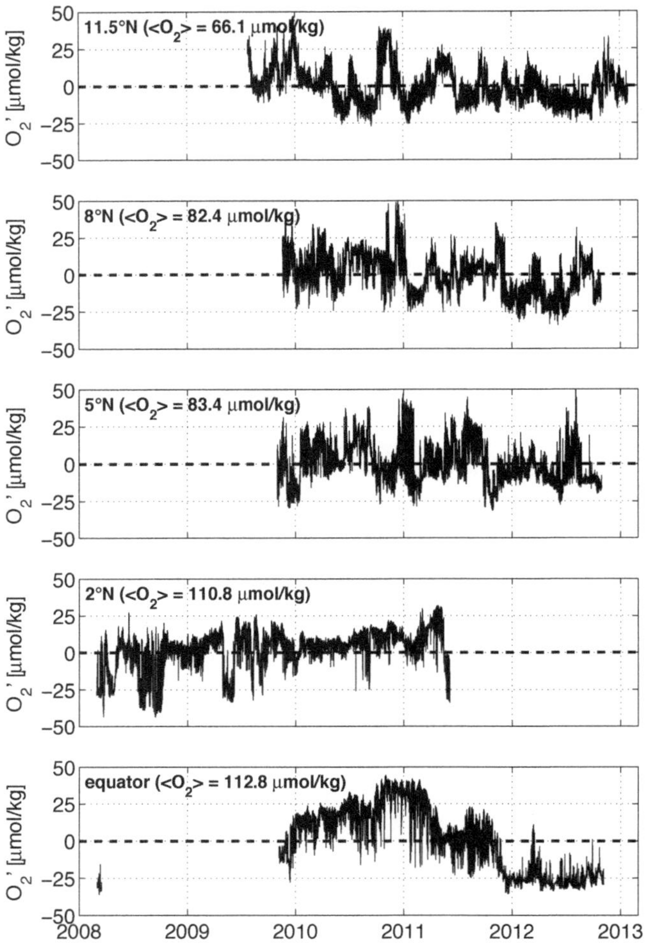

Figure 8. Time series of the oxygen anomaly at about 300 m in depth from moored observations along 23° W at different latitudes. Mean oxygen values at the different mooring locations are given in brackets.

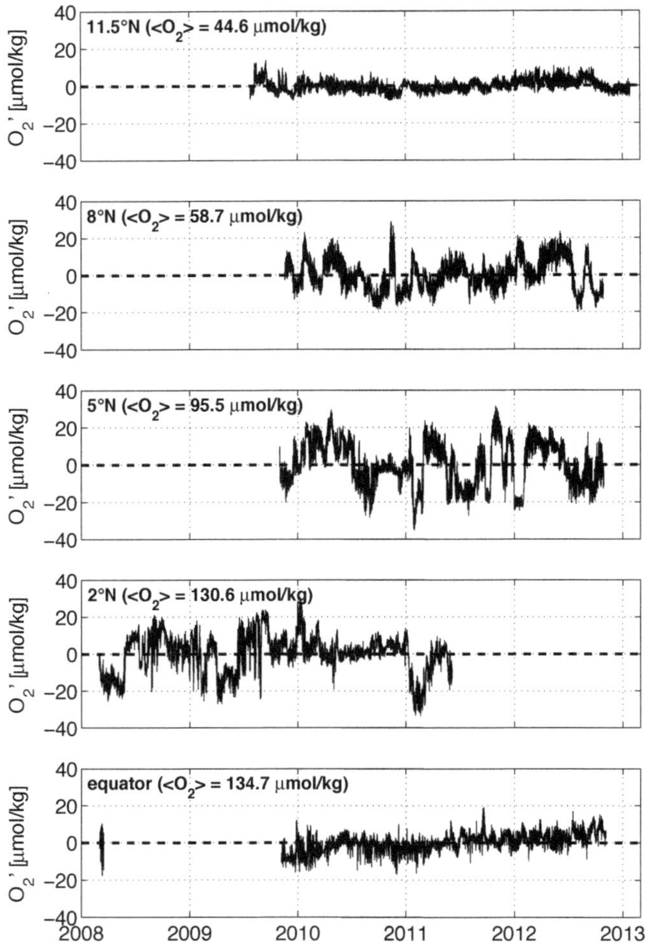

Figure 9. As Fig. 8, but at about 500 m in depth.

the deliberately released tracer during GUTRE, a mean diapycnal diffusivity of $(1.2 \pm 0.2) \times 10^{-5}\,\mathrm{m^2\,s^{-1}}$ was derived (Banyte et al., 2012). The tracer was injected on the isopycnal $\sigma_\theta = 26.88\,\mathrm{kg\,m^{-3}}$ (about 330 m), corresponding to the oxycline above the deep OMZ. GUTRE was accompanied by extensive microstructure and finescale shear measurements that delivered an estimate of $(1.0 \pm 0.2) \times 10^{-5}\,\mathrm{m^2\,s^{-1}}$ for K_ρ for the depth range between 150 and 500 m (Fischer et al., 2013). The value inferred from microstructure measurements only considers diapycnal mixing due to small-scale turbulence. However, double diffusive enhancement was found to be small ($\sim 0.1 \times 10^{-5}\,\mathrm{m^2\,s^{-1}}$) in this depth interval (Fischer et al., 2013), so the total diffusivities estimated by the two independent methods agree within the error bars. This estimate of diapycnal mixing is considerably larger than the expected background mixing at this latitude (e.g. Gregg et al., 2003), probably due to the presence of rough topography (e.g. the Sierra Leone Rise) in the southern part of the OMZ. Combining K_ρ with simultaneous profiles of the vertical oxygen gradient allows determination of the profile of the diapycnal oxygen flux. Its divergence represents the oxygen supply to the OMZ and amounted to about $1\,\mu\mathrm{mol\,kg^{-1}\,yr^{-1}}$ in the OMZ core, with the required oxygen transported downwards from the upper CW (Fischer et al., 2013).

Deeper-ranging microstructure profiles acquired during the most recent cruises to the ETNA OMZ (Table 1) allowed us to extend the analysis into the deeper water column down to 800 m in depth, i.e. allowed us to estimate the diapycnal oxygen flux from the AAIW below as well. In total, 200 microstructure profiles, 40 of them down to 800 m, were about equally partitioned into three subregions of the OMZ: a seamount subregion (7 % of OMZ area), an abyssal plain subregion (80 % of OMZ area), and a transition subregion (13 % of OMZ area). They served to estimate subregional mean profiles of the turbulent part of diapycnal diffusivity (Fig. 11). Double diffusive enhancement of K_ρ from simultaneous CTD profiles for each subregion following St. Laurent and Schmitt (1999) was accounted for to obtain subregional total K_ρ profiles (Fig. 11) and an area-weighted mean total K_ρ profile (Fig. 12). The mean diapycnal supply (Fig. 13) that, in the following, will be used in the oxygen budget, was

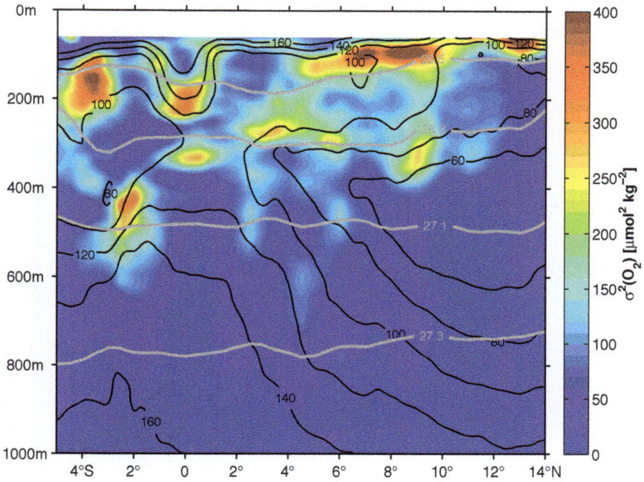

Figure 10. Oxygen variance along 23° W from repeat ship sections. The analysis was done on isopycnal surfaces and the results were projected back onto depth coordinates. Grey contours mark potential density (kg m^{-3}); black contours mark mean oxygen (µmol kg^{-1}).

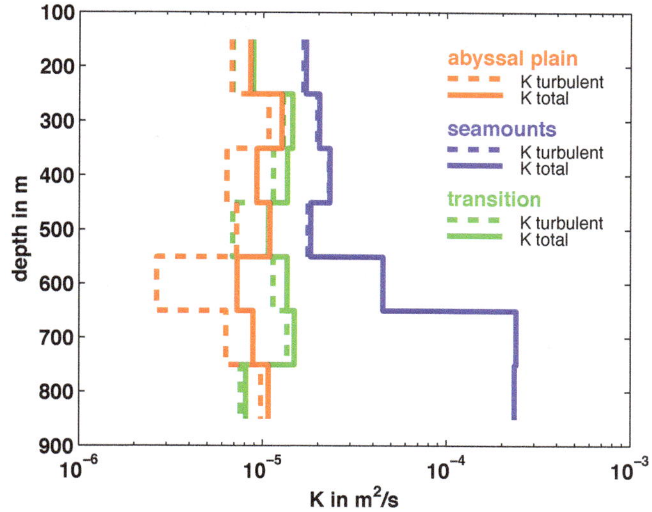

Figure 11. Profiles of the diapycnal eddy diffusivity as estimated from microstructure measurements (dashed lines) and by accounting for the effect of double diffusion (solid lines) for different regions: (red) abyssal plain, (blue) seamount region, and (green) transition region.

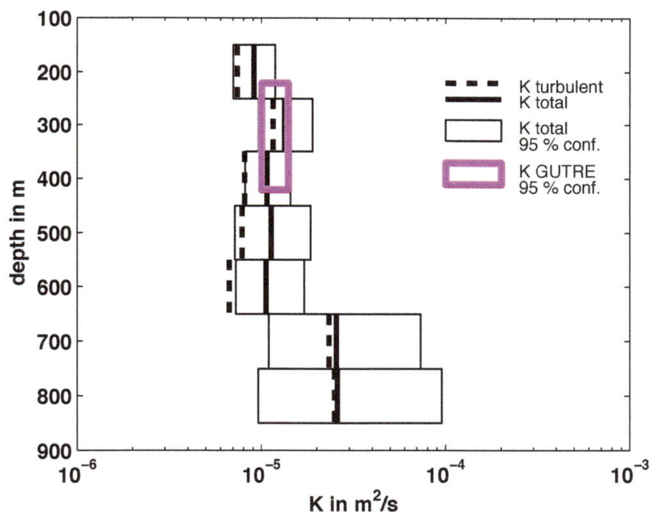

Figure 12. Profiles of diapycnal eddy diffusivity as estimated from microstructure measurements (dashed black line) and total diapycnal diffusivity after accounting for the effect of double diffusion (solid black line with 95 % confidence error level). The purple box represents the 95 % confidence error level of total diapycnal diffusity as estimated from the tracer release experiment.

then derived as the divergence of the low-pass filtered mean diapycnal flux. The mean flux profile was calculated as the area-weighted mean of the three flux profiles from the three subregions, which in turn were obtained by combining mean K_ρ with vertical oxygen gradient profiles from the three subregions. Error estimates are reported as 95% confidence limits and are based on standard errors of the mean of individual K_ρ and oxygen gradient profiles for each subregion. Subsequent error estimates for the mean total K_ρ profile (Fig. 12), flux profiles, and the mean supply profile (Fig. 13) were obtained from Gaussian error propagation (Ferrari and Polzin, 2005; Schafstall et al., 2010).

4.2 Lateral mixing

Lateral mixing is induced to first order by oceanic mesoscale activity, which predominantly acts along isopycnal surfaces. It effectively mixes oxygen in regions with strong isopycnal oxygen gradients, thus substantially contributing to the ventilation of an OMZ across its lateral boundaries (Luyten et al., 1983a; McCreary et al., 2013; Gnanadesikan et al., 2013; Hahn et al., 2014). For the ETNA OMZ, the eddy-driven meridional oxygen flux could be quantified along 23° W, both by a diffusive flux parameterisation and by eddy correlation (Hahn et al., 2014).

The diffusive flux parameterisation as the first method rests on the idea that the eddy-driven along-isopycnal oxygen flux can be expressed as a diffusive flux $F^d = K_e \nabla O_2$ which is down the mean oxygen gradient ∇O_2 with a horizontal eddy diffusivity K_e. Two independent estimates of the meridional component of K_e for the ETNA regime were derived. On the one hand, Banyte et al. (2013) analysed the lateral spreading of the tracer released at 330 m during GUTRE. On the other hand, Hahn et al. (2014) used hydrographic and velocity observations in the upper 1000 m from research vessels and moorings along 23° W during the last 15 years. Fundamentally, Hahn et al. (2014) based their analysis on the mixing length theory (as applied in Ferrari and Polzin, 2005) as well as on the theory of two-dimensional mesoscale turbulence on a β plane (Eden, 2007).

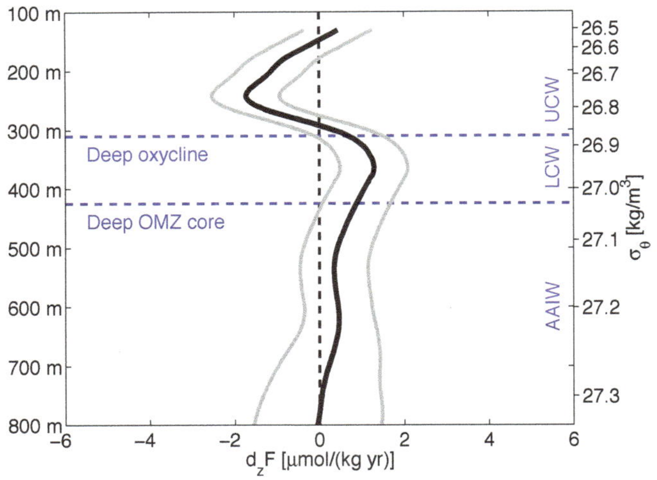

Figure 13. Mean oxygen supply due to diapycnal mixing (solid black line) for the open ocean ETNA OMZ and 95 % confidence error level (solid grey lines) as function of depth (left axis) or potential density (right axis). Blue dashed lines mark the depths of the deep oxycline and of the core of the deep OMZ that separate layers of upper and lower CW, and AAIW.

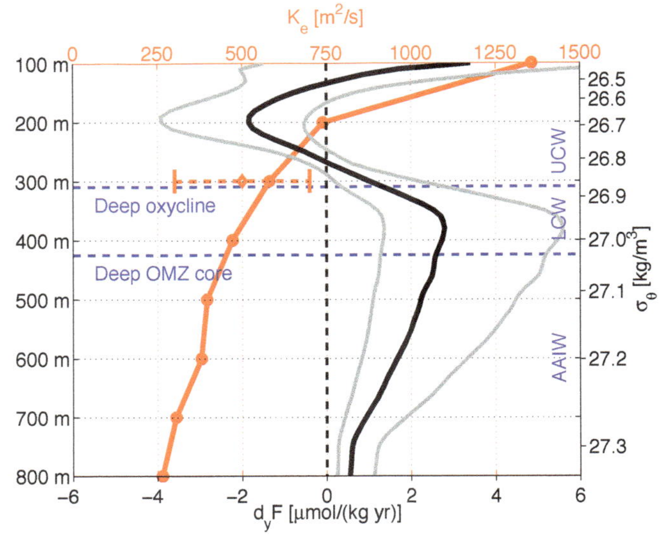

Figure 14. Eddy diffusivity as estimated from moored and shipboard observations (red circles, red line, upper axis) and from the tracer release experiment (red diamond with error bar, upper axis) as function of depth (left axis) or potential density (right axis). Also shown is the mean isopycnal meridional eddy-driven oxygen supply (black line, lower axis) for the open ocean ETNA OMZ with error levels (grey lines, lower axis) that were calculated from both the error of the curvature of the meridional oxygen distribution (95 % confidence) and the error of the eddy diffusivity (factor of 2 assumed). Blue dashed lines mark the depths of the deep oxycline and of the core of the deep OMZ that separate layers of upper and lower CW, and AAIW.

By comparing observed and simulated tracer distributions, Banyte et al. (2013) estimated a meridional eddy diffusivity of $500\,m^2\,s^{-1}$ at about 300 m with an uncertainty of $200\,m^2\,s^{-1}$ (Fig. 14). Hahn et al. (2014) estimated a meridional eddy diffusivity profile for the upper 1000 m with the range of uncertainty assumed as large as a factor of 2 following Ferrari and Polzin (2005). The profile shows maximum eddy diffusivity close to the surface and decreasing values with depth (Fig. 14). At 300 m, it yields $580\,m^2\,s^{-1}$, which is in good agreement with the estimate from Banyte et al. (2013). Together with the mean oxygen distribution, the obtained meridional eddy diffusivity was applied to derive the eddy-driven meridional oxygen flux along 23° W.

As a second method, the eddy correlation was used to calculate the eddy-driven meridional oxygen flux along isopycnal surfaces directly using mooring time series of oxygen and meridional velocity acquired at 5° N, 23° W and 8° N, 23° W in the years 2009–2012 and 2011–2012, respectively (see Hahn et al. (2014) for details). Although both estimation methods are accompanied by large uncertainties, a comparison of the results at the mooring sites reveals coherent profiles of the meridional oxygen flux (100–800 m). At the depth of the OMZ, they yield a northward oxygen flux towards the centre of the OMZ.

The oxygen that is meridionally supplied to the ETNA OMZ regime by lateral mixing can then be derived as the divergence of the eddy-driven meridional oxygen flux. The average profile of this eddy-driven meridional oxygen supply (6–14° N, 23° W) obtained using the diffusive flux parameterisation shows a substantial gain of oxygen at the depth of the OMZ and a loss of oxygen above (Fig. 14). The corresponding error was derived both from the error of the curvature of the meridional oxygen distribution (95 % confidence) and the error of the eddy diffusivity (factor 2 assumed following Ferrari and Polzin, 2005).

The tropical and subtropical oceans are generally assumed to be associated with an anisotropic horizontal eddy diffusivity (Banyte et al., 2013; Eden, 2007; Eden and Greatbatch, 2009; Kamenkovich et al., 2009) with larger horizontal eddy diffusivities in the zonal than in the meridional direction. Nevertheless, at the depth of the OMZ core, we consider the zonal eddy flux divergence small compared to the meridional eddy flux divergence, since the 2nd derivative of oxygen is an order of magnitude smaller in the zonal than the meridional direction.

4.3 Advection

We now turn to the remaining ventilation term in the budget, that is, the term associated with zonal advection (meridional advection is assumed to be negligible). We are only able to quantify this term as a residual. A rigorous determination of the advection term would require mean sections around a closed box to fulfil mass balance within the box. This cannot be achieved with the present observing system. However, our measurements along 23° W confirm that the advection term

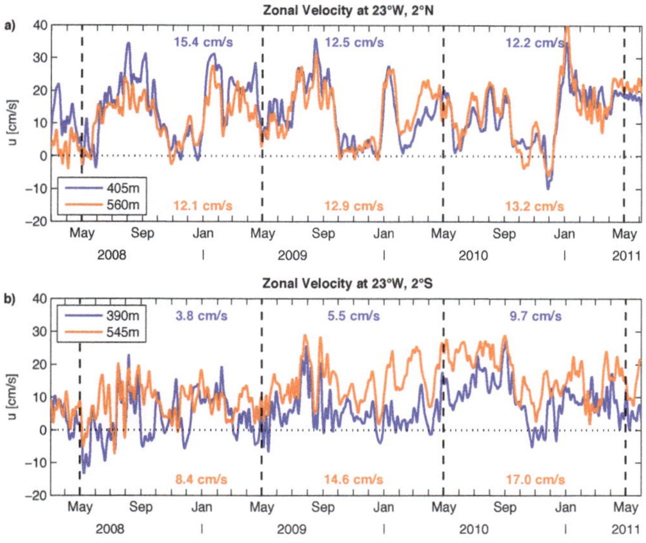

Figure 15. Zonal velocity from moored observations at 23° W, 2° N **(a)** and 23° W, 2° S **(b)** at about 400 m (blue lines) and 550 m (red lines). Blue and red numbers represent annual mean velocities at about 400 and 550 m in depth, respectively. Dashed vertical lines mark time periods used for the calculation of annual means; dotted horizontal line marks zero velocity.

is a major player in the ventilation of the OMZ, especially above 400 m in depth.

The key factor for carrying the relatively oxygen-rich waters eastwards from the western boundary is the presence of a series of latitudinally stacked zonal jets that are now known to be an ubiquitous feature of the tropical oceans (e.g. Maximenko et al., 2005; Qiu et al., 2013). Near the Equator in the Atlantic, these jets are confined below the Equatorial Undercurrent (EUC), but away from the Equator they extend to the surface, and at all latitudes they tend to have a strong depth-independent (barotropic) structure (Fig. 6). Brandt et al. (2010) suggested that a reduction in the strength of these jets north of the Equator was a factor in the recent reduction in oxygen within the OMZ. The influence these jets have on the meridional oxygen distribution can clearly be seen in Fig. 13 of Hahn et al. (2014) that shows the eddy-driven meridional oxygen supply. The red and blue alternating bands above 400 m in depth in that figure indicate a latitudinal redistribution of oxygen by mesoscale eddies, i.e. an oxygen gain for westward jets and an oxygen loss for eastward jets. It is clear from this figure that the zonal jets must play an important role in ventilating the upper 400 m of the water column in the latitude band (north of 6° N) of the OMZ. Looking at the equatorial region, the oxygen source (shown in blue) associated with the eastward jets, centred near 2° N and 2° S below 350 m in depth, can also be seen in Fig. 13 of Hahn et al. (2014). These so-called "flanking jets" are clearly an important ventilation pathway for maintaining the oxygen maximum at the Equator, as discussed further below.

The ventilation of the equatorial region, where there is a local oxygen maximum, has been studied by Brandt et al. (2012) using an advection–diffusion model. The role of the equatorial deep jets (EDJ; see Brandt et al., 2008, and Brandt et al., 2011) was also discussed in that paper. As can be seen from Fig. 6, the "flanking jets" are much stronger than the off-equatorial zonal jets noted earlier. Ascani et al. (2010) have suggested that these jets are maintained by Yanai waves, generated at the surface (possibly by instability of the energetic near-surface flow field forming tropical instability waves: von Schuckmann et al., 2008; Jochum et al., 2004), which break at depth. The jets show considerable variability (see Fig. 15) on monthly timescales, but are almost always unidirectional (especially in the Northern Hemisphere); their longitudinal dependence along the Equator is currently uncertain. The EDJ are also thought to be generated by downward propagating Yanai waves, in this case by barotropic instability of these waves as discussed by Hua et al. (2008) (see also d'Orgeville et al., 2007 and Ménesguen et al., 2009). The EDJ show downward phase propagation but upward energy propagation, consistent with the above theory, and lead to variability with a roughly 4.5 year period throughout the water column within 2 degrees latitude on either side of the Equator and above about 3000 m in depth (Brandt et al., 2011). As shown by Brandt et al. (2012), there is evidence of a corresponding 4.5 year variability in oxygen levels in the same region (Fig. 8, Equator), with variability at 300 m in depth at 23° W on the Equator having a range comparable to the range of the mean oxygen level along the Equator across the entire Atlantic.

5 Consumption

Oxygen consumption is a key mechanism for the formation of OMZs (Sverdrup, 1938; Wyrtki, 1962) and, although being a prominent part of the local oxygen budget of the OMZs, it is among the poorest constrained ones. We will consider here only the net consumption that is the combined effect of removal and production of oxygen. Removal of oxygen is related to the metabolism of marine life as well as to elementary chemical reactions, whereas production of oxygen is related to photosynthesis and as such is confined to the euphotic zone (e.g. Martz et al., 2008). We will focus in this section on pelagic oxygen consumption; removal of oxygen from the water column by uptake at the sediment–water interface will be discussed in Sect. 6.

Direct observations of oxygen in situ respiration are rare, primarily due to technical difficulties (e.g. Holtappels et al., 2014). The most commonly applied approach to quantifying time- and space-integrated oxygen removal and production processes is through an *apparent oxygen utilisation rate* (AOUR; e.g. Riley, 1951; Jenkins, 1982, 1998; Karstensen et al., 2008; Martz et al., 2008; Stanley et al., 2012). The AOUR is calculated as the ratio between the apparent oxygen utilisation (AOU) and age (τ). Hereby, AOU is determined as the difference between the air saturation value of dissolved oxygen (e.g. Weiss, 1970) at a given temperature and salinity (surface water saturation is commonly assumed to be 100 %) and the observed in situ oxygen concentration. The ageing of the water starts when a water parcel leaves the surface mixed layer ($\tau = 0$) and enters the ocean interior. As the ageing is closely linked to the ventilation process, the age is also called the ventilation age. In many cases the ventilation age is calculated from transient tracers (e.g. CFC-11, CFC-12, CCl_4, SF_6). Under the assumption of a given surface saturation (typically 100 %), the observed in situ tracer concentration is converted back to an equivalent atmospheric mixing ratio via its solubility function. The comparison with the respective tracer atmospheric time history finally yields the tracer age. For radioactive tracers (e.g. ^3He / ^3H, ^{39}Ar), a slightly different approach is used, but still requires the knowledge of surface concentrations or surface input functions (e.g. Roether et al., 2013; Lu et al., 2014). AOUR calculated using the tracer age follows an exponential decay with depth, at least for oceanic regions dominated by advection (Riley, 1951; Jenkins, 1982, 1998; Karstensen et al., 2008; Martz et al., 2008; Stanley et al., 2012). Such a structure supports the view that consumption is primarily a function of downward sinking particles (Martin et al., 1987).

The basic concept behind a ratio of along pathway oxygen removal and along pathway age increase assumes that the two quantities are consistently altered by ocean transport processes, e.g. it is assumed that as age increases so does AOU (Thiele and Sarmiento, 1990). While this seems to be a reasonable assumption for the ventilated gyre, it is questionable for the shadow zone where the OMZs are located. Here, diapycnal mixing (Fischer et al., 2013) and complex advection/lateral mixing patterns (Brandt et al., 2010; Hahn et al., 2014) have a strong influence on the local oxygen transport and water parcels from multiple source regions with different ventilation ages and along-path integrated oxygen consumption meet and mix.

Water mass composition and water ages can also be considered in a TTD approach (Haine and Hall, 2002), but limitations exist for non-steady state tracers (such as transient tracers). The TTD concept acknowledges the shortcomings in age calculations, which assign a single tracer age to a water parcel, and provides a framework to characterise the ventilation age more realistically (e.g. Waugh et al., 2004) by providing a mean age of the TTD. In a study using a transient tracer data set (up to 2009), Schneider et al. (2012) showed for the ETNA that the TTD obeys an inverse Gaussian function with the two moments Γ and Δ being equal ($\Delta / \Gamma = 1$), where Γ is the mean age and Δ defines the width of the TTD. In the limit of $\Delta / \Gamma = 0$, the mean age of the TTD equals the single tracer age.

Here an extended set of CFC-12, SF_6 and oxygen data collected in the ETNA OMZ is used to apply the TTD approach

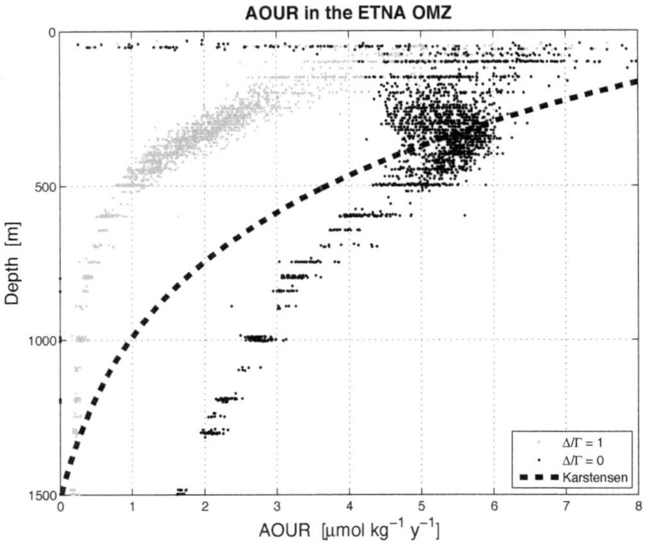

Figure 16. AOUR in the ETNA OMZ (between 4 and 14° N and east of 32° W). The AOUR was calculated using the TTD approach with two different assumptions about mixing: black dots corresponds to no mixing, $\Delta / \Gamma = 0$; grey dots to moderate mixing, $\Delta / \Gamma = 1$. The dashed line marks AOUR as obtained by Karstensen et al. (2008) using CFC-11 ages from the ventilated gyre.

for exploring the oxygen consumption rate. Using CFC-12 and SF_6 data (SF_6 preferentially used if available and CFC-12 if CFC-12 > 450 ppt, i.e. corresponding to atmospheric mixing ratios at about the end of the near-linear atmospheric increase) the AOUR is calculated using two different Δ / Γ ratios (Fig. 16). Note that the AOUR for $\Delta / \Gamma = 0$ is larger than values reported previously (Fig. 16) that were obtained by using a single tracer age concept applied to data collected in the ventilated gyre (e.g. Karstensen et al., 2008).

The two estimates for $\Delta / \Gamma = 0$ and $\Delta / \Gamma = 1$ represent an upper and lower limit of the AOUR within the ETNA OMZ, respectively. A shortcoming of the TTD concept in this region is its one-dimensionality (single water mass), i.e. it only considers the along-isopycnal mixing of parcels of a single source water mass, which might have encountered different advection and diffusion pathways and thus differ in age and AOU. The influence of diapycnal mixing (Fischer et al., 2013) and the mixing of two or more source waters (e.g. North and South Atlantic Central Water) (Kirchner et al., 2009; Brandt et al., 2010) is not considered by the TTD concept, which probably leads to a bias of the resulting AOUR. In fact our AOUR values for $\Delta / \Gamma = 1$ are lower than those calculated by Stanley et al. (2012) for the ventilated gyre region of the western North Atlantic close to Bermuda, where they used the TTD approach with $\Delta / \Gamma = 1$ on tritium (^3H) and ^3He measurements. They derived AOUR values close to $5\,\mu mol\,kg^{-1}\,yr^{-1}$ for the potential density level of $27.0\,kg\,m^{-3}$ that were similar to AOUR values obtained by Karstensen et al. (2008) using CFC-11 ages. For

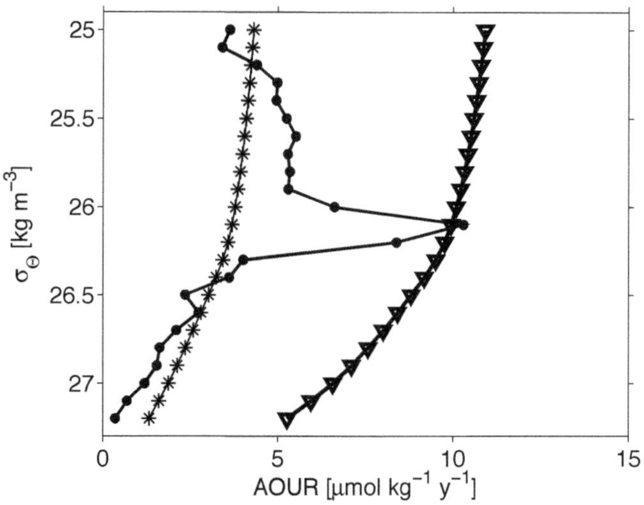

Figure 17. Three estimates of AOUR as a function of density: Schneider et al. (2012) used the TTD approach for the ETNA (stars), Karstensen et al. (2008) used CFC-11 water ages from the ventilated gyre only (triangles), and based on the ratio of North Atlantic mean AOU for isopycnal volumes and the corresponding reservoir ages (black dots, see further details, e.g. reservoir ages and volumes, in Karstensen et al., 2008; Figs. 9 and 10).

the same density level that is close to the OMZ core depth at roughly 400 m, we derived AOUR values of only about 1.5 μmol kg^{-1} yr^{-1} using the TTD approach with $\Delta / \Gamma = 1$. The main differences are that the waters off Bermuda are much better represented by a single water mass and that they are significantly younger with a TTD derived mean age of a few tens of years. Waters in the ETNA OMZ instead are a mixture of water masses from multiple sources, some of which might be rather old, resulting in a mean TTD age of 120–180 years (Fig. 7).

Another approach to estimating the large-scale AOUR is based on the reservoir age (Bolin and Rodhe, 1973), which is derived as the ratio of the total volume of the reservoir for an isopycnal range and the corresponding ventilating flux (that is, the subduction rate). The AOUR based on the reservoir age is then given by the ratio of the mean AOU of the isopycnal volume and the corresponding reservoir age. For the ETNA OMZ, the AOUR obtained using the reservoir ages of Karstensen et al. (2008) is for some density classes rather similar to the TTD approach with $\Delta / \Gamma = 0$, while for the well-ventilated isopycnal volumes (26.1 to 26.2 kg m^{-3}; see also Fig. 9 in Karstensen et al., 2008), it is closer to the AOUR from the tracer age approach (Fig. 17).

6 Processes at the continental margin

Processes contributing to the ventilation of OMZs at the continental margin are advective oxygen transport within the eastern boundary current system, upper-ocean diapycnal oxygen supply due to increased turbulent mixing on the continental slope and shelf, and eddy-driven isopycnal oxygen transport. In comparison to the open ocean OMZs, the consumption of oxygen at the continental margin is generally enhanced due to high pelagic primary production, which in turn results in an increased respiration associated with sinking particles in the water column and at the sediment–water interface. These processes are largely responsible for the regional oxygen distribution particularly defining the shape of the shallow OMZ in the ETNA. Along the eastern boundary, oxygen concentrations within the shallow OMZ decrease towards the north reaching a minimum at about 20° N (Fig. 4). For the deep OMZ, minimum oxygen levels at the continental margin are found south of 16° N (Machin and Pelegri, 2009).

6.1 Upwelling and circulation

The continental margin off Mauritania and Senegal is part of the Canary eastern boundary upwelling system that extends from the northern tip of the Iberian Peninsula at 43° N to south of Dakar at about 10° N (e.g. Mittelstaedt, 1991). Due to changes in wind forcing associated with the migration of the Intertropical Convergence Zone, coastal upwelling off Mauritania and Senegal exhibits a pronounced seasonality. Here winds favourable to upwelling prevail primarily from December to April. The seasonality in upwelling and associated primary production must be reflected in oxygen consumption and thus in water-column oxygen concentrations at the continental margin.

The ventilation of the waters above the continental margin occurs primarily through the Mauritania Current in the surface layer and the Poleward Undercurrent below. Both currents transport relatively oxygen-rich South Atlantic Central Water, which is supplied by the eastward flowing NECC and NEUC (Figs. 1, 4), northward into the upwelling region. Often, these two currents are not distinct from each other (e.g. Peña-Izquierdo et al., 2012). Usually, the Poleward Undercurrent is found attached to the continental slope between 50 and 300 m in depth, but it may extend as deep as 1000 m (Mittelstaedt, 1983; Barton, 1989; Hagen, 2001; Peña-Izquierdo et al., 2012). Average along-shore velocities from 18° N (Fig. 5) show a predominantly poleward flow in the upper 300 m over the continental slope of Mauritania exceeding 0.05 m s^{-1}. However, the effect of the eddy field and other variability on the mean flow is clearly not averaged out, due to the small number of available ship sections.

Previous studies showed that the Mauritania Current exhibits a seasonal behaviour (Mittelstaedt, 1991), which was found to be associated with the seasonality of the NECC, suggesting that the ventilation of the water masses above the continental margin also varies seasonally. In boreal winter and early boreal spring, when the NECC is weak, the Mauritania Current only reaches latitudes of about 14° N, while in boreal summer and early boreal autumn, due to the strengthening of the NECC and the relaxation of the northeast trade winds, the Mauritania Current reaches latitudes of about 20° N (Mittelstaedt, 1991; Stramma et al., 2008a). Besides the seasonal cycle, the flow variability off Mauritania and Senegal is influenced by intraseasonal coastal-trapped waves partly originating in the equatorial wave-guide (Polo et al., 2008). However, associated sea level anomalies are substantially weaker in the North Atlantic compared to the same latitude band in the South Atlantic. A strong influence of coastal-trapped waves on the oxygen distribution on the shelf of the ETNA as evidenced for the eastern boundary upwelling systems of the South Pacific and South Atlantic (Gutierrez et al., 2008; Monteiro et al., 2011) could so far not be shown.

Several studies have indicated that most of the water carried northward at the continental margin of Mauritania recirculates in the region off Cape Blanc at about 21° N within a cyclonic gyre (Mittelstaedt, 1983; Peña-Izquierdo et al., 2012). This circulation pattern is in agreement with the regional distribution of oxygen levels within the shallow oxygen minimum that exhibits the lowest oxygen concentrations at the continental margin and offshore just south of Cape Blanc (Peña-Izquierdo et al., 2012).

6.2 Benthic oxygen uptake

Oxygen uptake within the benthic region (i.e. the sediment and the immediately overlying water) is largely controlled by sediment oxygen consumption and can be a significant sink for oxygen from the water column above. In contrast to the difficulties of direct measurement of pelagic oxygen consumption, local measurements of sediment oxygen uptake are relatively straightforward for performing with a variety of techniques. Recent developments in measurement techniques include the use of benthic chambers, eddy-correlation techniques, multi-sensor microprofilers and benthic observatories (e.g. Glud, 2008). Total benthic oxygen uptake (TOU), which includes all processes consuming oxygen within the benthic region, is commonly measured by enclosure techniques such as benthic chambers. With these systems, the initial oxygen decrease of an overlying well-mixed water phase is approximately linear. TOU is then calculated based on the rate of oxygen decrease, accounting for the enclosed area and water volume. TOU rates have recently been measured in the upper 1000 m on the continental slope and shelf off Mauritania using benthic chambers attached to landers (Dale et al., 2014). The reported TOU rates that are quantified in terms of oxygen fluxes into the sediments were as high as 10 mmol $m^{-2} d^{-1}$ in depths between 50 and 100 m, and decreased quasi-exponentially to about 3 mmol $m^{-2} d^{-1}$ at a depth of 1000 m. To compare TOU rates to pelagic oxygen consumption, we have to apply the TOU to a water volume with a given in situ density: the consumption within a 1 m thick layer above the bottom due to TOU is 3 orders of magnitudes larger when compared to pelagic oxygen consumption occurring at similar depths. This is due to the volume-specific production and degradation of organic material in surface sediments, which supports high densities of microbes and metazoans (Glud, 2008). In shelf areas, it is estimated that 10 to 50 % of the pelagic primary production reaches the sediment (Canfield, 1993; Wollast, 1998), and benthic remineralisation plays a key role in this region for the recycling of nutrients and burial of carbon.

Although the benthic oxygen consumption due to TOU at the shelf greatly exceeds the pelagic oxygen consumption, benthic processes play a minor role for oxygen depletion within larger volumes such as that of the deep OMZ. To illustrate this, we assume that oxygen-depleted water masses are laterally exchanged between the shelf and the open ocean. Between 300 and 600 m in depth, the continental margin has a typical average topographic slope of about 4 %, corresponding to 25 m shelf width per 1 m depth change. Assuming a TOU of 5 mmol $m^{-2} d^{-1}$ results in an oxygen depletion by the sediments of 125 mmol d^{-1} per 1 m depth range and 1 m along-shelf distance. Using the range of pelagic oxygen consumption determined in Sect. 5 (1 to 5 μmol $kg^{-1} yr^{-1}$) and the corresponding in situ density, the equivalent water volume resulting in an oxygen depletion of 125 mmol d^{-1} would be 44×10^3 to 9×10^3 m^3, corresponding to distances from the shelf, where both processes have comparable influence, of 44 to 9 km. In other words, pelagic oxygen consumption within the deep OMZ, typically extending about 1000 km offshore, is 1 to 2 orders of magnitude larger than benthic oxygen consumption due to oxygen fluxes into the continental slope sediments. Reduced topographic slopes at shallower depths suggest a more important role of benthic oxygen uptake for the shallow OMZ, which is characterised by a minimum oxygen concentration close to the continental margin and is not as widespread as its deeper counterpart (cf. Figs. 3, 4).

6.3 Diapycnal oxygen fluxes at the continental margin

Diapycnal mixing on continental slopes and shelves is often found to be elevated due to tides interacting with topographic boundaries that accelerate an energy cascade from large-scale open ocean tides to small-scale turbulence (e.g. Sandstrom and Oakey, 1995). As shown by Schafstall et al. (2010), diapycnal mixing along the upper continental slope and lower shelf region off Mauritania is strongly elevated due to the presence of nonlinear internal waves that are boosted by the interaction of the barotropic tide with critically sloping topography (e.g. Holloway, 1985). Diapycnal nutrient fluxes calculated for the upwelling region are amongst the highest reported to date (Schafstall et al., 2010).

To assess the role of diapycnal mixing for ventilating the upper layer of the ocean above the continental slope, the diapycnal oxygen flux was calculated from 112 microstructure profiles collected over the continental slope between 500 and 100 m in water depth at 18° N along with CTD–O_2 profiles from two boreal winter cruises on the shelf of Mauritania (for details of the data set used, see Schafstall et al., 2010). Elevated mixing was found in a region with water depths shallower than 500 m (Schafstall et al., 2010). Within this region, the diapycnal flux of oxygen from the mixed layer into the stratified ocean is 73 mmol m^{-2} d^{-1}, with upper and lower 95 % confidence limits determined from Gaussian error propagation (Ferrari and Polzin, 2005; Schafstall et al., 2010) being 105 and 44 mmol m^{-2} d^{-1}, respectively. The diapycnal oxygen flux thus exceeds the benthic oxygen uptake by a factor of about 7. The diapycnal flux profile exponentially decays with depth, and the downward oxygen flux is reduced to less than 10 mmol m^{-2} d^{-1} at a depth of 60 m below the mixed layer, which has an average thickness of about 20 m. Diapycnal mixing is thus able to supply the oxygen fully that is required by the benthic oxygen uptake for water depths shallower than about 80 m. At about 150 m in depth, however, the diapycnal flux changes sign due to the presence of the shallow OMZ, and oxygen here is essentially fluxed upward, although at low rates. Thus, oxygen from the sea surface cannot contribute to ventilating the deeper water column via diapycnal mixing.

It should be noted that the diapycnal oxygen flux divergence from the mixed layer to 60 m below the mixed layer yields a diapycnal oxygen supply of about 400 μmol kg^{-1} yr^{-1}. In steady state other oxygen transport processes and consumption are required to balance this substantial oxygen supply. While vertical advection during the upwelling season might contribute to the balance, the oxygen supply due to other transport processes should be at least an order of magnitude lower in this region. The diapycnal oxygen supply to the upper thermocline can thus be used to define an upper limit of the oxygen consumption below the mixed layer. Such a consumption rate is, however, two orders of magnitude larger than the one estimated for the deep ocean as discussed above.

The results suggest that the high oxygen demand of the water column and the sediments within the upwelling region at shallow depths above the shallow OMZ may well be supplied from the surface via diapycnal mixing. At larger depths however, the continental slope must be ventilated via advective processes or isopycnal mixing. Nevertheless, although benthic oxygen uptake is an important local process decreasing oxygen levels in the bottom waters along the continental slope, it is negligible for the overall oxygen balance of the deep open ocean OMZ.

7 Long-term variability in ETNA OMZ

OMZs of the tropical oceans expanded and intensified during the last 50 years. Decreasing oxygen trends were found for the 300–700 m layers of selected regions, with the strongest decrease in the ETNA of -0.34 ± 0.13 μmol kg^{-1} yr^{-1} for the region 10–14° N, 20–30° W (Stramma et al., 2008b). The global analysis of observed changes in the oxygen content between 1960–1974 and 1990–2008 indicates a widespread and significant deoxygenation at about 200 m in depth in the tropical oceans (Stramma et al., 2010b). In the ETNA, this depth level corresponds to the intermediate oxygen maximum between the deep and shallow OMZs that is mainly ventilated by advection via zonal jets. A similar regional pattern of deoxygenation as for the 200 m level was found when vertically averaging oxygen changes over 200–700 m, albeit with a smaller amplitude (Stramma et al., 2010b).

One of the main questions regarding the observed oxygen trend is its possible relation to anthropogenic forcing that was suggested in a number of recent studies (Bopp et al., 2002; Keeling and Garcia, 2002; Plattner et al., 2002; Matear and Hirst, 2003; Oschlies et al., 2008; Schmittner et al., 2008; Frölicher et al., 2009; Keeling et al., 2010; Helm et al., 2011). Different mechanisms were suggested. Global warming results in decreasing oxygen solubility in surface waters, and due to the increasing upper ocean stratification, it might impact ocean circulation, subduction, and vertical mixing. Increased CO_2 levels and ocean acidification might impact biogeochemistry and oxygen consumption as well. However, up to now, current coupled climate–biogeochemistry models fail to reproduce the observed regional patterns of the oxygen trend, thus prohibiting a solid conclusion to be drawn about driving mechanisms of the observed ongoing deoxygenation (Stramma et al., 2012).

Similarly, there remain open questions as to how much of the observed oxygen changes are related to internal variability of the ocean and the climate system and what the dominant mechanisms are. The analysis of dissolved oxygen concentrations at 300 m in depth in the tropical and South Atlantic Ocean south of 20° N obtained from stations collected during the 1925–1927 *Meteor* expedition and the period 1990–2008 showed different and sometimes reversed trends compared to the mean oxygen trends found for the last 50 years, which indicates that the trend is not continuous, but that multidecadal variations are superimposed (Stramma et al., 2012). The oxygen trend along 23° W for the period 1972–2013 indicates a widespread oxygen decline, with the strongest oxygen reduction above the core of the deep OMZ and north of the Cape Verde archipelago (Fig. 18). However, oxygen anomalies within two boxes covering the region of relatively high oxygen above the deep oxycline (150–300 m) and the core region of the deep OMZ (350–700 m) (Fig. 19) show varying trends over the extended period (1900–2013) and the more recent period of enhanced measurements from 2006 to 2013. Note that the trend over the extended period is dominated by data taken during the 1970s, 1980s and the period 2006–2013. For the intermediate oxygen maximum (150–300 m), there is only a weak oxygen decline during the period 1900–2013 of -0.8 ± 0.5 μmol kg^{-1} $decade^{-1}$, while during the period 2006–2013, a much stronger decline of -14.3 ± 6.9 μmol kg^{-1} $decade^{-1}$ was observed. For the deep oxygen minimum (350–700 m), the long-term trend for the period 1900–2013 is -1.8 ± 0.3 μmol kg^{-1} $decade^{-1}$,

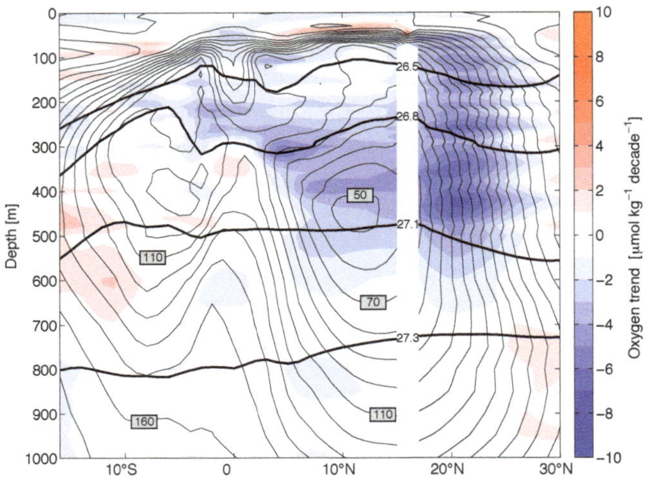

Figure 18. Oxygen trend along 23° W between 20 and 26° W and between 1972 and 2013 as obtained from the MIMOC climatology (Schmidtko et al., 2013). The trend was calculated on depth coordinates using oxygen anomalies relative to mean oxygen. Thin black contours mark mean oxygen (μmol kg^{-1}), thick black contours mark potential density (kg m^{-3}), both from the MIMOC climatology.

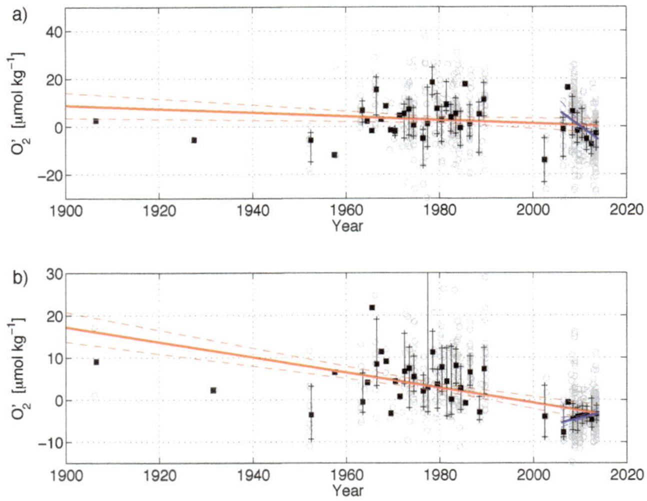

Figure 19. Oxygen anomalies for the region 9–15° N, 20–26° W and 150–300 m (intermediate oxygen maximum, upper panel) and 350–700 m (deep oxygen minimum, lower panel). Grey circles represent all available data, whiskers show interquartile range of data within each year and the black squares annual medians. Trends are calculated using annual medians weighted by the square root of available data within each year for the period 1900–2013 (solid red line) and 2006–2013 (solid blue line). The dashed lines mark the standard errors of the trends.

while during the period 2006–2013, oxygen increased by $2.7 \pm 1.9\,\mu$mol kg^{-1} decade^{-1}. These variations in the obtained trends that are related to different timescales and depth ranges may help to understand the underlying mechanism of long-term oxygen changes.

Different mechanisms might contribute to decadal to multidecadal oxygen variability: (1) Decadal to multidecadal AMOC changes would result in changes in the water mass distribution in the tropical North Atlantic, as identified for example in simulations with ocean–atmosphere general circulation models (Chang et al., 2008). Shifts in the boundaries between Northern and Southern Hemisphere water masses would likely affect oxygen distribution as well. (2) The transport of Indian Ocean CW toward the Atlantic via the Agulhas leakage might have increased during the last decades due to a poleward shift in the Southern Hemisphere westerlies. Such a change was observable in the NBUC as an increase in CW salinity (Biastoch et al., 2009) and might be associated with changes in the oxygen distribution as well. (3) Changes in the strength of latitudinally stacked zonal jets as derived by Brandt et al. (2010) result in changes in the advective pathways to the ETNA OMZ, with likely the strongest impact in the upper 300–400 m of the water column (Hahn et al., 2014). (4) Changes in the strength and location of the wind-driven gyres are a possible explanation for the long-term oxygen trends observed between 15° and 30° N in Fig. 18. (5) The variability of ventilation efficiency, either through dynamics (subduction) or saturation (warming), is able to produce oxygen anomalies that propagate into the ocean's interior (Karstensen et al., 2008).

In the North Atlantic, indications exist of a North Atlantic Oscillation (NAO) influence on multidecadal oxygen variations (Stendardo and Gruber, 2012). A similar influence of the NAO (e.g. due to associated changes in the northeast trade winds) on the water masses of the ETNA OMZ has not yet been shown. However, multidecadal changes in the strength of Atlantic STCs were detected in assimilation model runs. These changes include a minimum STC-layer (about 50–300 m) convergence in the early 1970s and a maximum in the early 1990s (Rabe et al., 2008), which would affect the supply of newly subducted oxygen-rich water masses from the subtropics to the tropics.

8 Similarities and differences between ETNA and ETSP OMZs

Similar to the hypoxic ETNA OMZ, the suboxic ETSP OMZ is located in the shadow zone equatorward of the subtropical gyre, with the lowest oxygen levels near the shelf break. The most prominent differences between the two OMZs are that the ETSP OMZ covers a much wider region and that oxygen values in its core region are close to zero (Karstensen et al., 2008), while the typical large-scale oxygen minimum in the ETNA only recently reached values slightly below $40\,\mu$mol kg^{-1} (Stramma et al., 2009). A continuation of the observed deoxygenation in the ETNA would turn the ETNA OMZ suboxic within a century; hence, it is worth looking at differences and similarities of the ETNA and the ETSP with regard to a possible shift of a hypoxic system to a suboxic system.

8.1 The large-scale distribution

Different from the ETNA with its Guinea Dome and the eastern tropical South Atlantic and eastern tropical North Pacific with similar domes, there is no dome in the ETSP (Kessler, 2006). Similar to the equatorial Atlantic, the equatorial Pacific is characterised by a local oxygen maximum and a system of eastward and westward currents (Figs. 2, 20). Near the Equator, the EUC, the NICC, and the SICC all carry water richer in oxygen than the adjacent westward flows (Stramma et al., 2010a). In the eastern Pacific, the Northern and Southern Subsurface countercurrents (NSCC and SSCC) are already low in oxygen and, different from the corresponding current bands in the Atlantic, do not provide oxygen-rich water to the OMZ. Near the Peruvian shelf, poleward and equatorward currents exist which supply equatorial and subtropical water to the eastern near-shelf regions (Fig. 2). The Chile–Peru Coastal Current (CPCC) and the Peru–Chile Current (PCC) flow equatorward in the near-surface layer close to the coast and farther than ~ 150 km from the coast, respectively, while the Peru–Chile Undercurrent (PCUC) flows poleward in subsurface layers along the outer continental shelf and inner slope (Chaigneau et al., 2013). Based on a hydrographic survey off Peru in January and February 2009 and in combination with float data and model results, Czeschel et al. (2011) prepared a schematic of the intermediate circulation of the ETSP and its link to the OMZ. The centre of the OMZ is a stagnant-flow area and the mean currents at 400 m in depth in the open ocean ETSP are weak. Along the $\sim 86°$ W section, the lowest oxygen is observed between 6° S and 10° S centred at about 400 m in depth and on the isopycnal $\sigma_\theta = 26.8\,\mathrm{kg\,m^{-3}}$. Along this isopycnal, the mean age is increased in the region of the low oxygen core, with a maximum mean age of about 300 yr at about 11° S, slightly poleward of the lowest oxygen concentration, and reduced near the Equator, with a mean age of about 200 yr (Fig. 20).

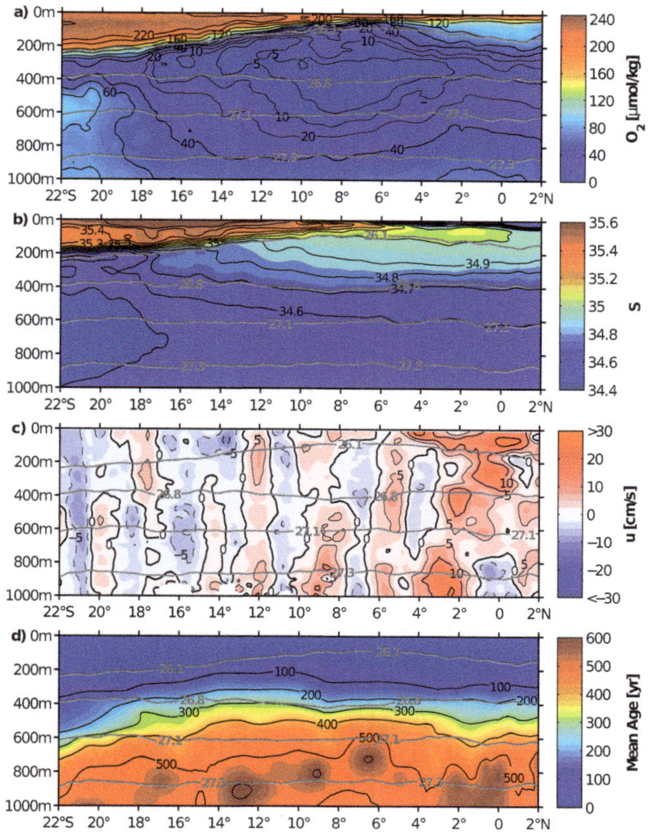

Figure 20. **(a)** Mean oxygen content, **(b)** salinity, **(c)** zonal velocity (positive eastward), and **(d)** mean age as obtained from meridional ship sections taken on three Pacific surveys along $\sim 86°$ W from 1993 to 2012. Grey contours mark potential density ($\mathrm{kg\,m^{-3}}$). The mean age is based solely on data from 1993. Eastward current bands, marked by reddish colours, are generally associated with elevated oxygen content.

8.2 Mesoscale processes

Mesoscale variability predominantly occurs as propagating Rossby waves and as nonlinear vortices or eddies. In particular, nonlinear vortices can trap and transport momentum, heat, mass and the chemical constituents of seawater, and therefore contribute to the large-scale water mass distribution (Chelton et al., 2007). Eddies are mainly generated by coastal flow instabilities that are influenced by remote equatorial forcing via coastal-trapped waves (Belmadani et al., 2012). They move westward from the coastal upwelling regions and hence carry shelf waters offshore. These eddies affect the regions' biogeochemical budgets, but also the primary productivity of the regions (Lachkar and Gruber, 2012), and seem to play an important role for the oxygen distribution on the poleward side of the OMZs. In global satellite observations of nonlinear mesoscale eddies by Chelton et al. (2011), it turned out that in the ETNA south and east of the Cape Verde Islands, almost no eddies with lifetimes of more than 16 weeks were present, while in the ETSP, a large number of such eddies could be identified. Their occurrence extends close to the Equator and the Peruvian shelf, as can be seen in Fig. 4a of Chelton et al. (2011). Despite the inferred weak eddy activity in the ETNA, water mass anomalies including local oxygen minima at shallow depths just below the mixed layer have been found in cyclonic as well as in anticyclonic mode water eddies in this region (see Fig. 4, showing a few profiles with oxygen concentrations below $40\,\mu\mathrm{mol\,kg^{-1}}$). In the ETSP, a region of high eddy production is located just off the shelf at 15–16° S, and strong eddies were described from a survey in November 2012. A strong anticyclonic mode water eddy located near the shelf of Peru at about 16° S showed a heat anomaly of 17.7×10^{18} J, a salt anomaly of 36.5×10^{10} kg (Stramma et al., 2013) and an oxygen anomaly of $-10.0 \times 10^{16}\,\mu$mol (Stramma et al., 2014). Even in a mooring at $\sim 20°$ S, 85° W, some 1500 km offshore, the passage of an anticyclonic mode water eddy carrying an oxygen anomaly of $-10.5 \times 10^{16}\,\mu$mol could be observed (Stramma et al., 2014). As eddies fall apart at the end of their lifetime, the anomalous hydrographic and biogeochemistry anomalies are redistributed in the ocean.

8.3 Oxygen budgets

A quantitative evaluation of the different terms of the oxygen budget of the tropical Pacific OMZ could not be performed so far. A rough estimate of the oxygen budget was instead given by Stramma et al. (2010a). They estimated the advective oxygen supply to the tropical Pacific OMZ from oxygen concentrations at, and zonal mass transport across, the 125° W meridian. The eastward mass transport associated with the EUC, SCCs and ICCs was estimated to be about 30×10^9 kg s^{-1}. It was assumed that this mass transport is returned by the adjacent westward currents with a typical relative oxygen difference between eastward and westward currents of about 20 μmol kg^{-1}. The resulting net advective molar oxygen supply across 125° W is 0.6×10^6 mol s^{-1} (Stramma et al., 2010a). The diffusive supply was estimated through the climatological 60 μmol kg^{-1} surface surrounding the tropical Pacific OMZ. Vertical and lateral oxygen gradients were evaluated at this surface and multiplied by a diapycnal diffusivity of 1×10^{-5} m^2 s^{-1} (Ledwell et al., 1998) and a horizontal eddy diffusivity of 500 m^2 s^{-1} characteristic of the off-equatorial regions (Davis, 2005), respectively. Integrating these products over the surface area resulted in a vertical diffusive molar oxygen supply of 0.4×10^6 mol s^{-1} mostly through the upper surface, where the gradients are large, and in a lateral diffusive molar oxygen supply of 0.8×10^6 mol s^{-1} (Stramma et al., 2010a). The mass of the tropical Pacific OMZs between 30° N and 30° S with oxygen concentrations lower than 60 μmol kg^{-1} is about 16×10^{18} kg. Dividing the estimates of molar supply by the mass leads to an advective oxygen supply of about 1.2 μmol kg^{-1} yr^{-1}, a lateral diffusive oxygen supply of 1.6 μmol kg^{-1} yr^{-1} and a vertical diffusive oxygen supply of 0.8 μmol kg^{-1} yr^{-1}. The oxygen utilisation rate calculated to balance the net oxygen supply resulted in about 3.6 μmol kg^{-1} yr^{-1}. These rough estimates of the oxygen budget are far from being a reliable result; however, it points to an allocation of about 33 % by advection, 45 % by eddy mixing and 22 % by vertical mixing. The calculation of the tropical Pacific oxygen budget differs from the calculation of the ETNA oxygen budget presented above: while advection along the Equator is included in the oxygen supply to the tropical Pacific OMZ, it is not in the ETNA OMZ. The budget of the ETNA OMZ included only the advective supply by zonal jets in the latitude range of the ETNA OMZ, while eddy mixing results in a meridional oxygen transport from the subtropical gyre in the north and the well-ventilated equatorial region in the south into the ETNA OMZ.

8.4 Trends in oxygen

As the ETSP OMZ is extremely low in oxygen, a decreasing trend is much more difficult to determine. Furthermore, data are sparse to investigate the trend. However, for the eastern Pacific equatorial region (5° S to 5° N, 105–115° W), a decrease of $0.13 \pm 0.32\,\mu$mol kg^{-1} yr^{-1} was described (Stramma et al., 2008b) for the 300–700 m depth layer for the last 50 years. The stronger decrease in oxygen in the ETNA compared to the ETSP is also visible from a global compilation of the trends of the last 50 years at 300 m in depth (Stramma et al., 2012).

On interannual to multidecadal timescales, oxygen variability in the ETSP is expected to be influenced by similar processes as those influencing the ETNA (see end of Sect. 7), albeit in response to the different large-scale climate modes that impact each ocean basin. In the Pacific, the multidecadal variability of the Pacific Decadal Oscillation (PDO) has the strongest influence on long timescales, while the El Niño–Southern Oscillation (ENSO), which mainly influences the upper 350 m of the ETSP, is superimposed on long-term changes (Czeschel et al., 2012). The variability of the Pacific STCs exhibits an ENSO signature with strong meridional transport occurring during La Niña and weak meridional transport during El Niño, and hence is a possible mechanism for oxygen variability (Zilberman et al., 2013). Model runs indicate a control of decadal and bidecadal climate variability in the tropical Pacific by the off-equatorial South Pacific Ocean triggered by changes in wind stress curl in the South Pacific extratropics (Tatebe et al., 2013) as an additional mechanism for oxygen variability. Besides decadal to multidecadal changes in the ventilation processes, variations in the oxygen consumption have been suggested to result in changes in the suboxic and hypoxic volumes of the tropical and subtropical Pacific on similar timescales (Deutsch et al., 2011; Ito and Deutsch, 2013). From three sediment cores along the North American margin, Deutsch et al. (2014) proposed that centennial changes in the North Pacific anoxia are linked to changes in tropical trade winds and their effect on upwelling and biological production.

9 Summary and discussion

The aim of the present paper is to provide a synthesis of the results from recent efforts to understand the physical mechanisms underpinning the functioning of the OMZs in the eastern tropical oceans, with a focus on the ETNA. The paper is mainly based on observations in the ETNA and the ETSP. The ETNA was selected to perform a dedicated observational programme consisting of a large number of research cruises, continuous moored observations, and TREs to better

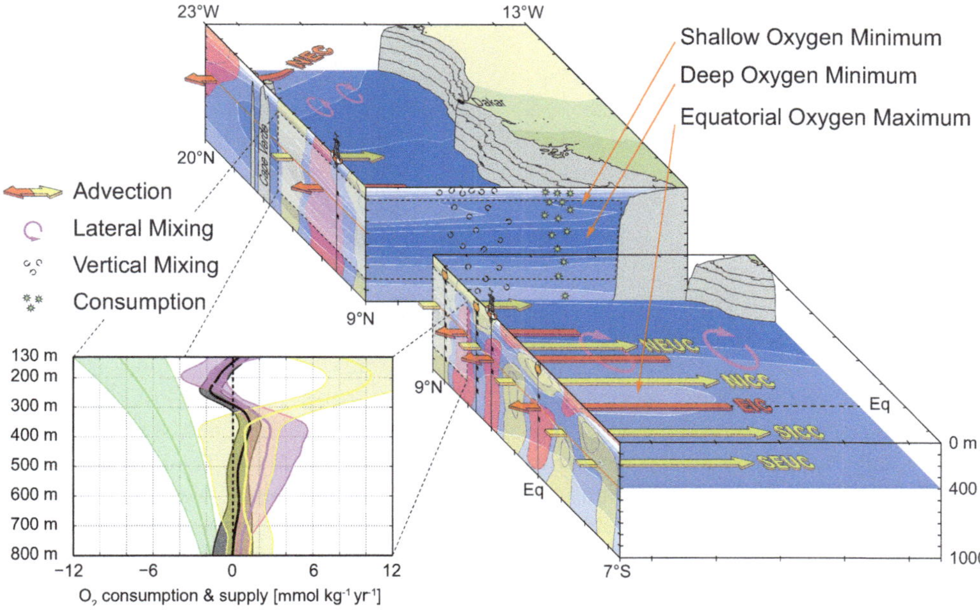

Figure 21. Schematic of the functioning of the ETNA OMZ and its oxygen budget. In the upper box, the oxygen distribution (bluish colours with dark/light blue corresponding to low/high oxygen) is shown at the sections along 23° W and 9° N and at a depth of 400 m; in the lower right box it is shown at the section along 23° W and at a depth of 400 m. Red and yellow areas at the 23° W section correspond to westward and eastward flow also marked by red and yellow arrows, respectively. The oxygen budget (lower left panel) includes physical supply by meridional (violet curve) and vertical mixing (black curve) as well as consumption after Karstensen et al. (2008) (green curve). The yellow curve is the residual of the other three terms, which is dominated by zonal advection. All error estimates (coloured shadings) are referred to a 95 % confidence (except the isopycnal meridional eddy supply, where the error was estimated from both the error of the oxygen curvature (95 % confidence) and the error of the eddy diffusivity (factor 2 assumed)) (see further details in the text and in Hahn et al., 2014).

understand the role of circulation and mixing in the ventilation of the OMZ. Results are summarised in the schematic Fig. 21. The ETSP was selected to allow a comparison of a hypoxic and suboxic OMZ. There are substantial differences in the dynamics of the thermocline, with a dominance of seasonal over interannual modes of variability in the Atlantic as opposed to interannual modes dominating in the Pacific as well as in the ocean–shelf interactions, which are possibly associated with different climate sensitivities of the OMZs in both oceans. However, a comparison of the factors of deoxygenation in the Atlantic and the Pacific might help to assess the possibility of a shift in the ETNA from hypoxic to suboxic.

One of the main results of the recent efforts is a first quantification of the oxygen budget of the deep ETNA OMZ (Brandt et al., 2010; Fischer et al., 2013; Hahn et al., 2014) that is extended here to 800 m in depth (Fig. 21). Integrating the different terms of the oxygen budget of the ETNA OMZ (Hahn et al., 2014) into the depth range below the deep oxycline from 350 to 570 m yields a consumption (after Karstensen et al., 2008) mainly balanced by the divergence of the meridional eddy flux (about 60 %) and the divergence of the diapycnal flux (20 %). The obtained residual of about 20 % can be ascribed in equal parts to the zonal advection and the long-term oxygen tendency as taken from Brandt et al. (2010). However, these are rough estimates. Most of the terms in the oxygen budget are associated with significant error, which particularly is the case for consumption and meridional eddy flux. Due to the TRE (Banyte et al., 2012) and repeated microstructure measurements (Fischer et al., 2013), the error in the diapycnal oxygen supply is comparatively small. The diapycnal oxygen supply is strongest slightly above the deep OMZ core, where it accounts for about one third of the oxygen supply required to balance consumption. There are, however, indications of regional variations in the diapycnal eddy diffusivity, with higher values over the seamount region (up to 1 order of magnitude) compared to the abyssal plains (Fig. 11), resulting also in a general increase in the diapycnal eddy diffusivity with depth (Fig. 12).

The contribution of the mean advection to the oxygen budget of the OMZ cannot be quantified from observational data. Instead, idealised advection–diffusion models were used to estimate this contribution (Brandt et al., 2010; Brandt et al., 2012). For these calculations, a basin-wide mean velocity field has to be prescribed based mainly on our knowledge of the mean flow along 23° W. However, the zonal extent of the zonal jets, their deviation from a purely zonal flow, and their connection to the well-ventilated western boundary regime are crucial in this calculation, but are not well constrained by observations, which leads again to a large uncertainty in the contribution of the mean advection to the oxygen budget of the ETNA OMZ.

Consumption as the main oxygen sink in the oxygen budget of the OMZ is currently best estimated as the net consumption along a water mass path from the subduction region toward the OMZ (Haine and Hall, 2002; Karstensen et al., 2008; Schneider et al., 2012). The different methods presented here yield a range of possible net consumption rates differing by a factor of 2–4 (Fig. 17). Besides this uncertainty, AOUR represents a large-scale net consumption rate that cannot account for the regional inhomogeneity in consumption, for example due to higher productivity in coastal, equatorial or open ocean upwelling regions compared to the oligotrophic ocean. For a local oxygen budget as presented here, the local oxygen consumption within the OMZ is required, which could substantially differ from values representing an integrated oxygen consumption along pathways from the subduction regions, through the oligotrophic ocean (often including the western boundary regime) into the OMZs. Additionally, the assumption of a consumption profile decreasing exponentially with depth (Martin et al., 1987) might be invalid. Lutz et al. (2002) noted the inability to fit sediment trap data to a single exponential function. Due to vertical changes in the liability of organic matter, sinking rate, and mineral ballast effect, they therefore suggested using the sum of two exponential functions with different decays. Processes that would also contribute to a deviation from a single exponential profile include respiration associated with the daily vertical migration cycle of zooplankton (Bianchi et al., 2013) or oxygen consumption at the sediment–ocean interface and associated lateral spreading of low-oxygen waters. To tackle the problem of regional and temporal consumption variability, new targeted data–model approaches are required, including observations of sinking particles or incubations for estimating pelagic oxygen consumption.

The relative importance of the different terms affecting the oxygen budgets of the ETNA and ETSP OMZs appear to be similar. For both OMZs, the eastward advection of oxygen-rich waters from the well-ventilated western boundary was found to be a dominant ventilation process. As the zonal currents are of similar strength in the tropical Pacific and Atlantic, the difference in the basin width of both oceans consequently results in lower oxygen concentrations and larger water mass ages in the eastern tropical Pacific (Fig. 20) compared to the eastern tropical Atlantic (Fig. 6).

Processes contributing to the oxygen budget at the eastern boundary include diapycnal mixing locally elevated due to tide–topography interaction, advective oxygen supply associated with (seasonally varying) eastern boundary circulation and coastal-trapped waves, mesoscale eddies favouring and redistributing oxygen anomalies, pelagic consumption and consumption at the sediment–ocean interface. Due to the high variability of most of these processes, both in space and time, the mean oxygen budget at the shelf is much less constrained compared to the open ocean. Often these processes are characterised by strong physical–biogeochemical interaction. For example, the downward oxygen flux from the mixed layer due to elevated diapycnal mixing at the shelf (Schafstall et al., 2010) must be balanced at least partly by local consumption. The extremely large vertical oxygen gradient at the shelf in the ETSP (from saturated oxygen levels in the mixed layer to zero oxygen within a few metres below) suggests extremely high consumption rates just below the mixed layer. Other examples are isolated eddies generated by the instability of the eastern boundary current. Such eddies transfer shelf water properties toward the open ocean while transforming these properties (particularly oxygen) by enhanced physical–biogeochemical interactions during their westward migration (Stramma et al., 2014). Their influence on the mean distribution of the shallow and deep OMZ could so far not be quantified. Dedicated process studies using mooring arrays, shipboard and multiple glider observations may help to elucidate the role of different processes in the eastern boundary oxygen budget.

The increase in the resolution of ocean circulation models improves the tropical circulation and associated oxygen distribution in the Atlantic (Duteil et al., 2014) and Pacific OMZs (Montes et al., 2014), suggesting that deficiencies in model physics largely contribute to the oxygen bias in coarser-resolution models. However, particularly the intermediate circulation (below 250 m) is still underestimated by these high-resolution simulations in realistic settings. To identify the physical mechanism responsible for the mean and variable zonal jets, idealised high-resolution models have been employed (Ménesguen et al., 2009; Ascani et al., 2010; Qiu et al., 2013). Such idealised models could furthermore be used, by including oxygen in the simulations, to study the roles of mean and variable advection in maintaining the tropical OMZs and to identify the mechanisms driving oxygen variability on interannual to multidecadal timescales.

The oxygen decline in the ETNA OMZ during the last decades corresponds to about 10 % of the oxygen sink due to consumption not balanced by ventilation processes. This is a substantial imbalance in the oxygen budget of the ETNA OMZ. The regional pattern along the 23° W section indicates the strongest oxygen reduction above the core of the deep OMZ and north of the Cape Verde archipelago (Fig. 18). Such a regional pattern is most likely due to changes in the circulation pattern associated with forced ocean dynamics as well as with internal ocean dynamics. Time series of all available oxygen data of the ETNA OMZ (Fig. 19) indicate variations on interannual, decadal, and multidecadal timescales; the long-term trend of deoxygenation associated with anthropogenic climate change might not be the dominant signal on such a regional scale. Improvements in model ventilation physics by increased resolution and/or improved parameterisations will reduce errors in the simulated mean oxygen distribution and its variability, but at the same time will help to understand the climate sensitivity of OMZs better with regard to anthropogenic climate change.

Oxygen data from shipboard repeat hydrography and moored observations show substantial interannual variability (Fig. 8) and trend-like changes (Fig. 19). The continuation of such measurements is essential to be able to test different hypotheses for the driving mechanisms of oxygen changes in the ocean. Using idealised or process models, distinct observed variability patterns might be reproduced and attributed to circulation changes and/or changes in the water mass distribution associated with the AMOC, STCs, PDO, or ENSO. For ocean circulation models, the acquired data provide the basis for improving the physical system in coupled climate–biogeochemistry simulations to make projections of future oxygen evolution more reliable.

Acknowledgements. This study was funded by the Deutsche Forschungsgemeinschaft as part of Sonderforschungsbereich 754 “Climate-Biogeochemistry Interactions in the Tropical Ocean”, through several research cruises with RV *Meteor*, RV *Maria S. Merian*, and RV *L'Atalante*, and by the Deutsche Bundesministerium für Bildung und Forschung (BMBF) as part of projects NORDATLANTIK (03F0605B, 03F0443B), RACE (03F0651B), SOPRAN (03F0462A, 03F0611A, 03F0662A), and AWA (01DG12073E). Moored velocity and oxygen observations were partly acquired in cooperation with the PIRATA project, and we would like to thank B. Bourlès, R. Lumpkin, C. Schmid, and G. Foltz for their help with mooring work and data sharing. We also thank J. Lübbecke, L. D. Bryant, and B. Dewitte for helpful discussions and comments on an earlier version of the manuscript. We thank the captains and crew of the RV *Maria S. Merian*, RV *Meteor*, RV *Poseidon*, and RV *L'Atalante* as well as our technical group for their help with the fieldwork.

Edited by: L. Levin

References

Ascani, F., Firing, E., Dutrieux, P., McCreary, J. P., and Ishida, A.: Deep equatorial ocean circulation induced by a forced-dissipated Yanai beam, J. Phys. Oceanogr., 40, 1118–1142, 2010.

Banyte, D., Tanhua, T., Visbeck, M., Wallace, D. W. R., Karstensen, J., Krahmann, G., Schneider, A., Stramma, L., and Dengler, M.: Diapycnal diffusivity at the upper boundary of the tropical North Atlantic oxygen minimum zone, J. Geophys. Res.-Oceans, 117, C09016, doi:10.1029/2011jc007762, 2012.

Banyte, D., Visbeck, M., Tanhua, T., Fischer, T., Krahmann, G., and Karstensen, J.: Lateral diffusivity from tracer release experiments in the tropical North Atlantic thermocline, J. Geophys. Res.-Oceans, 118, 2719–2733, 2013.

Barton, E. D.: The poleward undercurrent on the eastern boundary of the subtropical North Atlantic, in: Poleward flows along Eastern Ocean Boundaries, Lecture Note Series ed., edited by: Neshyba, S. J., Smith, R. L., and Mooers, C. N. K., Springer-Verlag, 82–95, 1989.

Belmadani, A., Echevin, V., Dewitte, B., and Colas, F.: Equatorially forced intraseasonal propagations along the Peru-Chile coast and their relation with the nearshore eddy activity in 1992–2000: A modeling study, J. Geophys. Res.-Oceans, 117, C04025, doi:10.1029/2011jc007848, 2012.

Bianchi, D., Galbraith, E. D., Carozza, D. A., Mislan, K. A. S., and Stock, C. A.: Intensification of open-ocean oxygen depletion by vertically migrating animals, Nat. Geosci., 6, 545–548, 2013.

Biastoch, A., Böning, C. W., Schwarzkopf, F. U., and Lutjeharms, J. R. E.: Increase in Agulhas leakage due to poleward shift of Southern Hemisphere westerlies, Nature, 462, 495–498, 2009.

Bolin, B. and Rodhe, H.: Note on Concepts of Age Distribution and Transit-Time in Natural Reservoirs, Tellus, 25, 58–62, 1973.

Bopp, L., Le Quere, C., Heimann, M., Manning, A. C., and Monfray, P.: Climate-induced oceanic oxygen fluxes: Implications for the contemporary carbon budget, Global Biogeochem. Cy., 16, doi:10.1029/2001gb001445, 2002.

Bourles, B., Lumpkin, R., McPhaden, M. J., Hernandez, F., Nobre, P., Campos, E., Yu, L. S., Planton, S., Busalacchi, A., Moura, A. D., Servain, J., and Trotte, J.: The PIRATA program: History, accomplishments, and future directions, B Am. Meteorol. Soc., 89, 1111–1125, 2008.

Brandt, P., Hormann, V., Bourles, B., Fischer, J., Schott, F. A., Stramma, L., and Dengler, M.: Oxygen tongues and zonal currents in the equatorial Atlantic, J. Geophys. Res.-Oceans, 113, C04012, doi:10.1029/2007jc004435, 2008.

Brandt, P., Hormann, V., Körtzinger, A., Visbeck, M., Krahmann, G., Stramma, L., Lumpkin, R., and Schmid, C.: Changes in the ventilation of the oxygen minimum zone of the tropical North Atlantic, J. Phys. Oceanogr., 40, 1784–1801, 2010.

Brandt, P., Funk, A., Hormann, V., Dengler, M., Greatbatch, R. J., and Toole, J. M.: Interannual atmospheric variability forced by the deep equatorial Atlantic Ocean, Nature, 473, 497–500, 2011.

Brandt, P., Greatbatch, R. J., Claus, M., Didwischus, S. H., Hormann, V., Funk, A., Hahn, J., Krahmann, G., Fischer, J., and Körtzinger, A.: Ventilation of the equatorial Atlantic by the equatorial deep jets, J. Geophys. Res.-Oceans, 117, C12015, 2012.

Canfield, D. E.: Organic Matter Oxidation in Marine Sediments, in: Interactions of C, N, P and S Biogeochemical Cycles and Global Change, edited by: Wollast, R., Mackenzie, F. T., and Chou, L., NATO ASI Series, Springer Berlin Heidelberg, 333–363, 1993.

Chaigneau, A., Dominguez, N., Eldin, G., Vasquez, L., Flores, R., Grados, C., and Echevin, V.: Near-coastal circulation in the Northern Humboldt Current System from shipboard ADCP data, J. Geophys. Res.-Oceans, 118, 5251–5266, 2013.

Chang, P., Zhang, R., Hazeleger, W., Wen, C., Wan, X. Q., Ji, L., Haarsma, R. J., Breugem, W. P., and Seidel, H.: Oceanic link between abrupt changes in the North Atlantic Ocean and the African monsoon, Nat. Geosci., 1, 444–448, 2008.

Chelton, D. B., Schlax, M. G., Samelson, R. M., and de Szoeke, R. A.: Global observations of large oceanic eddies, Geophys. Res. Lett., 34, L15606, doi:10.1029/2007gl030812, 2007.

Chelton, D. B., Schlax, M. G., and Samelson, R. M.: Global observations of nonlinear mesoscale eddies, Prog. Oceanogr., 91, 167–216, 2011.

Czeschel, R., Stramma, L., Schwarzkopf, F. U., Giese, B. S., Funk, A., and Karstensen, J.: Middepth circulation of the eastern tropical South Pacific and its link to the oxygen minimum zone, J. Geophys. Res.-Oceans, 116, C01015, doi:10.1029/2010jc006565, 2011.

Czeschel, R., Stramma, L., and Johnson, G. C.: Oxygen decreases and variability in the eastern equatorial Pacific, J. Geophys. Res.-Oceans, 117, C11019, doi:10.1029/2012jc008043, 2012.

d'Orgeville, M., Hua, B. L., and Sasaki, H.: Equatorial deep jets triggered by a large vertical scale variability within the western boundary layer, J. Mar. Res., 65, 1–25, 2007.

Dale, A. W., Sommer, S., Ryabenko, E., Noffke, A., Bohlen, L., Wallmann, K., Stolpovsky, K., Greinert, J., and Pfannkuche, O.: Benthic nitrogen fluxes and fractionation of nitrate in the Mauritanian oxygen minimum zone (Eastern Tropical North Atlantic), Geochim. Cosmochim. Ac., 134, 234–256, 2014.

Davis, R. E.: Intermediate-depth circulation of the Indian and South Pacific Oceans measured by autonomous floats, J. Phys. Oceanogr., 35, 683–707, 2005.

Deutsch, C., Brix, H., Ito, T., Frenzel, H., and Thompson, L.: Climate-Forced Variability of Ocean Hypoxia, Science, 333, 336–339, 2011.

Deutsch, C., Berelson, W., Thunell, R., Weber, T., Tems, C., McManus, J., Crusius, J., Ito, T., Baumgartner, T., Ferreira, V., Mey, J., and van Geen, A.: Centennial changes in North Pacific anoxia linked to tropical trade winds, Science, 345, 665–668, 2014.

Duteil, O., Schwarzkopf, F. U., Böning, C. W., and Oschlies, A.: Major role of the equatorial current system in setting oxygen levels in the eastern tropical Atlantic Ocean: A high- resolution model study, Geophys. Res. Lett., 41, 2033–2040, 2014.

Eden, C.: Eddy length scales in the North Atlantic Ocean, J. Geophys. Res.-Oceans, 112, C06004, doi:10.1029/2006jc003901, 2007.

Eden, C. and Greatbatch, R. J.: A diagnosis of isopycnal mixing by mesoscale eddies, Ocean Model, 27, 98–106, 2009.

Ferrari, R. and Polzin, K. L.: Finescale structure of the T-S relation in the eastern North Atlantic, J. Phys. Oceanogr., 35, 1437–1454, 2005.

Fischer, T., Banyte, D., Brandt, P., Dengler, M., Krahmann, G., Tanhua, T., and Visbeck, M.: Diapycnal oxygen supply to the tropical North Atlantic oxygen minimum zone, Biogeosciences, 10, 5079–5093, doi:10.5194/Bg-10-5079-2013, 2013.

Frölicher, T. L., Joos, F., Plattner, G. K., Steinacher, M., and Doney, S. C.: Natural variability and anthropogenic trends in oceanic oxygen in a coupled carbon cycle-climate model ensemble, Global Biogeochem. Cy., 23, GB1003, doi:10.1029/2008gb003316, 2009.

Glud, R. N.: Oxygen dynamics of marine sediments, Mar. Biol. Res., 4, 243–289, 2008.

Gnanadesikan, A., Bianchi, D., and Pradal, M. A.: Critical role for mesoscale eddy diffusion in supplying oxygen to hypoxic ocean waters, Geophys. Res. Lett., 40, 5194–5198, 2013.

Gregg, M. C., Sanford, T. B., and Winkel, D. P.: Reduced mixing from the breaking of internal waves in equatorial waters, Nature, 422, 513–515, 2003.

Gutierrez, D., Enriquez, E., Purca, S., Quipuzcoa, L., Marquina, R., Flores, G., and Graco, M.: Oxygenation episodes on the continental shelf of central Peru: Remote forcing and benthic ecosystem response, Prog. Oceanogr., 79, 177–189, 2008.

Hagen, E.: Northwest African upwelling scenario, Oceanol. Acta, 24, 113–128, 2001.

Hahn, J., Brandt, P., Greatbatch, R. J., Krahmann, G., and Körtzinger, A.: Oxygen variance and meridional oxygen supply in the Tropical North East Atlantic oxygen minimum zone, Clim. Dynam., 43, 2999–3024, 2014.

Haine, T. W. N. and Hall, T. M.: A generalized transport theory: Water-mass composition and age, J. Phys. Oceanogr., 32, 1932–1946, 2002.

Helm, K. P., Bindoff, N. L., and Church, J. A.: Observed decreases in oxygen content of the global ocean, Geophys. Res. Lett., 38, L23602, doi:10.1029/2011gl049513, 2011.

Holloway, P. E.: A comparison of semidiurnal internal tides from different bathymetric locations on the Australian North-West Shelf, J. Phys. Oceanogr., 15, 240–251, 1985.

Holtappels, M., Tiano, L., Kalvelage, T., Lavik, G., Revsbech, N. P., and Kuypers, M. M. M.: Aquatic Respiration Rate Measurements at Low Oxygen Concentrations, Plos One, 9, e89369, doi:10.1371/journal.pone.0089369, 2014.

Hua, B. L., d'Orgeville, M., Fruman, M. D., Menesguen, C., Schopp, R., Klein, P., and Sasaki, H.: Destabilization of mixed Rossby gravity waves and the formation of equatorial zonal jets, J. Fluid Mech., 610, 311–341, 2008.

Ito, T. and Deutsch, C.: Variability of the oxygen minimum zone in the tropical North Pacific during the late twentieth century, Global Biogeochem. Cy., 27, 1119–1128, 2013.

Jenkins, W. J.: Oxygen Utilization Rates in North-Atlantic Sub-Tropical Gyre and Primary Production in Oligotrophic Systems, Nature, 300, 246–248, 1982.

Jenkins, W. J.: Studying subtropical thermocline ventilation and circulation using tritium and He-3, J. Geophys. Res.-Oceans, 103, 15817–15831, 1998.

Jochum, M., Malanotte-Rizzoli, P., and Busalacchi, A.: Tropical instability waves in the Atlantic ocean, Ocean Model, 7, 145–163, 2004.

Kamenkovich, I., Berloff, P., and Pedlosky, J.: Anisotropic material transport by eddies and eddy-driven currents in a model of the North Atlantic, J. Phys. Oceanogr., 39, 3162–3175, 2009.

Karstensen, J., Stramma, L., and Visbeck, M.: Oxygen minimum zones in the eastern tropical Atlantic and Pacific oceans, Prog. Oceanogr., 77, 331–350, 2008.

Keeling, R. F. and Garcia, H. E.: The change in oceanic O-2 inventory associated with recent global warming, P. Natl. Acad. Sci. USA, 99, 7848–7853, 2002.

Keeling, R. F., Körtzinger, A., and Gruber, N.: Ocean deoxygenation in a warming world, Annu. Rev. Mar. Sci., 2, 199–229, 2010.

Kessler, W. S.: The circulation of the eastern tropical Pacific: A review, Prog. Oceanogr., 69, 181–217, 2006.

Kirchner, K., Rhein, M., Hüttl-Kabus, S., and Böning, C. W.: On the spreading of South Atlantic Water into the Northern Hemisphere, J. Geophys. Res.-Oceans, 114, C05019, doi:10.1029/2008JC005165, 2009.

Lachkar, Z. and Gruber, N.: A comparative study of biological production in eastern boundary upwelling systems using an artificial neural network, Biogeosciences, 9, 293–308, doi:10.5194/Bg-9-293-2012, 2012.

Ledwell, J. R., Watson, A. J., and Law, C. S.: Mixing of a tracer in the pycnocline, J. Geophys. Res.-Oceans, 103, 21499–21529, 1998.

Lu, Z. T., Schlosser, P., Smethie Jr, W. M., Sturchio, N. C., Fischer, T. P., Kennedy, B. M., Purtschert, R., Severinghaus, J. P.,

Solomon, D. K., Tanhua, T., and Yokochi, R.: Tracer applications of noble gas radionuclides in the geosciences, Earth-Sci. Rev., 138, 196–214, doi:10.1016/j.earscirev.2013.09.002, 2014.

Lutz, M., Dunbar, R., and Caldeira, K.: Regional variability in the vertical flux of particulate organic carbon in the ocean interior, Global Biogeochem. Cy., 16, 1037, doi:10.1029/2000gb001383, 2002.

Luyten, J., Pedlosky, J., and Stommel, H.: Climatic Inferences from the Ventilated Thermocline, Clim. Change, 5, 183–191, 1983a.

Luyten, J. R., Pedlosky, J., and Stommel, H.: The ventilated thermocline, J. Phys. Oceanogr., 13, 292–309, 1983b.

Machin, F. and Pelegri, J. L.: Northward penetration of Antarctic Intermediate Water off Northwest Africa, J. Phys. Oceanogr., 39, 512–535, 2009.

Malanotte-Rizzoli, P., Hedstrom, K., Arango, H., and Haidvogel, D. B.: Water mass pathways between the subtropical and tropical ocean in a climatological simulation of the North Atlantic ocean circulation, Dynam. Atmos. Oceans., 32, 331–371, 2000.

Martin, J. H., Knauer, G. A., Karl, D. M., and Broenkow, W. W.: Vertex – carbon cycling in the Northeast Pacific, Deep-Sea Res. Pt. A, 34, 267–285, 1987.

Martz, T. R., Johnson, K. S., and Riser, S. C.: Ocean metabolism observed with oxygen sensors on profiling floats in the South Pacific, Limnol. Oceanogr., 53, 2094–2111, 2008.

Matear, R. J. and Hirst, A. C.: Long-term changes in dissolved oxygen concentrations in the ocean caused by protracted global warming, Global Biogeochem. Cy., 17, 1125, doi:10.1029/2002gb001997, 2003.

Maximenko, N. A., Bang, B., and Sasaki, H.: Observational evidence of alternating zonal jets in the world ocean, Geophys. Res. Lett., 32, L12607, doi:10.1029/2005gl022728, 2005.

McCreary, J. P., Yu, Z. J., Hood, R. R., Vinaychandran, P. N., Furue, R., Ishida, A., and Richards, K. J.: Dynamics of the Indian-Ocean oxygen minimum zones, Prog. Oceanogr., 112, 15–37, 2013.

Ménesguen, C., Hua, B. L., Fruman, M. D., and Schopp, R.: Dynamics of the combined extra-equatorial and equatorial deep jets in the Atlantic, J. Mar. Res., 67, 323–346, 2009.

Mittelstaedt, E.: The upwelling area off Northwest Africa – a description of phenomena related to coastal upwelling, Prog. Oceanogr., 12, 307–331, 1983.

Mittelstaedt, E.: The ocean boundary along the Northwest African Coast – circulation and oceanographic properties at the sea-surface, Prog. Oceanogr., 26, 307–355, 1991.

Monteiro, P. M. S., Dewitte, B., Scranton, M. I., Paulmier, A., and van der Plas, A. K.: The role of open ocean boundary forcing on seasonal to decadal-scale variability and long-term change of natural shelf hypoxia, Environ. Res. Lett., 6, 025002, doi:10.1088/1748-9326/6/2/025002, 2011.

Montes, I., Dewitte, B., Gutknecht, E., Paulmier, A., Dadou, I., Oschlies, A., and Garcon, V.: High-resolution modeling of the Eastern Tropical Pacific oxygen minimum zone: Sensitivity to the tropical oceanic circulation, J. Geophys. Res.-Oceans, 119, 5515–5532, 2014.

Ollitrault, M., Lankhorst, M., Fratantoni, D., Richardson, P., and Zenk, W.: Zonal intermediate currents in the equatorial Atlantic Ocean, Geophys. Res. Lett., 33, L05605, doi:10.1029/2005gl025368, 2006.

Oschlies, A., Schulz, K. G., Riebesell, U., and Schmittner, A.: Simulated 21st century's increase in oceanic suboxia by CO_2-enhanced biotic carbon export, Global Biogeochem. Cy., 22, GB4008, doi:10.1029/2007gb003147, 2008.

Peña-Izquierdo, J., Pelegri, J. L., Pastor, M. V., Castellanos, P., Emelianov, M., Gasser, M., Salvador, J., and Vazquez-Dominguez, E.: The continental slope current system between Cape Verde and the Canary Islands, Sci. Mar., 76, 65–78, 2012.

Plattner, G. K., Joos, F., and Stocker, T. F.: Revision of the global carbon budget due to changing air-sea oxygen fluxes, Global Biogeochem. Cy., 16, 1096, doi:10.1029/2001gb001746, 2002.

Polo, I., Lazar, A., Rodriguez-Fonseca, B., and Arnault, S.: Oceanic Kelvin waves and tropical Atlantic intraseasonal variability: 1. Kelvin wave characterization, J. Geophys. Res.-Oceans, 113, C07009, doi:10.1029/2007jc004495, 2008.

Qiu, B., Chen, S. M., and Sasaki, H.: Generation of the North Equatorial Undercurrent jets by triad baroclinic Rossby wave interactions, J. Phys. Oceanogr., 43, 2682–2698, 2013.

Rabe, B., Schott, F. A., and Kohl, A.: Mean circulation and variability of the tropical Atlantic during 1952–2001 in the GECCO assimilation fields, J. Phys. Oceanogr., 38, 177–192, 2008.

Riley, G. A.: Oxygen, phosphate, and nitrate in the Atlantic Ocean, Bull. Bingham. Oceanogr. Coll., 12, 1–126, 1951.

Roether, W., Jean-Baptiste, P., Fourre, E., and Sultenfuss, J.: The transient distributions of nuclear weapon-generated tritium and its decay product He-3 in the Mediterranean Sea, 1952–2011, and their oceanographic potential, Ocean Sci., 9, 837–854, 2013, http://www.ocean-sci.net/9/837/2013/.

Sandstrom, H. and Oakey, N. S.: Dissipation in internal tides and solitary waves, J. Phys. Oceanogr., 25, 604–614, 1995.

Schafstall, J., Dengler, M., Brandt, P., and Bange, H.: Tidal-induced mixing and diapycnal nutrient fluxes in the Mauritanian upwelling region, J. Geophys. Res.-Oceans, 115, C10014, doi:10.1029/2009jc005940, 2010.

Schmidtko, S., Johnson, G. C., and Lyman, J. M.: MIMOC: A global monthly isopycnal upper-ocean climatology with mixed layers, J. Geophys. Res.-Oceans, 118, 1658–1672, 2013.

Schmittner, A., Oschlies, A., Matthews, H. D., and Galbraith, E. D.: Future changes in climate, ocean circulation, ecosystems, and biogeochemical cycling simulated for a business-as-usual CO_2 emission scenario until year 4000 AD, Global Biogeochem. Cy., 22, GB1013, doi:10.1029/2007gb002953, 2008.

Schneider, A., Tanhua, T., Körtzinger, A., and Wallace, D. W. R.: An evaluation of tracer fields and anthropogenic carbon in the equatorial and the tropical North Atlantic, Deep-Sea Res. Pt. I, 67, 85–97, 2012.

Schott, F. A., Stramma, L., and Fischer, J.: The warm water inflow into the western tropical Atlantic boundary regime, spring 1994, J. Geophys. Res.-Oceans, 100, 24745–24760, 1995.

Schott, F. A., Fischer, J., and Stramma, L.: Transports and pathways of the upper-layer circulation in the western tropical Atlantic, J. Phys. Oceanogr., 28, 1904–1928, 1998.

Schott, F. A., McCreary, J. P., and Johnson, G. C.: Shallow overturning circulations of the tropical-subtropical oceans, in: Earth Climate: The Ocean-Atmosphere Interaction, edited by: Wang, C., Xie, S.-P., and Carton, J. A., Geophysical Monograph 147, American Geophysical Union, Washington, DC, 261–304, 2004.

Schott, F. A., Dengler, M., Zantopp, R., Stramma, L., Fischer, J., and Brandt, P.: The shallow and deep western boundary circulation of the South Atlantic at 5°–11° S, J. Phys. Oceanogr., 35, 2031–2053, 2005.

St. Laurent, L. and Schmitt, R. W.: The contribution of salt fingers to vertical mixing in the North Atlantic Tracer Release Experiment, J. Phys. Oceanogr., 29, 1404–1424, 1999.

Stanley, R. H. R., Doney, S. C., Jenkins, W. J., and Lott, D. E.: Apparent oxygen utilization rates calculated from tritium and helium-3 profiles at the Bermuda Atlantic Time-series Study site, Biogeosciences, 9, 1969–1983, doi:10.5194/Bg-9-1969-2012, 2012.

Stendardo, I. and Gruber, N.: Oxygen trends over five decades in the North Atlantic, J. Geophys. Res.-Oceans, 117, C11004, doi:10.1029/2012jc007909, 2012.

Stramma, L. and England, M. H.: On the water masses and mean circulation of the South Atlantic Ocean, J. Geophys. Res.-Oceans, 104, 20863–20883, 1999.

Stramma, L., Brandt, P., Schafstall, J., Schott, F., Fischer, J., and Körtzinger, A.: Oxygen minimum zone in the North Atlantic south and east of the Cape Verde Islands, J. Geophys. Res.-Oceans, 113, C04014, doi:10.1029/2007jc004369, 2008a.

Stramma, L., Johnson, G. C., Sprintall, J., and Mohrholz, V.: Expanding oxygen-minimum zones in the tropical oceans, Science, 320, 655–658, 2008b.

Stramma, L., Visbeck, M., Brandt, P., Tanhua, T., and Wallace, D.: Deoxygenation in the oxygen minimum zone of the eastern tropical North Atlantic, Geophys. Res. Lett., 36, L20607, doi:10.1029/2009gl039593, 2009.

Stramma, L., Johnson, G. C., Firing, E., and Schmidtko, S.: Eastern Pacific oxygen minimum zones: Supply paths and multidecadal changes, J. Geophys. Res.-Oceans, 115, C09011, doi:10.1029/2009jc005976, 2010a.

Stramma, L., Schmidtko, S., Levin, L. A., and Johnson, G. C.: Ocean oxygen minima expansions and their biological impacts, Deep-Sea Res. Pt. I, 57, 587–595, 2010b.

Stramma, L., Oschlies, A., and Schmidtko, S.: Mismatch between observed and modeled trends in dissolved upper-ocean oxygen over the last 50 yr, Biogeosciences, 9, 4045–4057, doi:10.5194/Bg-9-4045-2012, 2012.

Stramma, L., Bange, H. W., Czeschel, R., Lorenzo, A., and Frank, M.: On the role of mesoscale eddies for the biological productivity and biogeochemistry in the eastern tropical Pacific Ocean off Peru, Biogeosciences, 10, 7293–7306, doi:10.5194/Bg-10-7293-2013, 2013.

Stramma, L., Weller, R. A., Czeschel, R., and Bigorre, S.: Eddies and an extreme water mass anomaly observed in the eastern south Pacific at the Stratus mooring, J. Geophys. Res. Oceans, 119, 1068–1083, 2014.

Suga, T. and Talley, L. D.: Antarctic intermediate water circulation in the tropical and subtropical South-Atlantic, J. Geophys. Res.-Oceans, 100, 13441–13453, 1995.

Sverdrup, H. U.: On the Explanation of the Oxygen Minima and Maxima in the Oceans, Journal du Conseil, 13, 163–172, 1938.

Tatebe, H., Imada, Y., Mori, M., Kimoto, M., and Hasumi, H.: Control of Decadal and Bidecadal Climate Variability in the Tropical Pacific by the Off-Equatorial South Pacific Ocean, J. Climate, 26, 6524–6534, 2013.

Thiele, G. and Sarmiento, J. L.: Tracer Dating and Ocean Ventilation, J. Geophys. Res.-Oceans, 95, 9377–9391, 1990.

Tsuchiya, M.: Thermostats and circulation in the upper layer of the Atlantic Ocean, Prog. Oceanogr., 16, 235–267, 1986.

Tsuchiya, M., Talley, L. D., and Mccartney, M. S.: An eastern Atlantic section from Iceland southward across the equator, Deep-Sea Res. Pt. A, 39, 1885–1917, 1992.

von Schuckmann, K., Brandt, P., and Eden, C.: Generation of tropical instability waves in the Atlantic Ocean, J. Geophys. Res.-Oceans, 113, C08034, doi:10.1029/2007jc004712, 2008.

Wattenberg, H.: Die Verteilung des Sauerstoffs im Atlantischen Ozean, Wissenschaftliche Ergebnisse der Deutschen Atlantischen Expedition auf dem Forschungs- und Vermessungsschiff Meteor 1925–1927, 9.1, edited by: Defant, A., de Gruyter, Berlin, Leipzig, 132 pp., 1938.

Waugh, D. W., Haine, T. W. N., and Hall, T. M.: Transport times and anthropogenic carbon in the subpolar North Atlantic Ocean, Deep-Sea Res. Pt. I, 51, 1475–1491, 2004.

Weiss, R. F.: Solubility of Nitrogen, Oxygen and Argon in Water and Seawater, Deep-Sea Res., 17, 721–735, 1970.

Wollast, R.: Evaluation and comparison of the global carbon cycle in the coastal zone and in the open ocean, in: The Sea, edited by: Robinson, A., Brink, K. H., Wiley, New York, 213–252, 1998.

Wüst, G.: Die Stratosphäre des Atlantischen Ozeans, Deutsche Atlantische Exped, Meteor 1925–1927, Wiss. Erg., 6, 288 pp., 1935.

Wyrtki, K.: The oxygen minima in relation to ocean circulation, Deep-Sea Res., 9, 11–23,1962.

Zhang, D. X., McPhaden, M. J., and Johns, W. E.: Observational evidence for flow between the subtropical and tropical Atlantic: The Atlantic subtropical cells, J. Phys. Oceanogr., 33, 1783–1797, 2003.

Zilberman, N. V., Roemmich, D. H., and Gille, S. T.: The Mean and the Time Variability of the Shallow Meridional Overturning Circulation in the Tropical South Pacific Ocean, J. Climate, 26, 4069–4087, 2013.

Preface: Towards a full greenhouse gas balance of the biosphere

L. Merbold[1], G. Wohlfahrt[2,3], K. Butterbach-Bahl[4,5], K. Pilegaard[6], T. DelSontro[7,8], P. Stoy[9], and D. Zona[10,11]

[1]Institute of Agricultural Sciences, ETH Zurich, 8092 Zurich, Switzerland
[2]Institute of Ecology, University of Innsbruck, 6020 Innsbruck, Austria
[3]European Academy of Bolzano, 39100 Bolzano, Italy
[4]Karlsruhe Institute of Technology, Institute for Meteorology and Climate Research, Atmospheric Environmental Research (IMK-IFU), 82467 Garmisch-Partenkirchen, Germany
[5]International Livestock Research Institute, Old Naivasha Road, Nairobi 00100, Kenya
[6]Center for Ecosystems and Sustainability, Department of Chemical and Biochemical Engineering, Technical University of Denmark, 4000 Roskilde, Denmark
[7]Institute of Biogeochemistry and Pollutant Dynamics, ETH Zurich, 8092 Zurich, Switzerland
[8]Department of Biological Sciences, University of Quebec at Montreal, Montreal, QC, H2J 2K9, Canada
[9]Department of Land Resources and Environmental Sciences, Montana State University, Bozeman, MT 59717-3120, USA
[10]Department of Animal and Plant Science, University of Sheffield, Western Bank, Sheffield, S10 2TN, UK
[11]Department of Biology, San Diego State University, 5500 Campanile Drive, San Diego, CA 92182, USA

Correspondence to: G. Wohlfahrt (georg.wohlfahrt@uibk.ac.at)

Ecosystem greenhouse gas (GHG) emissions (CO_2, CH_4, and N_2O) represent a major driver of global environmental change (IPCC, 2014). While there exists an emerging understanding on the net exchange of CO_2 across terrestrial and aquatic ecosystems due in part to the existence of large measurement and modeling networks (Baldocchi et al., 2001; Friend et al., 2007; Raymond et al., 2013; Tranvik et al., 2009), similar information on the biosphere–atmosphere exchange of non-CO_2 greenhouse gases (i.e., CH_4 and N_2O) is sparsely available in comparison. To date, a strong focus has been given to so-called high-emission ecosystems, such as wetlands, rivers and lakes, rice cultivations and ruminants for CH_4 (Nisbet et al., 2014) and agricultural ecosystems for N_2O emissions (Butterbach-Bahl et al., 2013; Sutton et al., 2007). Even though CO_2, CH_4, and N_2O emissions have been increasing during the last century, a combined quantification of the exchange of these three major greenhouse gases from a wide range of terrestrial and aquatic ecosystems is still missing. Therefore, approaches to develop full greenhouse gas monitoring networks, as currently undertaken in larger environmental research infrastructures such as ICOS (Integrated Carbon Observation System) and NEON (National Ecological Observation Network), are highly valuable.

The recently finished COST Action ES0804 – Advancing the integrated monitoring of trace gas exchange Between Biosphere and Atmosphere (ABBA) – evaluated the current state of greenhouse gas flux monitoring stations in Europe as a first step and initiated multi-flux monitoring sites in a second stage. In a third step, an initiative on scaling data collected at the site level to larger scales was undertaken. The fourth goal of the project was to train future researchers on experimental flux measurements as well as to introduce the next generation of researchers to modeling approaches. Under the auspices of Working Group 2 of ABBA, which focused on multi-flux monitoring sites, a full day session during the European Geosciences Union (EGU) General Assembly was organized in both 2012 and 2013 to exchange knowledge and discuss advantages and limitations of ecosystem-scale CH_4 and N_2O exchange observations. The session further focused on bridging the terrestrial and the aquatic research communities, who often act independently despite having similar research questions. In a synthesis sub-session, researchers discussed how to bring the knowledge gained at plot or ecosystem level to larger regional and continental scales via modeling approaches. Both sessions resulted in this Special Issue with which we aim to answer the following three questions:

1. What is the magnitude of N_2O and CH_4 emissions in different ecosystems and what are the respective driving forces?

2. What are the difficulties in measuring and modeling their exchange rates and how do these rates relate to ecosystem C and N turnover?

3. How do non-CO_2 greenhouse gases (CH_4 and N_2O) contribute to the total global warming potential of terrestrial and aquatic ecosystems?

The manuscripts included in this Special Issue investigate GHG emissions from a variety of aquatic and terrestrial ecosystems, ranging from freshwater rivers (Maeck et al., 2014) and lakes (Sturm et al., 2014) to grasslands (Hörtnagl and Wohlfahrt, 2014, Imer et al., 2013, Merbold et al., 2013), crops (Savage et al., 2014), forest ecosystems (Peichl et al., 2014; Wang et al., 2013), and peatlands (Tupek et al., 2015). The climate zones covered ranged from the boreal region (Hanis et al., 2013) via the temperate zone (Imer et al., 2013) to the tropics (Jamali et al., 2013; Werner et al., 2014). Multi-site studies, which comprised approximately half of the Special Issue contributions, compared different ecosystem types (Luo et al., 2013), investigated effects of gradients in elevation (Teh et al., 2014) and soil moisture (Tupek et al., 2015), ecosystem management (Imer et al., 2013, Schrier-Uijl et al., 2014), and forest age (Peichl et al., 2014). Hiller et al. (2014) scaled up anthropogenic and natural CH_4 emissions to the country scale, and Tian et al. (2013) and Kim and Kirschbaum (2014) conducted global analyses. Around one third of the papers covered all three major greenhouse gases (CO_2, CH_4 and N_2O), while most studies discussed at least two of them. The first major result of these papers is that the contribution of the CH_4 and N_2O exchange to the total GHG budget varied considerably, from small and intermediate (Merbold et al., 2013; Peichl et al., 2014) to up to 50 % (Hörtnagl and Wohlfahrt, 2014). Thus, this Special Issue drives the existing knowledge base substantially towards a more comprehensive quantification of the total ecosystem greenhouse gas balance.

A mix of experimental methods was employed amongst the contributed studies. Traditionally, chambers were the method of choice for flux measurements, and were also employed in many of the contributing manuscripts (Imer et al., 2013; Podgrajsek et al., 2014). Chamber measurements, however, are now increasingly complemented with or replaced by eddy-covariance flux measurements (Podgrajsek et al., 2014; Wang et al., 2013), which are facilitated by the technical advance of fast-response gas analyzers (Peltola et al., 2014; Savage et al., 2014). Challenges in measuring reliable fluxes with the various methods were discussed, in particular regarding CH_4 and N_2O and the spatiotemporal heterogeneity of their exchange (Maeck et al., 2014; Podgrajsek et al., 2014; Savage et al., 2014; Werner et al., 2014). A great example of this is the role of CH_4 ebullition from aquatic ecosystems, which are not only difficult to accurately quantify but also tend to dominate total emissions from aquatic systems where ebullition is common (Maeck et al., 2014; Sturm et al., 2014). This highlights the second major insight from this Special Issue, the importance of hot spots and hot moments that need to be considered when designing chamber flux sampling schemes and interpreting eddy-covariance flux data (Hörtnagl and Wohlfahrt, 2014; Imer et al., 2013; Savage et al., 2014).

Finally, as the third major result the papers in this Special Issue show that N_2O fluxes tend to be explained poorly by micrometeorological and hydrological variables (Luo et al., 2013; Merbold et al., 2013), suggesting that substrate availability and/or microbial community dynamics may play important roles. In contrast, abiotic drivers, and in particular soil water content, exhibited considerably more skill in predicting the sign and magnitude of CH_4 exchange (Tupek et al., 2015; Wang et al., 2013).

Thereby, this Special Issue emphasizes the necessity of concerted collaboration between different disciplines and the research communities for developing more comprehensive approaches. Ultimately, continued exchanges similar to those demonstrated herein will lead to a better understanding of the feedbacks that biosphere–atmosphere greenhouse gas exchange exerts on climate.

Acknowledgements. We gratefully acknowledge the financial support of the European Cooperation in Science and Technology (COST) for the COST Action ES0804 – Advancing the integrated monitoring of trace gas exchange Between Biosphere and Atmosphere (ABBA).

References

Baldocchi, D., Falge, E., Gu, L. H., Olson, R., Hollinger, D., Running, S., Anthoni, P., Bernhofer, C., Davis, K., Evans, R., Fuentes, J., Goldstein, A., Katul, G., Law, B., Lee, X. H., Malhi, Y., Meyers, T., Munger, W., Oechel, W., U, K. T. P., Pilegaard, K., and Schmid, H. P., Valentini, R., Verma, S., Vesala, T., Wilson, K., and Wofsy, S.: FLUXNET: A new tool to study the temporal and spatial variability of ecosystem-scale carbon dioxide, water vapor, and energy flux densities, B. Am. Meteorol. Soc., 82, 2415–2434, 2001.

Butterbach-Bahl, K., Baggs, E. M., Dannenmann, M., Kiese, R., and Zechmeister-Boltenstern, S.: Nitrous oxide emissions from soils: how well do we understand the processes and their controls?, Philos. Trans. Roy. Soc. B, 368, 20130122, doi:10.1098/rstb.2013.0122, 2013.

Friend, A. D., Arneth, A., Kiang, N. Y., Lomas, M., Ogee, J., Roedenbeckk, C., Running, S. W., Santaren, J.-D., Sitch, S., Viovy, N., Woodward, F. I., and Zaehle, S.: FLUXNET and modelling the global carbon cycle, Glob. Change Biol., 13, 610–633, 2007.

Hanis, K. L., Tenuta, M., Amiro, B. D., and Papakyriakou, T. N.: Seasonal dynamics of methane emissions from a subarctic fen in the Hudson Bay Lowlands, Biogeosciences, 10, 4465–4479, doi:10.5194/bg-10-4465-2013, 2013.

Hiller, R. V., Bretscher, D., DelSontro, T., Diem, T., Eugster, W., Henneberger, R., Hobi, S., Hodson, E., Imer, D., Kreuzer, M., Künzle, T., Merbold, L., Niklaus, P. A., Rihm, B., Schellenberger, A., Schroth, M. H., Schubert, C. J., Siegrist, H., Stieger, J., Buchmann, N., and Brunner, D.: Anthropogenic and natural methane fluxes in Switzerland synthesized within a spatially explicit inventory, Biogeosciences, 11, 1941–1959, doi:10.5194/bg-11-1941-2014, 2014.

Hörtnagl, L. and Wohlfahrt, G.: Methane and nitrous oxide exchange over a managed hay meadow, Biogeosciences, 11, 7219–7236, doi:10.5194/bg-11-7219-2014, 2014.

Imer, D., Merbold, L., Eugster, W., and Buchmann, N.: Temporal and spatial variations of soil CO_2, CH_4 and N_2O fluxes at three differently managed grasslands, Biogeosciences, 10, 5931–5945, doi:10.5194/bg-10-5931-2013, 2013.

IPCC: Summary for Policymakers, in: Climate Change 2014: Impacts, Adaptation, and Vulnerability, Part A: Global and Sectoral Aspects. Contribution of Working Group II to the Fifth Assessment Report of the Intergovernmental Panel on Climate Change, edited by: Field, C. B., Barros, V. R., Dokken, D. J., Mach, K. J., Mastrandrea, M. D., Bilir, T. E., Chatterjee, M., Ebi, K. L., Estrada, Y. O., Genova, R. C., Girma, B., Kissel, E. S., Levy, A. N., Maccracken, S., Mastrandrea, P. R., White, L. L., Cambridge, United Kingdom, and New York, NY, USA, Cambridge University Press, 1–32, 2014.

Jamali, H., Livesley, S. J., Hutley, L. B., Fest, B., and Arndt, S. K.: The relationships between termite mound CH_4/CO_2 emissions and internal concentration ratios are species specific, Biogeosciences, 10, 2229–2240, doi:10.5194/bg-10-2229-2013, 2013.

Kim, D.-G. and Kirschbaum, M. U. F.: The effect of land-use change on the net exchange rates of greenhouse gases: a meta-analytical approach, Biogeosciences Discuss., 11, 1053–1095, doi:10.5194/bgd-11-1053-2014, 2014.

Luo, G. J., Kiese, R., Wolf, B., and Butterbach-Bahl, K.: Effects of soil temperature and moisture on methane uptake and nitrous oxide emissions across three different ecosystem types, Biogeosciences, 10, 3205–3219, doi:10.5194/bg-10-3205-2013, 2013.

Maeck, A., Hofmann, H., and Lorke, A.: Pumping methane out of aquatic sediments – ebullition forcing mechanisms in an impounded river, Biogeosciences, 11, 2925–2938, doi:10.5194/bg-11-2925-2014, 2014.

Merbold, L., Steinlin, C., and Hagedorn, F.: Winter greenhouse gas fluxes (CO_2, CH_4 and N_2O) from a subalpine grassland, Biogeosciences, 10, 3185–3203, doi:10.5194/bg-10-3185-2013, 2013.

Nisbet, E. G., Dlugokencky, E. J., and Bousquet, P.: Methane on the Rise-Again, Science, 343, 493–495, 2014.

Peichl, M., Arain, A. M., Moore, T. R., Brodeur, J. J., Khomik, M., Ullah, S., Restrepo-Coupé, N., McLaren, J., and Pejam, M. R.: Carbon and greenhouse gas balances in an age sequence of temperate pine plantations, Biogeosciences, 11, 5399–5410, doi:10.5194/bg-11-5399-2014, 2014.

Peltola, O., Hensen, A., Helfter, C., Belelli Marchesini, L., Bosveld, F. C., van den Bulk, W. C. M., Elbers, J. A., Haapanala, S., Holst, J., Laurila, T., Lindroth, A., Nemitz, E., Röckmann, T., Vermeulen, A. T., and Mammarella, I.: Evaluating the performance of commonly used gas analysers for methane eddy covariance flux measurements: the InGOS inter-comparison field experiment, Biogeosciences, 11, 3163–3186, doi:10.5194/bg-11-3163-2014, 2014.

Podgrajsek, E., Sahlée, E., Bastviken, D., Holst, J., Lindroth, A., Tranvik, L., and Rutgersson, A.: Comparison of floating chamber and eddy covariance measurements of lake greenhouse gas fluxes, Biogeosciences, 11, 4225–4233, doi:10.5194/bg-11-4225-2014, 2014.

Raymond, P. A., Hartmann, J., Lauerwald, R. et al.: Global carbon dioxide emissions from inland waters, Nature, 503, 355–359, 2013.

Savage, K., Phillips, R., and Davidson, E.: High temporal frequency measurements of greenhouse gas emissions from soils, Biogeosciences, 11, 2709–2720, doi:10.5194/bg-11-2709-2014, 2014.

Schrier-Uijl, A. P., Kroon, P. S., Hendriks, D. M. D., Hensen, A., Van Huissteden, J., Berendse, F., and Veenendaal, E. M.: Agricultural peatlands: towards a greenhouse gas sink – a synthesis of a Dutch landscape study, Biogeosciences, 11, 4559–4576, doi:10.5194/bg-11-4559-2014, 2014.

Sturm, K., Yuan, Z., Gibbes, B., Werner, U., and Grinham, A.: Methane and nitrous oxide sources and emissions in a subtropical freshwater reservoir, South East Queensland, Australia, Biogeosciences, 11, 5245–5258, doi:10.5194/bg-11-5245-2014, 2014.

Sutton, M. A., Nemitz, E., Erisman, J. W., Beier, C., Bahl, K. B., Cellier, P., de Vries, W., Cotrufo, F., Skiba, U., Di Marco, C., Jones, S., Laville, P., Soussana, J. F., Loubet, B., Twigg, M., Famulari, D., Whitehead, J., Gallagher, M. W., Neftel, A., Flechard, C. R., Herrmann, B., Calanca, P. L., Schjoerring, J. K., Daemmgen, U., Horvath, L., Tang, Y. S., Emmett, B. A., Tietema, A., Penuelas, J., Kesik, M., Brueggemann, N., Pilegaard, K., Vesala, T., Campbell, C. L., Olesen, J. E., Dragosits, U., Theobald, M. R., Levy, P., Mobbs, D. C., Milne, R., Viovy, N., Vuichard, N., Smith, J. U., Smith, P., Bergamaschi, P., Fowler, D., and Reis, S.: Challenges in quantifying biosphere-atmosphere exchange of nitrogen species, Environ. Pollut., 150, 125–139, 2007.

Teh, Y. A., Diem, T., Jones, S., Huaraca Quispe, L. P., Baggs, E., Morley, N., Richards, M., Smith, P., and Meir, P.: Methane and nitrous oxide fluxes across an elevation gradient in the tropical Peruvian Andes, Biogeosciences, 11, 2325–2339, doi:10.5194/bg-11-2325-2014, 2014.

Tian, H., Chen, G., Lu, C., Xu, X., Ren, W., Banger, K., Zhang, B., Tao, B., Pan, S., Liu, M., and Zhang, C.: Global land-atmosphere exchange of methane and nitrous oxide: magnitude and spatiotemporal patterns, Biogeosciences Discuss., 10, 19811–19865, doi:10.5194/bgd-10-19811-2013, 2013.

Tranvik, L. J., Downing, J. A., Cotner, J. B., Loiselle, S. A., Striegl, R. G., Ballatore, T. J., Dillon, P., Finlay, K., Fortino, K., Knoll, L. B., Kortelainen, P. L., Kutser, T., Larsen, S., Laurion, I., Leech, D. M., McCallister, S. L., McKnight, D. M., Melack, J. M., Overholt, E., Porter, J. A., Prairie, Y., Renwick, W. H., Roland, F., Sherman, B. S., Schindler, D. W., Sobek, S., Tremblay, A., Vanni, M. J., Verschoor, A. M., von Wachenfeldt, E., and Weyhenmeyer, G. A.: Lakes and reservoirs as regulators of carbon cycling and climate, Limnol. Oceanogr., 54, 2298–2314, 2009.

Tupek, B., Minkkinen, K., Pumpanen, J., Vesala, T., and Nikinmaa, E.: CH_4 and N_2O dynamics in the boreal forest–mire ecotone, Biogeosciences, 12, 281–297, doi:10.5194/bg-12-281-2015, 2015.

Wang, J. M., Murphy, J. G., Geddes, J. A., Winsborough, C. L., Basiliko, N., and Thomas, S. C.: Methane fluxes measured by eddy covariance and static chamber techniques at a temperate forest in central Ontario, Canada, Biogeosciences, 10, 4371–4382, doi:10.5194/bg-10-4371-2013, 2013.

Werner, C., Reiser, K., Dannenmann, M., Hutley, L. B., Jacobeit, J., and Butterbach-Bahl, K.: N_2O, NO, N_2 and CO_2 emissions from tropical savanna and grassland of northern Australia: an incubation experiment with intact soil cores, Biogeosciences, 11, 6047–6065, doi:10.5194/bg-11-6047-2014, 2014.

8

Components of near-surface energy balance derived from satellite soundings – Noontime net available energy

K. Mallick[1], A. Jarvis[2], G. Wohlfahrt[3,*], G. Kiely[4], T. Hirano[5], A. Miyata[6], S. Yamamoto[7], and L. Hoffmann[1]

[1]Environmental Research and Innovation (ERIN), Luxembourg Institute of Science and Technology (LIST), L4422, Belvaux, Luxembourg
[2]Lancaster Environment Centre, Lancaster University, Lancaster LA1 4YQ, UK
[3]Institute of Ecology, University of Innsbruck, 6020 Innsbruck, Austria
[4]Hydrometeorology Research Group, Department of Civil and Environmental Engineering, University College Cork, Ireland
[5]Division of Environmental Resources, Research Faculty of Agriculture, Hokkaido University, Hokkaido, Japan
[6]National Institute for Agro-Environmental Sciences, Tsukuba, Japan
[7]Graduate School of Environmental Science, Okayama University Tsushimanaka3-1-1, Okayama 700-8530, Japan
[*]now at: European Academy of Bolzano, 39100 Bolzano, Italy

Correspondence to: K. Mallick (kaniska.mallick@gmail.com)

Abstract. This paper introduces a relatively simple method for recovering global fields of monthly midday (13:30 LT) near-surface net available energy (the sum of the sensible and latent heat flux or the difference between the net radiation and surface heat accumulation) using satellite visible and infrared products derived from the AIRS (Atmospheric Infrared Sounder) and MODIS (MODerate Resolution Imaging Spectroradiometer) platforms. The method focuses on first specifying net surface radiation by considering its various shortwave and longwave components. This was then used in a surface energy balance equation in conjunction with satellite day–night surface temperature difference to derive 12 h discrete time estimates of surface system heat capacity and heat accumulation, leading directly to retrieval for surface net available energy. Both net radiation and net available energy estimates were evaluated against ground truth data taken from 30 terrestrial tower sites affiliated with the FLUXNET network covering 7 different biome classes. This revealed a relatively good agreement between the satellite and tower data, with a pooled root-mean-square deviation of 98 and 72 W m^{-2} for monthly 13:30 LT net radiation and net available energy, respectively, although both quantities were underestimated by approximately 25 and 10 %, respectively, relative to the tower observation. Analysis of the individual shortwave and longwave components of the net radiation revealed the downwelling shortwave radiation to be main source of this systematic underestimation.

1 Introduction

An important manifestation of climate change is widespread alteration of the composition of the energy balance at the Earth's surface (Trenberth et al., 2009; Wild et al., 2013). Given the importance of being able to predict the consequences of climate change, both measurement and modelling of the components of surface energy balance attract significant attention from a broad range of related scientific disciplines (Stephens et al., 2012). Two such disciplines are hydrology and meteorology, which share a common interest in resolving the balance between sensible, H, and latent, λE, heat fluxes over a broad range of spatial and temporal scales (Anderson et al., 2012).

Net available energy, Φ, is a core variable used to predict the magnitude of H and λE given that it is defined as the sum of these two fluxes (Wright et al., 1992; Migletta et al., 2009; Anderson et al., 2012),

$$\Phi = \lambda E + H. \tag{1}$$

The utility of this definition arises from being able to also specify Φ as the difference between the net broadband radiation, R_N, and the rate of heat accumulation, G, below the plain across which R_N is specified,

$$\Phi = R_N - G. \tag{2}$$

Given that R_N is routinely measured using net radiometers, this affords an opportunity to specify Φ and hence either H or λE. For example, in modelling studies λE is invariably specified as a function of Φ using the ubiquitous equations such as those of Penman (1948) for open water or Monteith (1965) for land surfaces (Mu et al., 2011; Mallick et al., 2014a). Despite being the rate of change of heat stock in terrestrial environments, G is often interpreted as the "ground heat flux", and attempts to measure this using heat flux plates are commonplace (Mayocchi and Bristow, 1995; Sauer and Horton, 2005; Heitman et al., 2010). These measurements prove somewhat less reliable than R_N due to greater spatial heterogeneity in ground heat uptake (Gao et al., 1998; Tittebrand and Berger, 2009; Verhoef et al., 2012) allied to the fact that significant heat capacity resides in other elements of the land surface (Ochsner et al., 2007). As a result, G proves problematic in surface energy balance studies and is either ignored (Foken et al., 2006; Foken, 2008) or treated somewhat superficially (Choudhury, 1987), despite being significant under a broad range of conditions (Santanello and Friedl, 2003; Ochsner et al., 2007). Large-scale estimates of G are useful in the context of regional and global evapotranspiration modelling and for verification of regional and global circulation models (Kergoat et al., 2011).

The arrival of satellite retrievals for many of the components of R_N has opened up opportunities to develop large-scale estimates of this variable and hence λE (Batra et al., 2006; Mu et al., 2007, Anderson et al., 2012). For example, retrievals for the components of R_N have been available through the International Satellite Cloud Climatology Project (ISCCP) (Pinker and Laszlo, 1992; Stephens et al., 2012), the Earth Radiation Budget Experiment (ERBE) (Priestley et al., 2011), and the Clouds and the Earth's Radiant Energy System (CERES) (Mlynczak et al., 2011; Chen et al., 2013) onboard NASA's Earth Observing System (EOS) and Tropical Rainfall Measuring Mission (TRMM) satellites (Wielicki et al., 1998). Several studies have reported the estimation of R_N using a combination of MODIS (MODerate Resolution Imaging Spectroradiometer) atmospheric and land products over the USA, China and India (Cai et al., 2007; Mallick et al., 2009; Bisht and Bras, 2010, 2011) or NOAA-14 (National Oceanic and Atmospheric Administration) data over the Tibetan Plateau (Ma et al., 2002).

Unfortunately, in the absence of direct observations of G at spatial scales and coverage of satellite R_N, retrievals for Φ have had to rely on the parameterisation of G using surface temperature, albedo and vegetation index information (Bastiaanssen et al., 1998; Batra et al., 2006) or by assigning some fixed proportion of R_N (Choudhury, 1987; Humes et al., 1994) in satellite-based surface energy balance models (Mecikalski et al., 1999; Anderson et al., 2012). However studies by Murray and Verhoef (2007), Hsieh et al. (2009) and recently Verhoef et al. (2012) also demonstrated that G is, by definition, a highly dynamic quantity, and that the ratio G / R_N can range anywhere from 0.05 to 0.50 depending on the time of day, soil moisture and thermal properties, and vegetation density. Therefore, methods that are able to provide defensible estimates of G in conjunction with R_N would clearly be of great benefit to this area for determining Φ directly from satellite data and without relying unduly on any offline calibration. In this paper we present a method for retrieving R_N and Φ based on exploring both satellite radiance data and day–night surface temperature difference. The approach is necessarily simple in order to avoid over-reliance on models in the pre-processing and to reflect the fact that the focus of this work is the production of satellite estimates of monthly midday (13:30; all times listed are in local time, LT) Φ for use in a simple Bowen ratio (Bowen, 1926) λE specification framework as detailed in a companion paper by Mallick et al. (2014b) (we refer to this as M2 hereafter). Taking advantage of the extensive network of terrestrial eddy covariance tower sites (Baldocchi et al., 2001) which record direct measurements of R_N, H and λE, we use these measurements to derive independent non-radiative estimates of Φ in order to critically evaluate our satellite estimates of this quantity.

The method we present here for estimating R_N contrasts with more sophisticated model-based approaches which attempt to accommodate the complexity of atmospheric radiative transfer explicitly (e.g. Fouquart and Bonnel, 1980; Mlawer et al., 1997; Bisht and Bras, 2010, 2011; Hou et al., 2014). There are several reasons for adopting this stance. Firstly, the estimates of R_N need to be in agreement with the simple dynamic energy balance used to accommodate G when specifying Φ. Secondly, we believe it to be important that the complexity of the methods used here is commensurate with those methods used in the simple Bowen ratio approach as described in M2. Related to this, we have tried to restrict the approach to largely using only AIRS (Atmospheric InfraRed Sounder) data which provide the satellite soundings required for the Bowen ratio estimates. This single-platform approach is adopted to ensure the estimates do not suffer unduly from blending different data sources. Finally, as is the case with this method, complex radiative transfer approaches are also prone to the effects of uncertainty (Betts et al., 1993; Morcrette, 2002; Seidel et al., 2010), and therefore the parsimony implicit in the methods used here may be seen as advantageous.

2 Methodology

2.1 Satellite data sets

In the present study, two different data sources were used for the estimation of R_N and Φ, AIRS and MODIS. The AIRS sounder is carried by NASA's Aqua satellite, which was launched into a Sun-synchronous low Earth orbit on 4 May 2002 as part of NASA's Earth Observing System. It gives near-global coverage twice daily at 01:30–13:30 LT from an altitude of 705 km. Level 3 standard monthly day–night data products of air temperature and relative humidity profiles, cloud cover fraction, surface emissivity, near-surface air temperature, and surface-skin temperature and columnar total precipitable water at $1° \times 1°$ spatial resolution were obtained for 2003 from the online data archive of AIRS, distributed through NASA Mirador data holdings (http://mirador.gsfc.nasa.gov/). The monthly products are simply the arithmetic mean, weighted by counts, of the daily data of each grid box. The multi-day merged products have been used here because the IR retrievals are not cloud-proof and the multi-day product gave decent spatial cover in light of the missing cloudy-sky data. The data products were obtained in hierarchical data format (HDF4) along with their latitude–longitude projection. It is also important to mention that the daily 1° data contain orbital gaps and cloud contamination. In the 8-day data the co-incident land surface temperature in both the day and night pass was missing, and the atmospheric soundings were also missing in many places. It is the monthly data set where the soundings as well as both the day–night land surface temperatures were available and the data have complete global coverage.

We have used the MODIS Aqua atmospheric product data sets (MYD08_D3) (http://modis-atmos.gsfc.nasa.gov/index.html) at $1° \times 1°$ spatial resolution for extracting the solar zenith angle field. AIRS data do not contain any surface albedo field. For generating the surface albedo fields we used narrowband surface reflectances from combined MODIS Terra–Aqua 16-day data (MCD43C4) products acquired from the MODIS data archive (http://ladsweb.nascom.nasa.gov/data/search.html). The native spatial resolution of the MCD43C4 data sets is 0.05°. Therefore, all the narrowband surface reflectances were first resized into 1° by 1° to make them compatible with the AIRS spatial resolution and then the broadband surface albedo was generated from the narrowband reflectances following Liang et al. (1999) (presented in next section). It is important to mention that MODIS global albedo product (MCD43C3) contains bi-hemispherical reflectance (white-sky albedo) and directional hemispherical reflectance (black-sky albedo). Blue-sky albedo can be determined by weighting the white- and black-sky albedo with diffuse skylight fraction, which is a function of the aerosol optical depth and solar zenith angle. Look-up-table-based aerosol information and parameters are needed to convert the reflectances into the blue-sky albedo. But there are established formulations (Liang et al., 1999; Liang et al., 2002) to directly convert the narrowband reflectances into the broadband visible albedo that does not depend on any atmospheric variables and look-up tables, and therefore narrowband surface reflectances are used in the present study.

One of the core objectives of the work is to explore the potential of atmospheric sounding data. AIRS is the only dedicated sounder available which can be explored to address the objectives in the paper. MODIS also has soundings, but it was not designed for this and only has low-quality air temperature soundings. Coarse spatial resolution of AIRS would introduce many difficulties when it comes to the evaluation, but the most important aspect of the two companion papers (we refer to the current one as M1) is to introduce the possibility of using atmospheric sounding data as a means of observing surface energy fluxes (the companion paper on latent and sensible heat flux, M2). We have restricted Φ derivation to (largely) AIRS data (we used MODIS albedo because AIRS does not contain any albedo field) in order to exploit a single platform for the entire framework. We would also emphasise that the Φ retrievals are on one time slot per day for 13:30 LT, which is a standard for the studies that use polar-orbiting satellites.

2.2 Net radiation

The approach for estimating R_N uses the AIRS radiation products, although we have also made use of the MODIS surface reflectance and solar zenith angle products where necessary. R_N is generated by considering the following balance between net shortwave (R_{NS}) and longwave (R_{NL}) radiation at or near the Earth's surface,

$$R_N = R_{NS} + R_{NL} = (1-\alpha)R_{S\downarrow} + R_{L\downarrow} - R_{L\uparrow}, \tag{3}$$

where α is the surface albedo, $R_{L\downarrow}$ and $R_{L\uparrow}$ are the downwelling and upwelling thermal radiative fluxes, and $R_{S\downarrow}$ is the downwelling shortwave radiative flux (all fluxes specified in W m^{-2}). Our chosen reference level for R_N is the near surface given that this corresponds to the flux-based tower estimates we used in the evaluation. Therefore, surface $R_{S\downarrow}$ was estimated from its top-of-atmosphere clear-sky counterpart $R_{S0\downarrow}$ and AIRS cloud cover fraction (f) following Hildebrandt et al. (2007),

$$R_{S\downarrow} = (1-f)\tau_A R_{S0\downarrow}, \tag{4}$$

where τ_A is the clear-sky transmissivity of the atmosphere, which we assume is 0.75 (Cano et al., 1986; Thornton and Running, 1999; Hildebrandt et al., 2007; Gubler et al., 2012). Although clearly a simplification, a constant clear-sky transmissivity is widely used (e.g. Massaquoi, 1988; Bindi et al., 1992; Choudhury, 2001; Hildebrandt et al., 2007; Mallick et al., 2009) in recognition of the absence of robust alternatives. In addition, exploiting the AIRS cloud cover fraction data in

Eq. (2) should help accommodate the effects of variations in both the aerosol optical depth (Kaufman and Koran, 2006; Quass et al., 2010) and atmospheric water vapour (Adhikari et al., 2006).

The terrestrial surface albedo was generated using the MODIS Aqua–Terra surface reflectances r_i following Liang et al. (1999),

$$\alpha = \sum_{i=1}^{N} p_i r_i + 0.0036, \tag{5}$$

where r_i are the mid-point reflectances within the 0.62–0.67, 0.841–0.876, 0.459–0.479, 1.230–1.250, 1.628–1.653, 1.628–1.653, and 2.105–2.155 μm wavelength bands and p_i are the weightings for each wavelength bands taken as p_i = [0.3973; 0.2382; 0.3489; −0.2655; 0.1604; −0.0138; 0.0682] (Liang et al., 1999, 2002). The albedo of the ocean varies according to the cosine of solar zenith angle (Jin et al., 2004). Given that the oceanic surface reflectances are not available in either MODIS or AIRS, a constant albedo of 0.04 was assumed for oceans as satellite radiances are nadir.

Many of the longwave components of the radiative balance are very closely related to the raw IR radiances being measured by AIRS. Because these are not in the public domain, we have attempted to recover them as follows, although in future we would anticipate using the raw IR radiances more directly if possible. R_{NL} was calculated as

$$R_{\mathrm{NL}} = R_{\mathrm{L}\downarrow} - R_{\mathrm{L}\uparrow} = \varepsilon_{\mathrm{C}}\varepsilon_{\mathrm{S}}\sigma T_{\mathrm{C}}^4 - \varepsilon_{\mathrm{S}}\sigma T_{\mathrm{S}}^4, \tag{6}$$

where σ is the Stefan–Boltzmann constant (5.67×10^{-8} W m^{-2} K^{-4}), T_{c} is the columnar air temperature, and ε_{C} and ε_{S} are the column and surface emissivities. Among the different schemes for calculating ε_{C} we have used the formulation proposed by Prata (1996) as this appears to be the most reliable (Niemela et al., 2001; Bisht and Bras, 2010, 2011). This scheme uses AIRS total precipitable water (ξ) (cm) information to estimate ε_{C} as

$$\varepsilon_{\mathrm{C}} = 1 - (1+\xi)e^{-(1.2+3\xi)^{0.5}}. \tag{7}$$

The columnar air temperature T_{C} in Eq. (6) is taken as the average of the 2 m and 1000 hPa pressure level AIRS temperatures in an attempt to reflect a weighting toward the lower troposphere when specifying $R_{\mathrm{L}\downarrow}$. T_{S} and ε_{S} are taken directly from the AIRS skin temperature and surface emissivity products.

2.3 Surface heat capacity, ground heat flux and net available energy

The definition of G stems from consideration of the non-steady-state surface energy balance,

$$c\frac{\mathrm{d}T_{\mathrm{S}}(t)}{\mathrm{d}t} = R_{\mathrm{N}}(t) - \lambda E(h) - H(t) = G(t), \tag{8}$$

where c is the aggregate surface system heat capacity (MJ m^{-2} K^{-1}). The AIRS sounder platform samples twice daily at 01:30 and 13:30 LT. Despite being somewhat coarse, if a discrete time is taken, backward difference approximation of Eq. (8) with a sample interval of $\Delta t = 12$ h equivalent to that of the AIRS pass gives

$$\Delta T_{\mathrm{S}}(t) = b_1 R_{\mathrm{N}}(t) + b_2, \tag{9}$$

where ΔT_{s} is the day–night surface temperature change, $b_1 = \Delta t/c$ and $b_2 = -\Phi(t)\Delta t/c$. If we assume that the system is approximately in equilibrium over a 24 h cycle, and that $\Phi \approx 0$ at 01:30 LT (for all 30 sites analysed in this study Φ (01:30 LT) < 0.05Φ (13:30 LT); see also Tamai et al., 1998; Mamadou et al., 2014), then this gives the following simultaneous equations:

$$\Delta T_{\mathrm{S}}(13{:}30) = b_1 R_{\mathrm{N}}(13{:}30) + b_2, \tag{10a}$$

$$-\Delta T_{\mathrm{S}}(13{:}30) = b_1 R_{\mathrm{N}}(01{:}30), \tag{10b}$$

which can be solved analytically to derive b_1 and b_2 and hence Φ and c for each grid cell in the AIRS global array.

Equation (10) is a coarse approximation of Eq. (8) and hence potentially suffers from a number of deficiencies. Firstly, diurnal symmetry in ΔT_{s} is only appropriate when one considers weekly or monthly average behaviour, and that there are no additional heat losses to or gains from stores beyond the domain defined by the single heat capacity c. In this study we examined the monthly average behaviour because AIRS only gives partial global coverage on the daily timescale due to both cloud effects and the non-overlapping swath width of the sensor. Interactions with additional long-term heat stores is an issue in systems such as the oceans, where there can be a persistent heat loss/gains to/from deeper water over timescales of weeks to months, although relative to the diurnal fluctuation of stored surface heat this tends to be small (Stramma et al., 1986). Secondly, Φ can be either positive or negative at 01:30 LT, although it tends to be only a fraction (it never exceeds 5 % of afternoon Φ) of its 13:30 LT value due to the supply of relatively low amount of energy at night compared to the day. This may be less true for areas of land in the height of winter with cloud-covered days and over the sea where significant daytime heat accumulation could in part be re-released as night-time latent and sensible heat. Thirdly, Eqs. (9 and 10) attribute the magnitude of the daytime G to the night-time net longwave radiative balance, which is obviously rather uncertain. Fourthly, the air emissivity computation using Prata's equation was developed for the daytime and using it for the night-time emissivity may introduce errors. Finally, all the terms in Eq. (8) are highly dynamic and yet are treated as constant or varying linearly over the 12 h sample interval. It is difficult to predict what the consequences of this are, as it depends on the pattern of radiative forcing throughout the day, which can vary significantly in both time and space. Some illustrative examples of the theoretical assumptions of Eq. (10a, b) are depicted in

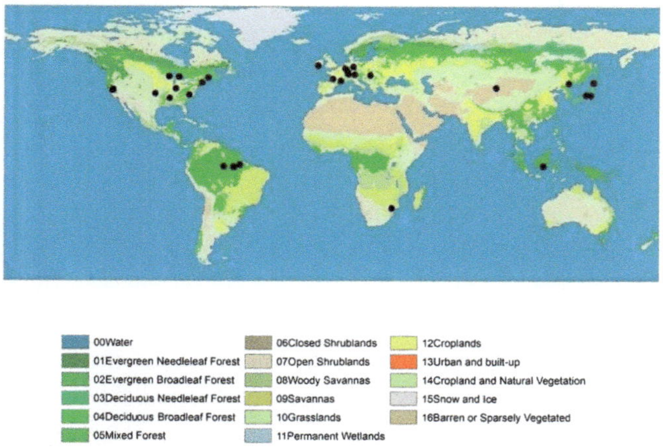

Figure 1. The distribution of the 30 eddy covariance tower sites used for evaluating R_N and Φ.

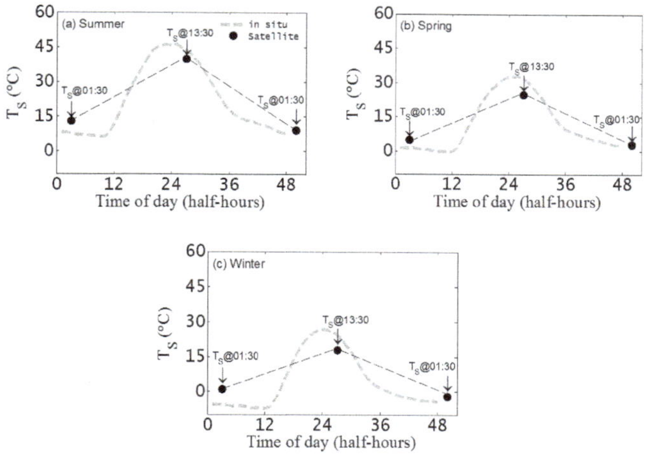

Figure 2. Illustrative examples of in situ monthly diurnal surface temperature (T_S) and the saw-tooth pattern of monthly satellite midday (13:30 LT) to night-time (01:30 LT) T_S (as hypothesised in Eq. 10) for three different seasons of a year. This shows a linear rise and fall of day–night T_S (dotted black line) or vice versa and indicates ΔT_S symmetry. This also shows how well the two T_S samples capture the dynamic range of the day and hence the discretisation is representative of the daily energy balance. The examples in Fig. 2 are shown with half-hourly data.

Figs. 2, 3 and 4. Figure 2 shows the examples of the diurnal symmetry of T_S for clear days in three different seasons where the saw-tooth pattern between noon (13:30 LT) and night (01:30 LT) T_S is evident. This clearly shows how well these two T_S samples capture the dynamic range of the day and hence the discretisation is representative of the daily energy balance. Figure 3a to d illustrate the diurnal evolution of Φ during three different times of the year (spring, summer and winter) for four broad biome categories (grassland, cropland, forest and savanna), which clearly indicates $\Phi \cong 0$ at 01:3 LT0 and within less than 5 % of afternoon Φ. Lastly, Fig. 4 highlights the two-dimensional relationship between the noontime G (13:30 LT) and night-time R_{NL} (01:30 LT) for the above-mentioned four biomes and the correlation between the two varied between 0.32 and 0.60, having high correlation over grassland and savanna and moderate correlation over forest and cropland. Despite large differences in the footprint size between G and R_{NL} measurements, the inverse relationship between the two variables in Fig. 4 clearly indicates the dependence of noontime G on the night-time longwave radiation balance. Therefore, although the theoretical approximations in Eqs. (8) and (10a, b) seem to be somewhat coarse and might have a few limitations (as described earlier), but Figs. 2, 3 and 4 indicate strong connectivity between ΔT_S, 01:30 LT R_N (or R_{NL}) and 13:30 LT Φ. However, given the structure of the atmosphere and the very small energy fluxes involved, high-latitude Φ estimates from this method are likely to be problematic in any case.

2.4 Sensitivity analysis

A general sensitivity analysis was carried out in order to assess the effects of the propagation of uncertainty onto the estimates of G, R_N and Φ. For this analysis the input terms were assigned uniform prior distributions of $\pm 10\,\%$ for all parameters other than temperatures for which ± 1 K uniform prior distributions were assumed. These assumed ranges resemble the stated uncertainties as given in the AIRS support literature (Aumann et al., 2003; Hearty et al., 2014). The sensitivity of each output to each input was calculated assuming an average, locally linear sensitivity. These were expressed as the change in output per unit change in input, normalised by the median value of each. Only absolute sensitivities >0.1 were considered significant. The standard deviations of the estimated distributions of G, R_N and Φ were used as the summary statistic for the measurement uncertainty of the proposed methodology.

2.5 Evaluation of R_N and Φ

To evaluate the satellite values of R_N and Φ, we have made use of the extensive FLUXNET terrestrial tower network (Baldocchi et al., 2001). Clearly, there is a scale conflict here, with the satellite retrievals being 1° whilst the individual tower observations are for scales of the order of 1 km or less. The tower R_N are from the broadband net radiometer sensors located on each tower. In the absence of reliable measures of G at the tower scale, and in order to derive genuinely independent measures of Φ against which to evaluate the satellite data, we have taken the tower net available energy as the sum of the measured sensible and latent heat flux, i.e. Eq. (1). Thereby we have assumed that the eddy covariance flux measurements are able to close the energy balance (i.e. $R_N - G = \lambda E + H$), the implications of which will be discussed below. We have chosen 30 sites covering a broad range of geographical locations selected from 7 land cover types, including evergreen broadleaf forest (EBF),

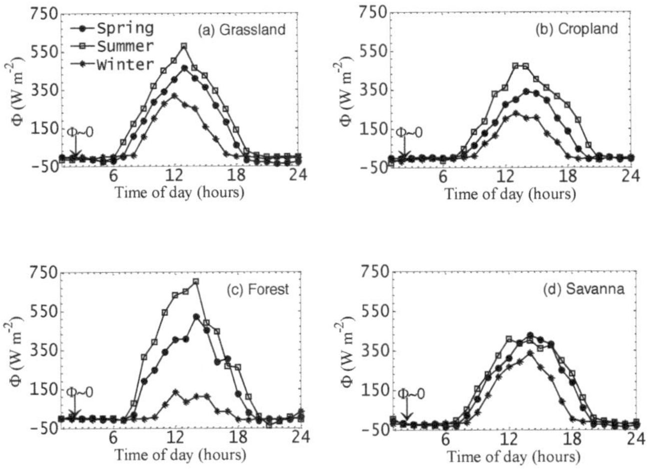

Figure 3. Examples of the diurnal evolution of net available energy (Φ) ($= \lambda E + H$) during three different times of year over four representative biome types. This shows $\Phi \cong 0$ around 01:30 LT.

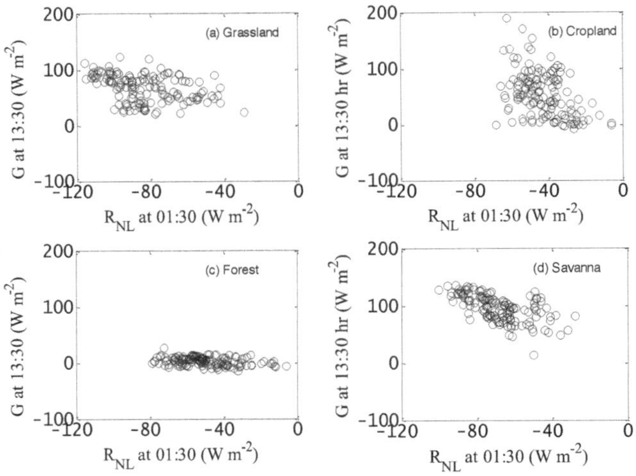

Figure 4. Observational relationship night-time (01:30 LT) net longwave radiation (R_{NL}) and midday (13:30 LT) G measurements. Despite the scale mismatch between the two measurements, moderate ($r = 0.32$–0.44) (forest, cropland) to high ($r = 0.60$) (grassland and savanna) relationship is notable.

mixed forest (MF), evergreen needleleaf forest (ENF), deciduous broadleaf forest (DBF), savanna (SAV), grassland (GRA) and cropland (CRO). A comprehensive list of the site characteristics are provided in Table 1. Each tower evaluation data set is comprised of the 13:30 LT time samples of R_N, H and λE which correspond to the satellite overpass. Again, the evaluation is based on pooling these data into weighted monthly average values. For the evaluation we have elected to compare all 12 months of data for 2003, as this year had the best overlap between the FLUXNET and AIRS databases. However before directly validating the satellite-retrieved Φ, the proposed Φ retrieval method is first evaluated using high-temporal-frequency ground-based observations of R_N and T_S over some eddy covariance sites representing four broad biome categories (grassland, cropland, forest and savanna). Both R_N and T_S at 13:30 LT and 01:30 LT were extracted from half-hourly measurements, and Φ at 13:30 LT was determined using Eq. (10a, b). The retrieved midday Φ was validated against tower-observed latent and sensible heat fluxes.

3 Results

Table 2 shows the results from the sensitivity analysis. For G we see the importance of the longwave specification and in particular ε_C. The standard deviation of the estimate of G from the ensemble is 18 W m^{-2}, giving approximate 95 % confidence detection limits of ± 36 W m^{-2} on the estimates. R_N is sensitive to all components of the radiation balance calculation as expected (Table 2). The standard deviation of the estimate of R_N from the ensemble is 40 W m^{-2}, giving approximate 95 % detection limits of ± 80 W m^{-2} on the estimates (Table 2). Not surprisingly, the sensitivity results for Φ mirror those of R_N, albeit with a marginally higher ensemble standard deviation of 44 W m^{-2} (Table 2).

The locations of the 30 terrestrial evaluation sites are marked in Fig. 1. Figure 5 shows annual average, global satellite scenes for 13:30 LT R_N, c, G and Φ for the year 2003. Missing data in the images are mainly due to missing data in the AIRS soundings at high latitudes or over the mountain belts, where it is difficult to profile air temperature and relative humidity reliably. In addition, persistent cloudy conditions also prevent reliable retrieval and hence are rejected, although these will be less evident in the monthly or annual average data.

Figure 5a shows the global distribution of R_N, which generally decreases with latitude, as expected. R_N also decreases over land due to the generally higher albedo, resulting in reduced absorption of the net shortwave radiation (Giambelluca et al., 1997; Gao and Wu, 2014) or relatively higher surface temperature increasing the net longwave component, especially over the drier regions (Liang et al., 1998; Trenberth, 2011). As a result the magnitude of R_N was around 200–300 W m^{-2} over the dry desert regions, whereas the oceanic values of R_N were 450–700 W m^{-2}.

Figure 5b shows the global distribution of c (surface heat capacity). The oceanic values of 4 to 8 MJ m^{-2} K^{-1} are equivalent to 1 to 2 m of seawater, which appears reasonable on the daily time step to which they relate (Stramma et al., 1986; Schwartz, 2007). These oceanic values are somewhat noisy due to the small day–night temperature differences observed for the oceans giving a relatively poor signal-to-noise ratio. However, behind this noise the pattern of oceanic c appears relatively uniform as one might expect. Over land c varies between 0.05 and 0.5 MJ m^{-2} K^{-1} with wetter tropical and high-latitude areas showing significantly

Table 1. Eddy covariance sites used for the evaluation of the satellite-derived R_N and Φ.

Biome type	Site name, country	Latitude (°)	Longitude (°)	Reference
Evergreen broadleaf forest (EBF)	Palangkaraya, Indonesia	−2.35	114.04	Hirano et al. (2007)
	Puechabon, France	43.74	3.6	Reichstein et al. (2003)
	Caxiuana Forest – Almeirim, Brazil	−1.72	−51.46	Carswell et al. (2002)
	Manaus – ZF2 K34, Brazil	−2.61	−60.21	de Araújo et al. (2002)
	Santarem – Km67, Brazil	−2.86	−54.96	Hutyra et al. (2007)
	Santarem – Km83, Brazil	−3.02	−54.58	Goulden et al. (2004)
Mixed forest (MF)	Vielsalm, Belgium	50.31	5.99	Aubinet et al. (2001)
	Tomakomai National Forest, Japan	42.73	141.52	Hirano et al. (2003)
	Changbaishan, China	42.4	128.09	Zhang et al. (2006)
Grassland (GRA)	Oensingen1 grass, Switzerland	47.29	7.73	Ammann et al. (2007)
	Neustift/Stubai Valley, Austria	47.12	11.32	Hammerle et al. (2008)
	Goodwin Creek, USA	34.25	−89.87	Unpublished
	Bugacpuszta, Hungary	46.69	19.61	Gilmanov et al. (2007)
	Dripsey, Ireland	51.99	−8.75	Jaksic et al. (2006)
Cropland (CRO)	ARM Southern Great Plains, USA	36.61	−97.49	Fischer et al. (2007)
	Bondville, USA	40.01	88.29	Meyers and Hollinger (2004)
	Tsukuba, Japan	36.05	140.03	Saito et al. (2005)
Evergreen needleleaf forest (ENF)	Le Bray, France	44.72	−0.77	Granier et al. (2000a)
	Duke Forest – loblolly pine, USA	35.98	−79.09	Katul et al. (2003)
	Blodgett Forest, USA	38.89	−120.63	Goldstein et al. (2000)
	Howland Forest, USA	45.2	−68.74	Hollinger et al. (1999)
Deciduous broadleaf forest (DBF)	Harvard Forest EMS Tower (HFR1), USA	42.54	−72.17	Urbanski et al. (2007)
	Univ. of Michigan Biological Station, USA	45.56	−84.71	Gough et al. (2009)
	Willow Creek, USA	45.81	−90.08	Cook et al. (2004)
	Hesse Forest – Sarrebourg, France	48.67	7.06	Granier et al. (2000b)
	Hainich, Germany	51.08	10.45	Anthoni et al. (2004)
	Morgan Monroe State Forest, USA	39.32	−86.41	Baldocchi et al. (2001)
	Takayama, Japan	36.15	137.42	Saigusa et al. (2002)
Savanna (SAV)	Tonzi Ranch, USA	38.43	−120.97	Baldocchi et al. (2004)
	Skukuza, South Africa	−25.02	31.49	Scholes et al. (2001)

higher values than the drier, less vegetated areas, as expected. The soil equivalent depth of this heat capacity is approximately 0.01 m, which again appears reasonable for a daily time step (Li and Islam, 1999), although in heavily vegetated areas c is obviously comprised of a more complex aggregation.

Figure 5c shows the global distribution of G. These are the 13:30 LT values, hence their being net positive as an annual average. Between 20° north and south, G is approximately 10 to 20 % of R_N, and this rises to more than 40 % above 50° north and south (Hsieh et al., 2009). Given that this opposes the pattern of R_N, one would conclude that there are either some deficiencies in the way G is specified here or that R_N partitions into latent heat far more effectively than surface heating in these warm wet environments (Liu et al., 2005). Again, terrestrial values are lower than their oceanic equivalents, mainly due to the lower heat capacity as well as reduced R_N as discussed above. This also highlights the role of the vegetation layer in preventing ground heating (Baker and Baker, 2002; Bounoua et al., 2010). The Sahara appears particularly prominent in this scene, with high rates of midday heat accumulation, which appears to be associated with a combination of moderate net radiation and relatively high heat capacity. The heterogeneity in this region appears to be related to the pattern of bare darker rock.

Figure 5d shows the global distribution of Φ, which follows a similar pattern to R_N as expected, although the pattern of G shown in Fig. 5c dictates that the north–south gradients in Φ are somewhat stronger than those of R_N. Before discussing these results, we consider their evaluation. In the first step, we validated the new method of Φ retrieval at representative FLUXNET sites using ground observations of the surface radiation components (T_S, R_N and G) as input before directly evaluating the satellite-based retrievals. Tower-scale evaluation of daily midday (13:30 LT) Φ is illustrated in Fig. 6a, b, c and d for four broad biome categories, which shows a modest correlation [$r = 0.91$ (± 0.03)

Table 2. Sensitivity analysis results. The forcing data are taken for midsummer in the Southern Great Plains, USA. Sensitivities are locally linear, averaged across the ensemble response and expressed as dimensionless relative changes. Only absolute sensitivities >0.1 are shown. $N = 10^5$ realisations.

		G (W m^{-2})	R_N (W m^{-2})	Φ (W m^{-2})
x	Sample range	dG/dx	dR_N/dx	dΦ/dx
τ_A	±10 %	–	1.30	1.58
f	±10 %	–	−0.77	−0.94
α	±10 %	–	−0.25	−0.31
ε_S	±10 %	1.00	−0.31	−0.37
ε_C	±10 %	−4.60	0.98	1.19
T_S	±1 K	0.75	−0.17	−0.21
T_{1000}	±1 K	−0.33	–	–
		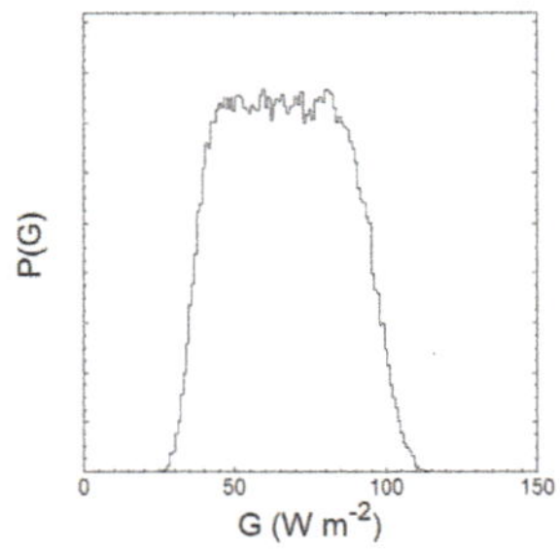	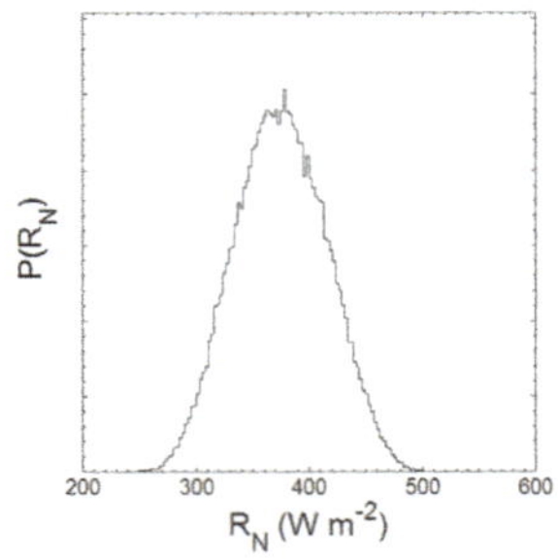	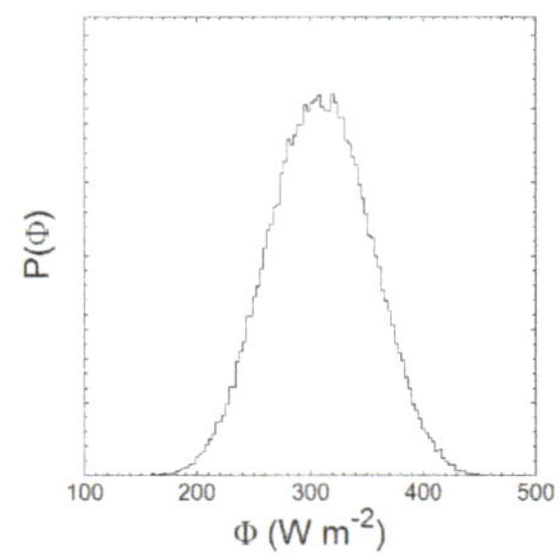
Standard deviation		18 W m^{-2}	40 W m^{-2}	44 W m^{-2}

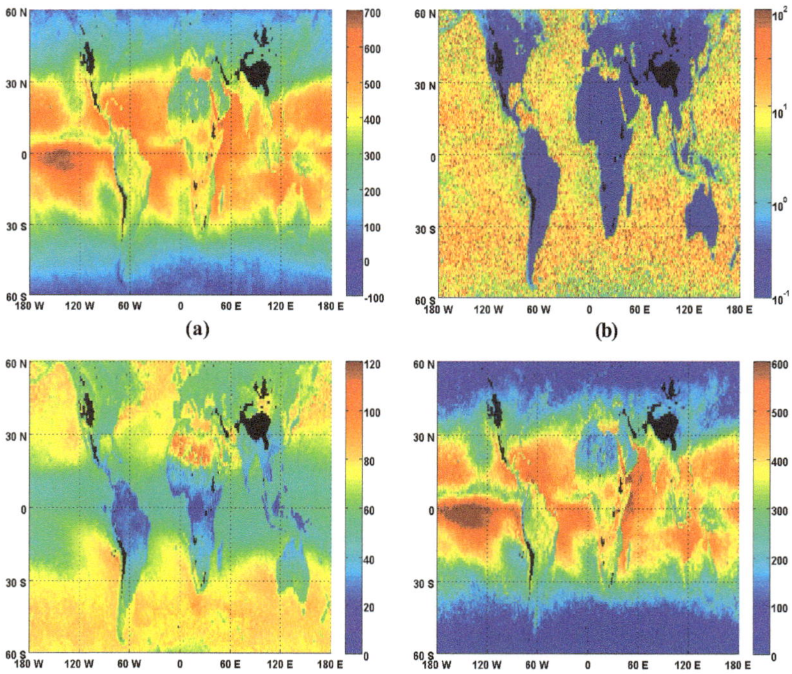

Figure 5. Global fields for annual average (year 2003) 13:30 LT: **(a)** net radiation, R_N (W m^{-2}); **(b)** surface heat capacity, c (MJ m^{-2} K^{-1}); **(c)** surface heat accumulation rate, G (W m^{-2}); **(d)** net available energy, Φ (W m^{-2}).

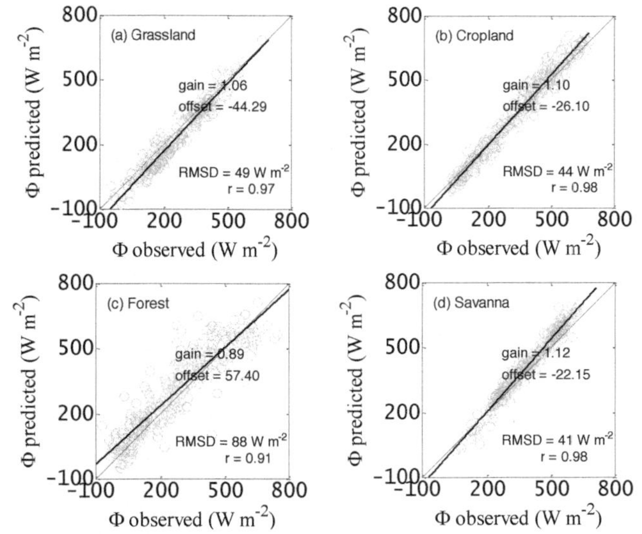

Figure 6. Validation of net available energy (Φ) ($= \lambda E + H$) using high-temporal-frequency (daily) observations of midday (13:30 LT) and night (01:30 LT) R_N and T_S (solving Eq. 10a and b) at the eddy covariance tower sites over four representative biomes.

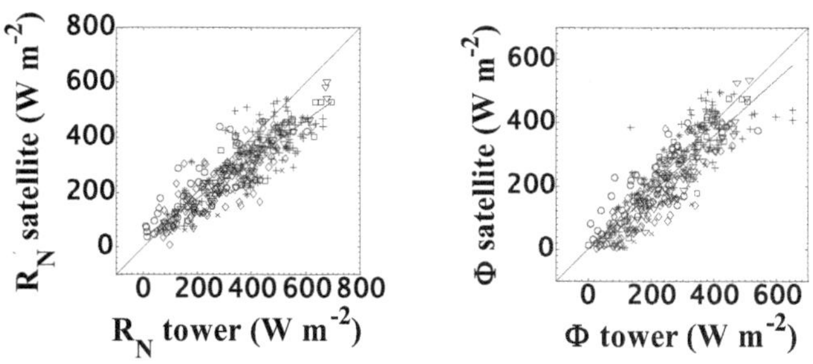

Figure 7. Comparison of satellite and tower monthly average 13:30 LT **(a)** R_N and **(b)** Φ [$\Phi = (H + \lambda E)$ (for the tower sites)]. For details of the site characteristics see Table 1. For the comparative statistics see Table 3. The solid line is the pooled linear regression given in Table 3. (+ EBF; × MF; ○ GRA; * CRO; ∇ ENF; ◇ DBF; □ SAV.)

to 0.98 (±0.04)][1] between observed and predicted Φ across all the biomes with regression statistics ranging between 0.89 (±0.07) and 1.12 (±0.05) for the gain and −44.29 (±20.15) to 59.40 (±29.03) for the offset (Fig. 6). The root-mean-square deviation (RMSD) varied between 41 (savanna) and 88 W m^{-2} (forest).

Figure 7a shows the pooled evaluation of satellite R_N, which produced an overall correlation of $r = 0.88$ (±0.03). Assuming both tower and satellite observations are linearly related through some "true" value, then the pooled values are co-related by $R_N(\text{satellite}) = 0.75(\pm 0.02) R_N(\text{tower}) + 23.37(\pm 8.20)$, i.e. a small but significant underestimation in R_N(satellite) relative to R_N(tower). The RMSD between the two was 98 W m^{-2}. The biome-specific statistics for R_N are given in Table 3, which reveals correlations ranging between 0.65 (EBF) and 0.96 (ENF), RMSD ranging between 74 (GRA) and 127 W m^{-2} (EBF), and regression statistics ranging between 0.58 (±0.08) and 0.87 (±0.04) for the gain and −32.40 (±23.73) and 107.45 (±39.93) for the offset.

Figure 7b shows the evaluation for Φ, which produced pooled statistics of $r = 0.87$ (±0.03) and an RMSD of 72 W m^{-2}, and the regression between the satellite-predicted and tower-observed Φ produced a regression line of $\Phi(\text{satellite}) = 0.90(\pm 0.03)\Phi(\text{tower}) - 2.43$ (±8.19). The biome-specific statistics for Φ are also given in Table 3 showing correlations ranging from 0.70 (EBF) to 0.95 (ENF), RMSD ranging between 62 (GRA & SAV) and 88 W m^{-2} (EBF) and regression coefficients ranging between 0.66 (±0.08) and 1.01 (±0.05) and between −65.25 (±27.07) and 108.71 (±32.10) for the gain and offset, respectively.

Figure 8 shows a sample of monthly time series for Φ for both the satellite and the towers. The sites were selected to represent the biome classes considered here, as well as ones for which complete annual data sets for 2003 were available. These results show that the satellite estimates generally track the trends in the tower data and hence that the pooled statistics are not masking the within site variability. Again, the site-wise comparative statistics for these data are given in Table 3.

4 Discussion

For R_N the statistics relating the satellite and tower data are different with to results of the following authors: Bisht et al. (2005), who obtained 74 W m^{-2} RMSD when evaluating MODIS Terra geophysical land products over the Southern Great Plains of the USA (our RMSD in grassland is only comparable here, while other biomes show larger error); Jacobs et al. (2004), who obtained a 14–46 W m^{-2} RMSD (12.2 % relative RMSD) when determining hourly R_N using GOES (Geostationary Operational Environmental Satellite) data over wetlands in southern Florida; Cai et al. (2007), who obtained 13.7 % error when evaluating MODIS Terra–Aqua data over China; Bisht and Bras (2010), who obtained 39–51 W m^{-2} RMSD over the central USA using MODIS Terra atmospheric data at 5–10 km spatial resolution; and Hwang et al. (2013) and Hou et al. (2014), who reported RMSD in instantaneous and daily R_N to be 58–142 and 37–40 W m^{-2} over Southeast Asia and China, respectively, using MODIS Terra data products. Stackhouse et al. (2000) evaluated the International Satellite Cloud Climatology Project (ISCCP) data and found errors in the range of 10 to 15 W m^{-2} in monthly average shortwave and longwave radiative fluxes. Other studies reported 33–60 W m^{-2} RMSD in daily R_N using 5 km MODIS Terra optical and thermal

[1] All uncertainties are expressed as ± 1 standard deviation unless otherwise stated.

data (a comprehensive list of relevant studies is given in Table 4). It is important to emphasise that, in the present study, the RMSD is being impacted in two ways: due to spatial scale mismatch and due to the time integration. When these errors are compounded in the derivation of R_N and compared with tower data, an RMSD of the order of 98 W m^{-2} appears reasonable considering the coarse spatial resolution of the AIRS data ($1° \times 1°$).

There have been very few attempts to retrieve satellite estimates of Φ and compare these with ground truth data, although the statistics from our attempt appear to be parallel to the results of Stisen et al. (2008), who studied a single grassland site in the Senegal River basin using moderate (high) spatio-temporal resolution (5 km spatial resolution, 15 min temporal resolution) MSG (Meteosat Second Generation) geostationary satellite data and obtained a correlation of $r = 0.71$ and an RMSD of 43 W m^{-2} in comparison to the surface measurements. While estimating evapotranspiration over Indian agroecosystems, Bhattacharya et al. (2010) obtained an RMSD of 56 W m^{-2} for noontime Φ using 8 km resolution Indian geostationary satellite data. In another study with MODIS Aqua data over semi-arid agroecosystems in India, Bhattacharya et al. (2011) reported an RMSD of 34 W m^{-2} in daily average Φ, which was associated with a significant tendency to underestimate Φ. As was common to all these studies, the ground heat flux was either modelled as an empirical approximation employing remotely sensed surface variables (albedo, vegetation index and T_S) or as a fixed fraction of R_N. Murray and Verhoef (2007) argued that these empirical approaches do not generalise well. In particular, prescribing G as a fixed fraction of R_N overlooks the role played by the thermal inertia of the land surface (Santanello and Friedl, 2003), leading to an underestimation of G in the morning and overestimation during the afternoon (Gentine et al., 2007). The retrieval of G proposed here using day–night surface temperature information attempts to account for this thermal inertia effect, and the results appear to support this approach especially when considering the scale mismatch between the tower and satellite observations.

As seen in Fig. 7a, there is a systematic underestimation of R_N relative to the tower values which exceeds the typical accuracy of net radiometer measurements of 20 W m^{-2} quoted by Foken (2008). We examined this underestimation in more detail wherever possible by evaluating three of the individual radiation components of R_N ($R_{S\downarrow}$, $R_{L\downarrow}$ and $R_{L\uparrow}$). All tower sites provided measurements of $R_{S\downarrow}$ (but not $R_{S\uparrow}$). Figure 9a shows $R_{S\downarrow}$ is systematically underestimated at the satellite scale, with $R_{S\downarrow}(\text{satellite}) = 0.70(\pm 0.02) R_{S\downarrow}(\text{tower}) + 68(\pm 12.24)$, which accounts for the mismatch of $R_N(\text{satellite}) \approx 0.75 R_N(\text{tower})$. Before attempting to account for the various reasons for this underestimation, it is important to realise that, unlike the IR components, the shortwave components are all-sky retrievals, i.e. like the tower data they do not omit cloudy-sky conditions. As a result, any bias in the shortwave

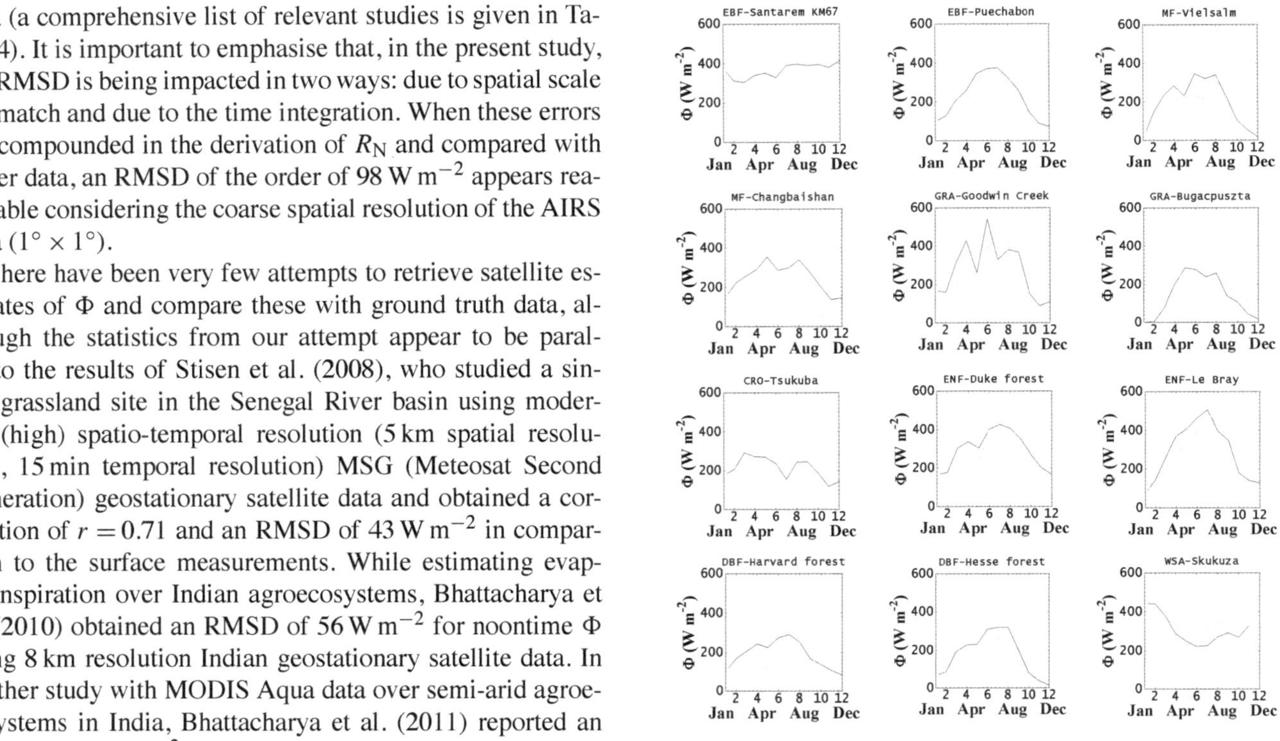

Figure 8. Satellite (grey) and tower (black) time series of monthly average 13:30 LT net available energy Φ [$\Phi = (H + \lambda E)$ (for the tower sites)] for a selection of sites for 2003. The numbers on the x axis are the month numbers, i.e. January is month number 1 and December is month number 12.

is not as a result of biased sampling when comprising the monthly average. Furthermore, the omission of non-clear-sky data would tend to lead to $R_{S\downarrow}(\text{satellite}) > R_{S\downarrow}(\text{tower})$.

Clearly, the retrieval of atmospheric shortwave transmissivity (τ_A) using cloud cover fraction is the principal reason for $R_{S\downarrow}(\text{satellite}) < R_{S\downarrow}(\text{tower})$ (Fig. 9a). The sensitivity analysis presented in Table 2 also indicates the significant sensitivity of R_N and Φ to the cloud cover fraction and atmospheric transmissivity. This shows that the method presented in the manuscript to estimate $R_{S\downarrow}$ needs further improvements. If we assume τ_A to be the principal reason for $R_{S\downarrow}(\text{satellite}) < R_{S\downarrow}(\text{tower})$, then a global value of 0.75 would be, on average, too low (Gueymard, 2003). A recent study of Longman et al. (2012) for the Mauna Loa Observatory (MLO) demonstrated the clear-sky τ_A could go up to 0.90. Given the relatively well defined relationship between $R_{S\downarrow}(\text{satellite})$ and $R_{S\downarrow}(\text{tower})$ seen in Fig. 3c, one would imagine that a more sophisticated dynamic representation of τ_A would offer substantial improvements in $R_{S\downarrow}(\text{satellite})$. Retrieval of τ_A including other atmospheric (e.g. cloud optical depth, aerosol optical depth, total precipitable water etc.) and surface (for example, single scattering albedo) variables in addition to the cloud cover fraction would offer a potential possibility of refining the $R_{S\downarrow}$ estimates (Chen et al., 2014; Longman et al., 2012; Kim and

Table 3. Comparative statistics for the satellite- and tower-derived monthly midday (13:30 LT) R_N and Φ for a range of biomes. Values in parenthesis are ± 1 standard deviation.

Biome	R_N RMSD ($W m^{-2}$)	R_N Gain	R_N Offset	R_N r	R_N N	Φ RMSD ($W m^{-2}$)	Φ Gain	Φ Offset	Φ r	Φ N
EBF	126.67	0.58 (±0.08)	107.45 (±39.93)	0.65 (±0.09)	69	87.67	0.66 (±0.08)	108.71 (±32.10)	0.70 (±0.09)	65
MF	104.21	0.82 (±0.07)	−32.40 (±23.73)	0.89 (±0.08)	36	87.29	0.97 (±0.10)	−65.25 (±27.07)	0.86 (±0.09)	32
GRA	74.29	0.73 (±0.05)	51.37 (±15.88)	0.88 (±0.06)	59	61.51	0.83 (±0.07)	15.71 (±16.21)	0.86 (±0.07)	53
CRO	89.13	0.73 (±0.08)	35.62 (±28.57)	0.84 (±0.09)	36	53.31	0.99 (±0.10)	−0.23 (±23.98)	0.87 (±0.09)	36
ENF	85.45	0.87 (±0.04)	−26.83 (±14.78)	0.96 (±0.04)	48	66.7	1.01 (±0.05)	−56.56 (±15.27)	0.95 (±0.05)	46
DBF	92.77	0.71 (±0.05)	21.74 (±15.15)	0.85 (±0.06)	84	71.57	0.88 (±0.06)	−16.70 (±14.23)	0.85 (±0.06)	80
SAV	103.98	0.69 (±0.08)	56.28 (±36.08)	0.87 (±0.11)	23	61.98	0.97 (±0.11)	−14.42 (±37.68)	0.88 (±0.25)	32
Pooled	98.21 (28 %)	0.75 (±0.02)	23.37 (±8.20)	0.88 (±0.03)	355	72.26 (22 %)	0.90 (±0.03)	−2.43 (±8.19)	0.87 (±0.03)	335

N: number of data points falling under individual biomes. $\Phi = (H + \lambda E)$ (for the tower sites).
EBF: evergreen broadleaf forest; MF: mixed forest; GRA: grassland; CRO: cropland; ENF: evergreen needleleaf forest; DBF: deciduous broadleaf forest; SAV: savanna.

Table 4. Summary of errors and characteristics of some of the dedicated satellite-based R_N and $R_{S\downarrow}$ retrieval studies.

Radiative flux variables	Reference	Sensor used	Spatial resolution	Temporal resolution	RMSD ($W m^{-2}$)
R_N	Verstraeten et al. (2005)	NOAA AVHRR	1 km	Instantaneous	5–5
	Bisht et al. (2005)	MODIS Terra	5 km	Instantaneous	74
	Cai et al. (2007)	MODIS Terra and Aqua synergy	5 km	Instantaneous	19
	Bisht and Bras (2010)	MODIS Terra	5 km	Instantaneous	23–39
	Hwang et al. (2013)	MODIS Terra	5 km	Instantaneous	58–142
	Hou et al. (2014)	MODIS Terra	5 km	Daily	37–40
	Bisht et al. (2005)	MODIS Terra	5 km	Daily	60
	Peng et al. (2013)	MODIS Terra	5 km	Daily	33
	Jin et al. (2011)	MODIS Terra	5 km	Monthly average	44
	Mallick et al. (2014) (current study)	AIRS and MODIS Terra	1°	Instantaneous monthly average	74–126
$R_{S\downarrow}$	Chen et al. (2014)	MODIS Terra	1°	Daily	39
	Sun et al. (2013)	In situ observations	Tower footprint	Instantaneous	36–89
	Huang et al. (2012)	MODIS Terra	5 km	Instantaneous	54–83 (in R_{NS})
	Huang et al. (2011)	MODIS Terra	5 km	Instantaneous	60–137
	Wang and Pinker (2009)	MODIS Terra and Aqua	1°	Instantaneous	77–158
	Kim and Hogue (2008)	MODIS Terra	5 km	Instantaneous	76
	Tang et al. (2006)	MODIS Terra	5 km	Daily	20–35 (in R_{NS})
	Mallick et al. (2014) (current study)	AIRS	1°	Instantaneous monthly average	110

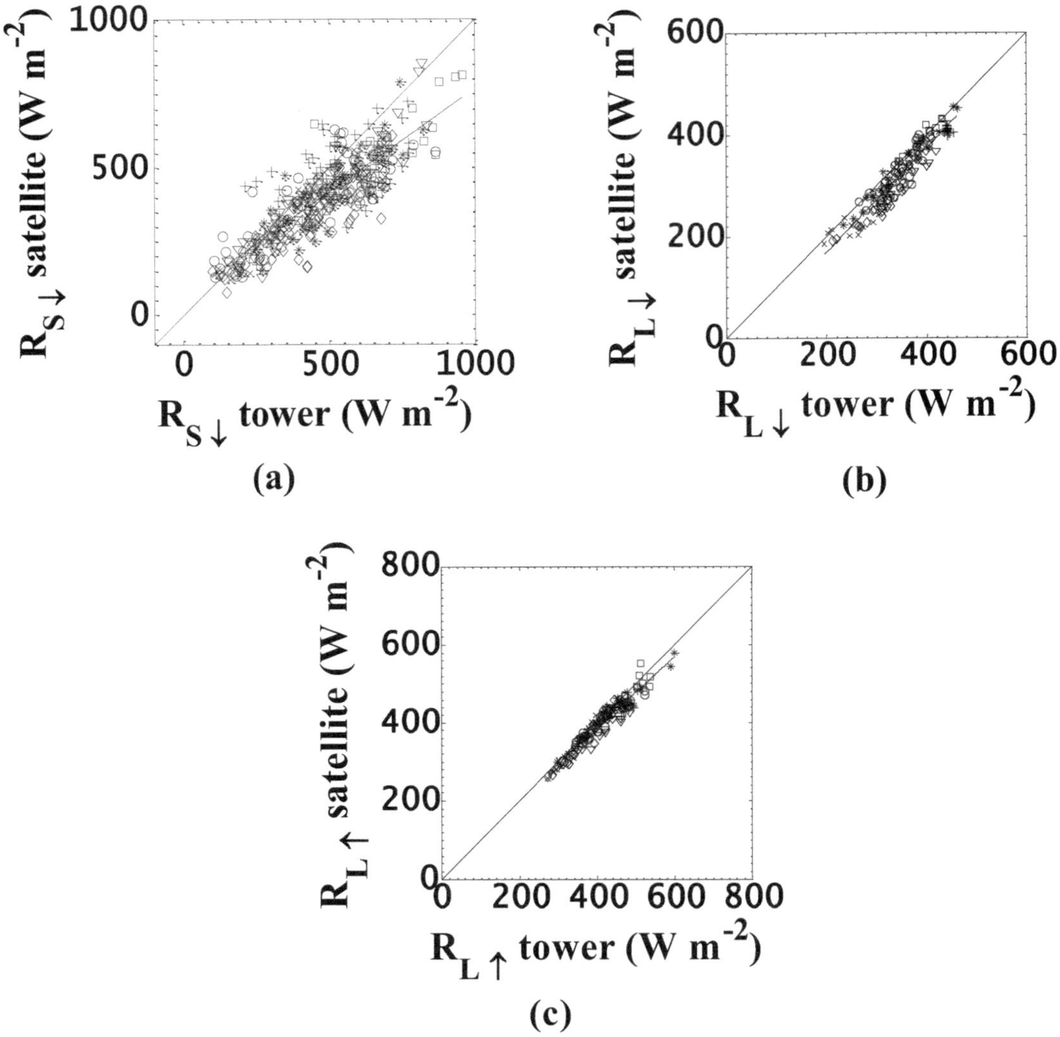

Figure 9. Comparison of satellite and tower monthly average 13:30 LT **(a)** $R_{S\downarrow}$, **(b)** $R_{L\downarrow}$ and **(c)** $R_{L\uparrow}$ for a selection of sites for which tower data for $R_{S\downarrow}$ (360 data points), $R_{L\downarrow}$ (159 data points) and $R_{L\uparrow}$ (159 data points) were available. The linear fit (solid line) between the two sources of $R_{S\downarrow}$ is $R_{S\downarrow}$(AIRS) = 0.70 (±0.02)$R_{S\downarrow}$(tower) +67.68 (±12.24); r = 0.84 (±0.03). The linear fit (solid line) between the two sources of $R_{L\downarrow}$ is $R_{L\downarrow}$(AIRS) = 1.03 (±0.03)$R_{L\downarrow}$(tower)–36.91 (±10.05); r = 0.95 (±0.03). The linear fit (solid line) between the two sources of $R_{L\uparrow}$ is $R_{L\uparrow}$(AIRS) = 0.91(±0.02)$R_{L\uparrow}$(tower) + 20.43 (±8.77); r = 0.96 (±0.02). The dashed lines are 1 : 1 in all cases. (+ EBF; × MF; ○ GRA; * CRO; ∇ ENF; ◇ DBF; □ SAV.)

Hogue, 2008). The exo-atmospheric shortwave radiation frequently interacts with the clouds, aerosols and water vapour during the transmission towards the Earth's surface. This interaction is wavelength-dependent over the entire shortwave spectrum (Chen et al., 2014; Kim and Hogue, 2008; Gueymard, 2003), and therefore a spectrally resolved τ_A scheme will be valuable to accurately determine $R_{S\downarrow}$. Recalibration of τ_A using the tower data is also a possibility, although we have avoided this given that the AIRS cloud cover fraction and scale mismatch between the satellite and tower could also be involved in the observed bias. For example, the diffuse fraction of $R_{S\downarrow}$(tower) can become enriched by surface-reflected solar radiation, particularly in undulating terrain (Dubayah and Loechel, 1997; Sultan et al., 2014). Nonlinear scaling effects of surface albedo (Oliphant et al., 2003; Salomon et al., 2006) can also be included in this because surface albedo interacts nonlinearly with surface characteristics such as surface wetness and land surface temperature (Ryu et al., 2008) or the leaf area index (Hammerle et al., 2008). Although, the RMSD of instantaneous $R_{S\downarrow}$ obtained in the present study (110 W m^{-2}) is different to other studies where $R_{S\downarrow}$ retrieval was based on either using parametric (radiative transfer) models or look-up tables derived from high-spatial-resolution MODIS data, it is worth comparing it with the statistics of some of those studies. Table 4 summarises the characteristics and associated errors of some of the recent $R_{S\downarrow}$ estimation studies, and shows an RMSD of 36–89 W m^{-2} at flux tower footprint, 54–137 W m^{-2} at 5 km

Table 5. The mean and standard deviation of monthly midday (13:30 LT) surface energy balance closure (C_{EB}) of different biome types for the 21 (out of 30) FLUXNET sites (N) under study. Nine out of 30 sites had missing ground heat flux. C_{EB} is computed according to Stoy et al. (2013).

Biome types	N	C_{EB}
EBF	2	0.92 ± 0.15
MF	3	0.76 ± 0.12
GRA	4	0.84 ± 0.08
CRO	3	0.77 ± 0.10
ENF	3	0.80 ± 0.02
DBF	4	0.66 ± 0.05
SAV	2	0.79 ± 0.08

spatial resolution and 77–158 W m^{-2} at 1° spatial resolution for the instantaneous $R_{S\downarrow}$ estimates and 20–39 W m^{-2} for the daily $R_{S\downarrow}$ (and net shortwave, R_{NS}) estimates. Considering the simplicity of the current approach and the large spatial scale of the AIRS data, an RMSD of the order of 110 W m^{-2} appears reasonable.

To probe the specification of R_N further, we investigated the individual longwave radiation components in relation to measures of theses fluxes available for a limited subset (14) of tower sites where the longwave radiative flux components were directly measured by pyrgeometers. From Fig. 9b and c it appears that there is quite good agreement between the satellite and tower data for both $R_{L\downarrow}$ and $R_{L\uparrow}$ and that any mismatch is insufficient to explain the discrepancy in R_N. This is somewhat surprising for two reasons. Firstly, unlike the shortwave component, R_L(tower) is all-sky whilst R_L(satellite) is only from clear-sky conditions where IR retrieval is possible. As a result, one would anticipate very significant differences in the monthly average values of the longwave components. However, it is difficult to predict the effect of this biased sampling on R_{NL}(satellite) given that cloud interacts with both $R_{L\downarrow}$ and $R_{L\uparrow}$ in complex ways. Secondly, one would anticipate significant scaling effects from the T^4 nonlinearity in Eq. (6), which can result in a disproportionate contribution of warmer elements within the system to both $R_{L\downarrow}$ and $R_{L\uparrow}$ (Kustas and Norman, 2000; Lakshmi and Zehrfuhs, 2002, Corbari et al., 2010). The fact that these effects are not seen to any significant degree could point to compensating errors in the analysis but does not distract from the central message of the importance of the bias in the shortwave when accounting for R_N(satellite) < R_N(tower).

Figure 7b and Table 3 show that Φ(satellite) $\approx 0.90\Phi$(tower), suggesting a slight compensation for the underspecification of $R_{S\downarrow}$ through the underspecification of G from the satellite data. However, this evaluation assumes the energy balance to be closed in the tower data (i.e. $R_N - G = \lambda E + H$), which typically is not the case, with $\lambda E + H$ often falling short of $R_N - G$ by 20 % (Wilson et al., 2002). Because the causes of this energy imbalance remain controversial (see Foken, 2008, for a review), it is difficult to estimate how much the tower values of $\lambda E + H$ are actually biased and hence the extent to which this bias affects our evaluation. Stoy et al. (2013) recently found a systematic relationship of the surface energy balance closure with landscape heterogeneity over 173 FLUXNET tower sites and reported an energy imbalance of 9–30 % over diverse biomes. In another study, Amiro et al. (2009) found relatively better fulfilment of energy balance closure by averaging data over longer periods. The monthly averages of (AIRS overpass time) 13:30 LT surface energy balance closure of the 30 sites used here (Table 5) show an average energy imbalance of ~20 % (ranging from 8 to 34 %). The errors in the tower data are also believed to be associated with different footprint characteristics for the different instruments used (Lin et al., 2008). For example, R_N observations typically have a footprint size of ~10 m^2, whilst air properties (e.g. air temperature, humidity) have footprint sizes of > 1 km^2. By way of illustration, if, in the worst case, the entire energy imbalance were to be attributed exclusively to $\lambda E + H$ (i.e. $R_N - G$ are quantified correctly), then the true midday $\lambda E + H$ could be some 20 % greater (Wilson et al., 2002). As a result, the present bias seen in Table 3 would change to Φ(satellite) $\approx 0.72\Phi$(tower) , again suggesting $R_{S\downarrow}$ to be the main source of bias in the satellite retrievals for both R_N and Φ.

We have only evaluated the satellite retrievals using data from terrestrial sites, and clearly it would be worthwhile repeating this for the ocean retrievals if possible. We have held back on this evaluation here because of the lack of an extensive network of instantaneous latent and sensible heat flux or radiative flux data over the oceans, although we note that the SEAFLUX project within the Global Energy and Water Experiment (GEWEX) initiative should give rise to such a database in the near future. From the terrestrial evaluation we would argue that the methodology employed here shows promise for specifying both noontime R_N and Φ, although the results suggest the need for improvements, particularly in the specification of $R_{S\downarrow}$. More detailed studies evaluating the representativeness of each tower site footprint in relation to the 1° scale within which it is situated could prove useful in this regard as would methods for cloud-proofing the satellite retrievals under persistent cloudy-sky conditions. Similarly, evaluation under extreme conditions (e.g. high altitudes and latitudes) is also required.

5 Conclusions

We have demonstrated a novel retrieval for midday (13:30 LT) surface net available energy (Φ) by blending the monthly atmospheric and land surface variables of AIRS and MODIS sensors. We have attempted to structure the method such that the Φ retrieval does not depend on any offline calibration. We performed a two-step evaluation of the retrieved values for Φ at some representative surface radiation measurement sites and also over 30 FLUXNET sites from all over the globe.

Current Φ estimates performed well when compared against high spatio-temporal in situ measurements and coarse-spatial-resolution satellite retrievals. Consistent underestimation of R_N was noted due to the underestimation of shortwave radiation, and in tropical latitudes the R_N (and Φ) agreement was relatively weaker. Combination of high-frequency cloud dynamics and relatively low seasonal variability of $R_{S\downarrow}$ makes it difficult to accurately model both Φ and R_N. One of the key challenges in modelling R_N (and Φ) in the tropics is to account for fast-changing cloud cover fraction and cloud optical properties throughout the day.

With the availability of high-spatial-resolution (1–5 km) MODIS day–night optical and thermal data, our present approach could be extended to derive high-spatial-resolution Φ estimates at the global scale. This could be achieved by estimating MODIS-based day–night R_N and combining day–night R_N with day–night T_S observations. At the same time, the current methodology could also be used on high-temporal-frequency observations of geostationary satellites (e.g. GOES and METEOSAT). Once surface heat uptake and heat capacity (through Eq. 10) has been estimated, hourly G and Φ could be determined from hourly R_N and T_S observations of geostationary satellites by assuming conservation of heat capacity over a particular day. Operational generation of satellite-based Φ product would be a valuable resource for a variety of investigations, such as estimating latent and sensible heat, evaluation of Earth system model outputs and quantifying the land–atmosphere coupling strength. Given that we have resorted to the minimal amount of calibration in deriving Φ, it would appear sensible to adopt a similar philosophy in developing satellite-based schemes for latent and sensible heat fluxes as proposed in M2.

In addition to opportunities to specify large-scale surface heat and water vapour fluxes, the heat capacity estimates made here clearly carry information on variations in terrestrial properties such as surface moisture storage, and we envisage that studies to develop this concept further could prove fruitful, particularly because of the emergence of satellite microwave data against which the results could be compared.

Acknowledgements. We would like to acknowledge the Goddard Earth Sciences Data and Information Services Center (GES DISC) and staff involved with Level 1 and Atmosphere Archive and Distribution System (LAADS), NASA, for making the AIRS and MODIS data available. We kindly acknowledge all the site PIs who provided terrestrial flux data through the FLUXNET data archive. The AmeriFlux regional network component of this archive is supported with funding from the US Department of Energy through its Terrestrial Carbon project. K. Mallick would like to thank Loise Wandera for helping with map preparation. The authors declare no conflict of interests. This work was originally developed under NERC grant number NEE0191531.

Edited by: P. Stoy

References

Adhikari, L., Wang, Z., and Whiteman, D.: Cloudy Assessment within an Atmospheric Infrared Sounder Pixel by Combining Moderate Resolution Spectroradiometer and Atmospheric Radiation Measurement Program Ground-Based Lidar and Radar Measurements, Sixteenth ARM Science Team Meeting Proceedings, Albuquerque, NM, March 27–31, 2006.

Amiro, B. D.: Measuring boreal forest evapotranspiration using the energy balance residual, J. Hydrol., 366, 112–118, 2009.

Ammann, C., Flechard, C. R., Leifeld, J., Neftel, A., and Fuhrer, J.: The carbon budget of newly established temperature grassland depends on management intensity, Agric. Ecosys. Environ., 121, 5–20, 2007.

Anderson, M. C., Allen, R. G., Morse, A., and Kustas, W. P.: Use of Landsat thermal imagery in monitoring evapotranspiration and managing water resources, Remote Sens. Environ., 122, 50–65, 2012.

Anthoni, P. M., Knohl, A., Rebmann, C., Freibauer, A., Mund, M., Ziegler, W., Kolle, O., and Schulze, E. D.: Forest and agricultural land-use-dependent CO_2 exchange in Thuringia, Germany, Glob. Change Biol., 10, 2005–2019, 2004.

Aubinet, M., Chermanne, B., Vandenhaute, M., Longdoz, B., Yernaux, M., and Laitat, F.: Long term carbon dioxide exchange above a mixed forest in the Belgian Ardennes, Agr. Forest Meteorol., 108, 293–315, 2001.

Aumann, H. H., Chahine, M. T., Gautier, C., Goldberg, M. D., Kalnay, E., McMillin, L. M., Revercomb, H., Rosenkranz, P. W., Smith, W. L., Staelin, D. H., Strow, L., L., and Susskind, J.: AIRS/AMSU/HSB on the aqua mission: design, science objectives, data products and processing systems, IEEE T. Geosci. Remote, 41, 253–264, 2003.

Baker, J. M. and Baker, D. G.: Long-term ground heat flux and heat storage at a mid-latitude site, Climatic Change, 54, 295–303, 2002.

Baldocchi, D. D., Falge, E., Gu, L., Olson, R., Hollinger, D., Running, S., Anthoni, P., Bernhofer, C., Davis, K., Evans, R., Fuentes, J., Goldstein, A., Katul, G., Law, B., Lee, X., Malhi, Y., Meyers, T., Munger, W., Oechel, W., Paw U, K. T., Pilegaard, K., Schmid, H. P., Valentini, R., Verma, S., Vesala, T., Wilson, K., and Wofsy, S.: Fluxnet: a new tool to study the temporal and spatial variability of ecosystem-scale carbon dioxide, water vapor, and energy flux densities, B. Am. Meteorol. Soc., 82, 2415–3434, 2001.

Baldocchi, D. D., Xu, L. K., and Kiang, N.: How plant functional-type, weather, seasonal drought, and soil physical properties alter water and energy fluxes of an oak-grass savanna and an annual grassland, Agr. Forest Meteorol., 123, 13–39, 2004.

Bastiaanssen, W. G. M., Menenti, M., Feddes, R. A., and Holtslag, A. A. M.: The Surface Energy Balance Algorithm for Land (SEBAL): Part 1 formulation, J. Hydrol., 212–213, 198–212, 1998.

Batra, N., Islam, S., Venturini, V., Bisht, G., and Jiang L.: Estimation and comparison of evapotranspiration from MODIS and

AVHRR sensors for clear sky days over the southern great plains, Remote Sens. Environ., 103, 1–15, 2006.

Betts, A. K., Ball, J. H., and Beljaars, A. C. M.: Comparison between the land surface response of the ECMWF model and the FIFE-1987 data, Q. J. Roy. Met. Soc., 119, 975–1001, 1993.

Bhattacharya, B. K., Mallick, K., Patel, N. K., and Parihar, J. S.: Regional clear sky evapotranspiration over agricultural land using remote sensing data from Indian geostationary meteorological satellite, J. Hydrol., 387, 65–80, 2010.

Bhattacharya, B. K., Mallick, K., Nigam, R., Dakore, K. K, and Sheikh, A. M.: Efficiency based wheat yield prediction in a semi-arid climate using surface energy budgeting with satellite observations, Agr. Forest Meteorol., 151, 1394–1408, 2011.

Bindi, M., Miglietta, F., and Zipoli, G.:, Different methods for separating diffuse and direct components of solar radiation and their application in crop growth models, Climate Res., 2, 47 -54, 1992.

Bisht, G., Venturini, V., Islam, S., and Jiang, L.: Estimation of Net Radiation Using MODIS (Moderate Resolution Imaging Spectroradiometer) Terra data for clear sky days, Remote Sens. Environ., 97, 52–67, 2005.

Bisht, G. and Bras, R.: Estimation of net radiation from the MODIS data under all sky conditions: Southern Great Plains case study, Remote Sens. Environ., 114, 1522–1534, 2010.

Bisht, G. and Bras, R. L.: Estimation of net radiation from the Moderate Resolution Imaging Spectroradiometer over the Continental United States, IEEE T. Geosci. Remote, 49, 2448–2462, 201, doi:10.1109/TGRS.2010.2096227, 2011.

Bounoua, L., Hall, F. G., Sellers, P. J., Kumar, A., Collatz, G. J., Tucker, C. J., and Imhoff, M. L.: Quantifying the negative feedback of vegetation to greenhouse warming: A modeling approach, Geophys. Res. Lett., 37, L23701, doi:10.1029/2010GL045338, 2010.

Bowen, I. S.: The ratio of heat losses by conduction and by evaporation from any water surface, Phy. Rev., 27, 779–787, 1926.

Cai, G., Xue, Y., Hu, Y., Guo, J., Wang, Y., and Qi, S.: Quantitative study of net radiation from MODIS data in the lower boundary layer in Poyang Lake area of Jiangxi Province, China, Int. J Remote Sens., 28, 4381–4389, 2007.

Cano, D., Monget, J. M., Albuisson, M., Guillard, H., Regas, N., and Wald, L.: A method for the determination of the global solar radiation from meteorological satellite data, Sol. Energy, 37, 31–39, 1986.

Carswell, F. E., Costa, A. L., Palheta, M., Malhi, Y., Meir, P., Costa, J. D. R., Ruivo, M. D., Leal, L. D. M., Costa, J. M. N., Clement, R. J., and Grace, J.: Seasonality in CO_2 and H_2O flux at an eastern Amazonian rain forest, J. Geophys. Res.-Atmos., 107, D208076, doi:10.1029/2000JD000284, 2002.

Chen, M., Zhuang, Q., and He, Y.: An Efficient Method of Estimating Downward Solar Radiation Based on the MODIS Observations for the Use of Land Surface Modeling, Remote Sens., 6, 7136–7157, doi:10.3390/rs6087136, 2014.

Chen, X., Huang, X., Loeb, N.G., and Wei, H.: Comparisons of Clear-Sky Outgoing Far-IR Flux Inferred from Satellite Observations and Computed from the Three Most Recent Reanalysis Products, J. Climate, 26, 478–494, doi:10.1175/JCLI-D-12-00212.1, 2013.

Choudhury, B. J.: Relationships between vegetation indices, radiation absorption, and net photosynthesis evaluated by a sensitivity analysis, Remote Sens. Environ., 22, 209–233, 1987.

Choudhury, B. J.: Estimating gross photosynthesis using satellite and ancillary data: approach and preliminary results, Remote Sens. Environ., 75, 1–21, 2001.

Cook, B. D., Davis, K. J., Wang, W., Desai, A., Berger, B. W., Teclaw, R. M., Martin, J. G., Bolstad, P. V., Bakwin, P. S., Yi, C., and Heilman, W.: Carbon exchange and venting anomalies in an upland deciduous forest in northern Wisconsin, USA, Agr. Forest Meteorol., 126, 271–295, 2004.

Corbari, C., Sobrino, J. A., Mancini, M., and Hidalgo, V.: Land surface temperature representativeness in a heterogeneous area through a distributed energy-water balance model and remote sensing data, Hydrol. Earth Syst. Sci., 14, 2141–2151, doi:10.5194/hess-14-2141-2010, 2010.

de Araújo, A. C., Nobre, A. D., Kruijt, B., Elbers, J. A., Dallarosa, R., Stefani, P., von Randow, C., Manzi, A. O., Culf, A. D., Gash, J. H. C., Valentini, R., and Kabat., P.: Comparative measurements of carbon dioxide fluxes from two nearby towers in a central Amazonian rainforest: the Manaus LBA site, J. Geophys. Res. Atmos., 107, 8090, doi:10.1029/2001JD000676, 2002.

Dubayah, R. and Loechel, S.: Modeling Topographic solar radiation using GOES data, J. Appl. Meteorol., 36, 141–154, 1997.

Fischer, M. L., Billesbach, D. P., Berry, J. A., Riley, W. J., and Torn, M. S.: Spatiotemporal variations in growing season exchanges of CO_2, H_2O, and sensible heat in agricultural fields of the Southern Great Plains, Earth Int., 11, 1–21, 2007.

Foken, T.: The energy balance closure problem: an overview, Ecol. Appl., 18, 1351–1367, 2008.

Foken, T., Wimmer, F., Mauder, M., Thomas, C., and Liebethal, C.: Some aspects of the energy balance closure problem, Atmos. Chem. Phys., 6, 4395–4402, doi:10.5194/acp-6-4395-2006, 2006.

Fouquart, Y. and Bonnel, B.: Computations of solar heating of the earth's atmosphere: A new parameterization, Beitr. Phys. Atmos., 53, 35–62, 1980.

Gao, J. and Wu, S.: Simulated Effects of Land Cover Conversion on the Surface Energy Budget in the Southwest of China, Energies, 7, 1251–1264, doi:10.3390/en7031251, 2014.

Gao, W., Coulter, R. L., Lesht, B. M., Qiu, J., and Wesely, M. L.: Estimating clear-sky regional surface fluxes in the southern great plains atmospheric radiation measurement site with ground measurements and satellite observations, J. Appl. Meteorol., 37, 5–22, 1998.

Gentine, P., Entekhabi, D., Chehbouni, A., Boulet, G., and Duchemin, B.: Analysis of evaporative fraction diurnal behaviour, Agr. Forest Meteorol., 143, 13–29, 2007.

Giambelluca, T. W., Hölscher, D., Bastos, T. X., Frazão, R. R., Nullet, M. A., and Zeigler, A. D.: Observations of albedo and radiation balance over postforest land surfaces in the eastern Amazon Basin, J. Climate, 10, 919–928, 1997.

Gilmanov, T. G., Soussana, J. F., Aires, L., Allard, V., Ammann, C., Balzarolo, M., Barcza, Z., Bernhofer, C., Campbell, C. L., Cernusca, A., Cescatti, A., Brown, J. C., Dirks, O. M., Dore, S., Eugster, W., Fuhrer, J., Gimeno, C., Gruenwald, T., Haszpra, L., Hensen, A., Ibrom, A., Jacobs, A. F. G., Jones, M. B., Lanigan, G., Laurila, T., Ohila, A., Manca, G., Marcolla, B., Nagy, Z., Pilegaard, K., Pinter, K., Pio, C., Raschi, A., Rogiers, N., Sanz, M. J., Stefani, P., Sutton, M., Tuba, Z., Valentini, R., Williams, M. L., and Wohlfahrt, G.: Partitioning European grassland net ecosystem CO_2 exchange into gross primary productivity and ecosystem respiration using light response function analysis, Agric. Ecosys. Environ., 121, 93–120, 2007.

Goldstein, A. H., Hultman, N. E., Fracheboud, J. M., Bauer, M. R., Panek, J. A., Xu, M., Qi, Y., Guenther, A. B., and Baugh, W.: Effects of climate variability on the carbon dioxide, water, and sensible heat fluxes above a ponderosa pine plantation in the Sierra Nevada (CA), Agr. Forest Meteorol., 101, 113–129, 2000.

Gough, C. M., Flower, C. E., Vogel, C. S. Dragoni, D., and Curtis, P. S.: Whole-ecosystem labile carbon production in a north temperate deciduous forest, Agr. Forest Meteorol., 149, 1531–1540, 2009.

Goulden, M. L., Miller, S. D., and da Rocha, H. R., Menton, M. C., De Freitas, H. C., Figuera, A. M. E. S., and De Sousa, C. A. D.: Diel and seasonal patterns of tropical forest CO_2 exchange, Ecol. Appl., 14, S42–S54, 2004.

Granier, A., Loustau, D., and Breda, N.: A generic model of forest canopy conductance dependent on climate, soil water availability and leaf area index, Ann. For. Sc., 57, 755–765, 2000a.

Granier, A., Biron, P., and Lemoine, D.: Water balance, transpiration and canopy conductance in two beech stands, Agr. Forest Meteorol., 100, 291–308, 2000b.

Gubler, S., Gruber, S., and Purves, R. S.: Uncertainties of parameterized surface downward clear-sky shortwave and all-sky longwave radiation., Atmos. Chem. Phys., 12, 5077–5098, doi:10.5194/acp-12-5077-2012, 2012.

Gueymard, C.: Direct solar transmittance and irradiance predictions with broadband models. Part I: detailed theoretical performance assessment, Sol. En., 74, 355–379, 2003.

Hammerle, A., Haslwanter, A., Tappeiner, U., Cernusca, A., and Wohlfahrt, G.: Leaf area controls on energy partitioning of a temperate mountain grassland, Biogeosciences, 5, 421–431, doi:10.5194/bg-5-421-2008, 2008.

Hearty, T. J., Savtchenko, A., Tian, B., Fetzer, E., Yung, Y. L., Theobald, M., Vollmer, B., Fishbein, E., and Won, Y.: Estimating sampling biases and measurement uncertainties of AIRS/AMSU-A temperature and water vapor observations using MERRA reanalysis, J. Geophys. Res. Atm., 119, 2725–2741, doi:10.1002/2013JD021205, 2014.

Heitman, J. L., Horton, R., Sauer, T J., Ren, T. S., and Xiao, X.: Latent heat in soil heat flux measurements, Agr. Forest Meteorol., 150, 1147–1153, 2010.

Hildebrandt, A., Aufi, M. A., Amerjeed, M., Shammas, M., and Eltahir, E. A. B.: Ecohydrology of a seasonal cloud forest in Dhofar: 1. Field experiment, Water Resour. Res., 43, W10411, doi:10.1029/2006WR005261, 2007.

Hirano, T., Hirata, R., Fujinuma, Y., Saigusa, N., Yamamoto, S., Harazono, Y., Takada, M., Inukai, K., and Inoue, G.: CO_2 and water vapor exchange of a larch forest in northern Japan, Tellus B, 55, 244–257, 2003.

Hirano, T., Segah, H., Harada, T., Limin, S., June, T., Hirata, R., and Osaki, M.: Carbon dioxide balance of a tropical peat swamp forest in Kalimantan, Indonesia, Glob. Change Biol., 13, 412–425, 2007.

Hollinger, D. Y., Goltz, S. M., Davidson, E. A., Lee, J. T., Tu, K., and Valentine, H. T.: Seasonal patterns and environmental control of carbon dioxide and water vapour exchange in an ecotonal boreal forest, Glob. Change Biol., 5, 891–902, 1999.

Hou, J., Jia, G., Zhao, T., Wang, H., and Tang, B.: Satellite based estimation of daily average net radiation under clear sky conditions, Adv. Atmos. Sci., 31, 705–720, 2014.

Huang, G. H., Liu, S. M., and Liang, S. L.: Estimation of net surface shortwave radiation from MODIS data, Int. J. Remote Sens., 33, 804–825, 2012.

Huang, G., Ma, M., Liang, S., Liu, S., and Li, X.: A LUT® approach to estimate surface solar irradiance by combining MODIS and MTSAT data, J. Geophys. Res., 116, D22201, doi:10.1029/2011JD016120, 2011.

Humes, K. S., Kustas, W. P., and Moran, M. S.: Use of remote sensing and reference site measurements to estimate instantaneous surface energy balance components over a semiarid rangeland watershed, Water Resour. Res., 30, 1363–1373, 1994.

Hutyra, L. R., Munger, J. W., Saleska, S. R., Gottlieb, E., Daube, B. C., Dunn, A. L., Amaral, D. F., de Camargo, P. B., and Wofsy, S. C.: Seasonal controls on the exchange of carbon and water in an Amazonian rain forest, J. Geophys. Res., 112, G03008, doi:10.1029/2006JG000365, 2007.

Hsieh C. I., Huang, C. W., and Kiely, G.: Long term estimation of soil heat flux by single layer soil temperature, Int. J. Biometeorol., 53, 113–123, 2009.

Hwang K., Choi, M., Lee, S. O., and Seo, J.-W.: Estimation of instantaneous and daily net radiation from MODIS data under clear sky conditions: a case study in East Asia, Irrig. Sci., 31, 1173–1184, doi:10.1007/s00271-012-0396-3, 2013.

Jacobs, J. M., Anderson, M. C., Friess, L. C., and Diak, G. R.: Solar radiation, longwave radiation and emergent wetland evapotranspiration estimates from satellite data in Florida, USA, Hydrol. Sc., 49, 461–176, 2004.

Jaksic, V., Kiely, G., Albertson, J., Katul, G., and Oren, R.: Net ecosystem exchange of grassland in contrasting wet and dry years, Agr. Forest Meteorol., 139, 323–334, 2006.

Jin, Y., Randerson, J. T., and Goulden, M. L.: Continental-scale net radiation and evapotranspiration estimated using MODIS satellite observations, Remote Sens. Environ., 115, 2302–2319, 2011.

Jin, Z., Charlock, T. P., Smith Jr., W. L., and Rutledge, K.: A parameterization of ocean surface albedo, Geophys. Res. Lett., 31, L22301, doi:10.1029/2004GL021180, 2004.

Katul, G. G., Leuning, R., and Oren, R.: Relationship between plant hydraulic and biochemical properties derived from a steady-state coupled water and carbon transport model, Pl. Cell Environ., 26, 339–350, 2003.

Kaufmann, Y. J. and Koran, I.: Smoke and Pollution Aerosol Effect on Cloud Cover, Science, 313, 655–658, 2006.

Kergoat, L., Grippa, M., Baille, A., Eymard, L., Lacaze, R., Mougin, E., Ottlé, C., Pellarin, T., Polcher, J. Rosnay, P. de., Roujean, J. -L., Sandholt, I., Taylor, C. M., Zin, I., and Zribi, M.: Remote sensing of the land surface during the African Monsoon Multidisciplinary Analysis (AMMA), Atmos. Sci. Lett., 12, 129–134, 2011.

Kim, J. and Hogue, T. S., Evaluation of a MODIS-based potential evapotranspiration product at the point-scale, J. Hydrometeorol., 9, 444–460, 2008.

Kustas, W. P. and Norman, J. M.: Evaluating the Effects of Subpixel Heterogeneity on Pixel Average Fluxes, Remote Sens. Environ., 74, 327–342, 2000.

Lakshmi, V. and Zehrfuhs, D.: Normalization and comparison of surface temperatures across a range of scales, IEEE T. Geosci. Remote 40, 2636–2646, 2002.

Li, J. and Islam, S.: On the estimation of soil moisture profile and surface fluxes partitioning from sequential assimilation of surface layer soil moisture, J. Hydrol., 220, 86–103, 1999.

Liang, S., Strahler, A. H., and Walthall, C.: Retrieval of land surface albedo from satellite observations: A simulation study, J. Appl. Meteorol., 38, 712–725, 1999.

Liang, S., Shuey, C., Russ, A., Fang, H., Chen, M., Walthall, C., and Daughtry, C.: Narrowband to Broadband Conversions of Land Surface Albedo: II. Validation, Remote Sens. Environ., 84, 25–41, 2002.

Liang, X., Wood, E. F., Lettenmaier, D. P., Lohmann, D., Boone, A., Chang, S., Chen, F., Dai, Y., Desborough, C., Dickinson, R. F., Duan, Q., Ek, M., Gusev, Y. M., Habets, F., Irannejad, P., Koster, R., Mitchell, K. E., Nasonova, O. N., Noilhan, J., Schaake, J., Schlosser, A., Shao, Y., Shmakin, A. B., Verseghy, D., Warrach, K., Wetzel, P., Xue, Y., Yang, Z. -L., and Zeng, Q.: The Project for Intercomparison of Land-surface Parameterization Schemes (PILPS) phase 2 (c) Red-Arkansas River basin experiment: 2. Spatial and temporal analysis of energy fluxes, Glob. Planet. Change, 19, 137–159, 1998.

Lin, B., Stackhouse Jr., P. W., Minnis, P., Wielicki, B. A., Hu, Y., Sun, W., Fan, T.-F., and Hinkelman, L. M.: Assessment of global annual atmospheric energy balance from satellite observations, J. Geophys. Res., 113, D16114, doi:10.1029/2008JD009869, 2008.

Liu, Z., Vavrus, S., He, F., Wen, N., and Zhong, Y.: Rethinking tropical ocean response to global warming: the enhanced equatorial warming, J. Climate, 18, 4684–4700, 2005.

Longman, R. J., Giambelluca, T. W., and Frazier, A. G.: Modeling clear-sky solar radiation across a range of elevations in Hawai'i: Comparing the use of input parameters at different temporal resolutions, J. Geophys. Res., 117, D02201, doi:10.1029/2011JD016388, 2012.

Ma, Y., Su, Z., Li, Z., Koike, T., and Menenti, M.: Determination of regional net radiation and soil heat flux over a heterogeneous landscape of the Tibetan Plateau, Hydrol. Process., 16, 2963–2971, 2002.

Mallick, K., Bhattacharya, B. K., Rao, V. U. M., Reddy, D. R., Banerjee, S., Hoshali, V., Pandey, V., Kar, G., Mukherjee, J., Vyas, S. P., Gadgil, A. S., and Patel, N. K.: Latent heat flux estimation in clear sky days over Indian agroecosystems using noontime satellite remote sensing data, Agr. Forest Meteorol., 149, 1646–1665, 2009.

Mallick, K., Jarvis, A. J., Boegh, E., Fisher, J. B., Drewry, D. T., Tu, K. P., Hook, S. J., Hulley, G., Ardo, J., Beringer, J., Arain, A., and Niyogi, D.: A surface temperature initiated closure (STIC) for surface energy balance fluxes, Remote Sens. Environ., 141, 243–261, 2014a.

Mallick, K., Jarvis, A., Wohlfahrt, G., Kiely, G., Hirano, T., Miyata, A., Yamamoto, S., and Hoffmann, L.: Components of near-surface energy balance derived from satellite soundings – Part 2: Noontime latent heat flux, Biogeosciences, 11, 7369–7382, doi:10.5194/bg-11-7369-2014, 2014b.

Mamadou, O., Cohard, J. M., Galle, S., Awanou, C. N., Diedhiou, A., Kounouhewa, B., and Peugeot, C.: Energy fluxes and surface characteristics over a cultivated area in Benin: daily and seasonal dynamics, Hydrol. Earth Syst. Sci., 18, 893–914, doi:10.5194/hess-18-893-2014, 2014.

Massaquoi, J. G. M.: Global solar radiation in Sierra Leone (West Africa), Solar Wind Tech., 5, 281–283, 1988.

Mayocchi, C. L. and Bristow, K. L.: Soil surface heat flux: some general questions and comments on measurements, Agr. Forest Meteorol., 75, 43–50, 1995.

Mecikalksi, J. R., Diak, G. R., Anderson, M. C., and Norman, J. M.: Estimating fluxes on continental scales using remotely-sensed data in an atmospheric-land exchange model, J. Appl. Meteorol., 35, 1352–1369, 1999.

Meyers, T. P. and Hollinger, S. E.: An assessment of storage terms in the surface energy balance of maize and soybean, Agr. Forest Meteorol., 125, 105–115, 2004.

Miglietta, F., Gioli, B., Brunet, Y., Hutjes, R. W. A., Matese, A., Sarrat, C., and Zaldei, A.: Sensible and latent heat flux from radiometric surface temperatures at the regional scale: methodology and evaluation, Biogeosciences, 6, 1975–1986, doi:10.5194/bg-6-1975-2009, 2009.

Mlawer, E. J., Taubman, S. J., Brown, P. D., Iacono, M. J., and Clough, S. A.: Radiative transfer for inhomogeneous atmosphere: RRTM, a validated correlated-*k* model for the longwave, J. Geophys. Res., 120D, 16663–16682, 1997.

Mlynczak, P. E., Smith, G. L., and Doelling, D. R.: The annual cycle of Earth radiation budget from Clouds and the Earth's Radiant Energy System (CERES) data, J. Appl. Meteorol. Clim., 50, 2490–2503, doi:10.1175/JAMC-D-11-050.1, 2011.

Monteith, J. L.: Evaporation and environment, Symp. Soc. Exp. Biol., 19, 205–234, 1965.

Morcrette, J. J.: Assessment of the ECMWF model cloudiness and surface radiation fields at the ARM SGP site, Month. Weather Rev., 130, 257–277, 2002.

Mu, Q., Heinsch, F. A., Zhao, M., and Running, S. W.: Development of a global evapotranspiration algorithm based on MODIS and global meteorology data, Remote Sens. Environ., 111, 519–536, 2007.

Mu, Q., Zhao M., and Running S. W.: Improvements to a MODIS Global Terrestrial Evapotranspiration Algorithm, Remote Sens. Environ., 115, 1781–1800, 2011.

Murray, T. and Verhoef, A.: Moving towards a more mechanistic approach in the determination of soil heat flux from remote measurements – I. A universal approach to calculate thermal inertia, Agr. Forest Meteorol., 147, 80–87, doi:10.1016/j.agrformet.2007.07.004, 2007.

Niemela, S., Raisanen, P., and Savijarvi, H.: Comparison of surface radiative flux parameterizations Part I: Longwave radiation, Atmos. Res., 58, 1–18, 2001.

Ochsner, T. E., Sauer, T. J., and Horton, R.: Soil heat storage measurements in energy balance studies, Agron. J., 99, 311–319, 2007.

Oliphant, A. J., Spronken-Smith, R. A., Sturman, A. P., and Owens, I. F.: Spatial Variability of Surface Radiation Fluxes in Mountainous Terrain, J. Appl. Meteorol., 42, 113–128, 2003.

Peng, J., Liu, Y., Zhao, X., and Loew, A.: Estimation of evapotranspiration from MODIS TOA radiances in the Poyang Lake basin, China, Hydrol. Earth Syst. Sci., 17, 1431–1444, doi:10.5194/hess-17-1431-2013, 2013.

Penman, H. L.: Natural evaporation from open water, bare soil and grass, Proc. Roy. Soc. London, A193, 120–145, 1948.

Pinker, R. T. and Laszlo, I.: Modeling Surface Solar irradiance for satellite applications on a global scale, J. Appl. Meteorol., 31, 194–211, 1992.

Prata, A. J.: A new long-wave formula for estimating downward clear-sky radiation at the surface, Q. J. R. Meteorol. Soc., 122, 1127–1151, 1996.

Priestley, K. J., Smith, G. L., Thomas, S., Cooper, D., Lee, R. B., Walikainen, D., Hess, P., Szewczyk, Z. P., and Wilson, R.: Radiometric Performance of the CERES Earth Radiation Budget Climate Record Sensors on the EOS Aqua and Terra Spacecraft through April 2007, J. Atmos. Ocean. Tech., 28, 3–21, doi:10.1175/2010JTECHA1521.1, 2011.

Quaas, J., Stevens, B., Stier, P., and Lohmann, U.: Interpreting the cloud cover – aerosol optical depth relationship found in satellite data using a general circulation model, Atmos. Chem. Phys., 10, 6129–6135, doi:10.5194/acp-10-6129-2010, 2010.

Reichstein, M., Tenhunen, J., Roupsard, O., Ourcival, J. M., Rambal, S., Miglietta, F., Peressotti, A., Pecchiari, M., Tirone, G., and Valentini, R.: Inverse modeling of seasonal drought effects on canopy CO_2/H_2O exchange in three Mediterranean ecosystems, J. Geophys. Res.-Atmos., 108, D23, doi:10.1029/2003JD003430, 2003.

Ryu, Y., Kang, S., Moon, S. K., and Kim, J.: Evaluation of land surface radiation balance derived from moderate resolution imaging spectroradiometer (MODIS) over complex terrain and heterogeneous landscape on clear sky days, Agr. Forest Meteorol., 148, 1538–1552, 2008.

Saigusa, N. Yamamoto, S., Murayama, S., Kondo, H., and Nishimura, N.: Gross primary production and net ecosystem exchange of a cool- temperate deciduous forest estimated by the eddy covariance method, Agr. Forest Meteorol., 112, 203–215, 2002.

Saito, M., Miyata, A., Nagai, H., and Yamada, T.: Seasonal variation of carbon dioxide exchange in rice paddy field in Japan, Agr. Forest Meteorol., 135, 93–109, 2005.

Salomon, J. G., Schaff, C. B., Strahler, A. H., Gao, F., and Jin, Y. F.: Validation of the MODIS bidirectional reflectance distribution function and albedo retrievals using combined observations from the Aqua and Terra platforms, IEEE T. Geosci. Remote, 44, 1555–1565, 2006.

Santanello, Jr., J. A. and Friedl, M. A.: Diurnal Covariation in Soil Heat Flux and Net Radiation, J. Appl. Meteorol., 42, 851–862, 2003.

Sauer, T. J. and Horton R.: Soil heat flux, in: Micrometeorology in Agricultural Systems, edited by: Hatfield, J. L. and Baker, J. M., Am. Soc. Agr., Madison, Wisconsin, USA, 131–154, 2005.

Scholes, R. J., Gureja, N., Giannecchinni, M., Dovie, D., Wilson, B., Davidson, N., Piggott, K., McLoughlin, C., van der Velde, K., Freeman, A., Bradley, S., Smart, R., and Ndala, S.:. The environment and vegetation of the flux measurement site near Skukuza, Kruger National Park, Koedoe, 44, 73–83, 2001.

Schwartz, S. E.: Heat capacity, time constant, and sensitivity of Earth's climate system, J. Geophys. Res., 112, D24S05, doi:10.1029/2007JD008746, 2007.

Seidel, F. C., Kokhanovsky, A. A., and Schaepman, M. E.: Fast and simple model for atmospheric radiative transfer, Atmos. Meas. Tech., 3, 1129–1141, doi:10.5194/amt-3-1129-2010, 2010.

Stackhouse, P. W., Gupta, S. K., Cox, S. J., Chiacchio, M., and Mikovitz, J. C.: The WCRP/GEWEX Surface Radiation Budget Project Release 2: An assessment of surface fluxes at 1° resolution, edited by: Smith, W. L. and Timofeyev, Y. M., IRS 2000: Current problems in atmospheric radiation, International Radiation Symposium, St. Petersburg, Russia, 24–29, 2000.

Stephens, G. L., Wild, M., Stackhouse, P. W., L'Ecuyer, T., Kato, S., and Henderson, D. S.: The Global Character of the Flux of Downward Longwave Radiation, J. Climate, 25, 2329–2340, doi:10.1175/JCLI-D-11-00262.1, 2012.

Stisen, S., Sandholt, I., Nørgaard, A., Fensholt, R., and Jensen, K. H.: Combining the triangle method with thermal inertia to estimate regional evapotranspiration applied to MSG-SEVIRI data in the Senegal river basin, Remote Sens. Environ., 112, 1242–1255, 2008.

Stoy, P. C., Mauder, M., Foken, T., Marcolla, B., Boegh, E., Ibrom, A., Arain, M. A., Arneth, A., Aurela, M., Bernhofer, C., Cescatti, A., Dellwik, E., Duce, P., Gianelle, D., van Gorsel, E., Kiely, G., Knohl, A., Margolis, H., McCaughey, H., Merbold, L., Montagnani, L., Papale, D., Reichstein, M., Saunders, M., Serrano-Ortiz, P., Sottocornola, M., Spano, D., Vaccari, F., Varlagin, A.: A data-driven analysis of energy balance closure across FLUXNET research sites: The role of landscape scale heterogeneity, Agr. Forest Meteorol., 171–172, 137–152, doi:10.1016/j.agrformet.2012.11.004, 2013.

Stramma, L., Cornillon, P., Weller, R. A., Price, J. F., and Briscoe, M. G.: Large diurnal sea surface temperature variability: satellite and in situ measurements, J. Geophys. Res., 16, 827–837, 1986.

Sultan, S., Wu, R., and Ahmed, I..: Impact of Terrain and Cloud Cover on the Distribution of Incoming Direct Solar Radiation over Pakistan, J. Geogr. Inform. Sys., 6, 70–77, doi:10.4236/jgis.2014.61008, 2014.

Sun, Z., Gebremichael, M., Wang, Q., Wang, J., Sammis, T. W., and Nickless, A.: Evaluation of Clear-Sky Incoming Radiation Estimating Equations Typically Used in Remote Sensing Evapotranspiration Algorithms, Remote Sens., 5, 4735–4752; doi:10.3390/rs5104735, 2013.

Tang, B., Li, Z.-L., and Zhang, R.: A direct method for estimating net surface shortwave radiation from MODIS data, Remote Sens. Environ., 103, 115–126, doi:10.1016/j.rse.2006.04.008, 2006.

Tamai, K., Abe, T., Araki, M., and Ito, H.: Radiation budget, soil heat flux and latent heat flux at the forest floor in warm, temperate mixed forest, Hydrol. Proc., 12, 2105–2114, 1998.

Thornton, P. E. and Running, S. W.: An improved algorithm for estimating incident daily solar radiation from measurements of temperature, humidity, and precipitation, Agr. Forest Meteorol., 93, 211–228, 1999.

Tittebrand, A. and Berger, F. H.: Spatial heterogeneity of satellite derived land surface parameters and energy flux densities for LITFASS-area, Atmos. Chem. Phys., 9, 2075–2087, doi:10.5194/acp-9-2075-2009, 2009.

Trenberth, K.: Changes in precipitation with climate change, Clim. Res., 47, 123–138, doi:10.3354/cr00953, 2011.

Trenberth, K., Fasullo, J., and Kiehl, J.: Earth's global energy budget, Bull. Am. Meteor. Soc., 90, 311–323, doi:10.1175/2008BAMS2634.1, 2009.

Urbanski, S., Barford, C., Wofsy, S. Kucharik, C., Pyle, E., Budney, J., Fitzjarrald, D., Czikowsky, M., and Munger, J. W.: Factors Controlling CO_2 Exchange at Harvard Forest on Hourly to Annual Time Scales, J. Geophys. Res., 112, G02020, doi:10.1029/2006JG000293, 2007.

Verhoef, A., Ottle, C., Cappelaere, B., Murray, T., Saux-Picart, S., Zribi, M., Maignan, F., Boulain, N., Demarty, J., and Ramier, D.: Spatio-temporal surface soil heat flux estimates from satellite data: results for the AMMA experiment at the Fakara (Niger) supersite, Agr. Forest Meteorol., 154–155, 55–66, doi:10.1016/j.agrformet.2011.08.003, 2012.

Verstraeten, W., Veroustraete, F., and Feyen, J.: Estimating evapotranspiration of European forests from NOAA-imagery at satellite overpass time: Towards an operational processing chain for integrated optical and thermal sensor data products, Remote Sens. Environ., 96, 256–276, 2005.

Wang, H. and Pinker, R. T., Shortwave radiative fluxes from MODIS: Model development and implementation, J. Geophys. Res., 114, D20201, doi:10.1029/2008JD010442, 2009.

Wielicki, B. A., Barkstrom, B. R., Baum, B. A., Charlock, T. P., Green, R. N., Kratz, D. P., Lee, R. B., Minnis, P., Smith, G. L., Wong, T. M., Young, D. F., Cess, R. D., Coakley, J. A., Crommelynck, D. A. H., Donner, L., Kandel, R., King, M. D., Miller, A. J., Ramanathan, V., Randall, D. A., Stowe, L. L., and Welch, R. M.: Clouds and the Earth's Radiant Energy System (CERES): Algorithm overview, IEEE Trans. Geosc. Remote Sens., 36, 1127–1141, 1998.

Wild, M., Folini, D., Schar, C., Loeb, N., Dutton, E. G., and Langlo, G. K.: The global energy balance from a surface perspective, Clim. Dynam., 40, 3107–3134, doi:10.1007/s00382-012-1569-8, 2013.

Wilson, K. B., Goldstein, A. H., Falge, E., Aubinet, M., Baldocchi, D., Berbigier, P., Bernhofer, Ch., Ceulemans, R., Dolman, H., Field, C., Grelle, A., Law, B., Meyers, T., Moncrieff, J., Monson, R., Oechel, W., Tenhunen, J., Valentini, R., and Verma, S.: Energy balance closure at FLUXNET sites, Agr. Forest Meteorol., 113, 223–243, 2002.

Wright, I. R., Gash, J. H. C., da Rocha, R., Shuttleworth, W. J., Nobre, C. A., Maitelli, G. T., Zamparoni, C. A. G. P., and Carvalho, P. R. A.: Dry season micrometeorology of central Amazonian ranchland, Q. J. Roy. Meteorol. Soc., 118, 1083–1099, 1992.

Zhang, J. H., Han, S. J., and Yu, G. R.: Seasonal variation in carbon dioxide exchange over a 200-year-old Chinese broad-leaved Korean pine mixed forest, Agr. Forest Meteorol., 137, 150–165, 2006.

9

Recent trends and drivers of regional sources and sinks of carbon dioxide

S. Sitch[1], P. Friedlingstein[1], N. Gruber[2], S. D. Jones[3], G. Murray-Tortarolo[1], A. Ahlström[4], S. C. Doney[5], H. Graven[6], C. Heinze[7,8,9], C. Huntingford[10], S. Levis[11], P. E. Levy[12], M. Lomas[13], B. Poulter[14], N. Viovy[15], S. Zaehle[16], N. Zeng[17], A. Arneth[18], G. Bonan[11], L. Bopp[15], J. G. Canadell[19], F. Chevallier[15], P. Ciais[15], R. Ellis[10], M. Gloor[20], P. Peylin[15], S. L. Piao[21], C. Le Quéré[3], B. Smith[4], Z. Zhu[22,23], and R. Myneni[24]

[1]University of Exeter, Exeter EX4 4QF, UK
[2]Institute of Biogeochemistry and Pollutant Dynamics, ETH Zurich, Zurich, Switzerland
[3]Tyndall Centre for Climate Change Research, University of East Anglia, Norwich NR4 7TJ, UK
[4]Lund University, Department of Physical Geography and Ecosystem Science, Sölvegatan 12, 223 62 Lund, Sweden
[5]Marine Chemistry and Geochemistry Department, Woods Hole Oceanographic Institution, 266 Woods Hole Road, Woods Hole, MA 02543, USA
[6]Department of Physics and Grantham Institute for Climate Change, Imperial College London, London SW7 2AZ, UK
[7]Geophysical Institute, University of Bergen, Bergen, Norway
[8]Bjerknes Centre for Climate Research, Bergen, Norway
[9]Uni Climate, Uni Research AS, Bergen, Norway
[10]Centre for Ecology and Hydrology, Benson Lane, Wallingford OX10 8BB, UK
[11]National Center for Atmospheric Research, Boulder, Colorado, USA
[12]Centre for Ecology and Hydrology, Bush Estate, Penicuik, Midlothian EH26 0QB, UK
[13]Department of Animal & Plant Sciences, University of Sheffield, Sheffield S10 2TN, UK
[14]Institute on Ecosystems and Department of Ecology, Montana State University, Bozeman, MT 59717, USA
[15]Laboratoire des Sciences du Climat et de l'Environnement, CEA CNRS UVSQ, 91191 Gif-sur-Yvette, France
[16]Biogeochemical Integration Department, Max Planck Institute for Biogeochemistry, P.O. Box 10 01 64, 07701 Jena, Germany
[17]Department of Atmospheric and Oceanic Science, University of Maryland, College Park, MD 20740, USA
[18]Karlsruhe Institute of Technology, Garmisch-Partenkirchen, Germany
[19]Global Carbon Project, CSIRO Oceans and Atmosphere Flagship, Canberra, Australia
[20]University of Leeds, School of Geography, Woodhouse Lane, Leeds LS9 2JT, UK
[21]College of Urban and Environmental Sciences, Peking University, Beijing 100871, China
[22]State Key Laboratory of Remote Sensing Science, Institute of Remote Sensing and Digital Earth, Chinese Academy of Sciences, Beijing 100101, China
[23]Center for Applications of Spatial Information Technologies in Public Health, Beijing 100101, China
[24]Department of Geography and Environment, Boston University, 675 Commonwealth Avenue, Boston, MA 02215, USA

Correspondence to: S. Sitch (s.a.sitch@exeter.ac.uk)

Abstract. The land and ocean absorb on average just over half of the anthropogenic emissions of carbon dioxide (CO_2) every year. These CO_2 "sinks" are modulated by climate change and variability. Here we use a suite of nine dynamic global vegetation models (DGVMs) and four ocean biogeochemical general circulation models (OBGCMs) to estimate trends driven by global and regional climate and atmospheric CO_2 in land and oceanic CO_2 exchanges with the atmosphere over the period 1990–2009, to attribute these trends to underlying processes in the models, and to quantify the uncertainty and level of inter-model agreement. The models were forced with reconstructed climate fields and observed global atmospheric CO_2; land use and land cover changes are not included for the DGVMs. Over the period 1990–2009, the DGVMs simulate a mean global land carbon sink of $-2.4 \pm 0.7\,\mathrm{Pg\,C\,yr^{-1}}$ with a small significant trend of $-0.06 \pm 0.03\,\mathrm{Pg\,C\,yr^{-2}}$ (increasing sink). Over the more limited period 1990–2004, the ocean models simulate a mean ocean sink of $-2.2 \pm 0.2\,\mathrm{Pg\,C\,yr^{-1}}$ with a trend in the net C uptake that is indistinguishable from zero ($-0.01 \pm 0.02\,\mathrm{Pg\,C\,yr^{-2}}$). The two ocean models that extended the simulations until 2009 suggest a slightly stronger, but still small, trend of $-0.02 \pm 0.01\,\mathrm{Pg\,C\,yr^{-2}}$. Trends from land and ocean models compare favourably to the land greenness trends from remote sensing, atmospheric inversion results, and the residual land sink required to close the global carbon budget. Trends in the land sink are driven by increasing net primary production (NPP), whose statistically significant trend of $0.22 \pm 0.08\,\mathrm{Pg\,C\,yr^{-2}}$ exceeds a significant trend in heterotrophic respiration of $0.16 \pm 0.05\,\mathrm{Pg\,C\,yr^{-2}}$ – primarily as a consequence of widespread CO_2 fertilisation of plant production. Most of the land-based trend in simulated net carbon uptake originates from natural ecosystems in the tropics ($-0.04 \pm 0.01\,\mathrm{Pg\,C\,yr^{-2}}$), with almost no trend over the northern land region, where recent warming and reduced rainfall offsets the positive impact of elevated atmospheric CO_2 and changes in growing season length on carbon storage. The small uptake trend in the ocean models emerges because climate variability and change, and in particular increasing sea surface temperatures, tend to counteract the trend in ocean uptake driven by the increase in atmospheric CO_2. Large uncertainty remains in the magnitude and sign of modelled carbon trends in several regions, as well as regarding the influence of land use and land cover changes on regional trends.

1 Introduction

Soon after the first high-precision measurements of atmospheric CO_2 started in the late 1950s, it became clear that the global-mean CO_2 growth rate is substantially lower than expected if all anthropogenic CO_2 emissions remained in the atmosphere (e.g. Keeling et al., 1976). The search for this "missing" carbon and the identification of the processes driving carbon sinks has been one of the dominating questions for carbon cycle research in the past decades (e.g. Tans et al., 1990; Sarmiento and Gruber, 2002; and others). While much progress has been achieved (e.g. Prentice et al., 2001; Sabine et al., 2004; Denman et al., 2007; Le Quéré et al., 2009), and estimates have converged considerably (Sweeney et al., 2007; Khatiwala et al., 2013; Wanninkhof et al., 2013), the spatial attribution of recent sink rates for the ocean and land, and particularly their changes through time, remain uncertain. To balance the global carbon budget, the combined sinks by land and ocean must have increased over recent decades (Keeling et al., 1995; Canadell et al., 2007; Raupach et al., 2008; Sarmiento et al., 2010; Gloor et al., 2010; Ballantyne et al., 2012). Sarmiento et al. (2010) showed that some of the increasing sinks are driven by the ocean, but also identified an even more substantial increase in the net uptake by the land biosphere between the 1980s and the 1990s. This increase in the global land and ocean sink has been sustained to date (Ballantyne et al., 2012).

There are several studies on the trends in carbon exchanges at the regional level based on atmospheric CO_2 observations (top-down approach) (Angert et al., 2005; Buermann et al., 2007; Chevallier et al., 2010; Sarmiento et al., 2010) and changes in high-latitude greenness on land (Nemani et al., 2003; Myneni et al., 1997) and changes in sea surface temperature in the ocean (Park et al., 2010). Atmospheric CO_2-based top-down approaches provide large-scale constraints on the land and ocean surface processes, but they cannot unambiguously identify the underlying processes or the regions driving these changes. Bottom-up studies using dynamic global vegetation models (DGVMs) or ocean biogeochemical general circulation models (OBGCMs) mechanistically represent many of the key land (Prentice et al., 2007) and ocean processes (Le Quéré et al., 2005), and offer the opportunity to investigate how changes in the structure and functioning of land ecosystems and the ocean in response to changing environmental conditions affect biogeochemical cycles. Therefore DGVMs and OBGCMs potentially allow for a more comprehensive analysis of surface carbon trends and provide insight into possible mechanisms behind regional trends in the carbon cycle.

There is a growing literature on regional carbon budgets for different parts of the world (Ciais et al., 1995; Phillips et al., 1998; Fan et al., 1998; Pacala et al., 2001; Janssens et al., 2003; Stephens et al., 2007; Piao et al., 2009; Lewis et al., 2009a; Ciais et al., 2010; Pan et al., 2011; Tjiputra et al., 2010; Roy et al., 2011; Schuster et al., 2013; Lenton et al., 2013), using bottom-up (inventory, carbon cycle models) and top-down methodologies, although they typically cover different time intervals. To date, no globally consistent attribution has been attempted for regional sources and sinks of atmospheric CO_2. This paper attempts to fill this gap by combining top-down and bottom-up approaches discussed in the regional syntheses of the REgional

Carbon Cycle Assessment and Processes project (RECCAP; Canadell et al., 2013) and by using factorial simulations to elucidate the processes that drive trends in the sources and sinks of atmospheric CO_2.

This study has two major aims. The first of these is to estimate the regional trends in the carbon exchange over the period 1990–2009, associated with changes in climate and atmospheric CO_2 concentration, for three land regions (northern land, tropical land, and southern land) and seven ocean regions (North Pacific, equatorial Pacific, South Pacific, North Atlantic, equatorial/South Atlantic, Indian Ocean, and Southern Ocean) (Fig. 1). The second aim is to determine which factors and processes among those included in the models are driving the modelled/observed trends in the regional land/ocean to atmosphere net CO_2 fluxes. For the land models, those factors and processes included are the CO_2 fertilisation effect on productivity and storage, as well as climate effects on productivity, respiration, and climate-caused natural disturbances (see Table S1 in the Supplement for details represented in individual models). A particular focus is on the impacts of climate variation and change on land ecosystems at the regional scale, as extreme climate events occurred during the period of 1990–2009 across many regions of the world, including North America (southwestern USA, 2000–2002), Europe (2003), Amazonia (2005), and eastern Australia (2001–2008), raising considerable attention in the ecological community regarding the consequences of recent climate variability on ecosystem structure and function (Allen et al., 2010) and the carbon cycle (Ciais et al., 2005; Van der Molen et al., 2011; Reichstein et al., 2013).

This study addresses the changes in the magnitude of the global carbon sink but does not discuss the efficiency of the sinks, which is widely discussed elsewhere (Raupach et al., 2014; Gloor et al., 2010; Ciais et al., 2013). These DGVMs have been extensively evaluated against observation-based gross primary production (GPP), land to atmosphere net CO_2 flux, and CO_2 sensitivity of net primary production (NPP) compared to results from free-air CO_2 enrichment (FACE) experiments (Piao et al., 2013).

Consideration of land use and land cover change (LULCC) on regional trends is beyond the scope of the present study, and therefore models assume a fixed present-day land use throughout the simulation period. Thus our results presented should be interpreted with this caveat in mind. There are large uncertainties in the global LULCC flux and its change through time, with an estimated decrease from $1.6 \pm 0.5\,\mathrm{Pg\,C\,yr^{-1}}$ (1990–1999) to $1.0 \pm 0.5\,\mathrm{Pg\,C\,yr^{-1}}$ (2000–2009) (LeQuéré et al., 2013). In addition, the net land use (LU) flux for the period 1990–2009 will be influenced by earlier LULCC (i.e. legacy fluxes), confounding the analysis. The response of the large fluxes associated with net primary productivity and heterotrophic respiration to climate variability and CO_2 are the focus of this study. Other companion papers investigate ecosystem response to interannual and seasonal timescales (Piao et al., 2013), and the carbon balance

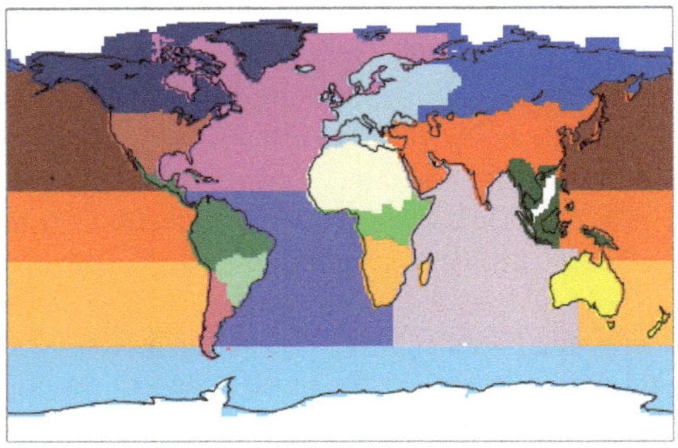

Figure 1. Land and ocean regions. The three land regions: northern land, tropical land, and southern land. Northern land comprises boreal North America (navy blue), Europe (light blue), boreal Asia (blue), temperate North America (pale red), and temperate Asia (red). Tropical land comprises tropical South American forests (sea green), northern Africa (sand), equatorial Africa (green), and tropical Asia (dark green). Southern land comprises South American savanna (pale green), temperate South America (violet), southern Africa (orange), and Australia and New Zealand (yellow). Ocean regions comprise North Pacific (dark red), equatorial Pacific (orange-red), South Pacific (orange), North Atlantic (orchid), equatorial/South Atlantic (slate blue), Indian Ocean (thistle), Southern Ocean (sky blue), and Arctic Ocean and Antarctica (white).

for individual land and ocean regions over the period 1990–2009 (see RECCAP special issue; Canadell et al., 2013, http://www.biogeosciences.net/special_issue107.html).

Trends and variability in the air–sea CO_2 fluxes simulated by the employed OBGCMs are driven by the increase in atmospheric CO_2 and by variability and change in ocean temperature, circulation, winds, and biology largely governed by climate variability. The air–sea CO_2 flux arising from the increase in atmospheric CO_2 is often referred to as the flux of anthropogenic CO_2, while the remainder, induced by changes in the natural cycling of carbon in the ocean–atmosphere system, is called the "natural" CO_2 component (e.g. Gruber et al., 2009). Although this conceptual separation has its limits (McNeill and Matear, 2013), it provides for a powerful way to understand how different forcings affect the net ocean sink.

DGVM results are compared with estimates of the residual land sink (RLS) and remote sensing products indicating trends of greening and browning in the northern region. Regional sources and sink trends are attributed to processes based on factorial simulations.

2 Methods

2.1 Dynamic global vegetation models

Following the studies of Le Quéré et al. (2009) and Sitch et al. (2008), a consortium of DGVM groups set up a project to investigate further the spatial trends in land–atmosphere flux and agreed to perform a factorial set of DGVM simulations over the historical period, 1901–2009. These simulations have contributed to the RECCAP activity (Canadell et al., 2011, 2013). There are now a variety of DGVMs with origins in different research communities that typically contain alternative parameterisations and a diverse inclusion of processes (Prentice et al., 2007; Piao et al., 2013). DGVMs have emerged from the land surface modelling (LSM), forest ecology, global biogeography, and global biogeochemical modelling communities. Representative of these research strands are the following nine DGVMs, which are applied here: Hyland (Levy et al., 2004), JULES (Cox, 2001; Clark et al., 2011), LPJ (Sitch et al., 2003), LPJ-GUESS (Smith et al., 2001), NCAR-CLM4 (Thornton et al., 2007, 2009; Bonan and Levis, 2010; Lawrence et al., 2011), ORCHIDEE (Krinner et al., 2005), OCN (Zaehle and Friend, 2010), SDGVM (Woodward et al., 1995; Woodward and Lomas, 2004), and VEGAS (Zeng, 2003; Zeng et al., 2005). In this study we focus on two aspects of land surface modelling: the carbon and the hydrological cycles. In the case of land surface models coupled to GCMs, energy exchange between the land surface and atmosphere is also simulated.

2.2 Ocean biogeochemical general circulation models

A total of four different groups have conducted the factorial simulations over the analysis period with three-dimensional OBGCMs and submitted their results to the RECCAP archive. These are MICOM-HAMOCCv1 (BER) (Assmann et al., 2010), CCSM-WHOI using CCSM3.1 (BEC) (Doney et al., 2009a, b), CCSM-ETH using CCSM3.0 (ETH) (Graven et al., 2012), and NEMO-PlankTOM5 (UEA) (Buitenhuis et al., 2010). Details of the models are given in the respective publications cited and in Table 2. Not all model simulations are independent of each other, as several of them share components. BEC and ETH employ the same OBGCM, but differ in their spin-up and surface forcing. The employed models have relatively similar horizontal resolution of the order of 1 to 3° in longitude and latitude, i.e. none of them is eddy-permitting or eddy-resolving. The four ecosystem/biogeochemical models are also of comparable complexity, i.e. including explicit descriptions of at least one phytoplankton and zooplankton group, with some models considering up to three explicitly modelled groups for phytoplankton and two for zooplankton. All models use the same gas exchange parameterisation of Wanninkhof (1992), although with different parameters. In particular, the ETH model used a lower value for the gas exchange coefficient than originally used in the CCSM standard configuration, yielding a global-mean gas transfer velocity that is more than 25 % lower than those of the other models (Graven et al., 2012). This reduction reflects the mounting evidence based on radiocarbon analyses that the original global-mean gas transfer velocity of Broecker et al. (1985) was too high (Peacock, 2004; Sweeney et al., 2007; Müller et al., 2008).

2.3 Data sets

2.3.1 Land

Climate forcing is based on a merged product of Climate Research Unit (CRU) observed monthly 0.5° climatology (v3.0, 1901–2009; New et al., 2000) and the high-temporal-resolution NCEP reanalysis. The merged product has a 0.5° spatial and 6 h temporal resolution. A coarse-resolution 3.75° × 2.5° version at monthly timescales was also produced (see Table 1 for spatial resolution of individual DGVMs). Global atmospheric CO_2 was derived from ice core and NOAA monitoring station data, and provided at annual resolution over the period 1860–2009. As land use and land cover change was not simulated in these model experiments, models assume a constant land use (invariant agricultural coverage) throughout the simulation period. Atmospheric nitrogen deposition data for CLM4CN and OCN were sourced from Jean-Francois Lamarque (personal communication, 2012) and Dentener et al. (2006), respectively.

Gridded fields of leaf area index (LAI) are used in the evaluation of DGVM northern greening trends. These LAI data sets were based on remote sensing data and were generated from the AVHRR GIMMS NDVI3g product using an artificial neural network (ANN)-derived model (Zhu et al., 2013). The data set has a temporal resolution of 15 days over the period 1981–2011, and a spatial resolution of 1/12°.

2.3.2 Ocean

Unlike how the land models simulations were set up, no common climatic forcing data set was used for the ocean model simulations. In fact, some models provided several simulation results obtained with different climatic forcings. Models were forced by the NCEP climatic data (Kalnay et al., 1996) in their original form, or in the modified CORE (Common Ocean-ice Reference Experiments – Corrected Normal Year Forcing (CORE-CNYF; Large and Yeager, 2004)) form (Table 2).

2.3.3 Atmospheric inversion

Simulated trends in land to atmospheric net CO_2 flux are compared with those from version 11.2 of the CO_2 inversion product from the Monitoring Atmospheric Composition and Climate – Interim Implementation (MACC-II) service (http://copernicus-atmosphere.eu/). The horizontal resolution of the inversion is 3.75 × 2.5 square degrees (longitude × latitude), and weekly temporal resolution, with night-time and daytime separated. The accuracy varies with the period and the location over the globe, depending on the density and the information content of the assimilated data, and usually decreases with increasing the resolution. Uncertainty

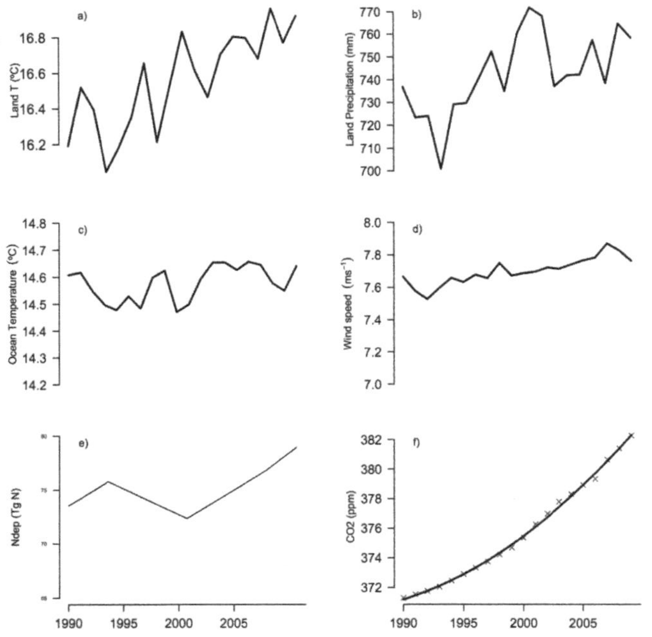

Figure 2. Global trends in environmental driving variables: (**a**) land temperature, (**b**) land precipitation, (**c**) ocean temperature, (**d**) wind speed, (**e**) N deposition, and (**f**) atmospheric [CO_2].

numbers at various scales can be found in Table 2 of Peylin et al. (2013). The inversion covers years 1979–2011, and a previous release has been documented by Chevallier et al. (2010). It uses a climatological prior without interannual variability, except for fossil fuel CO_2 emissions.

2.4 Experimental design

2.4.1 Land

Model spin-up consisted of recycling climate mean and variability from the early decades of the 20th century (1901–1920) with 1860 atmospheric CO_2 concentration of 287.14 ppm until carbon pools and fluxes were in steady state (zero mean annual land to atmospheric net CO_2 flux). The land models were then forced over the 1861–1900 transient simulation using varying CO_2 and continued recycling of climate as in the spin-up. The land models were then forced over the 1901–2009 period with changing CO_2, climate, and fixed present-day land use according to the following simulations:

- S_L1: changing CO_2 only (i.e. time-invariant present-day land use mask, fixed pre-industrial climate);

- S_L2: changing CO_2 and climate (i.e. time-invariant present-day land use mask).

For DGVMs including the N cycle, N deposition was a time-variant forcing in both simulations, such that the difference between S_L2 and S_L1 includes the synergistic effects of N deposition on CO_2 fertilisation (Zaehle et al., 2010).

Figure 2 shows the historical changes in climate, atmospheric CO_2 concentration, and nitrogen deposition over the period 1990–2009 used to force the DGVMs. A summary of DGVM characteristics is given in Table 1. A more detailed description of DGVM process representations is given in Table S1.

2.4.2 Ocean

The ocean models employed two different approaches for creating the initial conditions for the experiments. The first approach, followed by CCSM-ETH, CCSM-WHOI, and BER, involved first a multiple-century-long spin-up with climatological forcing and with atmospheric CO_2 held constant at its pre-industrial value, bringing these models very close to a climatological steady state for pre-industrial conditions (in some models ~ 1750; in others ~ 1850). In the second step, the models were then integrated forward in time through the historical period until 1948, with atmospheric CO_2 prescribed to follow the observed trend and a climatological forcing. The length of the spin-up varies from a few hundred years to several thousand years, resulting in differing global integrated drift fluxes, although their magnitudes are substantially smaller than 0.05 $Pg\,C\,yr^{-1}$ with essentially no rate of change. The second approach, followed by NEMO-PlankTOM5 (UEA), was to initialise the model with reconstructed initial conditions in 1920, and then also run it forward in time until 1948 with prescribed atmospheric CO_2, repeating the daily forcing conditions of a single year (1980). The modelled export production was tuned to obtain an ocean CO_2 sink of 2.2 $Pg\,C\,yr^{-1}$ in the 1990s. This second method offers the advantage that the model's carbon fields remain closer to the observations compared to the long spin-up approach, but it comes at the cost of generating a drift that affects the mean conditions and to a lesser extent the trend. Tests with the model runs of Le Quéré et al. (2010) suggest the drift in the mean CO_2 sink is about 0.5 $Pg\,C\,yr^{-1}$ and the drift in the trend is about 0.005 $Pg\,C\,yr^{-2}$ globally, and is largest in the Southern Ocean.

From ~ 1950 onward, the models performed two separate simulations:

- S_O1: CO_2 only, i.e. atmospheric CO_2 increases, but models are forced with climatological atmospheric boundary conditions (referred to as ACO2 in the RECCAP archive);

- S_O2: CO_2 and climate, i.e. as S_O1, but models are forced with "realistic" year-to-year variability in atmospheric boundary conditions (ANTH).

In these runs, both S_O1 and S_O2 are affected by the same drift, and their differences thus remove the drift. The CCSM-based models performed an additional experiment to better separate between the fluxes of natural and anthropogenic CO_2:

Table 1. Characteristics of the nine dynamic global vegetation models.

Model name	Abbreviation	Spatial resolution	Land surface model	Full nitrogen cycle	River export flux	Fire simulation	Harvest/grazing flux	Source
Community Land Model 4CN	CLM4CN	0.5° × 0.5°	Yes	Yes	No	Yes	No	Oleson et al. (2010); Lawrence et al. (2011)
Hyland	HYL	3.75° × 2.5°	No	No	No	No	Yes	Friend et al. (1997); Levy et al. (2004)
Lund–Potsdam–Jena	LPJ	0.5° × 0.5°	No	No	No	Yes	Yes	Sitch et al. (2003)
LPJ-GUESS	LPJ-GUESS	0.5° × 0.5°	No	No	No	Yes	No	Smith et al. (2001)
ORCHIDEE-CN	OCN	3.75° × 2.5°	Yes	Yes	No	No	Yes	Zaehle and Friend (2010); Zaehle et al. (2010)
ORCHIDEE	ORC	0.5° × 0.5°	Yes	No	No	No	No	Krinner et al. (2005)
Sheffield-DGVM	SDGVM	3.75° × 2.5°	No	No	Yes	Yes	No	Woodward et al. (1995)
TRIFFID	TRI	3.75° × 2.5°	Yes	No	No	No	No	Cox (2001)
VEGAS	VEGAS	0.5° × 0.5°	Yes	No	Yes	Yes	Yes	Zeng et al. (2005)

Table 2. Characteristics of the four ocean biogeochemical general circulation models (OBGCMs). All include NPZD-type ecosystem models and N, P, Si, and Fe nutrient components.

Model name	Abbreviation	Spatial resolution	Meteorological forcing	Gas transfer formulation	Years used	Source
MICOM-HAMOCCv1	BER	2.4° × 0.8–2.4°	NCEP	Wanninkhof (1992)	1990 to 2009	Assmann et al. (2010)
CCSM-WHOI	BEC	3.6° × 0.8–1.8°	NCEP	Wanninkhof (1992)	1990 to 2009	Doney et al. (2009a, b)
CCSM-ETH	ETH	3.6° × 0.9 × 1.9°	CORE	Wanninkhof (1992)	1990 to 2007	Graven et al. (2012)
NEMO-PlankTOM5	UEA	2° × 0.5–2°	NCEP	Wanninkhof (1992)	1990 to 2009	Buitenhuis et al. (2010)

– S_O3: pre-industrial CO_2 and climate, i.e. atmospheric CO_2 is fixed at its pre-industrial level, but atmospheric boundary conditions vary as in S_O2 (PIND).

From these simulations, only the results from 1990 through to 2009 were analysed. Only the UEA and CCSM-WHOI models made results available for the S_O1 and S_O2 simulations for the entire analysis time. The results for the BER model for 2009 are incomplete, and the CCSM-ETH simulations extend only to 2007. In order to maintain a sufficiently large set of models, we decided to focus our analysis primarily on the 1990–2004 period, but occasionally also include the results through to 2009, with the important caveat that the latter are based only on two models.

2.5 Output variables

2.5.1 Land

In this study we focus primarily on the simulated carbon cycle variables, net NPP, RH (heterotrophic respiration), and LAI, a measure of vegetation greenness. The land to atmosphere net CO_2 flux is

$$\text{land to atmosphere net}\,CO_2\,\text{flux} = -\text{NBP}$$
$$= \text{RH} + \text{wildfire flux riverine C flux} + \text{harvest} - \text{NPP},$$

where we have adopted the atmospheric perspective with regard to the sign of the fluxes, i.e. negative numbers indicate a sink for atmospheric CO_2 and a negative trend indicates an increasing sink or a decreasing source.

DGVMs typically do not represent all these processes; a list for each individual DGVM is given in Table 1. DGVM results for simulation S_L2 are compared against the global RLS, calculated as the annual anthropogenic CO_2 emissions (fossil fuel, cement manufacture, and land use C flux) minus the annual CO_2 growth rate and model mean ocean C sink as given by Friedlingstein et al. (2010). The ocean uptake is from the same OGGCMs as the ones used here, and the land use C flux is based on a book-keeping approach from Houghton (2010). Note the RLS depends on a LULCC model of emissions (the one of Houghton). Strictly speaking, comparison of model land to atmosphere net CO_2 flux with RLS is therefore inconsistent because these models treat areas affected by LUC as pristine ecosystems, and these areas are generally associated with a high land carbon sinks. Simulated net carbon flux from S2 is therefore likely to overestimate the RLS sink, by construction.

The regional analysis will focus on three large land regions (Fig. 1), and within these regions, trends at a finer spatial resolution, from multi-grid-cell to the sub-region, are analysed.

The comparison of DGVM simulated trends in the northern growing season against satellite-derived NDVI (normalised difference vegetation index) observations was based on eight models (JULES, LPJ, LPJ-GUESS, NCAR-CLM4, ORCHIDEE, OCN, SDGVM, VEGAS), which provided LAI outputs. The means and trends in the onset, end, and length of growing season were computed. Growing season variables were calculated using the methodology of Murray-Tortarolo et al. (2013). Leaf onset is defined as the day when LAI begins to increase above a critical threshold (CT), defined as

$$\text{CT} = \text{LAI}_{\text{min}} + 0.2 \cdot (\text{LAI}_{\text{max}} - \text{LAI}_{\text{min}}),$$

where LAI_{min} and LAI_{max} represent the minimum and maximum LAI over the annual cycle. Similarly, leaf senescence, or offset, or end of growing season, is defined as the day when LAI decreases below the CT. The length of the growing season in days is calculated as the end minus the onset. This calculation was made for each grid cell above 30° N (i.e. northern extratropics) from the models and the satellite data. In addition, any grid cell where LAI varied by less than 0.5 over the annual cycle from the satellite data was considered to be predominantly evergreen (e.g. boreal forest), and thus excluded from the analysis. We also masked out regions where LAI decreases in the summer (drought deciduous vegetation). In addition, when the growing season spans over the end of year (e.g. Mediterranean and some pixels particularly on the southern margin of the domain), we include the first 3 months of the second year in our analysis. Means and trends were calculated using a linear model over the period 1990–2009.

2.5.2 Ocean

The modelling groups provided output on a monthly basis for the years 1990 through to 2004 and 2009 at two levels of priority. Tier-one data included the surface ocean fields of the air–sea CO_2 flux, oceanic pCO_2, dissolved inorganic carbon (DIC), alkalinity (Alk), temperature (T), salinity (S), and mixed layer depth. The second-tier data included the biological export at 100 m, the vertically integrated net primary production, and the surface chlorophyll a concentration. Some models also supplied three-dimensional climatological fields of DIC, Alk, T, and S.

To determine the different factors contributing to the modelled trends and variations, we undertook two (linear) separations:

- The contribution of climate variability and change on the ocean carbon cycle: $X\_var = X(S\_O2) - X(S\_O1)$, X is any variable or flux, where the expression in parentheses represents the results of the corresponding simulation, and X_var represents the impact of climate change and variability on the ocean carbon cycle.
- The contribution of anthropogenic CO_2: $X\_ant = X(S\_O2) - X(S\_O3)$.

For each of the integrations, but particularly for the changing CO_2 and climate simulation S_O2, we analysed the factors contributing to the temporal change in the air–sea CO_2 flux F by a linear Taylor expansion (see e.g. Lovenduski et al., 2007 and Doney et al., 2009a):

$$\Delta F = \partial F/\partial \mathrm{ws} \cdot \Delta \mathrm{ws} + \partial F/\partial T \cdot \Delta T + \partial F/\partial \mathrm{ice} \cdot \Delta \mathrm{ice} + \partial F/\partial \mathrm{sDIC} \cdot \Delta \mathrm{sDIC} + \partial F/\partial \mathrm{sAlk} \cdot \Delta \mathrm{sAlk} + \partial F/\partial \mathrm{FS} \cdot \Delta S,$$

where ws is the wind speed, ice is the sea-ice fraction, sDIC and sAlk are the salinity normalised DIC and Alk concentrations, and $\partial F/\partial \mathrm{FS}$ is the change in the air–sea CO_2 flux in response to freshwater fluxes. This latter term includes not only the sensitivity of oceanic pCO_2 to changes in salinity but also the dilution effects of freshwater on DIC and Alk (see Doney et al., 2009a, for details). The partial derivatives were computed directly from the model equations for the mean conditions in each region. The changes in the driving components were derived from the trend computed via a linear regression of the model results and then multiplied by the length of the time series.

3 Results

3.1 Global Trends

3.1.1 Land

The ensemble mean global land to atmosphere net carbon dioxide flux from S_L2 is $-2.38 \pm 0.72\,\mathrm{Pg\,C\,yr^{-1}}$ over the period 1990–2009 ($P = 0.04$, where P is the probability of a trend statistically indistinguishable from zero; a significance level of 0.05 is selected) (Fig. 3, Fig. S1 in the Supplement, Table 3). The numbers behind ± signs are the 1 standard deviation of 20-year means for nine DGVMs. This compares to the global RLS of $-2.45 \pm 1.17\,\mathrm{Pg\,C\,yr^{-1}}$, inferred from the global carbon budget by Friedlingstein et al. (2010) over the same period. All DGVMs agree on an increasing land sink with a net flux trend over this period ranging between -0.02 and $-0.11\,\mathrm{Pg\,C\,yr^{-2}}$, corresponding to the OCN and Hyland DGVMs, respectively (Table 3). DGVMs simulate an increase in the land C sink with an ensemble mean trend of $-0.06 \pm 0.03\,\mathrm{Pg\,C\,yr^{-2}}$ ($P < 0.05$) over the period 1990–2009 (Table 3) in response to changes in climate and atmospheric CO_2 content. The two DGVMs with a fully coupled carbon and nitrogen cycle (CN) also simulate an increase in the land sink, at -0.02 ($P = 0.6$) for OCN and $-0.05\,\mathrm{Pg\,C\,yr^{-2}}$ ($P = 0.06$) for CLM4CN. DGVMs suggest the increase in global land sink between 1990 and 2009 is driven by increases in simulated global NPP (Fig. 3).

DGVMs simulate an ensemble mean global NPP of $62.9 \pm 8.73\,\mathrm{Pg\,C\,yr^{-1}}$ over the period 1990–2009 (Table 3). All DGVMs simulate an increase in NPP over this period, with an ensemble mean DGVM trend in NPP of $0.22 \pm 0.08\,\mathrm{Pg\,C\,yr^{-2}}$ ($P = 0.00$) (Table 3). Models with a higher NPP trend also produce a higher land to atmosphere net CO_2 flux trend (Fig. S2 in the Supplement). The ensemble mean NPP trend of $0.22 \pm 0.08\,\mathrm{Pg\,C\,yr^{-2}}$ ($P < 0.01$) from simulation S_L2 (CO_2 and climate forcing) contrasts with an ensemble trend of $0.19 \pm 0.08\,\mathrm{Pg\,C\,yr^{-2}}$ ($P < 0.01$) and $0.03 \pm 0.05\,\mathrm{Pg\,C\,yr^{-2}}$ ($P = 0.24$) over the same period for the S_L1 (CO_2 only) and S_L2–S_L1 (the climate effect), respectively (Tables S2, S3 in the Supplement). These results suggest that the simulated increase in global NPP is mainly in response to increasing atmospheric CO_2 (direct

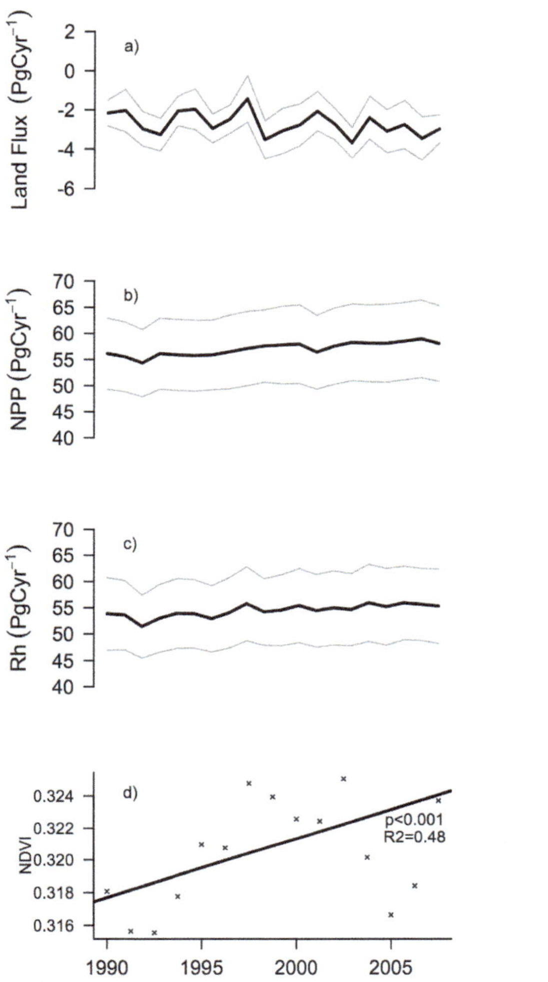

Figure 3. Global trends in ensemble land model responses. (**a**) DGVM mean model land to atmosphere net CO_2 flux and standard deviation (grey lines); (**b**) component fluxes, NPP; and (**c**) RH (= RH + wildfire + riverine C flux); and (**d**) remotely sensed trends in annual mean NDVI (crosses), a measure of vegetation greenness, and a linear regression through the data points (bold line).

CO_2 fertilisation of photosynthesis, in addition to the indirect benefits from an improved water balance in water-limited ecosystems due to the physiological effects of CO_2 on water use efficiency). VEGAS, CLM4CN, and OCN simulate the smallest positive trends in NPP among the DGVMs in response to elevated CO_2 forcing (Table S2). This suggests that the potential CO_2 fertilisation effect may be already strongly limited by present-day nitrogen availability in some ecosystems (Vitousek and Howarth, 1991). There is more uncertainty among models on the impact of climate changes on global NPP, with only two models simulating a significant positive trend (Table S3).

DGVMs simulate an ensemble mean global RH of 57.5 ± 9.8 Pg C yr^{-1} over the period 1990–2009 (Table 3). All DGVMs simulate an increase in RH for S_L2 (CO_2 and climate), with an ensemble mean trend of

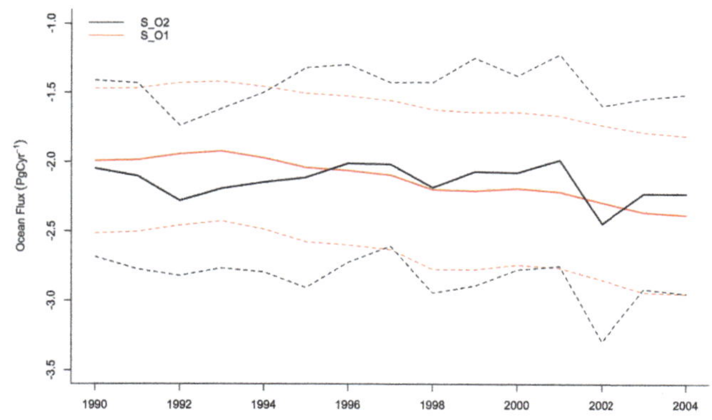

Figure 4. Global trends in ensemble ocean model fluxes. Black line: results from simulation S_O2 with variable "climate" and increasing CO_2. Red line: results from simulation S_O1 with constant "climate" and increasing CO_2. The dashed grey and dashed red lines indicate the ± uncertainty bands given by the four models that contribute to the ensemble mean.

0.16 ± 0.05 Pg C yr^{-2} ($P < 0.01$) over the period 1990–2009 (Table 3). This is lower than the trend in global NPP, resulting in a trend towards increasing net land carbon uptake. This is unsurprising as there is a lagged response in increases in RH relative to NPP, reflecting the turnover time of the newly incorporated plant material. The ensemble mean trend in RH is 0.12 ± 0.06 Pg C yr^{-2} ($P < 0.01$) and 0.04 ± 0.02 Pg C yr^{-2} ($P = 0.09$) over the same period for the S_L1 (CO_2 only) and S_L2–S_L1 (the climate effect), respectively (Tables S2, S3). This implies the dominant effect on RH is increased substrate for microbial respiration, with the additional litter input into soils, as a consequence of enhanced NPP, rather than enhanced rates of microbial decomposition with rising temperatures. Nevertheless, the simulated mean residence time (MRT = soil carbon / RH) of soil organic matter decreases, in response to warming, which is especially pronounced in high-latitude regions (Fig. S3 in the Supplement). The difference in land–atmosphere flux trend between the CN models OCN (-0.02 Pg C yr^{-2}) and CLM4CN (-0.05 Pg C yr^{-2}) is largely due their difference in RH trends at 0.14 and 0.11 Pg C yr^{-2}, respectively, rather than differential responses of simulated NPP to elevated CO_2 (Table 3).

Only four DGVMs simulated wildfire fluxes (CLM4CN, LPJ, LPJ-GUESS, SDGVM). No significant trends in the global wildfire flux were reported by any of the DGVMs.

3.1.2 Ocean

The global ocean is simulated to have acted as a very substantial sink for atmospheric CO_2 but one that has increased only slightly over the last two decades (see also discussion in Wanninkhof et al., 2013). The mean ocean sink in the four models (CCSM-ETH, CCSM-WHOI, UEA, and BER) increased from ~ -2.0 Pg C yr^{-1} in the early 1990s to

~ -2.1 Pg C yr^{-1} during the first 5 years of the 21st century (Fig. 4).

We separate the mean and variable components by using our factorial experiments, i.e. by using S_O1 results to identify the ocean uptake in the absence of climate variability and change, and the difference between S_O2 and S_O1 as measure of the impact of climate change. This separation reveals that, in the absence of climate variability and change, the global ocean uptake would have increased from about -1.98 ± 0.04 Pg C yr^{-1} for the 1990–1994 period to -2.3 ± 0.09 Pg C yr^{-1} for 2000–2004 (for the two models that provided S_O1 results up to 2009 (CCSM-WHOI and UEA), the uptake flux would have increased from -1.99 to -2.56 Pg C yr^{-1} for 2005–2009). This global net uptake flux and its substantial trend in time (-0.03 Pg C yr^{-2} for 1990–2004, and -0.04 Pg C yr^{-2} for 1990–2010) is entirely driven by the increase in atmospheric CO_2 and is – integrated globally – numerically equivalent to the ocean uptake flux of anthropogenic CO_2. Climate variability and change modified these fluxes, and particularly the trend in these models. The four models suggest an enhancement of the uptake in the early 1990s (1990–1994) of about -0.2 Pg C yr^{-1}, turning into a reduction of the uptake in the subsequent period (1995–1999), followed by a further reduction in the 2000–2004 period of $\sim +0.1$ Pg C yr^{-1}. This trend toward reduced uptake in response to climate variability and change of $+0.03$ Pg C yr^{-2} nearly completely compensates for the anthropogenic CO_2 driven increase in uptake, causing the overall uptake of CO_2 to have a nearly flat trend over the 1990–2004 period of <0.01 Pg C yr^{-2} The same tendencies are found for the two models that extend over the entire 1990–2009 period: in these models, climate change and variability reduces the CO_2-driven trend of -0.04 Pg C yr^{-2} by more than $+0.02$ Pg C yr^{-2}, to around -0.02 Pg C yr^{-2}.

With consideration of the different factors affecting the ocean carbon sink following our Taylor expansion, we find increasing sea surface temperature to be a globally important driver for the positive trends (reduced sinks) induced by climate change and variability. Over the 1990–2004 period, the surface ocean warmed, on average, by 0.004 °C yr^{-1} (0.005 °C yr^{-1} from 1990 through to 2009). Isochemically, this leads to an increase in the oceanic pCO_2 of ~ 0.06 μatm yr^{-1}, which appears small. However, it needs to be compared with the trend in the global-mean air–sea pCO_2 difference of about ~ 0.1 μatm yr^{-1} that is required in order to generate a trend in the ocean uptake of -0.03 Pg C yr^{-2} (see e.g. Matsumoto and Gruber, 2005; Sarmiento and Gruber, 2006). The overall sink is therefore largely a consequence of the increase in atmospheric CO_2 (i.e. it mostly corresponds to the uptake flux of anthropogenic CO_2), but it includes a substantial perturbation flux stemming from the impact of climate variability and change on the ocean carbon cycle.

3.2 Regional trends

3.2.1 Land

Northern land

All DGVMs agree on a land C sink over the northern land region, with a mean land–atmosphere flux of -1.03 ± 0.30 Pg C yr^{-1} over the period 1990–2009 (Fig. S4 in the Supplement, Table 3). The ensemble mean land–atmosphere flux trend is near zero for this region between 1990 and 2009 (Fig. S5 in the Supplement). Of particular interest are sub-regions with a simulated positive land–atmosphere flux trend (Fig. 5), implying a diminishing sink of atmospheric CO_2 or an increasing source of CO_2 to the atmosphere. At least six models out of nine agree on a decreasing regional land sink across some areas in temperate North America, eastern Europe, northeastern China, and Mongolia (Fig. 5). These largely correspond to regions with negative trends in precipitation (Fig. 6).

Over the northern region, which covers almost 50 % of the land surface, DGVMs simulate an ensemble mean NPP of 24.1 ± 4.48 Pg C yr^{-1}, which represents almost 40 % of the global total (Table 3). All DGVMs simulate an increase in northern NPP over this period, with a trend in NPP of 0.06 ± 0.02 Pg C yr^{-2} ($P < 0.01$) (Table 3). However, enhanced productivity in the northern land region accounts for only around 29 % of the simulated global trend in NPP. The ensemble mean NPP trend of 0.06 ± 0.02 Pg C yr^{-2} ($P < 0.01$) from simulation S2 (CO_2 and climate forcing) compares to a trend of 0.07 ± 0.03 Pg C yr^{-2} ($P < 0.01$) and -0.00 ± 0.04 Pg C yr^{-2} ($P = 0.85$) for the S_L1 (CO_2 only) and S_L2–S_L1 (the climate effect), respectively (Tables S2, S3). All DGVMs simulate a positive trend in NPP in response to elevated CO_2 across the northern land region, and trends are all significant at the 95 % confidence level with the exception of CLM4CN ($P = 0.21$).

Large areas in temperate North America and Asia experienced warming combined with reductions in precipitation over the period 1990–2009 (Fig. 5). Indeed, although DGVMs simulate larger mean NPP in temperate compared to boreal regions (Table S5 in the Supplement), they simulate significant positive trends in boreal North America and boreal Asia, whereas trends in both temperate North America and Asia are smaller and not significant at the 95 % confidence level (Table S5).

In response to warming, models simulate an earlier onset (ensemble mean model trend $= -0.078 \pm 0.131$ days yr^{-1}) and delayed termination of the growing season (0.217 ± 0.097 days yr^{-1}) based on LAI, and thus a trend towards a longer growing season in the northern extratropics (0.295 ± 0.228 days yr^{-1}) (Fig. 7). This is in broad agreement with observed greening trends (Zhu et al., 2013; Murray-Tortarolo et al., 2013): onset $= -0.11$ days yr^{-1}, offset $= 0.252$ days yr^{-1}, and

Table 3. Mean and trends in NPP, RH, and land–atmosphere flux as simulated by individual DGVMs and the ensemble mean.

MODEL	NPP ($Pg\,C\,yr^{-1}$)	Trend ($Pg\,C\,yr^{-2}$)	*P* value	RH ($Pg\,C\,yr^{-1}$)	Trend ($Pg\,C\,yr^{-2}$)	*P* value	Land–atm CO_2 flux ($Pg\,C\,yr^{-1}$)	Trend ($Pg\,C\,yr^{-1}$)	*P* value
Global_Land									
CLM4CN	51.508	0.148	0.000	47.668	0.106	0.000	−1.459	−0.052	0.059
HYLAND	73.422	0.319	0.000	68.835	0.203	0.000	−3.466	−0.109	0.000
LPJ	59.306	0.216	0.000	47.612	0.117	0.000	−2.251	−0.068	0.061
LPJ-GUESS	62.506	0.174	0.000	55.448	0.145	0.000	−1.802	−0.043	0.346
OCN	53.941	0.155	0.000	50.611	0.135	0.000	−2.272	−0.015	0.568
ORCHIDEE	75.516	0.293	0.000	72.037	0.208	0.000	−3.479	−0.086	0.046
SDGVM	60.965	0.240	0.000	53.778	0.190	0.000	−2.127	−0.044	0.170
TRIFFID	71.929	0.305	0.000	69.167	0.244	0.000	−2.762	−0.061	0.265
VEGAS	57.308	0.113	0.006	51.930	0.092	0.000	−1.783	−0.018	0.551
Ensemble	**62.934**	**0.218**	**0.000**	**57.454**	**0.160**	**0.000**	**−2.378**	**−0.055**	**0.048**
SD	**8.729**	**0.076**		**9.791**	**0.053**		**0.721**	**0.030**	
Northern_Land									
CLM4CN	17.523	0.043	0.003	16.215	0.036	0.000	−0.670	−0.007	0.612
HYLAND	19.139	0.098	0.000	17.591	0.080	0.000	−0.876	−0.014	0.311
LPJ	24.566	0.079	0.001	19.578	0.062	0.006	−1.168	−0.006	0.735
LPJ-GUESS	28.484	0.039	0.085	25.883	0.067	0.009	−0.634	0.023	0.521
OCN	21.008	0.044	0.035	19.264	0.047	0.008	−1.117	0.007	0.632
ORCHIDEE	30.337	0.070	0.007	29.112	0.063	0.000	−1.226	−0.006	0.740
SDGVM	25.144	0.063	0.006	22.598	0.065	0.006	−0.828	0.004	0.762
TRIFFID	28.476	0.088	0.009	27.006	0.103	0.001	−1.470	0.016	0.455
VEGAS	21.895	0.048	0.012	18.914	0.043	0.001	−1.322	−0.000	0.968
Ensemble	**24.064**	**0.063**	**0.001**	**21.796**	**0.063**	**0.001**	**−1.034**	**0.002**	**0.865**
SD	**4.484**	**0.022**		**4.562**	**0.020**		**0.295**	**0.012**	
Tropical_Land									
CLM4CN	26.400	0.090	0.000	24.464	0.058	0.000	−0.692	−0.039	0.110
HYLAND	34.489	0.112	0.000	32.695	0.067	0.000	−1.560	−0.044	0.001
LPJ	25.830	0.100	0.001	21.224	0.035	0.001	−0.817	−0.049	0.031
LPJ-GUESS	21.922	0.078	0.000	19.332	0.051	0.000	−0.785	−0.036	0.038
OCN	22.750	0.084	0.000	21.476	0.065	0.000	−0.982	−0.017	
ORCHIDEE	31.313	0.151	0.000	29.640	0.108	0.000	−1.673	−0.043	0.084
SDGVM	23.505	0.118	0.000	20.677	0.075	0.000	−0.984	−0.038	0.030
TRIFFID	29.801	0.141	0.000	28.925	0.096	0.000	−0.876	−0.045	0.218
VEGAS	23.472	0.041	0.061	21.994	0.033	0.004	−0.278	−0.010	0.527
Ensemble	**26.609**	**0.102**	**0.000**	**24.492**	**0.065**	**0.000**	**−0.961**	**−0.036**	**0.045**
SD	**4.350**	**0.034**		**4.752**	**0.025**		**0.428**	**0.013**	
Southern_Land									
CLM4CN	7.617	0.014	0.187	7.017	0.011	0.036	−0.098	−0.005	0.719
HYLAND	19.875	0.109	0.000	18.623	0.056	0.000	−1.035	−0.051	0.000
LPJ	8.940	0.037	0.074	6.833	0.021	0.004	−0.267	−0.013	0.355
LPJ-GUESS	12.124	0.058	0.003	10.255	0.026	0.001	−0.385	−0.031	0.192
OCN	10.222	0.027	0.165	9.909	0.023	0.053	−0.174	−0.004	0.744
ORCHIDEE	13.884	0.073	0.002	13.304	0.037	0.000	−0.581	−0.036	0.027
SDGVM	12.358	0.059	0.034	10.539	0.050	0.000	−0.317	−0.010	0.701
TRIFFID	13.707	0.077	0.020	13.290	0.045	0.000	−0.417	−0.032	0.269
VEGAS	11.971	0.024	0.382	11.049	0.016	0.140	−0.182	−0.009	0.656
Ensemble	**12.300**	**0.053**	**0.011**	**11.202**	**0.032**	**0.000**	**−0.384**	**−0.021**	**0.196**
SD	**3.528**	**0.031**		**3.597**	**0.016**		**0.285**	**0.017**	

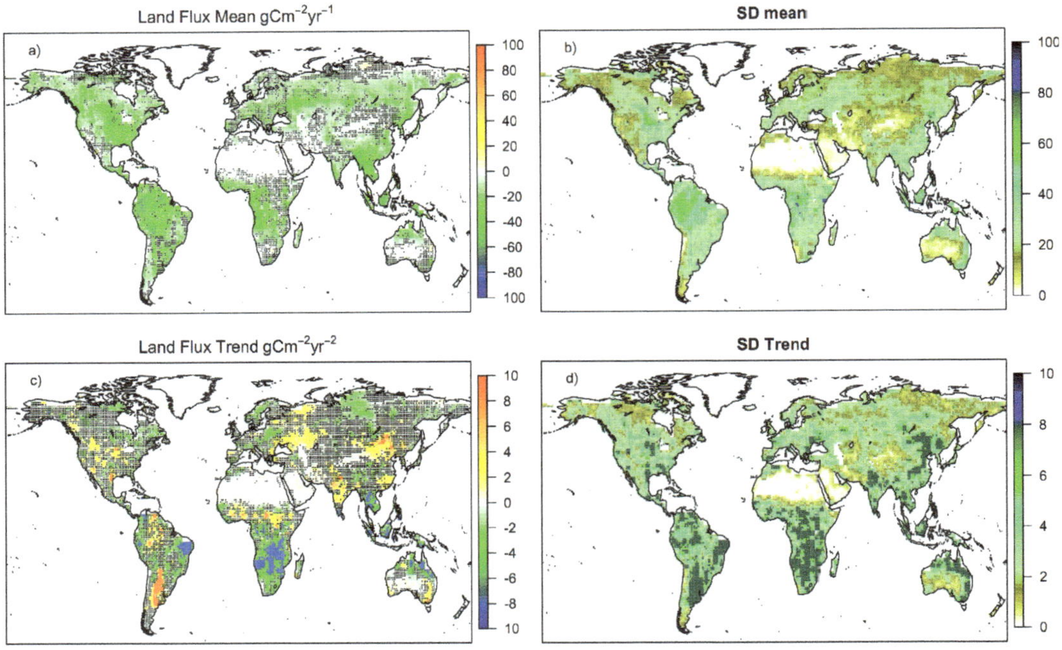

Figure 5. **(a)** Average land to atmosphere net CO_2 flux over the period 1990–2009 for the ensemble mean and model disagreement, with stippling representing agreement for <66 % of DGVMs , and **(b)** standard deviation across DGVMs. **(c)** The trend in land to atmosphere net CO_2 flux across the ensemble, and model disagreement, with stippling representing agreement of <66 % of the DGVMs , and **(d)** the standard deviation of the trend.

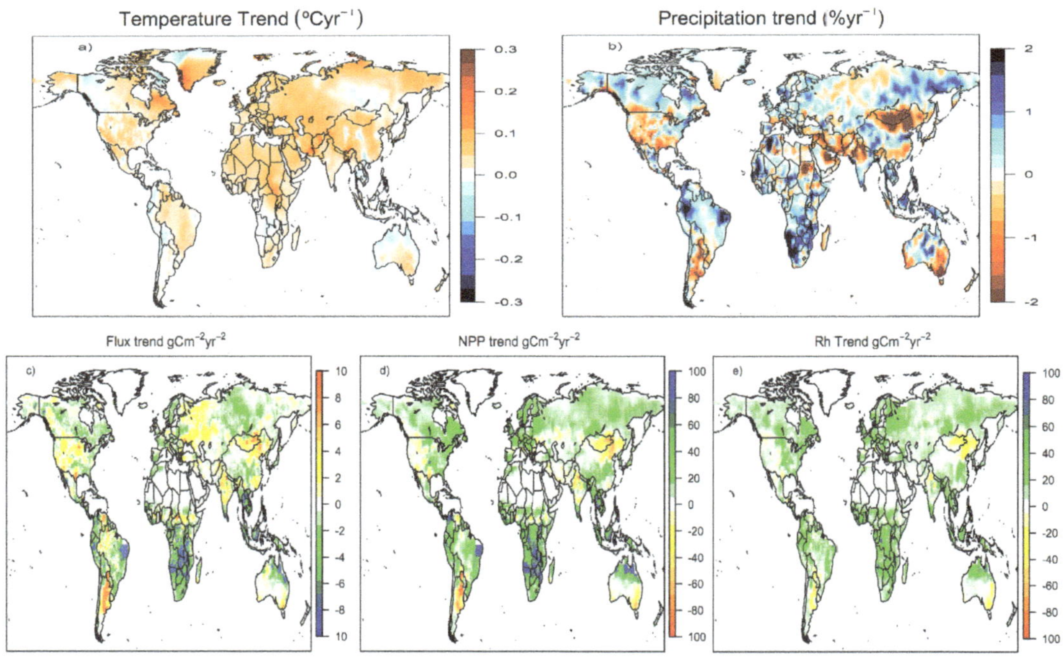

Figure 6. Trends in land climate drivers and process responses. **(a)** Trend in temperature ($°C\,yr^{-1}$), **(b)** trend in precipitation ($\%\ yr^{-1}$), **(c)** trend in land to atmosphere net CO_2 flux ($gC\,m^{-2}\,yr^{-2}$), **(d)** trend in NPP ($gC\,m^{-2}\,yr^{-2}$), and **(e)** trend in RH (= RH + wildfire + Riverine C flux) ($gC\,m^{-2}\,yr^{-2}$).

growing season length $= 0.361$ days yr^{-1}. There is less agreement among models on reproducing the observed browning trends in some regions of the boreal forest.

DGVMs simulate an ensemble mean RH of 21.8 ± 4.6 Pg C yr^{-1} across the northern land region (Table 3). All DGVMs simulate an increase in northern RH over the period 1990–2009, with a significant trend in RH of 0.063 ± 0.02 Pg C yr^{-1} ($P < 0.01$) (Table 3). DGVMs simulate larger mean RH in temperate compared to boreal regions, yet smaller positive trends for Asia (Table S6 in the Supplement). This is because of relatively smaller increases in substrate (i.e. NPP) in temperate regions and greater warming in boreal regions stimulating microbial decomposition, reducing mean residence time of carbon in soils (MRT = soil carbon / RH; see Fig. S3).

No significant trends in the wildfire flux were reported by any of the DGVMs for the northern land region. However, DGVMs agree on simulating a small negative trend in wildfire flux across boreal North America and tundra.

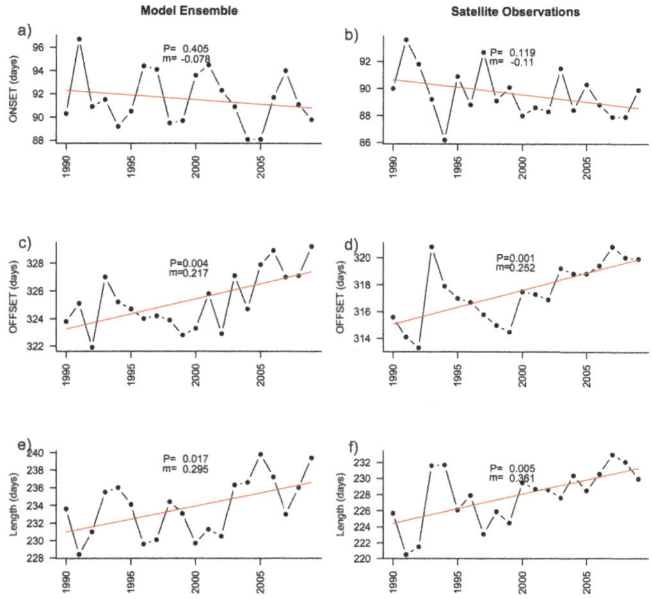

Figure 7. Ensemble-mean trends in the onset **(a, b)**, offset **(c, d)**, and length of growing season in days **(e, f)** for the ensemble mean (left) compared with satellite-derived estimates (right).

Tropical land

All DGVMs simulate an increasing land C sink over recent decades, in response to changes in climate and atmospheric CO_2 concentration over the tropical land region, with an ensemble mean land–atmosphere flux of -0.96 ± 0.43 Pg C yr^{-1} (Table 3, Fig. S4) and trend of -0.04 ± 0.01 Pg C yr^{-2} ($P = 0.05$), or -0.88 ± 0.33 g C m^{-2} yr^{-2} on an area basis (Table 3, Table S4 in the Supplement Fig. S5). This represents 65 % of the increase in global land sink over the last two decades across the tropical land, which covers 27 % of the land surface (Table S4). DGVMs simulate significant negative trends (i.e. increasing sinks) across tropical Asia and equatorial Africa (Table S4).

DGVMs simulate an ensemble mean NPP of 26.6 ± 4.35 Pg C yr^{-1} averaged over the tropical region, representing 42 % of the global total (Table 3). All DGVMs simulate a significant increase in tropical NPP over this period, with an ensemble mean trend in NPP of 0.10 ± 0.03 Pg C yr^{-2} ($P = 0.00$) for S_L2 (Table 3). This compares to a trend of 0.09 ± 0.03 Pg C yr^{-2} ($P < 0.01$) and 0.02 ± 0.02 Pg C yr^{-2} ($P = 0.33$) over the same period for the S_L1 (CO_2 only) and S_L2–S_L1 (the climate effect), respectively (Tables S2, S3). Again, the simulated trend in NPP is dominated by the simulated response of ecosystems to elevated atmospheric CO_2 content. DGVMs simulate positive NPP trends across tropical South American forests, tropical Asia, equatorial Africa, and North African savanna (Table S5). Nevertheless there are some areas within tropical South America and North African savanna regions with negative trends in NPP (Fig. 6).

All DGVMs simulate an increase in RH over the period 1990–2009, with an ensemble mean RH of 24.49 ± 4.75 Pg C yr^{-1} (Table 3) and trend of 0.065 ± 0.025 Pg C yr^{-2} ($P < 0.01$). This can be largely attributed to the response of ecosystems to elevated CO_2 (Table S2).

No significant trends in the wildfire flux were reported by any of the DGVMs for the tropical land region. However, DGVMs agree on simulating a negative trend in wildfire flux across equatorial Africa and tropical Asia.

Southern land

All DGVMs agree on a net land sink over the southern land during the last two decades, with an ensemble mean land–atmosphere flux of -0.38 ± 0.29 Pg C yr^{-1} (Table 3, Fig. S4). Although all DGVMs simulate an increase in the land sink over the southern extratropics, with an ensemble mean land–atmosphere trend of -0.02 ± 0.02 Pg C yr^{-2} ($P = 0.20$) (Fig. S5) or -0.58 ± 0.45 g C m^{-2} yr^{-2} on an area basis, only trends for HYL and ORC are significant at the 95 % confidence level (Table 3). Ensemble mean trends are significant for temperate South American and southern African regions at 0.005 ± 0.005 Pg C yr^{-2} ($P = 0.05$) and -0.022 ± 0.011 Pg C yr^{-2} ($P = 0.01$), respectively (Table S4). For southern Africa, all DGVMs simulate an increase in the land sink in response to climate variability and change over this period (five out of nine are significant at the 90 % confidence level) (Table S7 in the Supplement, Fig. 6). In contrast, the simulated decrease in land sink for temperate South America is associated with a decrease in precipitation over 1990–2009 (Table S8 in the Supplement).

DGVMs simulate an ensemble mean NPP of 12.3 ± 3.53 Pg C yr^{-1} over the southern extratropics, which represents ~20 % of the global total (Table 3) across 24 % of the land surface. All DGVMs simulate an increase in NPP over this period, with a significant ensemble mean trend of 0.05 ± 0.03 Pg C yr^{-2} ($P = 0.01$), i.e. the southern land region accounts for around 25 % of the simulated global trend in NPP. Southern Africa is the only southern sub-region with a significant trend in NPP of 0.041 ± 0.018 Pg C yr^{-2} ($P < 0.01$) (Table S5), due to a positive response of plant production to both CO_2 and climate, and is likely in response to increases in precipitation over the last two decades (Table S7, Fig. 5).

DGVMs simulate an ensemble mean RH of 11.20 ± 3.60 Pg C yr^{-1} over the southern land region (Table 3). All DGVMs simulate an increase in RH over the period 1990–2009, with a significant trend in the ensemble mean RH of 0.03 ± 0.02 Pg C yr^{-2} ($P < 0.01$). This is only partly explained by the response of ecosystems to elevated CO_2; over southern Africa the ensemble mean trend in RH from S_L1 is 0.01 ± 0.01 Pg C yr^{-2} ($P < 0.01$), and a climate-induced positive trend in RH of 0.01 ± 0.00 Pg C yr^{-1} ($P < 0.01$) (Table S2, S7).

No significant trends in the wildfire flux were reported by any of the DGVMs for the southern land region. However DGVMs agree on simulating a negative trend in wildfire flux across southern Africa.

In summary, the globally increasing trend in land carbon sink is about two-thirds due to tropical ecosystems and one-third due to the southern land region, with zero contribution from northern land. This partitioning in trend is quite different from the mean carbon sink fluxes themselves, which is more like 43 : 41 : 16 (northern : tropical : southern).

Qualitative change in processes

A qualitative assessment of the differential responses of the underlying land processes to changes in environmental conditions, and their contribution to the sink–source land–atmosphere flux trends is shown in Fig. 8. Many regions are simulated to have a negative land–atmosphere flux trend, with increases in NPP leading increases in RH. However there are locations with positive trends over the period 1990–2009, i.e. red colours in Fig. 8. In some regions models simulate a positive trend in NPP but an even larger positive trend in RH (eastern Europe, southeastern USA, Amazonia, southern China, North America tundra). Warming is likely to enhance both NPP and RH in high-latitude ecosystems, but primarily RH in low latitudes. Reduced precipitation may partially or fully offset the benefits of elevated atmospheric CO_2 abundance on NPP, and the response of RH to changes in precipitation is not obvious, as this is influenced by the initial soil moisture status. This is because microbial activity increases with increasing soil moisture at low moisture levels, before reaching a maximum activity, and then begins to decline as water completely fills the soil pore spaces and oxygen becomes more limiting to respiration. Locations in the western USA, southern Asia, northern boreal China, southeastern South America, and western and southern Australia are simulated to have negative NPP trends over the last two decades, as a result of reduced rainfall, and there is a less negative trend in RH, possibly due to a reduction in microbial respiration rates with increased soil dryness. The warming and drying in central Asia (northeastern China and Mongolia) and southern Australia is simulated to reduce the rate of microbial decomposition in these regions (Fig. S3), which partly opposes the NPP-driven lagged decrease in RH. The source trend in eastern Europe is simulated as a combination of a negative trend in NPP, as a result of a combination of elevated temperatures and reduced precipitation (i.e. soil drying), and a positive trend in RH driven by increasing temperature, despite reduced plant litter input.

.2 Ocean

Regional fluxes

The large-scale distribution of the modelled mean surface fluxes consists of strong outgassing in the tropical regions, especially in the Pacific, and broad regions of uptake in the mid-latitudes, with a few regions in the high latitudes of particularly high uptake, such as the North Atlantic (Fig. 9). This pattern is largely the result of the exchange flux of natural CO_2 that balances globally to a near-zero flux, but exhibits regionally strong variations (Gruber et al., 2009). Superimposed on this natural CO_2 flux pattern is the uptake of anthropogenic CO_2, which is taken up everywhere, but with substantial regional variation. Large anthropogenic CO_2 uptake fluxes occur in the regions of surface ocean divergence, such as the equatorial Pacific and particularly the Southern Ocean (Sarmiento et al., 1992; Gloor et al., 2003; Mikaloff Fletcher et al., 2006). This is a result of the divergence causing waters to upwell to the surface which have not been exposed to the atmosphere for a while, thereby permitting them to take up a substantial amount of anthropogenic CO_2. This reduces the outgassing that typically characterises these regions as a result of these upwelling waters also bringing with them high carbon loads from the remineralisation of organic matter.

Over the analysis period, the air–sea CO_2 fluxes exhibit only a remarkably small trend in most places, with some regions increasing in uptake, while others show a positive flux anomaly, i.e. lesser uptake. Thus the small global trend in ocean uptake over the 1990–2004 analysis period is a result of also the individual regions having relatively modest trends.

Process analysis

The regional flux trends are, however, much smaller than expected from an ocean with constant circulation that is only responding to increasing atmospheric CO_2, and hence would tend to increase its uptake of anthropogenic CO_2 through

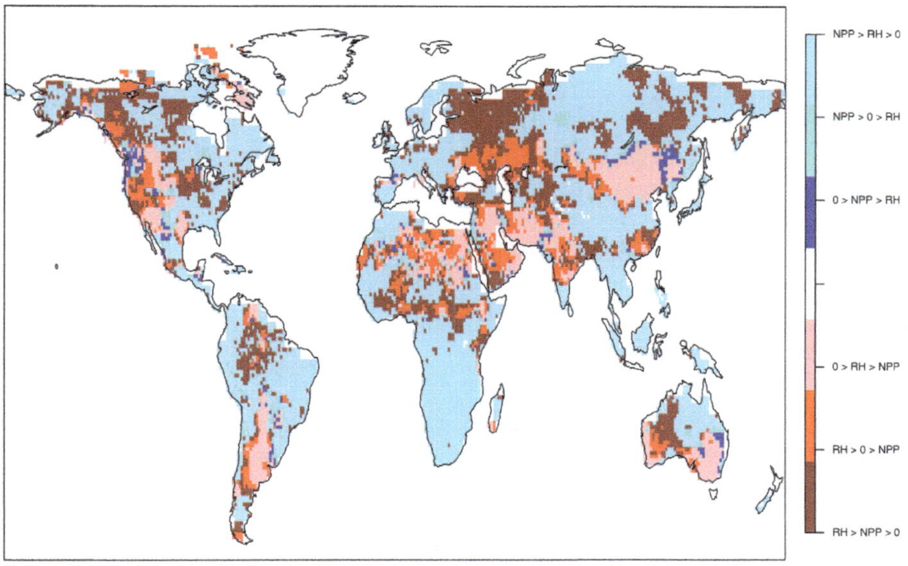

Figure 8. Qualitative change in processes over the period 1990–2009. Negative trend in land–atmosphere net CO_2 flux: enhanced NPP > enhanced RH (= RH + wildfire + riverine C flux) (pale blue); enhanced NPP, reduced RH (turquoise); and reduced NPP < reduced RH (dark blue). Positive trend in land–atmosphere net CO_2 flux: enhanced NPP < enhanced RH (dark red); reduced NPP, enhanced RH (red); and reduced NPP > reduced RH (pink).

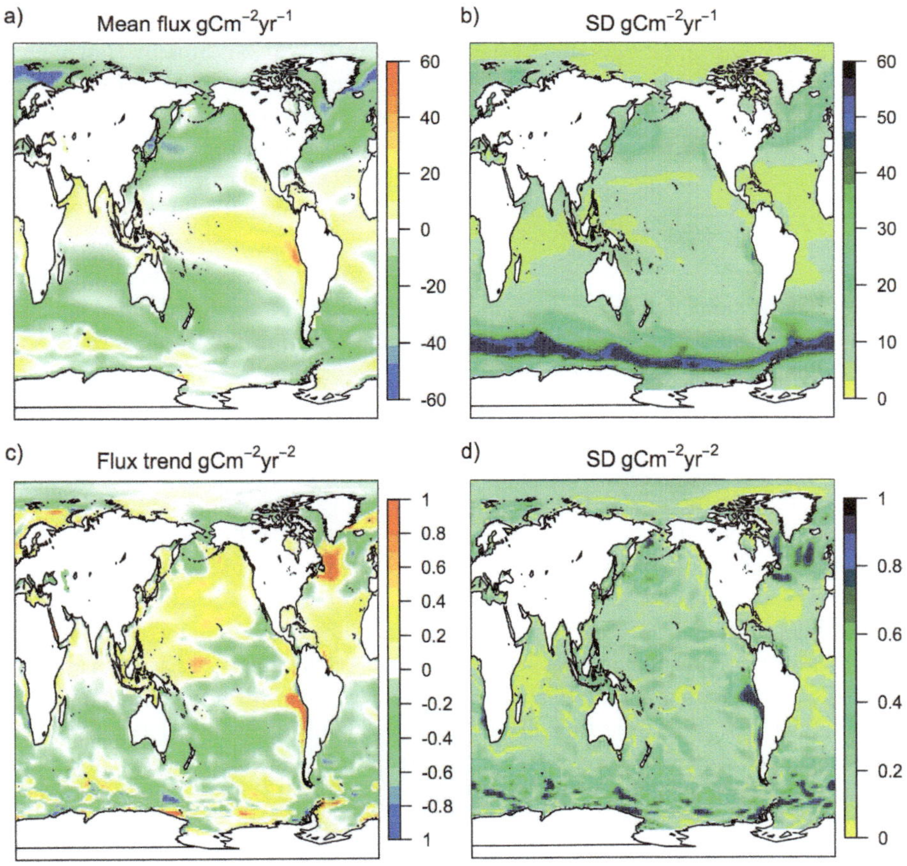

Figure 9. Gridded maps of the ensemble mean sea–air CO_2 flux over the period 1990–2004 **(a)**, standard deviation of the mean flux across the four OBGCMs **(b)**, the trend in the net flux across the ensemble **(c)**, and the standard deviation of the trend **(d)**.

time (Fig. 10). In the absence of climate variability and change, all regions would have flux density trends of more than $-0.05\,\mathrm{g\,C\,m^{-2}\,yr^{-2}}$, with some regions, such as the Southern Ocean, exceeding $-0.15\,\mathrm{g\,C\,m^{-2}\,yr^{-2}}$. However climate variability and change compensate for these negative trends in every single region by increasing them by $+0.04\,\mathrm{g\,C\,m^{-2}\,yr^{-2}}$ or more (with the exception of the South Pacific), such that the overall trends fluctuate from region to region around zero (Fig. 10). The largest reductions in trends are simulated to occur in the North and equatorial Pacific and in the North Atlantic, where they even cause a change in the sign of the overall trend. A similar, although slightly more moderate, pattern is seen if the analysis is undertaken for the entire 1990–2009 period with two models only. The most important difference is found in the North Atlantic, where the climate variability impact is substantially smaller, and not offsetting the anthropogenic CO_2 trend when analysed for 1990–2009.

The mechanisms driving the oceanic flux trends differ between the analysed regions. Attribution of regional trends to specific processes or changes in specific state variables in the different models is a work in progress, and is difficult to achieve with high confidence as yet. This is due to the antagonistic effect of ocean warming on CO_2 solubility and on dissociation of carbonic acid into bicarbonate and carbonate, as well as to the complex changes in ocean circulation and mixing, which themselves influence the biological carbon pumps of the ocean.

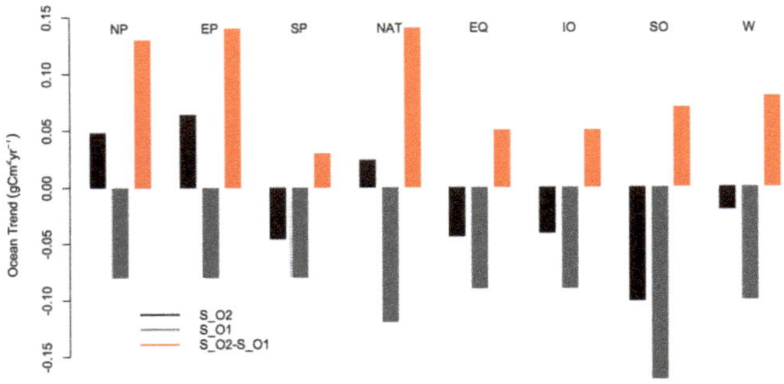

Figure 10. Regional ocean flux trends from 1990 through to 2004 for the standard case, i.e. variable climate and increasing CO_2 (simulation S_O2), and for the constant climate case (simulation S_O1), and their difference (S_O2–S_O1). Ocean regions comprise North Pacific (NP), equatorial Pacific (EP), South Pacific (SP), North Atlantic (NAT), equatorial/South Atlantic (EQ), Indian Ocean (IO), and Southern Ocean (SO), and world oceans (W).

4 Discussion

4.1 Land

The DGVMs used in this study simulate an increase in land carbon uptake over the period 1990–2009. The result is consistent with the earlier findings of Sarmiento et al. (2010), who suggested a large increase in the RLS between 1960 and 1988 and between 1989 and 2009 (Table S9 in the Supplement). The ensemble mean land–atmosphere flux increased by $-1.11\,\mathrm{Pg\,C\,yr^{-1}}$ for the same period, compared to the estimated RLS increase of $-0.88\,\mathrm{Pg\,C\,yr^{-1}}$ from Sarmiento et al. (2010). The DGVM ensemble trends in land uptake for the globe, northern, tropical, and southern land regions of -0.06 ± 0.03, 0.00 ± 0.01, -0.04 ± 0.01, and $-0.02\pm0.02\,\mathrm{Pg\,C\,yr^{-2}}$, respectively, compare favourably with the inversion estimates of -0.06 ± 0.04, -0.01 ± 0.01, -0.04 ± 0.02, and $-0.01\pm0.01\,\mathrm{Pg\,C\,yr^{-2}}$ over the period 1990–2009. Although encouraging, these results should be interpreted with caution because the inversion accounts for any trend in the land use change flux over this period, whereas DGVMs had fixed land use.

There is empirical evidence of a large increase in biomass in intact forest in tropical South America and Africa (Pan et al., 2011; Baker et al., 2004; Lewis et al., 2009a, b), which is consistent with the DGVM projections presented here. Lewis et al. (2009b) found broad agreement between biomass trends from observations and from a suite of carbon cycle models applied with 20th century forcing of climate and atmospheric CO_2 content, using a similar protocol to the current analysis. DGVMs suggest a large component of the uptake trend is associated with a positive NPP response to elevated CO_2, which is broadly consistent with the enhancement of forest production due to CO_2 observed in FACE experiments (Norby et al., 2005), although they are largely located in temperate forest ecosystems. However, recent studies have highlighted the role of nitrogen in limiting the long-term CO_2 response (Canadell et al., 2007; Norby et al., 2010) in these ecosystems. The long-term plant response to elevated CO_2 is likely affected by nutrients and its impact on plant C allocation (Zaehle et al., 2014), however only two out of the nine models used here (CLM4CN and OCN) include interactive nutrient cycling (see DGVM characteristics, Table S1).

In contrast to the large trend in net C uptake across the tropics, DGVMs simulate no statistically significant trend over the northern land region. In particular, trends in NPP over temperate regions are smaller than those in boreal regions, and are also not significant. Many temperate areas experienced a decrease in rainfall between 1990 and 2009, and suffered periods of prolonged and severe drought. Examples include the drought in the western USA of 2000–2004 (McDowell et al., 2008; Anderreg et al., 2012) and the 2003 summer heatwave in Europe (Ciais et al., 2005). Zscheischler et al. (2014) suggest that negative productivity extremes dominated interannual variability in productivity during the period 1982–2011; these extremes are evident particularly over temperate latitudes.

Satellite observations suggest a general greening trend in high latitudes, with an earlier onset and longer growing season in high-latitude ecosystems, which is reproduced by the DGVMs. Observations suggest a greening tundra and a slower greening and possible browning in some regions of the boreal forest (Tucker et al., 2001; Bhatt et al., 2010), especially in North America (Beck and Goetz, 2011). In tundra ecosystems, an earlier onset is attributed to warming and earlier snowmelt. In these ecosystems, the start of the growing season corresponds to near peak in radiation. Thus any temperature-induced earlier snowmelt (McDonald et al., 2004; Sitch et al., 2007a) is likely to enhance plant production. Warming may not have such a great effect on the end of the growing season in Arctic tundra ecosystems, as this may be driven primarily by radiation. DGVMs simulate a significant positive trend in NPP in boreal North America and boreal Asia and the circumpolar tundra. Nitrogen limitation is also likely to constrain the productivity at high latitudes, but it was not possible to quantify N-limitation effects on regional trends in this study.

DGVMs simulate decreasing NPP across northeastern China and Mongolia, contributing to the overall decreasing land uptake trend, in response to recent climate. In a regional study, Poulter et al. (2013) investigated the differential response of cool semi-arid ecosystems to recent warming and drying trends across Mongolia and northern China, using multiple sources of evidence, including the LPJ DGVM, FPAR remotely sensed data (derived from GIMMS NDVI3g), and tree-ring widths. They found coherent patterns of high sensitivity to precipitation across data sources, which showed some areas with warming-induced springtime greening and drought-induced summertime browning, and limitations to NPP explained mainly by soil moisture.

Browning has occurred as a consequence of regional drought, wildfire, and insect outbreak, and their interaction, especially in North America (Beck and Goetz, 2011). Disturbance plays a key role in the ecology of many global ecosystems. For example, wildfire plays a dominant role in the carbon balance of boreal forest in central Canada and other regions (Bond-Lamberty et al., 2007), and insect outbreaks like the mountain pine beetle epidemic between 2000 and 2006 in British Colombia, Canada, resulted in the transition of forests from a small carbon sink to a source (Kurz et al., 2008). In general, disturbance and forest management are inadequately represented by the current generation of DGVMs, even though several models include simple prognostic wildfire schemes (Table S1), while some are starting to include other disturbance types such as insect attacks (Jönsson et al., 2012) and windthrow (Lagergren et al., 2012). The extension of DGVMs to include representations of globally and regionally important disturbance types and their response to changing environmental conditions is a priority.

In Table 4, DGVM results are compared with the RECCAP synthesis papers documenting carbon sources and sinks for individual regions. Note that DGVMs provided one source of evidence for some regional papers. Over Russia, DGVMs agree on a sink yet underestimate that sink's magnitude, likely related to soil respiration (which is unsurprising, as many DGVMs have a limited representation of permafrost and active layer thickness) (Dolman et al., 2012). In South America, DGVMs agree with inventory-based estimates on a sink in natural forests (Gloor et al., 2012). DGVMs also agree with other data sources on the sign and magnitude of the natural land sink over Australia (Harverd et al., 2013). Over Europe DGVMs simulate a smaller mean land sink than the synthesis study suggests (Luyssaert et al., 2012). However, the regional synthesis was conducted over the shorter time period 2001–2005. For the Arctic, DGVMs tend to simulate a lower sink than regional process-based models (McGuire et al., 2012). However, over the 1990–2006 period, DGVMs are in line with observations and inversions on the magnitude and sign of the natural land sink, and DGVM results also suggest a sink trend in line with observations. DGVMs simulate a land sink over South Asia in agreement with inversions; however there were limited data to compare trends from DGVMs and other products (Patra et al., 2013). For East Asia, DGVM results agree remarkably well with remote sensing model–data fusion and inverse models on the magnitude of the land sink over the period 1990–2009. Finally, for Africa, DGVMs are broadly consistent with inventory- and flux-based estimates in simulating a land sink over Africa, albeit of lower magnitude (Valentini et al., 2014).

4.2 Ocean

The investigated OBGCMs consistently simulate an ocean characterised by a substantial uptake of CO_2 from the atmosphere, but with a global integrated trend in the last two decades ($-0.02 \pm 0.01\,\mathrm{Pg\,C\,yr^{-2}}$) that is substantially smaller than that expected based on the increase in atmospheric CO_2. Results based on the predictions from ocean inversion and ocean Green function methods (Mikaloff Fletcher et al., 2006; Gruber et al., 2009; Khatiwala et al., 2009) suggest an increase in ocean uptake with a trend of the order of $-0.04\,\mathrm{Pg\,C\,yr^{-2}}$ over the analysis period (see also Wanninkhof et al., 2013). These latter methods assume constant circulation, while our simulations here include the impact of climate variability and change.

Our analyses reveal that recent climate variability and change has caused the ocean carbon cycle to take up less CO_2 from the atmosphere than expected on the basis of the increase in atmospheric CO_2, i.e. it reduces the efficiency of the ocean carbon sink. Globally, we find that this efficiency reduction is primarily a result of ocean warming, while, regionally, many more processes (e.g. wind changes, alkalinity/DIC concentration changes) are at play.

Is this reduction in uptake efficiency over the analysis period the first sign of a positive feedback between global warming and the ocean carbon cycle – or, alternatively, could it just reflect natural decadal-scale variability in air–sea CO_2 fluxes? Without a formal attribution study, it is not possible to provide a firm answer. We suspect that the majority of the trend in the efficiency is due to

Table 4. Ensemble DGVM regional NBP mean comparison with RECCAP regional chapter analyses.

Region	DGVM mean NBP ($TgC\,yr^{-1}$)	Region processed-based models	Inventory-based	Flux-based	Inversion	Best estimate
Russia	−199		−761	−709	−653	
South America (forest)	−472 ± 211		−570 ± 170 (1990–94) −530 ± 140 (1995–99) −450 ± 250 (2000–04) −150 ± 230 (2005–09)			
Africa	−410 ± 310		−740 ± 1190	−1340 ± 1320	50 ± 280 (LULCC 510 ± 280)	
Australia & New Zealand	−70 ± 78	−36 ± 29 (LULCC 18 ± 7)				
Europe	−179 ± 92					−891 ± 155 (2001–05)
Arctic (1990–2006)	−86	−177				−96
South Asia	−210 ± 164				−35.4 (1997–06) −317 to −88.3 (2007–08)	
East Asia	−224 ± 141		−293 ± 33 combined inventory–EO-flux approach			270 ± 507

"natural" decadal-scale variability; however, largely based on the results of McKinley et al. (2011) and Fay and McKinley (2013), who showed that whereas trends in oceanic pCO_2 (and air–sea CO_2 fluxes) are variable on a decadal timescale, they do converge towards atmospheric pCO_2 trends when analysed over a longer 30-year period for most global regions. Nevertheless, they also show that warming (partly driven by anthropogenic climate change) in the permanently stratified subtropical gyre of the North Atlantic has started to reduce ocean uptake in recent years. In the Southern Ocean, where Le Quéré et al. (2007) and Lovenduski et al. (2008) used models to suggest a reduction in ocean carbon uptake efficiency over the past 25 years in response to increasing Southern Ocean winds, Fay and McKinley (2013) concluded that the data are insufficient to draw any conclusions.

We should note that the associated uncertainties remain large. Of particular concern is the moderate success of the models in simulating the time-mean ocean sinks and their long-term seasonal cycle (e.g. McKinley et al., 2006). Furthermore, some of the models underestimate the oceanic uptake of transient tracers such as anthropogenic radiocarbon (see e.g. Graven et al., 2012). Such a reduction in the oceanic uptake efficiency is also not suggested by independent measures of oceanic CO_2 uptake, such as the atmospheric O_2 / N_2 method (Manning and Keeling, 2006; Ishidoya et al., 2012), although the large uncertainties in these estimates make the determination of trends in uptake highly uncertain.

All the models have been tuned to reproduce data synthesis on ocean surface pCO_2 (Pfeil et al., 2013; Takahashi et al., 2009) and deep ocean (Key et al., 2004) reasonably well. Specific systematic data assimilation procedures, however, have not been applied. On decadal timescales, the ocean CO_2 flux feedback to climate change (change in hydrography and circulation) and rising ambient CO_2 (change in CO_2 buffering) reacts only slowly on the global average due to the long timescales of oceanic motion and marine CO_2 equilibration with the atmosphere. Changes in biogeochemical and ecosystem processes, such as locally varying gas exchange velocities, phytoplankton blooms, and associated particle flux pulses, can lead to regional interannual variations in air–sea CO_2 fluxes, but may partially cancel for averages over larger regions. With ocean observations only over about a two-decade time frame, it is difficult to quantify longer-term trends due to other proposed mechanisms: a gradual slowing-down of meridional overturning circulation due to a strengthening of density stratification; redissolution of $CaCO_3$ sediment from the seafloor associated with fossil fuel neutralization; and potential changes in biogenic particle fluxes due to carbon overconsumption and changing ballasting (cf. Keller et al., 2014). Whether more complex models will render better results will depend on how well the additional free parameters in more complex biogeochemical models can be constrained by measurements. So far, more complex – and hence potentially more realistic – models do not necessarily give better results than the present nutrient-phytoplankton-zooplankton-detritus (NPZD)-type models models as applied here (Le Quéré et al., 2005; Kriest et al., 2010).

4.3 Reducing uncertainty in regional sinks

In order to better quantify the regional carbon cycle and its trends, DGVM and ocean carbon cycle models need to improve both process representations and model evaluation and benchmarking (Luo et al., 2012). There is a need for up-to-date global climate and land use and cover change data sets to force the DGVMs, as well as a deeper investigation of the quality and differences between the different reanalysis products used to force ocean carbon cycle models. Also, techniques such as detection and attribution can be applied to elucidate trends in the regional carbon cycle and their drivers.

4.3.1 Model evaluation and benchmarking

Piao et al. (2013) evaluated the DGVM model results for their response to climate variability and to CO_2 trends, and the seasonal cycle of CO_2 fluxes were benchmarked in Peng et al. (2014). Piao et al. (2013) found DGVMs to simulate higher mean and interannual variations (IAVs) in gross primary production than a data-driven model (Jung et al., 2011), particularly in the tropics; however, this is the region where the data-driven model is most uncertain. DGVMs were able to capture the IAVs in RLS, although the simulated land–atmosphere net CO_2 flux appears too sensitive to variations in precipitation in tropical forests and savannas. However, Poulter et al. (2014) found an increase in the sensitivity of the net flux to precipitation over the last three decades across continental Australia. Piao et al. (2013) found that the simulated net CO_2 flux was more sensitive than productivity to temperature variations. When compared to ecosystem warming experiments the DGVMs tend to underpredict the response of NPP to temperature at temperate sites. DGVMs simulated an ensemble mean NPP enhancement comparable to FACE experiment observations (Piao et al., 2013). However, modelling of ecosystem function in water-stressed environments and changes in plant water use with elevated CO_2 remains a challenge for DGVMs (Morales et al., 2005; Keenan et al., 2009; De Kauwe et al., 2013).

There is a critical need for comprehensive model benchmarking, as a first step to attempt to reduce model uncertainty. Several prototype carbon cycle benchmarking schemes have been developed (Randerson et al., 2009; Cadule et al., 2010). A more in-depth evaluation and community benchmarking set needs to be agreed upon and implemented which also evaluates models for their implicit land response timescales (especially relevant in the discussion on future tipping elements and non-linear future responses) and for the simulated carbon, water, and nutrient cycles. New emerging frameworks now exist (Blyth et al., 2011; Abramowitz, 2012; Luo et al., 2012; Dalmonech and Zaehle, 2013; Harverd et al., 2013). One example within RECAPP is a multiple-constraint approach applied to reduce uncertainty in land carbon and water cycles over Australia (Haverd et al., 2013).

4.3.2 Model resolution

Simulated ocean carbon dynamics may be sensitive to horizontal resolution, particularly as model resolution improves sufficiently to adequately capture mesoscale eddies. Mesoscale turbulence influences the ocean carbon cycle in a variety of ways, and the present eddy parameterisations may not adequately capture the full range of effects and the responses to climate variability and change. For example, mesoscale processes are thought to modulate biological productivity by altering the supply of limiting nutrients (Falkowski et al., 1991; McGillicuddy et al., 1998; Gruber et al., 2011). A particularly crucial issue involves the wind-driven overturning circulation in the Southern Ocean, where non-eddy-resolving models indicate a strong sensitivity of the overturning circulation and ocean carbon uptake to surface wind stress (Le Quéré et al., 2007; Lovenduski et al., 2008). Some eddy-resolving models, in contrast, suggest that enhanced wind stress is dissipated by increased eddy activity, leading to only a small increase in overturning (Böning, et al., 2008), although more recent results indicate a larger response (Gent and Danabasoglu, 2011; Matear et al., 2013).

4.3.3 Model structure

There is a need for improved representation of ecological processes in land and ocean models, e.g. nutrient cycling (N, P), demographic dynamics, disturbance (wildfire, windthrow, insects), land use and land cover change in land models, and better representation of the key functional diversity in ocean and land biogeochemical models. DGVMs need to represent land use and land cover changes, forest management, and forest age in order to improve estimates of the regional and global land carbon budget. There have been recent developments to include nutrient dynamics, mostly nitrogen, in global land biosphere models (as reviewed by Zaehle and Dalmonech, 2011). Too few model simulations are available to date to allow for an ensemble model trend assessment. However, a few general trends appear robust: as evident from Table 3, CN models generally show less of a response to increasing atmospheric CO_2 due to nitrogen limitation of plant production. N dynamics further alter the climate–carbon relationship, which tend to reduce the C loss from temperate and boreal terrestrial ecosystems due to warming – but with a considerable degree of uncertainty (Thornton et al., 2009; Sokolov et al., 2008; Zaehle et al., 2010). Changes in the nitrogen cycle due to anthropogenic reactive nitrogen additions (both fertiliser to croplands and N deposition on forests and natural grasslands) further modify the terrestrial net C balance and contribute with -0.2 to $-0.5\,\mathrm{Pg\,C\,yr^{-1}}$ to the current land sink (Zaehle and Dalmonech, 2011). Zaehle et al. (2011), using the OCN model, estimated the 1995–2005 trend in land uptake due to N deposition to be $-1.1 \pm 1.7\,\mathrm{Tg\,C\,yr^{-2}}$, with strong regional differences depending on the regional trends in air pollution and reactive N loading of the atmosphere and the nitrogen status of the ecosystems, which are generally lower in less responsive ecosystems close to nitrogen saturation highly polluted regions. The DGVMs applied here do not consider the P cycle; P limitation on land carbon uptake may be particularly important in tropical forests and savannas (Edwards et al., 2005; Wang et al., 2010; Zhang et al., 2014).

There are several additional land processes that have not been considered in this current multi-model analysis. These include the effects of aerosols and tropospheric ozone on the carbon cycle. Unlike a global forcing agent such as CO_2, the effects of air pollutants (aerosols, NO_x, and O_3), with their shorter atmospheric lifetimes, are at the regional scale.

Aerosol-induced changes in radiation quantity and quality (i.e. the ratio of diffuse to direct) affect plant productivity and the land sink (Mercado et al., 2009). From around 1960 until the 1980s, radiation levels declined across industrialised regions, a phenomenon called "global dimming", followed by a recent brightening in Europe and North America with the adoption of air pollution legislation. Reductions in acid rain have been found to greatly influence trends in riverine DOC, vegetation health, and rates of soil organic matter decomposition. Tropospheric ozone is known to be toxic to plants and lead to reductions in plant productivity, and thus reduce the efficiency of the land carbon sink (Sitch et al., 2007b; Anav et al., 2011). Drivers of the land carbon sink related to air pollution – e.g. N deposition, acid precipitation, diffuse and direct radiation, and surface O_3 – have varied markedly in space and time over recent decades. Although likely important for regional carbon cycle trends, quantifying these effects is beyond the scope of the present study.

The Pinatubo eruption in 1991, at the start of the study period, had a major influence on many carbon cycle processes, leading to an enhanced land sink over the period 1991–1993. This has been attributed to a combination of cooling-induced reductions in high-latitude respiration and enhanced productivity associated with changes in diffuse radiation (Jones and Cox, 2001; Lucht et al., 2002; Peylin et al., 2005; Mercado et al., 2009; Frölicher et al., 2013). The direct effect of aerosols on climate drivers is implicitly included in this study (i.e. responses to high-latitude cooling, tropical drying, reduced net incoming solar radiation); however diffuse radiation effects are not included.

Similar gaps need to be addressed in ocean biogeochemical models. The ecosystem modules in the current generation of OBGCMs lack the ability to assess many of the suggested mechanisms by which climate and ocean acidification could alter marine biogeochemistry and ocean carbon storage. Proposed biological processes that could influence ocean carbon uptake and release involve, for example, decoupling of carbon and macronutrient cycling, changes in micronutrient limitation, variations in elemental stoichiometry in organic matter, and changes in the vertical depth scale for the respiration of sinking organic carbon particles (e.g. Boyd and Doney, 2003; Sarmiento and Gruber, 2006). Some advances have been made with the incorporation of dynamic iron cycling and iron limitation, multiple plankton groups, calcification, and nitrogen fixation (Le Quéré et al., 2005). However, the evaluation of these aspects of the models is currently hindered by both data- and process-level information limitations.

4.3.4 Climate and land use and cover data sets

In addition to model structure, the choice of climate forcing and model initial conditions can also contribute to differences in the simulated terrestrial carbon sink. At regional scales, differences in land cover can introduce $\sim 10\,\%$ uncertainty in simulated regional-scale GPP (Jung et al., 2007; Quaife et al., 2008) and $\sim 3.5\,\%$ uncertainty for global NPP. Climate forcing uncertainty tends to have larger effects on carbon flux uncertainty than land cover (Hicke, 2005; Poulter et al., 2011), with up to 25 % differences in GPP reported over Europe (Jung et al., 2007) and a 10 % difference for global NPP (Poulter et al., 2011). Climate forcing uncertainty and land cover (i.e. PFT distributions) can alter long-term trends in land to atmosphere net CO_2 flux and interannual variability of carbon fluxes to climate (Poulter et al., 2011). The DGVMs applied here did not consider LULCC. This is an active area of research; models need a consistent implementation of LULCC. Uncertainties in the simulated net land use flux are associated with assumptions on the implementation of LULCC gridded maps (e.g. whether conversion to cropland in a grid-cell is taken preferentially from grassland, forest, or both), simulated biomass estimates, and subsequent decomposition rates. However DGVMs offer the exciting prospect of disentangling the component fluxes associated with land use (e.g. direct emissions and legacy fluxes) and separating the environmental and direct human impacts on the net LU flux (Gasser and Ciais, 2013; Pongratz et al., 2014; Stocker et al., 2014).

5 Conclusions

Land models suggest an increase in the global land net C uptake over the period 1990–2009, with increases in tropical and southern regions and negligible increase in northern regions. The increased sink is mainly driven by trends in NPP, in response to increasing atmospheric CO_2 concentration, and modulated by change in climate. Over the same period, ocean models suggest a negligible increase in net ocean C uptake – a result of ocean warming counteracting the expected increase in ocean uptake driven by the increase in atmospheric CO_2. At the sub-regional level, trends vary both in sign and magnitude, particularly over land. Areas in temperate North America, eastern Europe, and northeastern China show a decreasing regional land sink trend, due to regional drying, suggesting a possibility for a transition to a net carbon source in the future if drying continues or droughts become more severe and/or frequent. In the ocean, the trends tend to be more homogeneous, but the underlying dynamics differ greatly, ranging from ocean warming, to winds, and to changes in circulation/mixing and ocean productivity, making simple extrapolations into the future difficult.

Our conclusions need to be viewed with several important caveats: only a few models include a fully coupled carbon–nitrogen cycle, and no model included land use and land cover changes. Ocean models tend to be too coarse in resolution to properly represent important scales of motions and mixing, such as eddies and other mesoscale processes, and coastal boundary processes. Furthermore, their representation of ocean ecosystem processes and their sensitivity to climate change and other stressors (e.g. ocean acidification, deoxygenation, etc.; Gruber, 2011; Boyd, 2011) is rather simplistic.

There is a need for detailed model evaluation and benchmarking in order to reduce the uncertainty in the sinks in the land and ocean and, particularly, in how these sinks have changed in the past and how they may change in the future. For land ecosystems, a concerted effort is needed in the DGVM community to incorporate nutrient cycling as well as land use and land cover change. For the oceans, models need to improve their representation of unresolved physical transport and mixing processes, and ecosystem models need to evolve to better characterise their response to global change.

Acknowledgements. S. Sitch acknowledges financial support by RCUK through NERC (grant no. NE/J010154/). N. Gruber and C. Heinze acknowledge financial support by the European Commission through the EU FP7 projects CARBOCHANGE (grant no. 264879) and GEOCARBON (grant no. 283080). N. Gruber was additionally supported through ETH Zurich. S. C. Doney acknowledges support from the US National Science Foundation (NSF AGS-1048827). P. Friedlingstein, A. Arneth, and S. Zaehle acknowledge support by the European Commission through the EU FP7 project EMBRACE (grant no. 282672). A. Arneth and S. Sitch acknowledge the support of the European Commission-funded project LUC4C (grant no. 603542). The research leading to these results received funding from the European Community's Seventh Framework Programme (FP7 2007–2013) under grant agreement no. 238366. A. Ahlström and B. Smith acknowledge funding through the Mistra Swedish Research Programme on Climate, Impacts and Adaptation (SWECIA). This study is a contribution to the Lund Centre for Studies of Carbon Cycle and Climate Interactions (LUCCI) and the Strategic Research Area Modelling the Regional and Global Earth System (MERGE). This is a contribution to the Bjerknes Centre for Climate Research (BCCR) and core project BIOFEEDBACK of the Centre for Climate Dynamics (SKD) at BCCR. C. Heinze acknowledges support from NOTUR/NorStore projects NN2980K and NS2980K. The authors wish to thank Jonathan Barichivich for discussions on greening trends.

Edited by: F. Joos

References

Abramowitz, G.: Towards a public, standardized, diagnostic benchmarking system for land surface models, Geosci. Model Dev., 5, 819–827, doi:10.5194/gmd-5-819-2012, 2012.

Allen, C. D., Macalady, A., Chenchouni, H., Bachelet, D., McDowell, N., Vennetier, M., Gonzales, P., Hogg, T., Rigling, A., Breshears, D., Fensham, R., Zhang, Z., Kitzberger, T., Lim, J., Castro, J., Allard, G., Running, S., Semerci, A., and Cobb, N.: A global overview of drought and heat-induced tree mortality reveals emerging climate change risks for forests, Forest Ecol. Manag., 4, 660–684, doi:10.1016/j.foreco.2009.09.001, 2010.

Anav, A., Menut, L., Khvorostyanov, D., and Viovy, N.: Impact of tropospheric ozone on the Euro-Mediterranean vegetation, Glob. Change Biol., 17, 2342–2359, 2011.

Anderegg, W. R. L., Berry J. A., Smith, D. D., Sperry J. S., Anderegg, L. D. L., and Field, C. B.: The roles of hydraulic and carbon stress in a widespread climate-induced forest die-off, P. Natl. Acad. Sci. USA, 109, 233–237, doi:10.1073/pnas.1107891109, 2012.

Angert, A., Biraud, S., Bonfils, C., Henning, C. C., Buermann, W., Pinzon, J., Tucker, C. J., and I. Fung: Drier summers cancel out the CO_2 uptake enhancement induced by warmer springs, P. Natl. Acad. Sci., 102, 10823–10827, 2005.

Assmann, K. M., Bentsen, M., Segschneider, J., and Heinze, C.: An isopycnic ocean carbon cycle model, Geosci. Model Dev., 3, 143–167, doi:10.5194/gmd-3-143-2010, 2010.

Baker, T. R., Phillips, O. L., Malhi, Y., Almeida, S., Arroyo, L., Di Fiore, A., Erwin, T., Higuchi, N., Killeen, T. J., Laurance, S. G., Laurance, W. F., Lewis, S. L., Monteagudo, A., Neill, D. A., Nunez Vargas, P., Pitman, N. C. A., Silva, J. N. M., and Vasquez Martinez, R.: Increasing biomass in Amazonian forest plots, Philos. T. R. Soc. B, 359, 353–365, 2004.

Ballantyne, A. P., Alden, C. B., Miller, J. B., Tans, P. P., and White, J. W. C.: Increase in observed net carbon dioxide uptake by land and oceans during the past 50 years, Nature, 488, 70–72, doi:10.1038/nature11299, 2012.

Beck, S. A. and Goetz, S. J.: Satellite observations of high northern latitude vegetation productivity changes between 1982 and 2008: ecological variability and regional differences, Environ. Res. Lett. 6, 045501, doi:10.1088/1748-9326/6/4/045501, 2011.

Bhatt, U. S., Walker, D. A., Raynolds, M. K., Comiso, J. C., Epstein, H. E., Jia, G., Gens, R., Pinzon, J. E., Tucker, C. J., Tweedie, C. E., and Webber, P. J.: Circumpolar Arctic tundra vegetation change is linked to sea ice decline, Earth Interact., 14, 1–20, 2010.

Blyth, E., Clark, D. B., Ellis, R., Huntingford, C., Los, S., Pryor, M., Best, M., and Sitch, S.: A comprehensive set of benchmark tests for a land surface model of simultaneous fluxes of water and carbon at both the global and seasonal scale, Geosci. Model Dev., 4, 255–269, doi:10.5194/gmd-4-255-2011, 2011.

Bonan, G. B. and Levis, S.: Quantifying carbon-nitrogen feedbacks in the Community Land Model (CLM4), Geophys. Res. Lett., 37, L07401, doi:10.1029/2010GL042430, 2010.

Bond-Lamberty, B., Peckham, S. D., Ahl, D. E., and Gower, S. T.: Fire as the dominant driver of central Canadian boreal forest carbon balance, Nature, 450, 89–92, 2007.

Böning, C. W., Dispert, A., Visbeck, M., Rintoul, S. R., and Schwarzkopf, F. U.: The response of the Antarctic Circumpolar Current to recent climate change, Nat. Geosci., 1, 864–869, 2008.

Boyd, P. and Doney, S. C.: The impact of climate change and feedback process on the ocean carbon cycle, Ocean Biogeochemistry, edited by: Fasham, M., Springer, 157–193, 2003.

Boyd, P. W.: Beyond ocean acidification, Nat. Geosci., 4, 273–274, 2011.

Broecker, W. S., Peng, T.-H., Ostlund, G., and Stuiver, M.: The distribution of bomb radiocarbon in the ocean, J. Geophys. Res., 90, 6953–6970, 1985.

Buermann, W., Lintner, B. R., Koven, C. D., Angert, A., Pinzon, J. E., Tucker, C. J., and Fung, I. Y.: The changing carbon cycle at Mauna Loa observatory, P. Natl. Acad. Sci., 104, 4249–4254, 2007.

Buitenhuis, E. T., Rivkin, R. B., Sailley, S., and Le Quéré, C.: Biogeochemical flues through microzooplankton, Global Biogeochem. Cy., 24, GB4015, doi:10.1029/2009GB003601, 2010.

Cadule, P., Friedlingstein, P., Bopp, L., Sitch, S., Jones, C. D., Ciais, P., Piao, S. L., and Peylin, P.: Benchmarking coupled climate-carbon models against long-term atmospheric CO_2 measurements, Global Biogeochem. Cy., 24, Gb2016, doi:10.1029/2009gb003556, 2010.

Canadell, J. G., Pataki D., Gifford, R., Houghton, R. A., Lou, Y., Raupach, M. R., Smith, P., and Steffen, W.: Saturation of the terrestrial carbon sink, in: Terrestrial Ecosystems in a Changing World, edited by: Canadell, J. G., Pataki, D., and Pitelka, L., 59–78, The IGBP Series, Springer-Verlag, Berlin Heidelberg, 59–78, 2007.

Canadell, J. G., Ciais, P., Gurney, K., Le Quéré, C., Piao, S., Raupach, M. R., and Sabine, C. L.: An international effort to quantify regional carbon fluxes, EOS T. Am. Geophys. Un., 92, 81–82, 2011.

Canadell, J. G., Ciais, P., Sabine, C., and Joos, F. (Eds.): REgional Carbon Cycle Assessment and Processes (RECCAP), Special issue, Biogeosciences, http://www.biogeosciences-discuss.net/special_issue83.html, 2013.

Chevallier, F., Ciais, P., Conway, T. J., Aalto, T., Anderson, B. E., Bousquet, P., Brunke, E. G., Ciattaglia, L., Esaki, Y., Fröhlich, M., Gomez, A., Gomez-Pelaez, A. J., Haszpra, L., Krummel, P. B., Langenfelds, R. L., Leuenberger, M., Machida, T., Maignan, F., Matsueda, H., Morguí, J. A., Mukai, H., Nakazawa, T., Peylin, P., Ramonet, M., Rivier, L., Sawa, Y., Schmidt, M., Steele, L. P., Vay, S. A., Vermeulen, A. T., Wofsy, S., and Worthy, D.: CO_2 surface fluxes at grid point scale estimated from a global 21-year reanalysis of atmospheric measurements, J. Geophys. Res., 115, D21307, doi:10.1029/2010JD013887, 2010.

Ciais, P., Tans, P. P., Trolier, M., White, J. W. C., and Francey, R. J.: A large northern hemisphere terrestrial CO_2 sink indicated by the $^{13}C/^{12}C$ ratio of atmospheric CO_2, Science, 269, 1098–1102, doi:10.1126/science.269.5227.1098, 1995.

Ciais, P., Reichstein, M., Viovy, N., Granier, A., Ogee, J., Allard, V., Aubinet, M., Buchmann, N., Bernhofer, C., Carrara, A., Chevallier, F., De Noblet, N., Friend, A. D., Friedlingstein, P., Grunwald, T., Heinesch, B., Keronen, P., Knohl, A., Krinner, G., Loustau, D., Manca, G., Matteucci, G., Miglietta, F., Ourcival, J. M., Papale, D., Pilegaard, K., Rambal, S., Seufert, G., Soussana, J. F., Sanz, M. J., Schulze, E. D., Vesala, T., and Valentini, R.: Europe-wide reduction in primary productivity caused by the heat and drought in 2003, Nature, 437, 529–533, doi:10.1038/nature03972, 2005.

Ciais, P., Canadell, J. G, Luyssaert, S., Chevallier, F., Shvidenko, A., Pousii, Z., Jonas, M., Peylin, P., Wayne King, A., Schulze, E.-D., Piao, S., Roedenbeck, C., Wouters, P., and Breon, F.-M.: Can we reconcile atmospheric estimates of the Northern terrestrial carbon sink with land-based accounting?, Current Opinion in Environmental Sustainability, 2, 225–230, 2010.

Ciais, P., Sabine, C., Bala, G., Bopp, L., Brovkin, V., Canadell, J., Chhabra, A., DeFries, R., Galloway, J., Heimann, M., Jones, C., Le Quéré, C., Myneni, R. B., Piao, S., and Thornton, P.: Carbon and Other Biogeochemical Cycles, in: Climate Change 2013: The Physical Science Basis, Contribution of Working Group I to the Fifth Assessment Report of the Intergovernmental Panel on Climate Change, edited by: Stocker, T. F., Qin, D., Plattner, G.-K., Tignor, M., Allen, S. K., Boschung, J., Nauels, A., Xia, Y., Bex, V., and Midgley, P. M., Cambridge University Press, Cambridge, United Kingdom and New York, NY, USA, 465–570, 2013.

Clark, D. B., Mercado, L. M., Sitch, S., Jones, C. D., Gedney, N., Best, M. J., Pryor, M., Rooney, G. G., Essery, R. L. H., Blyth, E., Boucher, O., Harding, R. J., Huntingford, C., and Cox, P. M.: The Joint UK Land Environment Simulator (JULES), model description – Part 2: Carbon fluxes and vegetation dynamics, Geosci. Model Dev., 4, 701–722, doi:10.5194/gmd-4-701-2011, 2011.

Cox, P. M.: Description of the "TRIFFID" dynamic global vegetation model, Hadley Centre Technical Note, 24, 1–16, 2001.

Dalmonech, D. and Zaehle, S.: Towards a more objective evaluation of modelled land-carbon trends using atmospheric CO_2 and satellite-based vegetation activity observations, Biogeosciences, 10, 4189–4210, doi:10.5194/bg-10-4189-2013, 2013.

De Kauwe, M. G., Medlyn, B. E., Zaehle, S., Walker, A. P., Dietze, M. C., Hickler, T., Jain, A. K., Luo, Y., Parton, W. J., Prentice, C., Smith, B., Thornton, P. E., Wang, S., Wang, Y-P., Wårlind, D., Weng, E. S., Crous, K, Y., Ellsworth, D. S., Hanson, P. J., Seok-Kim, H., Warren, J. M., Oren, R., and Norby, R. J.: Forest water use and water use efficiency at elevated CO_2: a model-data intercomparison at two contrasting temperate forest FACE sites, Glob. Change Biol., 19, 1759–1779, 2013.

Denman, K. L., Brasseur, G., Chidthaisong, A., Ciais, P., Cox, P. M., Dickinson, R. E., Hauglustaine, D., Heinze, C., Holland, E., Jacob, D., Lohmann, U., Ramachandran, S., da Silva Dias, P. L., Wofsy, S. C. , and Zhang, X.: Couplings Between Changes in the Climate System and Biogeochemistry, in: Climate Change 2007: The Physical Science Basis, Contribution of Working Group I to the Fourth Assessment Report of the Intergovernmental Panel on Climate Change, edited by: Solomon, S., Qin, D., Manning, M., Chen, Z., Marquis, M., Averyt, K. B., Tignor, M., and Miller, H. L., Cambridge University Press, Cambridge, United Kingdom and New York, NY, USA, 499–588, 2007.

Dentener, F., Stevenson, D., Ellingsen, K., van Noije, T., Schultz, M., Amann, M., Atherton, C., Bell, N., Bergmann, D., Bey, I., Bouwman, L., Butler, T., Cofala, J., Collins, B., Drevet, J., Doherty, R., Eickhout, B., Eskes, H., Fiore, A., Gauss, M., Hauglustaine, D., Horowitz, L., Isaksen, I. S. A., Josse, B., Lawrence, M., Krol, M., Lamarque, J. F., Montanaro, V., Muller, J. F., Peuch, V. H., Pitari, G., Pyle, J., Rast, S., Rodriguez, J., Sanderson, M., Savage, N. H., Shindell, D., Strahan, S., Szopa, S., Sudo, K., Van Dingenen, R., Wild, O., and Zeng, G.: The global atmospheric environment for the next generation, Environ. Sci. Technol., 40, 3586–3594, doi:10.1021/es0523845, 2006.

Dolman, A. J., Shvidenko, A., Schepaschenko, D., Ciais, P., Tchebakova, N., Chen, T., van der Molen, M. K., Belelli Marchesini, L., Maximov, T. C., Maksyutov, S., and Schulze, E.-D.: An estimate of the terrestrial carbon budget of Russia using inventory-based, eddy covariance and inversion methods, Biogeosciences, 9, 5323–5340, doi:10.5194/bg-9-5323-2012, 2012.

Doney, S. C., Lima, I., Feely, R. A., Glover, D. M., Lindsay, K., Mahowald, N., Moore, J. K., and Wanninkhof, R.: Mechanisms governing interannual variability in upper-ocean inorganic carbon system and air-sea CO_2 fluxes: Physical climate and atmospheric dust, Deep-Sea Res.-Pt. II, 56, 640–655, 2009a.

Doney, S. C., Lima, I., Moore, J. K., Lindsay, K., Behrenfeld, M. J., Westberry, T. K., Mahowald, N., Glover, D. M., and Takahashi, T.: Skill metrics for confronting global upper ocean ecosystem-biogeochemistry models against field and remote sensing data, J. Marine Syst., 76, 95–112, doi:10.1016/j.jmarsys.2008.05.015, 2009b.

Edwards E., S. McCaffery, S., and Evans, J.: Phosphorus status determines biomass response to elevated CO_2 in a legume: C-4 grass community, Glob. Change Biol., 11, 1968–1981, 2005.

Falkowski, P. G., Ziemann, D., Kolber, Z., and Bienfang, P. K.: Role of eddy pumping in enhancing primary production in the ocean, Nature, 352, 55–58, 1991.

Fan, S., Gloor, M., Mahlmam, J., Pacala, S., Sarmiento, J., Takahashi, T., and Tans, P.: A large terrestrial carbon sink in North America implied by atmospheric and oceanic carbon dioxide data and models, Science, 282, 442–446, 1998.

Fay, A. R. and McKinley, G. A.: Global trends in surface ocean *p*CO2 from in situ data, Global Biogeochem. Cy., 27, 541–557, doi:10.1002/gbc.20051, 2013.

Friedlingstein, P., Houghton, R. A., Marland, G., Hackler, J., Boden,T. A., Conway, T. J., Canadell, J. G., Raupach, M. R., Ciais, P., and Le Quéré, C.: Update on CO_2 emissions, Nat. Geosci., 3, 811–812, doi:10.1038/ngeo1022, 2010.

Friend, A. D., Stevens, A. K., Knox, R. G., and Cannell, M. G. R.: A process-based, terrestrial biosphere model of ecosystem dynamics (Hybrid v3.0), Ecol. Model., 95, 249–287, 1997.

Frölicher, T. L., Joos, F., Raible, C. C., and Sarmiento, J. L.: Atmospheric CO_2 response to volcanic eruptions: the role of ENSO, season, and variability, Global Biogeochem. Cy., 27, 239–251, 2013.

Gasser, T. and Ciais, P.: A theoretical framework for the net land-to-atmosphere CO_2 flux and its implications in the definition of "emissions from land-use change", Earth Syst. Dynam., 4, 171–186, doi:10.5194/esd-4-171-2013, 2013.

Gent, P. R. and Danabasoglu, G.: Response to Increasing Southern Hemisphere Winds in CCSM4, J. Climate, 24, 4992–4998, 2011.

Gloor, M., Gruber, N., Sarmiento, J. L., Sabine, C., Feely, D., and Rödenbeck, C.: A first estimate of present and preindustrial air-sea CO_2 flux patterns based on ocean interior carbon measurements and models, Geophys. Res. Lett., 30, 1010, doi:10.1029/2002GL015594, 2003.

Gloor, M., Sarmiento, J. L., and Gruber, N.: What can be learned about carbon cycle climate feedbacks from the CO_2 airborne fraction?, Atmos. Chem. Phys., 10, 7739–7751, doi:10.5194/acp-10-7739-2010, 2010.

Gloor, M., Gatti, L., Brienen, R., Feldpausch, T. R., Phillips, O. L., Miller, J., Ometto, J. P., Rocha, H., Baker, T., de Jong, B., Houghton, R. A., Malhi, Y., Aragão, L. E. O. C., Guyot, J.-L., Zhao, K., Jackson, R., Peylin, P., Sitch, S., Poulter, B., Lomas, M., Zaehle, S., Huntingford, C., Levy, P., and Lloyd, J.: The carbon balance of South America: a review of the status, decadal trends and main determinants, Biogeosciences, 9, 5407–5430, doi:10.5194/bg-9-5407-2012, 2012.

Graven, H. D., Gruber, N., Key, R., Khatiwala, S., and Giraud, X.: Changing controls on oceanic radiocarbon: New insights on shallow-to-deep ocean exchange and anthropogenic CO_2 uptake, J. Geophys. Res., 117, C10005, doi:10.1029/2012JC008074, 2012.

Gruber, N.: Warming up, turning sour, losing breath: Ocean biogeochemistry under global change, Philos. T. R. Soc. A, 369, 1980–1996, doi:10.1098/rsta.2011.0003, 2011.

Gruber, N., Gloor, M., Fletcher, S. E. M., Doney, S. C., Dutkiewicz, S., Follows, M. J., Gerber, M., Jacobson, A. R., Joos, F., Lindsay, K., Menemenlis, D., Mouchet, A., Muller, S. A., Sarmiento, J. L., and Takahashi, T.: Oceanic sources, sinks, and transport of atmospheric CO_2, Global Biogeochem. Cy., 23, GB1005, doi:10.1029/2008GB003349, 2009.

Gruber, N., Lachkar, Z., Frenzel, H., Marchesiello, P., Munnich, M., McWilliams, J., Nagai, T., and Plattner, G.-K.: Eddy-induced reduction of biological production in Eastern Boundary Upwelling Systems, Nat. Geosci., 4, 787–792, doi:10.1038/ngeo1273, 2011.

Haverd, V., Raupach, M. R., Briggs, P. R., J. G. Canadell., Davis, S. J., Law, R. M., Meyer, C. P., Peters, G. P., Pickett-Heaps, C., and Sherman, B.: The Australian terrestrial carbon budget, Biogeosciences, 10, 851–869, doi:10.5194/bg-10-851-2013, 2013.

Hicke, J. A.: NCEP and GISS solar radiation data sets available for ecosystem modeling: Description, differences, and impacts on net primary production, Global Biogeochem. Cy., 19, GB2006, doi:10.1029/2004GB002391, 2005.

Houghton, R. A.: How well do we know the flux of CO_2 from land-use change? Tellus B, 62, 337–351, doi:10.1111/j.1600-0889.2010.00473.x, 2010.

Ishidoya, S., Aoki, S., Goto, D., Nakazawa, T., Taguchi, S., and Patra, P. K.: Time and Space variations of O_2/N_2 ratio in the troposphere over Japan and estimation of the global CO_2 budget for the period 2000–2010, Tellus B, 64, 18964, doi:10.3402/tellusb.v64i0.18964, 2012.

Janssens, I. A., Freibauer, A., Ciais, P., Smith, P., Nabuurs, G.-J., Folberth, G., Schlamadinger, B., Hutjes, R. W. A., Ceulemans, R., Schulze, E.-D., Valentini, R., and Dolman, A. J.: Europe's terrestrial biosphere absorbs 7 to 12 % of European anthropogenic CO_2 emissions, Science 300, 1538–1542, 2003.

Jones, C. D. and Cox, P. M.: Modelling the volcanic signal in the atmospheric CO_2 record, Global Biogeochem. Cy., 15, 453–465, doi:10.1029/2000GB001281, 2001.

Jönsson, A. M., Schröder, M., Lagergren, F., Anderbrandt, O., and Smith, B.: Guess the impact of *Ips typographus* – an ecosystem modelling approach for simulating bark beetle outbreaks, Agr. Forest Meteorol., 166–167, 188–200, 2012.

Jung, M., Vetter, M., Herold, M., Churkina, G., Reichstein, M., Zaehle, S., Cias, P., Viovy, N., Bondeau, A., Chen, Y., Trusilova, K., Feser, F., and Heimann, M.: Uncertainties of modelling GPP over Europe: A systematic study on the effects of using different drivers and terrestrial biosphere models, Global Biogeochem. Cy., 21, GB4021, doi:10.1029/2006GB002915, 2007.

Jung, M., Reichstein, M., Margolis, H. A., Cescatti, A., Richardson A. D., Arain, M. A., Arneth, A., Bernhofer, C., Bonal, D., Chen J., Gianelle, D., Gobron, N., Kiely, G., Kutsch, W., Lasslop, G., Law, B. E., Lindroth, A., Merbold, L., Montagnani, L., Moors, E. J., Papale, D., Sottocornola, M., Vaccari, F., and Williams, C.: Global patterns of land-atmosphere fluxes of carbon dioxide, latent heat, and sensible heat derived from eddy covariance, satellite, and meteorological observations, J. Geophys. Res., 116, G00J07, doi:10.1029/2010JG001566, 2011.

Kalnay, E., Kanamitsu, M., Kistler, R., Collins, W., Deaven, D., Gandin, L., Iredell, M., Saha, S., Woollen, J., Zhu, Y., Chelliah, M., Ebisuzaki, W., Higgins, W., Janowiak, J., Mo, K. C., Ropelewski, C., Wang, J., Leetmaa, A., Reynolds, R., Jenne, R., and Joseph, D.: The NCEP/NCAR 40-year reanalysis project, B. Am. Meteorol. Soc., 77, 437–471, 1996.

Keeling, C. D., Bacastow, R. B., Bainbridge, A. E., Ekdahl Jr, C. A., Guenther, P. R., Waterman, L. S., and Chinm J. F. S.: Atmospheric carbon dioxide variations at Mauna Loa observatory, Hawaii, Tellus, 28, 538–551, 1976.

Keeling, C. D., Whorf, T. P., Wahlen, M., and van der Plicht, J.: Interannual extremes in the rate of rise of atmospheric carbon dioxide since 1980, Nature, 375, 666–670, 1995.

Keenan, T., García, R., Friend, A. D., Zaehle, S., Gracia, C., and Sabate, S.: Improved understanding of drought controls on seasonal variation in Mediterranean forest canopy CO_2 and water fluxes through combined in situ measurements and ecosystem modelling, Biogeosciences, 6, 1423–1444, doi:10.5194/bg-6-1423-2009, 2009.

Keller, K. M., Joos, F., and Raible, C. C.: Time of emergence of trends in ocean biogeochemistry, Biogeosciences, 11, 3647–3659, doi:10.5194/bg-11-3647-2014, 2014.

Key, R. M., Kozyr, A., Sabine, C. L., Lee, K., Wanninkhof, R., Bullister, J. L., Feely, R. A., Millero, F. J., Mordy, C., and Peng, T.-H.: A global ocean carbon climatology: results from Global Data Analysis Project (GLODAP), Global Biogeochem. Cy., 18, GB4031, doi:10.1029/2004GB002247, 2004.

Khatiwala, S., Primeau, F., and Hall, T.: Reconstruction of the history of anthropogenic CO_2 concentrations in the ocean, Nature, 462, 346–349, doi:10.1038/nature08526, 2009.

Khatiwala, S., Tanhua, T., Mikaloff Fletcher, S., Gerber, M., Doney, S. C., Graven, H. D., Gruber, N., McKinley, G. A., Murata, A., Ríos, A. F., and Sabine, C. L.: Global ocean storage of anthropogenic carbon, Biogeosciences, 10, 2169–2191, doi:10.5194/bg-10-2169-2013, 2013.

Kriest, I., Khatiwala, S., and Oschlies, A.: Towards an assessment of simple global marine biogeochemical models of different complexity, Progr. Oceanogr., 86, 337–360, 2010.

Krinner, G., Viovy, N., de Noblet-Ducoudre, N., Ogee, J., Polcher, J., Friedlingstein, P., Ciais, P., Sitch, S., and Prentice, I. C.: A dynamic global vegetation model for studies of the coupled atmosphere-biosphere system, Global Biogeochem. Cy., 19, Gb1015, doi:10.1029/2003gb002199, 2005.

Kurz, W. A., Dymond, C. C., Stinson, G., Rampley, G. J., Neilson, E. T., Carroll, A. L., Ebata, T., and Safranyik, L.: Mountain pine beetle and forest carbon feedback to climate change, Nature, 452, 987–990, doi:10.1038/nature06777, 2008.

Lagergren, F., Jönsson, A. M., Blennow, K., and Smith, B.: Implementing storm damage in a dynamic vegetation model for regional applications in Sweden, Ecol. Model., 247, 71–82, 2012.

Large, W. G. and Yeager, S.: Diurnal to decadal global forcing for ocean and sea-ice models: The data sets and flux climatologies, NCAR Tech. Rep. TN-460 _STR, 105 pp., 2004.

Lawrence, D. M., Oleson, K. W., Flanner, M. G., Thornton, P. E., Swenson, S. C., Lawrence, P. J., Zeng, X., Yang, Z.-L., Levis, S., Sakaguchi, K., Bonan, G. B., and Slater, A. G.: Parameterization improvements and functional and structural advances in version 4 of the Community Land Model, J. Adv. Model. Earth Syst., 3, M03001, doi:10.1029/2011ms000045, 2011.

Lenton, A., Tilbrook, B., Law, R. M., Bakker, D., Doney, S. C., Gruber, N., Ishii, M., Hoppema, M., Lovenduski, N. S., Matear, R. J., McNeil, B. I., Metzl, N., Mikaloff Fletcher, S. E., Monteiro, P. M. S., Rödenbeck, C., Sweeney, C., and Takahashi, T.: Sea-air CO_2 fluxes in the Southern Ocean for the period 1990–2009, Biogeosciences, 10, 4037–4054, doi:10.5194/bg-10-4037-2013, 2013.

Le Quéré, C. L., Harrison, S. P., Colin Prentice, I. C., Buitenhuis, E. T., Aumont, O., Bopp, L., Claustre, H., Cotrim Da Cunha, L., Geider, R., Giraud, X., Klaas, C., Kohfeld, K. E., Legendre, L., Manizza, M., Platt, T., Rivkin, R. B., Sathyendranath, S., Uitz, J., Watson, A. J., and Wolf-Gladrow, D.: Ecosystem dynamics based on plankton functional types for global ocean biogeochemistry models, Glob. Change Biol., 11, 2016–2040, doi:10.1111/j.1365-2486.2005.1004.x, 2005.

Le Quéré, C, Rödenbeck, C., Buitenhuis, E. T., Conway, T. J., Langenfelds, R., Gomez, A., Labuschagne, C., Ramonet, M., Nakazawa, T., Metzl, N., and Gillett, N.: Saturation of the Southern ocean CO_2 sink due to recent climate change, Science, 316, 1735–1738, doi:10.1126/science.1136188, 2007.

Le Quéré C, Raupach M. R., Canadell, J. G., Marland, G., Bopp, L., Ciais, P., Conway, T. J., Doney, S. C., Feely, R. A., Foster P, Friedlingstein, P., Gurney, K., Houghton, R. A., House, J. I., Huntingford, C., Levy, P. E., Lomas, M. R., Majkut, J., Metzl, N., Ometto, J. P., Peters, G. P., Prentice, I. C., Randerson, J. T., Running, S. W., Sarmiento, J. L., Schuster, U., Sitch, S., Takahashi, T., Viovy, N., van der Werf , G. R., and Woodward, F. I.: Trends in the sources and sinks of carbon dioxide, Nat. Geosci., 2, 831–836, 2009.

Le Quéré, C., Takahashi, T., Buitenhuis, C. E., Rödenbeck, C., and Sutherland, S.: Impact of climate change and variability on the global oceanic sink of CO_2, Global Biogeochem. Cy., 24, GB4007, doi:10.1029/2009GB003599, 2010.

Le Quéré, C., Andres, R. J., Boden, T., Conway, T., Houghton, R. A., House, J. I., Marland, G., Peters, G. P., van der Werf, G. R., Ahlström, A., Andrew, R. M., Bopp, L., Canadell, J. G., Ciais, P., Doney, S. C., Enright, C., Friedlingstein, P., Huntingford, C., Jain, A. K., Jourdain, C., Kato, E., Keeling, R. F., Klein Goldewijk, K., Levis, S., Levy, P., Lomas, M., Poulter, B., Raupach, M. R., Schwinger, J., Sitch, S., Stocker, B. D., Viovy, N., Zaehle, S., and Zeng, N.: The global carbon budget 1959–2011, Earth Syst. Sci. Data, 5, 165–185, doi:10.5194/essd-5-165-2013, 2013.

Levy, P. E., Cannell, M. G. R., and Friend, A. D.: Modelling the impact of future changes in climate, CO_2 concentration and land use on natural ecosystems and the terrestrial carbon sink, Global Environ. Chang., 14, 21–30, 2004.

Lewis, S. L., Lopez-Gonzalez, G., Sonké, B., Affum-Baffoe, K., Baker, T. R., Ojo, L. O., Phillips, O. L., Reitsma, J. M., White, L., James, A., Comiskey, K., Djuikouo, M.-N., Ewango, C. E. N., Feldpausch, T. R., Hamilton, A. C., Gloor, M., Hart, T., Hladik, A., Lloyd, J., Lovett, J. C., Makana, J.-R., Malhi, Y., Mbago, F. M., Ndangalasi, H. J., Peacock, J., Peh, K. S.-H., Sheil, D., Sunderland, T., Swaine, M. D., Taplin, J., Taylor, D., Thomas, S. C., and Wöll, R. V. H.: Increasing carbon storage in intact African tropical forests, Nature, 457, 1003–1006, 2009a.

Lewis, S. L., Lloyd, J., Sitch, S., Mitchard, E. T. A., and Laurance, W. F.: Changing Ecology of Tropical Forests: Evidence and Drivers, Annu. Rev. Ecol. Evol. Syst., 40, 529–49, 2009b.

Lovenduski, N. S., Gruber, N., Doney, S. C., and Lima, I. D.: Enhanced CO_2 outgassing in the Southern Ocean from a positive phase of the Southern Annular Mode, Global Biogeochem. Cy. 21, GB2026, doi:10.1029/2006GB002900, 2007.

Lovenduski, N. S., Gruber, N., and Doney, S. C.: Toward a mechanistic understanding of the decadal trends in the Southern Ocean carbon sink, Global Biogeochem. Cy., 22, 1–9, 2008.

Lucht, W., Prentice, I. C., Myneni, R. B., Sitch, S., Friedlingstein, P., Cramer, W., Bousquet, P., Buermann, W., and Smith, B.: Climate control of the high-latitude vegetation greening trend and Pinatubo effect, Science, 296, 1687–1689, 2002.

Luo, Y. Q., Randerson, J. T., Abramowitz, G., Bacour, C., Blyth, E., Carvalhais, N., Ciais, P., Dalmonech, D., Fisher, J. B., Fisher, R., Friedlingstein, P., Hibbard, K., Hoffman, F., Huntzinger, D., Jones, C. D., Koven, C., Lawrence, D., Li, D. J., Mahecha, M., Niu, S. L., Norby, R., Piao, S. L., Qi, X., Peylin, P., Prentice, I. C., Riley, W., Reichstein, M., Schwalm, C., Wang, Y. P., Xia, J. Y., Zaehle, S., and Zhou, X. H.: A framework for benchmarking land models, Biogeosciences, 9, 3857–3874, doi:10.5194/bg-9-3857-2012, 2012.

Luyssaert, S., Abril, G., Andres, R., Bastviken, D., Bellassen, V., Bergamaschi, P., Bousquet, P., Chevallier, F., Ciais, P., Corazza, M., Dechow, R., Erb, K.-H., Etiope, G., Fortems-Cheiney, A., Grassi, G., Hartmann, J., Jung, M., Lathière, J., Lohila, A., Mayorga, E., Moosdorf, N., Njakou, D. S., Otto, J., Papale, D., Peters, W., Peylin, P., Raymond, P., Rödenbeck, C., Saarnio, S., Schulze, E.-D., Szopa, S., Thompson, R., Verkerk, P. J., Vuichard, N., Wang, R., Wattenbach, M., and Zaehle, S.: The European land and inland water CO_2, CO, CH_4 and N_2O balance between 2001 and 2005, Biogeosciences, 9, 3357–3380, doi:10.5194/bg-9-3357-2012, 2012.

Manning, A. C. and Keeling, R. F.: Global oceanic and land biota sinks from the Scripps atmospheric oxygen flask sampling network, Tellus B, 58B, 95–116, 2006.

Matear, R. J., Chamberlain, M. A., Sun, C. and Feng, M.: Climate change projection of the Tasman Sea from an Eddy-resolving Ocean Model, J. Geophys. Res.-Oceans, 118, 2961–2976, doi:10.1002/jgrc.20202, 2013.

Matsumoto, K. and Gruber, N.: How accurate is the estimation of anthropogenic carbon in the ocean?, An evaluation of the DC* method, Global Biogeochem. Cy., 19, GB3014, doi:10.1029/2004GB002397, 2005.

McDonald, K. C., Kimball, J. S., Njoke, E., Zimmermann, R., and Zhao, M.: Variability in springtime thaw in the terrestrial high latitudes: monitoring a major control on the biospheric assimilation of atmospheric CO_2 with spaceborne microwave remote sensing, Earth Interact., 8, 1–23, 2004.

McDowell, N. G, Pockman, W. T., Allen, C. D., Breshears, D. D., Cobb, N., Kolb, T., Plaut, J., Sperry, J., West, A., Williams, D. G., and Yepez, E. A.: Mechanisms of plant survival and mortality during drought: why do some plants survive while others succumb to drought?. New Phytol., 178, 719–739, 2008.

McGillicuddy Jr, D. J., Robinson, A. R., Siegel, D. A., Jannasch, H. W., Johnson, R., Dickey, T. D., McNeil, J., Michaels, A. F., and Knap, A. H.: Influence of mesoscale eddies on new production in the Sargasso Sea, Nature, 394, 263–266, 1998.

McGuire, A. D., Christensen, T. R., Hayes, D., Heroult, A., Euskirchen, E., Kimball, J. S., Koven, C., Lafleur, P., Miller, P. A., Oechel, W., Peylin, P., Williams, M., and Yi, Y.: An assessment of the carbon balance of Arctic tundra: comparisons among observations, process models, and atmospheric inversions, Biogeosciences, 9, 3185–3204, doi:10.5194/bg-9-3185-2012, 2012.

McKinley, G. A., Takahashi, T., Buitenhuis, E., Chai, F., Christian, J. R., Doney, S. C., Jiang, M.-S., Le Quéré, C., Lima, I., Linday, K., Moore, J. K., Murtugudde, R., Shi, L., and Wetzel, P.: North Pacific carbon cycle response to climate variability on seasonal to decadal timescales. J. Geophys. Res., 111, C07S06, doi:10.1029/2005JC003173, 2006.

McKinley, G. A., Fay, A. R., Takahashi, T., and Metzl, N: Convergence of atmospheric and North Atlantic carbon dioxide trends on multidecadal timescales, Nat. Geosci, 4, 606–610, doi:10.1038/NGEO1193, 2011.

McNeil, B. I. and Matear, R. J.: The non-steady state oceanic CO_2 signal: its importance, magnitude and a novel way to detect it, Biogeosciences, 10, 2219–2228, doi:10.5194/bg-10-2219-2013, 2013.

Mercado, L. M., Bellouin, N., Sitch, S., Boucher, O., Huntingford, C., and Cox, P. M.: Impact of Changes in Diffuse Radiation on the Global Land Carbon Sink, Nature, 458, 1014–1017, 2009.

Mikaloff Fletcher, S. E., Gruber, N., Jacobson, A. R., Doney, S. C., Dutkiewicz, S., Gerber, M., Follows, M., Joos, F., Lindsay, K., Menemenlis, D., Mouchet, A., Müller, S. A., and Sarmiento, J. L.: Inverse estimates of anthropogenic CO_2 uptake, transport, and storage by the oceans, Global Biogeochem. Cy., 20, GB2002, doi:10.1029/2005GB002530, 2006.

Morales, P., Sykes, M. T., Prentice, I. C., Smith, P., Smith, B., Bugmann, H., Zierl, B., Friedlingstein, P., Viovy, N., Sabaté, S., Sánchez, A., Pla, E., Gracia, C. A., Sitch, S., Arneth, A., and Ogee, J.: Comparing and evaluating process-based ecosystem model predictions of carbon and water fluxes in major European forest biomes, Glob. Change Biol., 11, 2211–2233, doi:10.1111/j.1365-2486.2005.01036.x, 2005.

Myneni, R. B., Keeling, C. D., Tucker, C. J., Asrar, G., and Nemani, R. R.: Increased plant growth in the northern high latitudes from 1981 to 1991, Nature, 386, 698–702, doi:10.1038/386698a0, 1997.

Murray-Tortarolo, G., Anav, A., Friedlingstein, P., Sitch, S., Piao, S., Zhu, Z., Poulter, B., Zaehle, S., Ahlstrom, A., Lomas, M., Levis, S., Viovy, N., and Zeng, N.: Evaluation of DGVMs in reproducing satellite derived LAI over the Northern Hemisphere, Part I: Uncoupled DGVMs, Remote Sens., 5, 4819–4838; doi:10.3390/rs5104819, 2013.

Müller, S. A., Joos, F., Plattner, G.-K., Edwards, N. R., and Stocker, T. F.: Modeled natural and excess radiocarbon: Sensitivities to the gas exchange formulation and ocean transport strength, Global Biogeochem. Cy., 22, doi:10.1029/2007GB003065, 2008.

Nemani, R. R., Keeling, C. D., Hashimoto, H., Jolly, W. M., Piper, S. C., Tucker, C. J., Myneni, R. B., and Running, S. W.: Climate-driven increases in global terrestrial net primary production from 1982 to 1999, Science, 300, 1560–1563, 2003.

New, M. G., Hulme, M., and Jones, P. D.: Representing twentieth-century space-climate variability, Part II, Development of 1901–1996 monthly grids of terrestrial surface climate, J. Climate, 13, 2217–2238, 2000.

Norby, R. J., DeLucia, E. H., Gielen, B., Calfapietra, C., Giardina, C. P., King, J. S., Ledford, J., McCarthy, H. R., Moore, D. J. P., Ceulemans, R., De Angelis, P., Finzi, A. C., Karnosky, D. F., Kubiske, M. E., Lukac, M., Pregitzer, K. S., Scarascia-Mugnozza, G. E., Schlesinger, W. H., and Oren, R.: Forest response to elevated CO2 is conserved across a broad range of productivity, P. Natl. Acad. Sci. USA, 102, 18052–18056, 2005.

Norby, R. J., Warren, J. M., Iversen, C. M., Medlyn, B. E., and McMurtrie, R. E.: CO_2 enhancement of forest productivity constrained by limited nitrogen availability, P. Natl. Acad. Sci., 107, 19368–19373, 2010.

Oleson, K. W., Lawrence, D. M., Bonan, G. B., Flanner, M. G., Kluzek, E., Lawrence, P. J., Levis, S., Swenson, S. C., and Thornton, P. E.: Technical description of version 4.0 of the Community Land Model (CLM), NCAR Tech. Note NCAR/TN-478+STR, 257 pp., 2010.

Pacala, S. W., Hurtt, G. C., Baker, D., Peylin, P., Houghton, R. A., Birdsey, R. A., Heath, L., Sundquist, E. T., Stallard, R. F., Ciais, P., Moorcroft, P., Caspersen, J. P., Shevliakova, E., Moore B., Kohlmaier, G., Holland, E., Gloor, M., Harmon, M. E., Fan, S.-M., Sarmiento, J. L., Goodale, C. L., Schimel, D., and Field, C. B.: Consistent land- and atmosphere-based US carbon sink estimates, Science, 292, 2316–2320, 2001.

Pan, Y., Birdsey, R.A., Fang, J., Houghton, R., Kauppi, P. E., Kurz, W. A., Phillips, O. L., Shvidenko, A., Lewis, S. L., Canadell, J. G., Ciais, P., Jackson, R. B., Pacala, S., McGuire, A. D., Piao, S., Rautiainen, A., Sitch, S., Hayes, D., and Watson, C.: A large and persistent carbon sink in the World's forests, 1990–2007, Science, doi:10.1126/science.1201609, 2011.

Park, G.-H., Wanninkhof, R., Doney, S. C., Takahashi, T., Lee, K., Feely, R. A., Sabine, C. L., Triñanes, J., and Lima, I. D.: Variability of global net sea-air CO_2 fluxes over the last three decades using empirical relationships, Tellus B, 62, 352–368, 2010.

Patra, P. K., Canadell, J. G., Houghton, R. A., Piao, S. L., Oh, N.-H., Ciais, P., Manjunath, K. R., Chhabra, A., Wang, T., Bhattacharya, T., Bousquet, P., Hartman, J., Ito, A., Mayorga, E., Niwa, Y., Raymond, P. A., Sarma, V. V. S. S., and Lasco, R.: The carbon budget of South Asia, Biogeosciences, 10, 513–527, doi:10.5194/bg-10-513-2013, 2013.

Peacock, S.: Debate over the ocean bomb radiocarbon sink: Closing the gap, Global Biogeochem. Cy., 18, GB2022, doi:10.1029/2003GB002211, 2004.

Peng S., Ciais, P., Chevallier, F., Peylin, P., Cadule, P., Sitch, S., Piao, S., Ahlström, A., Levy, P., Lomas, M., Poulter, B., Viovy, N., Wang, T., Zaehle, S., Zeng, N., Zhao, F. and Zhao, H.: Benchmarking the seasonal cycle of CO_2 fluxes simulated by terrestrial ecosystem models, Global Biogeochem. Cy., 29, doi:10.1002/2014GB004931, 2014.

Peylin, P., Bousquet, P., Le Quéré, C., Sitch, S., Friedlingstein, P., McKinley, G., Gruber, N., Rayner, P., and Ciais, P.: Multiple constraints on regional CO_2 flux variations over land and oceans, Global Biogeochem. Cy., 19, GB1011, doi:10.1029/2003GB002214, 2005.

Peylin, P., Law, R. M., Gurney, K. R., Chevallier, F., Jacobson, A. R., Maki, T., Niwa, Y., Patra, P. K., Peters, W., Rayner, P. J., Rödenbeck, C., van der Laan-Luijkx, I. T., and Zhang, X.: Global atmospheric carbon budget: results from an ensemble of atmospheric CO_2 inversions, Biogeosciences, 10, 6699–6720, doi:10.5194/bg-10-6699-2013, 2013.

Pfeil, B., Olsen, A., Bakker, D. C. E., Hankin, S., Koyuk, H., Kozyr, A., Malczyk, J., Manke, A., Metzl, N., Sabine, C. L., Akl, J., Alin, S. R., Bates, N., Bellerby, R. G. J., Borges, A., Boutin, J., Brown, P. J., Cai, W.-J., Chavez, F. P., Chen, A., Cosca, C., Fassbender, A. J., Feely, R. A., González-Dávila, M., Goyet, C., Hales, B., Hardman-Mountford, N., Heinze, C., Hood, M., Hoppema, M., Hunt, C. W., Hydes, D., Ishii, M., Johannessen, T., Jones, S. D., Key, R. M., Körtzinger, A., Landschützer, P., Lauvset, S. K., Lefèvre, N., Lenton, A., Lourantou, A., Merlivat, L., Midorikawa, T., Mintrop, L., Miyazaki, C., Murata, A., Nakadate, A., Nakano, Y., Nakaoka, S., Nojiri, Y., Omar, A. M., Padin, X. A., Park, G.-H., Paterson, K., Perez, F. F., Pierrot, D., Poisson, A., Ríos, A. F., Santana-Casiano, J. M., Salisbury, J., Sarma, V. V. S. S., Schlitzer, R., Schneider, B., Schuster, U., Sieger, R., Skjelvan, I., Steinhoff, T., Suzuki, T., Takahashi, T., Tedesco, K., Telszewski, M., Thomas, H., Tilbrook, B., Tjiputra, J., Vandemark, D., Veness, T., Wanninkhof, R., Watson, A. J., Weiss, R., Wong, C. S., and Yoshikawa-Inoue, H.: A uniform, quality controlled Surface Ocean CO_2 Atlas (SOCAT), Earth Syst. Sci. Data, 5, 125–143, doi:10.5194/essd-5-125-2013, 2013.

Phillips, O. L., Malhi, Y., Higuchi, N., Laurance, W. F., Nunez, P. V., Vasquez, R. M., Laurance, S. G., Ferreira, L. V., Stern, M., Brown, S., and Grace, J.: Changes in the carbon balance of tropical forests: Evidence from long-term plots, Science, 282, 439–442, 1998.

Piao, S. L., Fang, J. Y., Ciais, P., Peylin, P., Huang, Y., Sitch, S., and Wang, T: The carbon balance of terrestrial ecosystems in China, Nature, 458, 1009–1014, 2009.

Piao, S. L., Sitch, S., Ciais, P., Friedlingstein, P., Peylin, P., Wang, X. H., Ahlström, A., Anav, A., Canadell, J. G., Huntingford, C., Jung, M., Levis, S., Levy, P. E., Li, J. S., Lin, X., Lomas, M. R., Lu, M ., Luo, Y. Q., Ma, Y. C., Myneni, R. B., Poulter, B., Sun, Z. Z., Wang, T., Viovy, N., Zaehle, S., and Zeng, N.: Evaluation of terrestrial carbon cycle models for their response to climate variability and to CO_2 trends, Glob. Change Biol., 19, 2117–2132, 2013.

Pongratz, J., Reick, C. H., Houghton, R. A., and House, J. I.: Terminology as a key uncertainty in net land use and land cover change carbon flux estimates, Earth Syst. Dynam., 5, 177–195, doi:10.5194/esd-5-177-2014, 2014.

Poulter, B., Frank, D. C., Hodson, E. L., and Zimmermann, N. E.: Impacts of land cover and climate data selection on understanding terrestrial carbon dynamics and the CO_2 airborne fraction, Biogeosciences, 8, 2027–2036, doi:10.5194/bg-8-2027-2011, 2011.

Poulter, B., Pederson, N., Liu, H., Zhu, Z., D'Arrigo, R., Ciais, P., Davi, N., Frank, D., Leland, C., Myneni, R., Piao, S., and Wang, T.: Recent trends in Inner Asian forest dynamics to temperature and precipitation indicate high sensitivity to climate change, Agr. Forest Meteorol., 178–179, 31–45, 2013.

Poulter, B., Frank, D., Ciais, P., Myneni, R., Andela, N., Bi, J., Broquet, G., Canadell, J. G., Chevallier, F., Liu, Y. Y., Running, S. W., Sitch, S. and van der Werf, G. R.: Contribution of semi-arid ecosystems to interannual variability of the global carbon cycle, Nature, doi:10.1038/nature13376, 2014.

Prentice I. C., Farquhar, G. D., Fasham, M. J. R., Goulden, M. L., Heimann, M., Jaramillo, V.J., Kheshgi, H. S., Le Quere, C., Scholes, R. J., Wallace, D. W. R., Contributing Authors Archer, D., Ashmore, M. R., Aumont, O., Baker, D., Battle, M., Bender,

M., Bopp, L. P., Bousquet, P., Caldeira, K., Ciais, P., Cox, P. M., Cramer, W., Dentener, F., Enting, I. G., Field, C. B., Friedlingstein, P., Holland, E. A., Houghton, R. A., House, J. I., Ishida, A., Jain, A. K., Janssens, I. A., Joos, F., Kaminski, T., Keeling, C. D., Keeling, R. F., Kicklighter, D. W., Kohfeld, K. E., Knorr, W., Law, R., Lenton, T. , Lindsay, K., Maier-Reimer, E., Manning, A. C., Matear, R. J., McGuire, A. D., Melillo, J. M., Meyer, R., Mund, M., Orr, J. C., Piper, S., Plattner, K., Rayner, P. J., Sitch, S., Slater, R., Taguchi, S. , Tans, P. P., Tian, H. Q., Weirig, M. F., Whorf, T., and Yool, A.: The carbon cycle and atmospheric carbon dioxide, in. Climate Change 2001: The Scientific Basis edited by: Houghton, J. T., Cambridge Univ. Press, New York, 183–237, 2001.

Prentice, I. C., Bondeau, A., Cramer, W., Harrison, S. P., Hickler, T., Lucht, W., Sitch, S., Smith, B., and Sykes, M. T.: Terrestrial Ecosystems in a changing world, IGBP Book Series, edited by: Canadell. J., Pitelka, L., and Pataki, D., Springer, Heidelberg, Germany, pp. 336, 2007.

Quaife, T., Quegan, S., Disney, M., Lewis, P., Lomas, M. R., and Woodward, F. I.: Impact of land cover uncertainties on estimates of biospheric carbon fluxes, Global Biogeochem. Cy., 22, GB4016, doi:10.1029/2007GB003097, 2008.

Randerson, J. T., Hoffman, F. M., Thornton, P. E., Mahowald, N. M., Lindsay, K., Lee, Y.-H., Nevison, C. D., Doney, S. C., Bonan, G., Stoeckli, R., Covey, C., Running, S. W., and Fung, I. Y.: Systematic assessment of terrestrial biogeochemistry in coupled climate-carbon models, Glob. Change Biol., 15, 2462–2484, 2009.

Raupach, M. R., Canadell, J. G., and Le Quéré, C.: Anthropogenic and biophysical contributions to increasing atmospheric CO2 growth rate and airborne fraction, Biogeosciences, 5, 1601–1613, doi:10.5194/bg-5-1601-2008, 2008.

Raupach, M. R., Gloor, M., Sarmiento, J. L., Canadell, J. G., Frölicher, T. L., Gasser, T., Houghton, R. A., Le Quéré, C., and Trudinger, C. M.: The declining uptake rate of atmospheric CO_2 by land and ocean sinks, Biogeosciences, 11, 3453–3475, doi:10.5194/bg-11-3453-2014, 2014.

Reichstein, M., Bahn, M., Ciais, P., Frank, D., Mahecha, M. D., Seneviratne, S. I., Zscheischler, J., Beer, C., Buchmann, N., Frank, D. C., Papale, D., Rammig, A., Smith, P., Thonicke, K., van der Velde, M., Vicca, S., Walz, A., and Wattenbach, M.: Climate extremes and the carbon cycle, Nature, 500, 287–295, 2013.

Roy, T., Bopp, L., Gehlen, M., Schneider, B., Cadule, P., Frölicher, T. L., Segschneider, J., Tjiputra, J., Heinze, C., and Joos, F.: Regional Impacts of Climate Change and Atmospheric CO_2 on Future Ocean Carbon Uptake: A Multimodel Linear Feedback Analysis, J. Climate, 24, 2300–2318, 2011.

Sabine, C. S., Feely, R. A.., Gruber, N. , Key, R. M., Lee, K., Bullister, J. L., Wanninkhof, R., Wong, C. S., Wallace, D. W. R., Tillbrook, B., Millero, F. J., Peng, T.-H., Kozyr, A., Ono, T., and Rios, A. F.: The oceanic sink for anthropogenic CO_2, Science, 305, 367–371, 2004.

Sarmiento, J. L. and Gruber, N.: Sinks for anthropogenic carbon, Phys. Today, 55, 30–36, 2002.

Sarmiento, J. L. and Gruber, N.: Ocean Biogeochemical Dynamics, Princeton Univ. Press, 526 pp., 2006.

Sarmiento, J. L., Orr, J. C., and Siegenthaler, U.: A perturbation simulation of CO_2 uptake in an ocean general circulation model, J. Geophys. Res., 97, 3621–3645, 1992.

Sarmiento, J. L., Gloor, M., Gruber, N., Beaulieu, C., Jacobson, A. R., Mikaloff Fletcher, S. E., Pacala, S., and Rodgers, K.: Trends and regional distributions of land and ocean carbon sinks, Biogeosciences, 7, 2351–2367, doi:10.5194/bg-7-2351-2010, 2010.

Schuster, U., McKinley, G. A., Bates, N., Chevallier, F., Doney, S. C., Fay, A. R., González-Dávila, M., Gruber, N., Jones, S., Krijnen, J., Landschützer, P., Lefèvre, N., Manizza, M., Mathis, J., Metzl, N., Olsen, A., Rios, A. F., Rödenbeck, C., Santana-Casiano, J. M., Takahashi, T., Wanninkhof, R., and Watson, A. J.: An assessment of the Atlantic and Arctic sea-air CO_2 fluxes, 1990–2009, Biogeosciences, 10, 607–627, doi:10.5194/bg-10-607-2013, 2013.

Sitch, S., Smith, B., Prentice, I. C., Arneth, A., Bondeau, A., Cramer, W., Kaplan, J. O., Levis, S., Lucht, W., Sykes, M. T., Thonicke, K., and Venevsky, S.: Evaluation of ecosystem dynamics, plant geography and terrestrial carbon cycling in the LPJ dynamic global vegetation model, Glob. Change Biol., 9, 161–185, doi:10.1046/j.1365-2486.2003.00569.x, 2003.

Sitch, S., McGuire, A. D., Kimball, J., Gedney, N., Gamon, J., Engstrom, R., Wolf, A., Zhuang, Q., Clein, J. S., and McDonald, K. C.: Assessing the carbon balance of circumpolar arctic tundra using remote sensing and process modeling, Ecol. Appl., 17, 213–234, 2007a.

Sitch, S., Cox, P. M., Collins, W. J., and Huntingford., C.: Indirect radiative forcing of climate change through ozone effects on the land-carbon sink, Nature, 448, 791–794, doi:10.1038/nature06059, 2007b.

Sitch, S., Huntingford, C., Gedney, N., Levy, P. E., Lomas, M., Piao, S., Betts, R., Ciais, P., Cox, P., Friedlingstein, P., Jones, C. D., Prentice, I. C., and Woodward, F. I.: Evaluation of the terrestrial carbon cycle, future plant geography, and climate-carbon cycle feedbacks using 5 Dynamic Global Vegetation Models (DGVMs), Glob. Change Biol., 14 1–25, doi:10.1111/j.1365-2486.2008.01626.x, 2008.

Smith, B., Prentice, I. C., and Sykes, M. T.: Representation of vegetation dynamics in the modelling of terrestrial ecosystems: comparing two contrasting approaches within European climate space, Global Ecol. Biogeogr., 10, 621–637, 2001.

Sokolov, A. P., Kicklighter, D. W., Melillo, J. M., Felzer, B. S., Schlosser, C. A., and Cronin, T. W.: Consequences of considering carbon-nitrogen interactions on the feedbacks between climate and the terrestrial carbon cycle, J. Climate, 21, 3776–3796, doi:10.1175/2008JCLI2038.1, 2008.

Stocker, B. D., Feissli, F., Strassmann, K. M., Spahni, R., and Joos, F.: Past and future carbon fluxes from land use change, shifting cultivation and wood harvest, Tellus B, 66, 23188, doi:10.3402/tellusb.v66.23188, 2014.

Stephens, B. B., Gurney, K. R., Tans, P. P., Sweeney, C., Peters, W., Bruhwiler, L., Ciais, P., Ramonet, M., Bousquet, P., Nakazawa, T., Aoki, S., Machida, T., Inoue, G., Vinnichenko, N., Lloyd, J., Jordan, A., Heimann, M., Shibistova, O., Langenfelds, R. L., Steele, L. P., Francey, R. J., and Denning, A. S.: Weak northern and strong tropical land carbon uptake from vertical profiles of atmospheric CO_2, Science, 316, 1732–1735, 2007.

Sweeney, C., Gloor, E., Jacobson, A. R., Key, R. M., McKinley, G., Sarmiento, J. L., and Wanninkhof, R.: Constraining global air-sea gas exchange for CO_2 with recent bomb ^{14}C measurements, Global Biogeochem. Cycl., 21, GB2015, doi:10.1029/2006GB002784, 2007.

Takahashi, T., Sutherland, S. C., Wanninkhof, R., Sweeney, C., Feely, R. A., Chipman, D., Hales, B., Friederich, G., Chavez, F., Sabine, C., Watson, A., Bakker, D. C. E., Schuster, U., Metzl, N., Yoshikawa-Inoue, H., Ishii, M., Midorikawa, T., Nojiri, Y., Kortzinger, A., Steinhoff, T., Hoppema, M., Olafsson, J., Arnarson, T. S., Tilbrook, B., Johannessen, T., Olsen, A., Bellerby, R., Wong, C. S., Delille, B., Bates, N. R., and de Baar, H. J. W.: Climatological mean and decadal changes in surface ocean pCO_2, and net sea-air CO2 flux over the global oceans, SOCOVV Symposium Volume, Deep Sea Res.-Pt. II, 810, 554–577, 2009.

Tans, P. P., Fung, I. Y., and Takahashi, T.: Observational constraints on the global atmospheric CO_2 budget, Science, 247, 1431–1439, doi:10.1126/science.247.4949.1431, 1990.

Thornton, P. E., Lamarque, J. F., Rosenbloom, N. A., and Mahowald, N. M.: Influence of carbon-nitrogen cycle coupling on land model response to CO_2 fertilization and climate variability, Global Biogeochem. Cy., 21, GB4018, doi:10.1029/2006GB002868, 2007.

Thornton, P. E., Doney, S. C., Lindsay, K., Moore, J. K., Mahowald, N., Randerson, J. T., Fung, I., Lamarque, J.-F., Feddema, J. J., and Lee, Y.-H.: Carbon-nitrogen interactions regulate climate-carbon cycle feedbacks: results from an atmosphere-ocean general circulation model, Biogeosciences, 6, 2099–2120, doi:10.5194/bg-6-2099-2009, 2009.

Tjiputra, J. F., Assmann, K., and Heinze, C.: Anthropogenic carbon dynamics in the changing ocean, Ocean Science, 6, 605–614, 2010.

Tucker, C. J., Slayback, D. A., Pinzon, J. E., Los, S. O., Myneni, R. B., and Taylor, M. G.: Higher northern latitude normalized difference vegetation index and growing season trends from 1982 to 1999, Int. J. Biometeorol., 45, 184–190, 2001.

Valentini, R., Arneth, A., Bombelli, A., Castaldi, S., Cazzolla Gatti, R., Chevallier, F., Ciais, P., Grieco, E., Hartmann, J., Henry, M., Houghton, R. A., Jung, M., Kutsch, W. L., Malhi, Y., Mayorga, E., Merbold, L., Murray-Tortarolo, G., Papale, D., Peylin, P., Poulter, B., Raymond, P. A., Santini, M., Sitch, S., Vaglio Laurin, G., van der Werf, G. R., Williams, C. A., and Scholes, R. J.: A full greenhouse gases budget of Africa: synthesis, uncertainties, and vulnerabilities, Biogeosciences, 11, 381–407, doi:10.5194/bg-11-381-2014, 2014.

Van der Molen, M. K., Dolman, A. J., Ciais, P., Eglin, T., Gobron, N., Law, B. E., Meir, P., Peters, W., Phillips, O. L., Reichstein, M., Chen, T., Dekker, S. C., Doubková, M., Friedl, M. A., Jung, M., van den Hurk, B. J. J. M., de Jeu, R. A. M., Kruijt, B., Ohta, T., Rebel, K. T., Plummer, S., Seneviratne, S. I., Sitch, S., Teuling, A. J., van der Werf, G. R., and Wang, G.: Drought and ecosystem carbon cycling, Agr. Forest Meteorol., 151, 765–773, doi:10.1016/j.agrformet.2011.01.018, 2011.

Vitousek, P. M. and Howarth, R. W.: Nitrogen limitation on land and in the sea: How can it occur?, Biogeochemistry, 13, 87–115, 1991.

Wang, Y. P., Law, R. M., and Pak, B.: A global model of carbon, nitrogen and phosphorus cycles for the terrestrial biosphere, Biogeosciences, 7, 2261–2282, doi:10.5194/bg-7-2261-2010, 2010.

Wanninkhof, R.: Relationship between wind speed and gas exchange over the ocean, J. Geophys. Res., 97, 7373–7382, 1992.

Wanninkhof, R., Park, G.-H., Takahashi, T., Sweeney, C., Feely, R., Nojiri, Y., Gruber, N., Doney, S. C., McKinley, G. A., Lenton, A., Le Quéré, C., Heinze, C., Schwinger, J., Graven, H., and Khatiwala, S.: Global ocean carbon uptake: magnitude, variability and trends, Biogeosciences, 10, 1983–2000, doi:10.5194/bg-10-1983-2013, 2013.

Woodward, F. I. and Lomas, M. R.: Vegetation-dynamics-simulating responses to climate change, Biological Reviews, 79, 643–670, 2004.

Woodward, F. I., Smith, T. M., Emanuel, W. R.: A global land primary productivity and phytogeography model. Global Biogeochem. Cy., 9, 471–490, 1995.

Zaehle, S. and Friend, A. D.: Carbon and nitrogen cycle dynamics in the O-CN land surface model, I: Model description, site- scale evaluation and sensitivity to parameter estimates, Global Biogeochem. Cy., 24, GB1005, doi:10.1029/2009GB003521, 2010.

Zaehle, S. and Dalmonech, D.: Carbon – nitrogen interactions on land at global scales: current understanding in modeling climate biosphere feedbacks, Current Opinion in Environmental Sustainability, 3, 311–320, 2011.

Zaehle, S., Friedlingstein, P., and Friend, A. D.: Terrestrial nitrogen feedbacks may acccelerate future climate change, Geophys. Res. Lett., 37, L01401, doi:10.1029/2009GL041345, 2010.

Zaehle, S., Ciais, P., Friend, A. D., and Prieur, V.: Carbon benefits of anthropogenic reactive nitrogen offset by nitrous oxide emissions, Nat. Geosci., 4, 601–605, doi:10.1038/ngeo1207, 2011.

Zaehle, S., Medlyn, B. E., De Kauwe, M. G., Walker, A. P., Dietze, M. C., Hickler, T., Luo, Y., Wany, Y.-P., El-Masri, B., Thornton, P., Jain, A., Wang, S., Warlind, D., Weng, E., Parton, W., Iversen, C. M., Gallet-Budynek, A., McCarthy, H., Finzi, A., Hanson, P. J., Prentice, I. C., Oren, R., and Norby, R. J.: Evaluation of eleven terrestrial carbon-nitrogen cycle models against observations from two temperate Free-Air CO_2 Enrichment Studies, New Phytologist, 202, 803–822, doi:10.1111/nph.12697, 2014.

Zeng, N.: Glacial-interglacial atmospheric CO_2 change – The glacial burial hypothesis, Adv. Atmos. Sci., 20, 677–673, 2003.

Zeng, N., Mariotti, A., and Wetzel, P.: Terrestrial mechanisms of interannual CO_2 variability, Global Biogeochem. Cy., 19, GB1016, doi:10.1029/2004gb002273, 2005.

Zhang, Q., Wang, Y. P., Matear, R. J., Pitman, A. J., and Dai, Y. J.: Nitrogen and phosphorous limitations significantly reduce future allowable CO_2 emissions, Geophys. Res. Lett, 41, 632–637, doi:10.1002/2013GL058352, 2014.

Zhu, Z., Bi, J., Pan, Y., Ganguly, S., Anav, A., Xu, L., Samanta, A., Piao, S., Nemani, R. R., and Myneni, R. B.: Global Data Sets of Vegetation Leaf Area Index (LAI)3g and Fraction of Photosynthetically Active Radiation (FPAR)3g Derived from Global Inventory Modeling and Mapping Studies (GIMMS) Normalized Difference Vegetation Index (NDVI3g) for the Period 1981 to 2011, Remote Sens., 5, 927–948, 2013.

Zscheischler, J., Mahecha, M. D., von Buttlar, J., Harmeling, S., Jung, M., Rammig, A., Randerson, J. T., Schoelkopf, B., Seneviratne, S. I., Tomelleri, E., Zaehle, S., and Reichstein, M.: A few extreme events dominate global interannual variability in gross primary production, Environ. Res. Lett., 9, 035001, doi:10.1088/1748-9326/9/3/035001, 2014.

10

Impacts of simulated herbivory on volatile organic compound emission profiles from coniferous plants

C. L. Faiola[1,*], B. T. Jobson[1], and T. M. VanReken[1]

[1]Laboratory for Atmospheric Research, Department of Civil and Environmental Engineering, Washington State University, Pullman, Washington, USA
[*]now at: Department of Applied Physics, University of Eastern Finland, Kuopio, Finland

Correspondence to: T. M. VanReken (vanreken@wsu.edu)

Abstract. The largest global source of volatile organic compounds (VOCs) in the atmosphere is from biogenic emissions. Plant stressors associated with a changing environment can alter both the quantity and composition of the compounds that are emitted. This study investigated the effects of one global change stressor, increased herbivory, on plant emissions from five different coniferous species: bristlecone pine (*Pinus aristata*), blue spruce (*Picea pungens*), western redcedar (*Thuja plicata*), grand fir (*Abies grandis*), and Douglas-fir (*Pseudotsuga menziesii*). Herbivory was simulated in the laboratory via exogenous application of methyl jasmonate (MeJA), a herbivory proxy. Gas-phase species were measured continuously with a gas chromatograph coupled to a mass spectrometer and flame ionization detector (GC–MS–FID). Stress responses varied between the different plant types and even between experiments using the same set of saplings. The compounds most frequently impacted by the stress treatment were alpha-pinene, beta-pinene, 1,8-cineol, beta-myrcene, terpinolene, limonene, and the cymene isomers. Individual compounds within a single experiment often exhibited a different response to the treatment from one another.

1 Introduction

The largest global source of volatile organic compounds (VOCs) in the atmosphere is emissions from vegetation (Guenther et al., 2000, 2012). These biogenic VOCs (BVOCs) oxidize in the atmosphere and can contribute significantly to the formation of secondary pollutants such as ozone and secondary organic aerosol (SOA) (Atkinson, 2000; Ehn et al., 2014; Hamilton et al., 2009; Kroll and Seinfeld, 2008), and thus play a key role in Earth's climate (Carslaw et al., 2010). Plants emit a wide range of organic compounds that will be classified here structurally into three categories: small oxygenated VOCs (OVOCs), terpenoids (isoprene, monoterpenes, sesquiterpenes, and their oxygenated derivatives), and aromatics (Herrmann and Weaver, 1999; Kesselmeier and Staudt, 1999). The regulation of BVOC emissions depends on both physiological and physicochemical controls that vary both between plant species and between different compounds produced within a single tree (Niinemets et al., 2004).

Some BVOCs are constitutive, meaning they are continuously synthesized and emitted by the plant while being regulated by the physiological and physicochemical mechanisms described above. Constitutive emissions can be either de novo or pooled depending on the absence or presence of storage structures. A single plant can emit both de novo and pooled emissions simultaneously (Loreto et al., 2000). In contrast to constitutive emissions, some BVOC emissions are inducible, meaning they are only synthesized and emitted when the plant is exposed to an abiotic or biotic stress that initiates their production. These stress-induced emission rates can make up a significant amount of total plant BVOC emissions (Blande et al., 2007; Brilli et al., 2009; Staudt and Lhoutellier, 2007). They also can increase or decrease the secondary organic aerosol formation potential of the BVOC emissions depending on the types of VOCs that are induced (Mentel et al., 2013).

Plant stress can significantly alter the BVOC emission profile both by inducing emissions of additional compounds and by changing the emissions of constitutive compounds (Arneth and Niinemets, 2010). This is an important consideration because different VOCs, even within the same class of compounds, can vary by orders of magnitude in their chemical reactivity (Atkinson and Arey, 1998). A variety of stress exposure studies have been performed investigating BVOC emission changes due to ozone exposure (Heiden et al., 1999; Vuorinen et al., 2004), salt stress (Loreto and Delfine, 2000; Teuber et al., 2008), increased CO_2 (Calfapietra et al., 2009; Constable et al., 1999), enhanced radiation (Harley et al., 1996), drought and/or high temperatures (Kleist et al., 2012; Niinemets, 2010; Niinemets et al., 2010), herbivory (Achotegui-Castells et al., 2013; Copolovici et al., 2011; Engelberth et al., 2004), and pathogen attack (Jansen et al., 2009a; Toome et al., 2010). A thorough review on this topic was presented by Peñuelas and Staudt (2010). Despite the numerous studies investigating this topic, most of these stress influences on BVOC emission rates are still not understood well enough to be included in the models used to develop emissions inventories (Guenther et al., 2012). This is in large part the result of two main factors: (1) the absence of enough quantitative experimental data to generate useful algorithms and (2) the large variability in stress response between trees and even between different compounds emitted by the same tree (Peñuelas and Staudt, 2010, and references therein).

Generally, a plant's response to stress depends on the longevity and severity of the stress exposure. Under mild to moderate abiotic stress, biochemical defense pathways are activated that induce and/or increase BVOC emissions, a response that protects the plant from both oxidative and thermal stress (Loreto and Schnitzler, 2010). However, the stress response changes for different types of compounds depending on the physicochemical properties of the compound. For example, emissions of small OVOCs (e.g., methanol, acetaldehyde, and acetone) are closely related to stomatal conductance whereas terpenes are not (Niinemets et al., 2004). Terpenes are hydrocarbons that can diffuse out of the plants into the atmosphere directly through the plant's membranes (Fall and Monson, 1992; Loreto et al., 1996). Consequently, stomatal conductance has no impact on the regulation of terpene emissions because of their chemical properties. In contrast, OVOCs cannot diffuse directly through plant membranes and easily dissolve in aqueous solutions, which further hinders volatilization. Thus, the effects of drought and/or heat stress impact OVOC emissions and terpene emissions differently because plants have evolved mechanisms to deal with these stressors by controlling their stomata. This stressor increases OVOC emissions in the short-term, but after prolonged exposure to the stressor, plants close their stomata to conserve water and a resulting drop in OVOC emissions occurs (Filella et al., 2007; Graus et al., 2013). This same threshold effect was not observed for terpene foliar concentrations and terpene emissions from Mediterranean tree species and C4 crops (Blanch et al., 2009; Graus et al., 2013). However, other studies have demonstrated that under severe enough drought stress, monoterpene emissions also begin to decrease (Ormeno et al., 2007; Simpraga et al., 2011). Presumably, at some extreme, the plant shuts down metabolic activity and terpene pools, if present, are depleted.

One important stressor in future climates will be the increased number of plant-eating pests, leading to increased herbivory (Bale et al., 2002). Plants have evolved to respond to herbivory stress by emitting BVOCs as a defense, using them for communication with other plants, and to signal natural predators of the herbivores (Engelberth et al., 2004). It is well established that herbivory can increase monoterpene, sesquiterpene, and small OVOC emission rates and substantially alter the BVOC profile (Achotegui-Castells et al., 2013; Hu et al., 2008; Laothawornkitkul et al., 2008; Semiz et al., 2012). The presence of herbivore infestation can increase BVOC emissions by 4- to 20-fold (Amin et al., 2012, 2013; Berg et al., 2013), and this response can last for several weeks (Priemé et al., 2000). These results suggest that herbivore stress could have a substantial impact on SOA formation in forest environments in the future. However, the number of plants studied using quantitative analytical techniques to measure compound-specific BVOC emission rates is not representative of all the major BVOC emitters in different environments. Furthermore, within the pool of plants that have been studied, large variation has been observed in responses. Emissions of different compounds from the same plant exhibit different temporal responses to herbivory stress (Copolovici et al., 2011). Additionally, the plant stress response varies depending on the type of biotic stress and/or the type of plant; other studies have shown increases in total terpene emission rates after herbivory exposure with no change in the VOC profile (Jansen et al., 2009b; Priemé et al., 2000) or different responses of the same plant to pathogen vs. herbivory stress (Vuorinen et al., 2007). Finally, extrapolating these results to natural environments is further complicated where simultaneous exposure to multiple stressors is likely the rule rather than the exception; multiple abiotic and biotic stressors can interact to significantly alter the plant's response relative to any single stressor (Holopainen and Gershenzon, 2010; Trowbridge et al., 2014; Winter et al., 2012).

This study adds to our knowledge of climate change stress impacts on BVOC emission rates by quantitatively investigating the impacts of an herbivore treatment on the VOC profile and emission rates from five different coniferous tree species that have not been the focus of other herbivory studies. This study was a component of a project that investigated the effects of herbivory stress on the composition of biogenic SOA generated from BVOC emissions (Faiola et al., 2014b). Published data on this topic is extremely limited, so one goal of this work was to identify key tree species that could produce a large herbivore-treatment effect on SOA composition. The herbivore treatment was an exogenous application of the plant hormone methyl jasmonate (MeJA). MeJA is a compound that plants use in nature to warn neighboring plants about the presence of herbivores; when plants are exposed to this compound, their emissions respond in a manner similar to if they were being attacked (Martin et al., 2003). This response is not plant species specific and allows even plants of different species to communicate with one another (Farmer and Ryan, 1990). The plant species used in this study are native to temperate coniferous forests in the mountainous regions of the western United States and Canada.

Responses to the simulated herbivory stress varied between tree types. Additionally, responses also varied between experiments using the same group of trees within a single tree species, and for different compounds within the same experiment. These results reinforce the necessity to obtain quantitative, compound-specific stress response measurements on a survey of representative trees in an area before stress-induced emissions can be integrated into biogenic emissions model inventories. We also identify a list of VOCs that showed similar stress responses across experiments and could significantly affect atmospheric chemical processes in future scenarios where increased herbivory is present.

2 Experimental approach

This research is a component of a larger project investigating plant stress impacts on biogenic secondary organic aerosol formation, using the Washington State University's Biogenic Aerosol Formation Facility. This facility is a dual chamber system with two separate Teflon® FEP bags: one a dynamic plant emission enclosure where sapling trees are stored and the other an aerosol growth chamber. This dual chamber system uses emissions from living vegetation as a precursor VOC source for SOA generation. The objective of this paper is to present impacts of plant stress on the BVOC emission profile from the subset of experiments where continuous gas-phase measurements were available from the plant chamber. Analysis of the impacts of the stress treatment on the composition of subsequently formed SOA are presented in a separate paper (Faiola et al., 2014b).

2.1 Tree description and treatment

Experiments were performed with saplings from five different coniferous species: bristlecone pine (*Pinus aristata*), blue spruce (*Picea pungens*), western redcedar (*Thuja plicata*), grand fir (*Abies grandis*), and Douglas-fir (*Pseudotsuga menziesii*). *Pinus aristata* and *Picea pungens* are found in the Rocky Mountains of Colorado. *Thuja plicata*, *Abies grandis*, and *Pseudotsuga menziesii* have wider latitudinal ranges and are found in the northern Rockies of the United States and Canada as well as the western mountain ranges of North America from Alaska to California. Emphasis in the experimental design was on the diversity of representative tree species included, with the goal of identifying species that responded strongly to stress treatment in ways that might affect SOA composition. This emphasis limited the number of replications that were possible.

Saplings were 1–3 years of age at the time of the experiments, and were purchased from the University of Idaho Forestry Nursery. Plants were cared for by greenhouse staff to ensure consistent watering and fertilization. They were stored outside of the greenhouse to be closer to their natural environmental conditions and prevent unnatural plant emission behavior that could occur within greenhouse conditions. This also meant the plants could have been exposed to natural stressors (e.g., heat or herbivory). These natural stressors were not controlled but would be representative of conditions encountered by the plants in nature because it is likely that exposure to multiple stressors is the rule rather than the exception in a forest environment (Holopainen and Gershenzon, 2010). Plant specimens were transported from the greenhouse to the laboratory plant chamber at least 2 days before treatment in order to capture a baseline VOC profile. Plants required 24–36 h to acclimate to the plant chamber after transportation. A summary of experiments is provided in Table 1.

Treatments using MeJA or jasmonic acid have been used to simulate herbivory response in plants (Filella et al., 2006; Rodriguez-Saona et al., 2001) and can change the terpene emission profile (Martin et al., 2003). The stress treatment used in these experiments was a foliar application of 200 mL of 10 mM MeJA solution in 18.2 MΩ purity water, based on previously reported methods (Martin et al., 2003). Negative control experiments were performed with each tree species, but only two (one from *Pinus aristata* and one from *Picea pungens*) were performed while the GC–MS–FID was in operation. The negative control treatment was a foliar application of 200 mL of 18.2 MΩ purity water.

2.2 Description of plant chamber and analytical instrumentation

Three to nine individual saplings were stored in the 0.9 m × 0.9 m × 0.9 m plant enclosure for each experiment; the number depended on the size and age of the trees. The plant enclosure was equipped with a lamp (Lumatek High-PAR Output HPS Lamp, 600W) set on a 12 h on/off cycle to simulate the day/night cycle. Photosynthetically active radiation (PAR) was continuously monitored with an Apogee model SQ-215 quantum sensor. Temperature and relative humidity were not controlled but were continuously monitored with a Vaisala model HMP110 humidity and temperature probe. The plant enclosure was continuously purged with zero air at 9.5 standard L min^{-1} (Aadco model 737 pure air generator).

Gas-phase emissions from the saplings were continuously monitored with a gas chromatograph coupled to a mass spectrometer and flame ionization detector (Agilent model

Table 1. Experiment Summary.

Plant scientific name	Common name	Experiment ID	Experiment type	Measurement dates	Treatment day and time	SOA generation experiments*
Picea pungens	Blue spruce	PP-E1	MeJA	12–17 May	15 May 11:40	PPu-1-Post
Picea pungens	Blue spruce	PP-C	Negative control	8–15 July	11 July 15:00	*none*
Picea pungens	Blue spruce	PP-E2	MeJA	15–19 July	17 July 10:40	PPu-2-Pre, PPu-2-Post
Pinus aristata	Bristlecone pine	PA-E	MeJA	19–24 May	22 May 11:30	PA-3-Pre, PA-3-Post
Pinus aristata	Bristlecone pine	PA-C	Negative control	26–31 May	29 May 11:00	PA-4-Pre
Abies grandis	Grand fir	AG-E	MeJA	23–28 June	26 June 11:30	AG-1-Pre, AG-1-Post
Thuja plicata	Western redcedar	TP-E	MeJA	16–23 September	22 September 08:30	TP-3-Pre1, TP-3-Pre2, TP-3-Post
Pseudotsuga menziesii	Douglas-fir	PM-E	MeJA	23–30 September	26 September 09:00	PM-2-Pre, PM-2-Post

* SOA composition results presented in Faiola et al. (2014b).

6890/5973 GC–MS–FID, DB-5MS column) with a time resolution of ~ 70 min. This instrument was equipped with a custom-built pre-concentration system described previously by Faiola et al. (2012, 2014a). The pre-concentration unit traps analytes on the Tenax© GR adsorbent and uses thermodesorption to inject compounds into the GC system. The FID is essentially a "carbon counter", meaning that the current produced from the detector is a function of the number of carbons in the molecule. Consequently, if the structure of the molecule is known, the concentration may be quantified using the effective carbon number concept with an upper-limit instrumental error of $\pm 10\,\%$ (Faiola et al., 2012). Identifications of the following compounds could be made based on retention times determined using commercial standards: 3-carene, terpinolene, limonene, alpha-pinene, beta-pinene, alpha-terpinene, beta-myrcene, and *o*-cymene. Molecular structures of other peaks were determined by interpreting the mass spectra acquired with the MS detector along with retention indices for monoterpenes. Integrated peak areas from the FID were converted to emission rates using Eq. (1):

$$E = \frac{A_a \chi_s N_s M_a F}{1000 A_s N_a B}. \tag{1}$$

Here, E is the emission rate normalized to plant biomass in units of μg-C g^{-1} h^{-1}, A_a and A_s are the integrated FID peak areas of the analyte and internal standard, respectively, χ_s is the mixing ratio of the internal standard (ppbV), N_a and N_s are the effective carbon numbers of the analyte and internal standard, respectively, M_a is the analyte molar mass of carbon (g-C mol^{-1}), F is the molar flow through the plant enclosure (mol-air h^{-1}), 1000 is a conversion factor to obtain the appropriate units, and B is the biomass of needles in the plant enclosure (g). Effective carbon numbers were estimated using the effective carbon number concept (Faiola et al., 2012; Sternberg et al., 1962). Biomass was estimated by collecting and weighing a subset of needles from each tree after they were removed from the plant chamber. Needles were dried for a minimum of 24 h in an oven before weighing; furthermore, dry needle weight was scaled up to the tree level by estimating the number of needles on each tree.

The GC–MS–FID used in this study was optimized to quantify monoterpenes. It can also quantitatively analyze aromatic emissions of a similar size. These emissions are dependent on temperature and were temperature normalized to 303 K using Eq. (2) (Guenther et al., 1993):

$$E(T) = E_s * e^{(\beta(T - T_s))}, \tag{2}$$

where $E(T)$ is the measured emission rate at a measured temperature (T), and E_s is the standardized basal emission rate (BER) at standard temperature (T_s). The activity adjustment factor, β (K^{-1}), was calculated for each experiment using measured emission rates between the post-acclimation period and treatment application. The number of points varied from experiment to experiment, but included a minimum of 24 h of measurements. Activity adjustment factors were calculated for terpenes and terpenoid aromatics separately because their chemical structures are slightly different and thus their chemical properties are expected to also differ. Results of these calculations are summarized in Table 2. The activity adjustment factors calculated here ranged from 0.15 to 0.59 K^{-1}, with most values ranging from 0.15 to 0.26 K^{-1}. Where a relationship between temperature and emission rate was observed and an activity adjustment factor could be calculated, nearly all values calculated for the terpenes were consistent with the ranges previously reported for coniferous tree species by Helmig et al. (2013) and Ortega et al. (2008) (0.08 to 0.28 K^{-1}) (0.00 to 0.23 K^{-1}). The one exception was the activity adjustment factor calculated for *Pseudotsuga menziesii*, which was much higher than any of the others, but which also had the highest temperature/emission rate (ER) correlation observed from any experiment ($r^2 = 0.91$ for monoterpenes and $r^2 = 0.89$ for aromatics). No aromatic compounds were observed above detection limit during the pre-treatment period for experiment PP-E1, so no activity adjustment factor could be calculated. Additionally, there was no relationship between temperature and emission rate during the pre-treatment period for the *Abies grandis* experiment. In this case, the average activity adjustment factor from the other experiments was used to temperature normalize the emissions for the *Abies grandis* experiment (excluding the apparent outlier from *Pseudotsuga menziesii*).

In addition to monoterpenoids, this analytical system could detect and identify isoprene and some small OVOCs. However, these compounds had low breakthrough volumes for the Tenax© adsorbent used, and so they were not quanti-

Table 2. Summary of activity adjustment factors for total monoterpenes and total aromatics that were calculated from pre-treatment emissions. Dashes indicate that no relationship could be established between temperature and emission rate for that experiment.

Experiment ID	MT β (K^{-1})	r^2	Aromatic β (K^{-1})	r^2	Temperature range (K)
PP-E1	0.21	0.87	–	–	293–300
PP-E2	0.17	0.82	0.21	0.76	298–305
PA-E	0.19	0.72	0.25	0.69	292–301
AG-E	–	–	–	–	–
TP-E	0.15	0.86	0.26	0.79	297–302
PM-E	0.52	0.91	0.59	0.89	297–301

tatively captured on the adsorbent trap. Thus, absolute emission rates are not reported for those compounds. Instead, the relative measured value could be analyzed to look at trends in changing emissions from day to day. Where used, these emissions were normalized to their maximum measured emission rate and presented as a unitless value.

2.3 Calculating atmospheric reactivity of BVOC emissions

One potential impact of stress-induced changes in the monoterpenoid profile is on the oxidative reactivity of the BVOC emissions. To evaluate this, it is necessary to isolate the impact of the changing terpenoid profile on reactivity and exclude any impacts from changes to absolute emission rates. To do this, the sum total monoterpenoid mixing ratio was normalized to 1 ppbV and the mixing ratio of each individual monoterpenoid was calculated from the relative terpenoid contribution. This reactivity will be referred to as the concentration-normalized reactivity of the BVOC emission profile. The total mixing ratio value of 1 ppbV was selected as a reasonable approximation of summertime afternoon monoterpene mixing ratios in the canopy of a forest environment (Bryan et al., 2012; Nölscher et al., 2012). The compounds used in the reactivity calculations and their corresponding OH and O_3 rate constants are presented in Table 3. Reaction rate constants were obtained from experimental results in the literature where available (Atkinson et al., 1990; Calvert et al., 2000; Corchnoy and Atkinson, 1990; Gai et al., 2013; Reissell et al., 2001; United States Environmental Protection Agency, 2014) or were calculated using the method described in Calvert et al. (2000). Ring strain was ignored for the ozone reaction rate constants. Concentration-normalized OH and O_3 reactivity of plant BVOC emission profiles were calculated from the sum of the individual BVOC reactivities, which were calculated as the product of the reaction rate constant and the normalized mixing ratio. The resulting total OH and O_3 reactivity is the inverse of the OH and O_3 lifetime. Only those compounds listed in Table 3 were included in the calculation. This list includes all the major VOCs that were identified in these experiments.

Table 3. Reaction rate constants for monoterpenoids at 298 ± 2 K. Units are cm^3 $molecule^{-1}$ s^{-1}.

Compound	OH rate constant	O_3 rate constant
santene	1.10×10^{-10}	1.10×10^{-15}
2-bornene	5.64×10^{-11}	1.20×10^{-16}
alpha-thujene	8.69×10^{-11}	4.00×10^{-16}
alpha-pinene	5.37×10^{-11}	8.66×10^{-17}
alpha-fenchene	5.14×10^{-11}	1.10×10^{-17}
camphene	5.33×10^{-11}	9.00×10^{-19}
2,4-thujadiene	1.08×10^{-10}	1.31×10^{-16}
beta-terpinene	1.44×10^{-10}	4.42×10^{-16}
beta-myrcene	2.15×10^{-10}	4.70×10^{-16}
alpha-phellandrene	3.13×10^{-10}	3.00×10^{-15}
3-carene	8.80×10^{-11}	3.70×10^{-17}
alpha-terpinene	3.63×10^{-10}	2.10×10^{-14}
limonene	1.70×10^{-10}	2.00×10^{-16}
beta-phellandrene	1.68×10^{-10}	4.70×10^{-17}
1,8-cineol	1.11×10^{-11}	1.50×10^{-19}
beta-ocimene	2.52×10^{-10}	5.40×10^{-16}
gamma-terpinene	1.77×10^{-10}	1.40×10^{-16}
terpinolene	2.25×10^{-10}	1.90×10^{-15}
m-cymene	1.51×10^{-11}	5.00×10^{-20}
p-cymene	1.51×10^{-11}	5.00×10^{-20}
o-cymene	1.51×10^{-11}	5.00×10^{-20}
o-cymenene	6.65×10^{-11}	5.00×10^{-20}
p-cymenene	6.65×10^{-11}	5.00×10^{-20}
2-carene	8.00×10^{-11}	2.30×10^{-16}
p-allylanisole	5.20×10^{-11}	1.03×10^{-17}
camphor	4.60×10^{-12}	7.00×10^{-20}
beta-pinene	7.89×10^{-11}	1.50×10^{-17}

References used to determine these reaction rate constants were Atkinson et al., 1990; Calvert et al., 2000; Corchnoy and Atkinson, 1990; Gai et al., 2013; Reissell et al., 2001; United States Environmental Protection Agency, 2014.

3 Results and discussion

In this section, pre-treatment BVOC profiles from each experiment are presented first and compared with previous reports of BVOC measurements from the same tree species. This was done to investigate whether the pre-treatment BVOC profiles were representative of trees in a natural setting. Then, the stress response from each tree type is described separately, including changes to the daily average monoterpenoid profiles and temporal trends in absolute emission rates. A summary of the main compounds that were affected by the stress treatment from each tree is presented. Finally, the concentration-normalized OH and O_3 reactivity are presented to investigate the impact of changing the BVOC profile before and after stress treatment.

3.1 Pre-treatment monoterpene profiles

Monoterpenoids were the dominant biogenic emissions that were quantitatively measured from each tree type in this study. These compounds have been the focus of numerous field measurements using the same species used in these experiments. Figure 1 summarizes the pre-treatment monoterpene profile for each experiment in this study. Values are presented as the percent of total monoterpenoid emission rates for each experiment. The same results are provided in absolute emission rates in Table 4. The profiles were calculated using all data from the end of the acclimation period until immediately before the stress treatment was applied. This time period varied from experiment to experiment but always included a minimum of 24 h of measurements. In total, 32 monoterpenoid chemical species were observed prior to treatment, including two oxygenated monoterpenes, camphor, and 1,8-cineol. Minor constituents were summed for inclusion in the profile. This group includes the following compounds: santene, 2-bornene, alpha-fenchene, 2,4-thujadiene, beta-terpinene, 2-carene, alpha-phellandrene, alpha-terpinene, gamma-terpinene, alpha-thujene, the aromatic cymenene isomers, acetophenone, two unidentified monoterpenes, and four unidentified aromatic compounds. Together, this category accounted for < 10 % of all pre-treatment monoterpenoid emissions. Toluene was also measured during some experiments but was not a major component and was not included in this analysis.

The pre-treatment monoterpene profile varied between the tree species (Fig. 1). However, despite differences in their distribution, the same seven compounds made up greater than 75 % of all monoterpene emissions from all trees: alpha-pinene, limonene, 3-carene, beta-pinene, beta-myrcene, camphene, and beta-phellandrene. For the two sets of *Picea pungens* experiments, the pre-treatment profiles were substantially different, even though the same four saplings were used in each of the three experiments. *Picea pungens* emissions in May (PP-E1) were dominated by alpha-pinene and limonene, while in July (PP-E2 and PP-C) they were dominated by limonene and beta-myrcene. Each of these profiles were consistent with previous measurements made in a field setting. The *Picea pungens* monoterpene profile presented by Helmig et al. (2013) had higher contributions from alpha-pinene in spring, but decreased in August and September in a manner similar to what we observed in July. Furthermore, we observed an increase in the contribution of

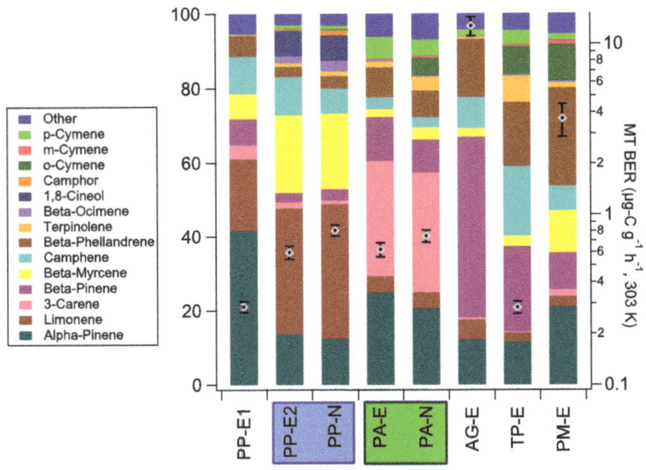

Figure 1. Pre-treatment monoterpenoid profiles for each experiment. PP-E1 represents *Picea pungens* stress experiment 1, PP-E2 is *Picea pungens* stress experiment 2, PP-N is *Picea pungens* negative control, PA-E is *Pinus aristata* stress experiment, PA-N is *Pinus aristata* negative control, AG-E is *Abies grandis* stress experiment, and PM-E is *Pseudotsuga menziesii* stress experiment. The two shaded boxes denote the paired stress/negative control experiments that were performed consecutively with the same set of saplings. The left axis shows the proportion of each compound emitted as a percent of total monoterpenoids. The diamonds associated with the right axis show the average pre-treatment basal emission rate (BER) of total monoterpenes normalized to a temperature of 303 K in units of μg-C g^{-1} h^{-1}. The x axis label is the experiment ID (Table 1). The average BER was calculating using all data from the end of the acclimation period until immediately before the stress treatment was applied (> 24 h of measurements). The error bars represent the standard deviation of the averaged value.

1,8-cineol in the July experiments vs. the May experiment, which Helmig et al. (2013) also described. The *Picea pungens* monoterpenoid BER in this study ranged from 0.29 to 0.81 μg-C g^{-1} h^{-1} (0.32–0.92 μg g^{-1} h^{-1}). Previous reports ranged from < 0.10 to 1.45 μg g^{-1} h^{-1} throughout the year, and during the months of May–July (the time period when our experiments were performed) the reported BER range was 0.87–1.45 μg g^{-1} h^{-1} (Helmig et al., 2013). Thus, the *Picea pungens* BER in our experiments was on the lower end of what has been reported from *Picea pungens* in the field.

The monoterpenoid profile of the Rocky Mountain bristlecone pine (*Pinus aristata*) has not been previously reported to our knowledge. A profile of the Great Basin bristlecone pine (*Pinus longaeva*) was presented by Helmig et al. (2013), and is used here for comparison. Both profiles were dominated by 3-carene, alpha-pinene, and beta-pinene. Within this study, the two *Pinus aristata* experiments exhibited nearly identical pre-treatment monoterpene emission profiles. These measurements were taken within 2 weeks of one another. The *Pinus aristata* monoterpenoid BER was 0.62–0.75 μg-C g^{-1} h^{-1} (0.70–0.85 μg g^{-1} h^{-1}), which is on the higher

Table 4. Summary of the temperature-normalized pre-treatment emission rates for the dominant compound emissions. Units are emission rates in μg-C g^{-1} h^{-1} normalized to 303 K. A dash indicates the compound was not detected and *bdl* indicates the compound was detected but it was below the calculated detection limit for quantification (detection limit = 0.003 μg-C g^{-1} h^{-1}). The average sum basal emission rate (BER) is provided at the bottom of the table for each experiment; σ denotes the standard deviation of the measurements used to calculate the pre-treatment average.

	PP-E1	PP-E2	PP-N	PA-E	PA-N	AG-E	TP-E	PM-E
alpha-pinene	0.119	0.081	0.100	0.154	0.153	1.537	0.033	0.769
limonene	0.056	0.204	0.293	0.027	0.033	0.682	0.007	0.102
3-carene	0.011	0.010	0.008	0.195	0.242	0.076	bdl	0.067
beta-pinene	0.020	0.015	0.025	0.074	0.067	6.203	0.066	0.363
beta-myrcene	0.020	0.125	0.165	0.014	0.025	0.297	0.008	0.422
camphene	0.028	0.061	0.053	0.019	0.021	1.054	0.053	0.244
beta-phellandrene	0.016	0.016	0.027	0.049	0.053	1.958	0.049	0.968
terpinolene	–	0.006	0.011	0.010	0.028	0.074	0.020	0.054
beta-ocimene	–	0.011	0.022	–	bdl	–	–	0.008
1,8-cineol	–	0.041	0.055	–	–	–	–	–
camphor	–	bdl	0.011	–	–	–	–	–
o-cymene	–	–	–	–	0.036	–	0.022	0.358
m-cymene	–	–	–	0.005	0.005	–	0.002	0.045
p-cymene	bdl	0.008	0.010	0.036	0.032	0.247	0.011	0.062
other	0.016	0.018	0.026	0.038	0.052	0.548	0.013	0.199
sum BER	0.286	0.597	0.806	0.621	0.746	12.675	0.284	3.661
σ	0.022	0.054	0.061	0.060	0.060	1.576	0.023	0.807

end of the range of *Pinus longaeva* BER values reported by Helmig et al. (2013) in May and June, 0.16–0.74 μg g^{-1} h^{-1}.

The *Abies grandis*, *Pseudotsuga menziesii*, and *Thuja plicata* monoterpene profiles each differed from what has been reported previously. The profile from *Abies grandis* in this study was dominated by beta-pinene, but no beta-pinene was observed by Ortega et al. (2008). This difference could be explained by natural genotypic variation because Ortega et al. (2008) also observed natural variation in the constitutive BVOC profiles between plants of the same tree species. However, the *Abies grandis* monoterpenoid pre-treatment BER measured in our experiment was 12.67 μg-C g^{-1} h^{-1}, substantially higher than any other pre-treatment monoterpenoid BER observed in this study and more than an order of magnitude greater than that reported by Ortega et al. (2008) for the same tree species. These high emission rates could suggest the *Abies grandis* saplings were likely exhibiting a stress response prior to treatment.

For *Pseudotsuga menziesii*, the dominant monoterpene emission measured in this study was beta-phellandrene (40 % of all monoterpenoid emissions). Helmig et al. (2013) observed alpha-pinene and beta-pinene comprising more than 50 % of all *Pseudotsuga menziesii* monoterpenoid emissions throughout an entire year of measurements, which was consistent with the profile presented in Geron et al. (2000). However, Ortega et al. (2008) observed variability in *Pseudotsuga menziesii* monoterpene profiles in the field, reporting that limonene and camphene were the dominant emissions during one set of measurements, while sabinene and alpha-pinene were for another. Furthermore, beta-pinene emissions were measured for one reported BVOC profile by Ortega et al. (2008) but not for the other. Thus, the pre-treatment profile in this laboratory study could still be representative of a natural baseline condition. The pre-treatment *Pseudotsuga menziesii* BER measured in our laboratory chamber was 3.66 μg-C g^{-1} h^{-1}. This was the second-highest observed BER value prior to treatment, and is consistent with previous reports where values as high as 3.40 μg-C g^{-1} h^{-1} were measured from *Pseudotsuga menziesii* branch enclosures by Ortega et al. (2008). However, our laboratory experiment was conducted in September when seasonal reports of emissions have shown decreasing emission trends. For example, the highest BER reported in the field by Helmig et al. (2013) was 2.51 μg-C g^{-1} h^{-1} in June, but they reported that by September the monoterpenoid BER had dropped back down to 0.12 μg-C g^{-1} h^{-1}. Thus, the BERs in our experiment were at the upper range of what would be expected in the natural environment from *Pseudotsuga menziesii* at this time of year.

Thuja plicata monoterpenoid emissions in this study were dominated by beta-pinene, camphene, and beta-phellandrene, whereas Ortega et al. (2008) found that 61 % of all monoterpenoid emissions were composed of the oxygenated compounds alpha- and beta-thujone. We did not observe any thujone emissions throughout the measurement period. The monoterpenoid pre-treatment BER from *Thuja plicata* was the lowest we observed from any species at 0.28 μg-C g^{-1} h^{-1}. This was consistent with the *Thuja plicata* BER reported by Ortega et al. (2008), 0.30 μg-C g^{-1} h^{-1}.

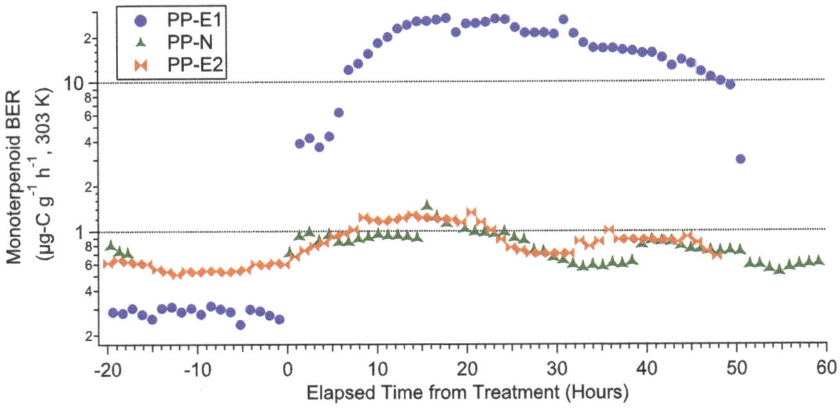

Figure 2. A summary of monoterpenoid emissions from all three *Picea pungens* experiment. The only experiment to exhibit a clear stress effect on monoterpenoid emission rates following treatment was the first methyl jasmonate (MeJA) experiment performed in May (PP-E1).

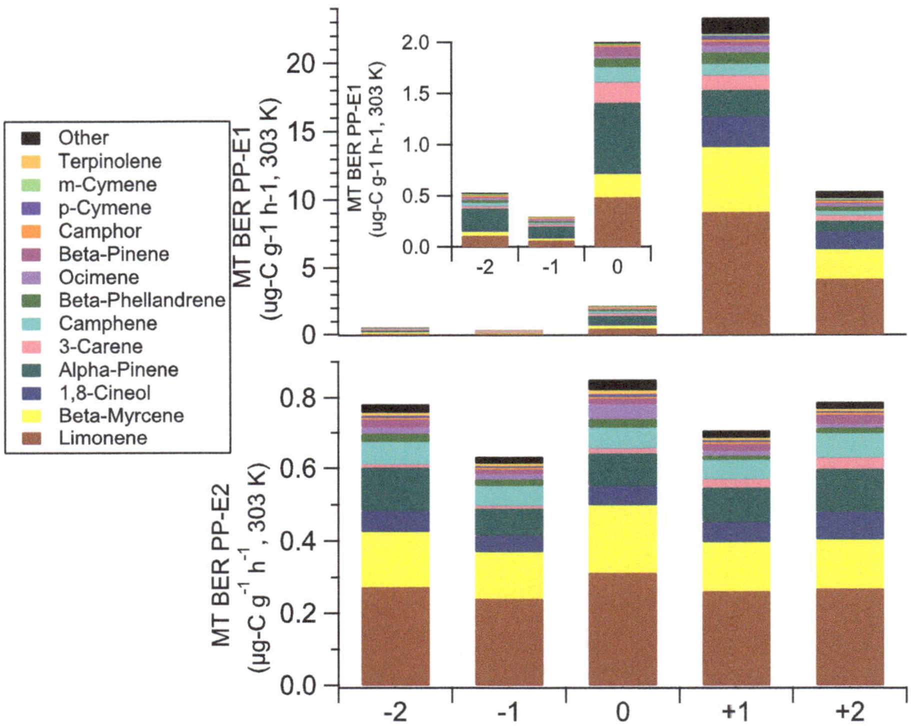

Figure 3. Summary of monoterpenoid profile for the two *Picea pungens* methyl jasmonate (MeJA) experiments. The x axis denotes the day relative to treatment, where treatment was performed on day 0. The y axis is the monoterpenoid (MT) basal emission rate normalized to 303 K. Results from the methyl jasmonate (MeJA) experiment performed in May are presented in the top plot and the results from the MeJA experiment performed in July are presented in the bottom plot. Note the difference in y axis scale for the top plot vs. the bottom plot. The inset in the top plot is provided as an enlargement of the profiles for days −2, −1, and 0 for experiment PP-E1.

3.2 Blue spruce (*Picea pungens*)

Three experiments were performed using *Picea pungens* saplings, two with MeJA treatments and one negative control. All three experiments were performed using the same four saplings, and the negative control experiment was performed the week prior to the July MeJA-treatment experiment. The two MeJA-treatment experiments did not produce consistent results. To illustrate this, a plot of the total monoterpenoid BER vs. elapsed time since treatment is shown in Fig. 2. The first treatment experiment performed in May exhibited a clear stress response where monoterpene emissions increased from 0.29 ± 0.2 to $23.27 \pm 2.15\,\mu$g-C g^{-1} h^{-1}. This represents an 80-fold increase after treatment. Emissions remained elevated above pre-treatment values over the next 50 h. In stark contrast, the monoterpene emissions from the

July MeJA experiment did not demonstrate a significantly different response to stress than did the negative control. There was a small increase in emissions for both PP-N and PP-E2 on the day of treatment. The short lived, slight emissions increase observed in these experiments could possibly be the result of an abiotic surface adsorption disruption effect; water displaces organic molecules previously adsorbed to the needle surfaces and produces a burst in measured emissions. This phenomenon has been observed in a natural forest environment where bursts of VOC emission were observed following rain (in a natural forest setting) or water application (in a laboratory setting) (Faiola et al., 2014a; Greenberg et al., 2012; Warneke et al., 1999). This would suggest that there was no significant stress treatment effect and that the small increase in some emissions observed on the treatment day could be a function of the treatment method itself rather than an actual stress response.

The difference in these results was also apparent when the complete BVOC profiles were examined (Fig. 3). These values are the average daytime emissions (6:00 to 18:00 LT). To simplify the presentation, BVOCs that individually constituted less than 1 % of all monoterpenoid emissions were summed and presented in the "Other" category. The pre-treatment aromatic emissions for the PP-E1 experiment were too low to calculate an aromatic activity adjustment factor, so the activity adjustment factor for aromatics calculated from PP-E2 data was used to normalize aromatic emission rates for both experiments.

In PP-E1, the maximum stress response for all classes of compounds was observed the day after treatment (day +1). The highest-emitted monoterpene before treatment was alpha-pinene (>40 % of all MT emissions, Fig. 1). After treatment, limonene, beta-myrcene, and 1,8-cineol dominated the emission profile. Limonene and beta-myrcene were constitutive emissions that were stimulated more than other constitutive emissions after treatment. In addition to enhancing constitutive emissions, the stress treatment also induced many new monoterpenoid emissions, including alpha-phellandrene, alpha-terpinene, 1,8-cineol, ocimene, gamma-terpinene, and terpinolene. Some of these induced compounds did not contribute significantly to the overall post-treatment emissions and were thus lumped into the other category, but they are worth noting because they were only observed after treatment had been applied. Specifically, 1,8-cineol and ocimene were emitted at rates well over 2 orders of magnitude higher than the detection limit after treatment, i.e., above the 80-fold increase in total emissions, which suggests these emissions were truly induced and not just emitted at rates below the detection limit prior to treatment. Negligible amounts of aromatic compounds were observed before treatment. After treatment, even though aromatics still made up a small relative proportion of overall emissions, the aromatic emissions (predominantly *p*-cymene) increased significantly to $0.5\,\mu\text{g-C}\,\text{g}^{-1}\,\text{h}^{-1}$, which was similar to the pre-treatment sum monoterpenoid BERs for many of the tree species presented in Fig. 1. Emissions of all classes of compounds began to decrease again within 48 h after treatment, but still remained elevated relative to pre-treatment values when measurements ceased.

In contrast to the May experiment, in the July *Picea pungens* experiment the monoterpenoid average profile did not significantly change after treatment (Fig. 3). This could be due to seasonal differences in the sensitivity of *Picea pungens* to herbivore treatment. This has been observed in other coniferous plant species. For example, monoterpene synthesis in *Pinus sylvestris* is more responsive to plant stressors during the spring when shoots are actively growing (Bäck et al., 2005). In the *Picea pungens* experiment presented here, there were small increases in terpinolene and ocimene emissions on the day of treatment, but they quickly returned to pre-treatment levels. Furthermore, results from the May experiment suggested that 1,8-cineol was a stress-induced compound that was only observed after treatment, but this same compound constituted a significant proportion of the pre-treatment BVOC emission profile in the July experiment. This could be a natural seasonal effect; field measurements have demonstrated seasonal changes in 1,8-cineol emission rates from *Picea pungens* (Helmig et al., 2013). However, it is also possible that the 1,8-cineol emission rate fluctuations observed in the field were due to the presence of some natural stressor. Thus, the pre-treatment profile for the July experiment could indicate that the trees' metabolic stress pathways had been activated prior to experimental treatment. This hypothesis is further supported by the higher percentage of beta-myrcene and limonene emissions present in the July pre-treatment profile that more closely resemble the post-treatment stress profile from the May experiment. This combined with the low emission rate values could suggest that the trees had been exposed to an external stressor for an adequate length of time to cause the plant to begin shutting down metabolic processes. If this was the case, the application of an additional stress treatment did not produce a stress response under those conditions.

Averaging emission rates over each day provides a clean picture of the overall VOC profiles, but any patterned variability that may occur through the day would be hidden by this approach. Another way to investigate changing VOC profiles is to compare the emission rate data for different compounds to evaluate their covariance. If paired compounds co-vary, then their relative emissions are consistent over time. If their correlation is weaker, it suggests that the profile is changing, possibly due to differences in the factors regulating the compounds' emissions.

Constitutive emissions co-varied throughout the negative control experiment (PP-N). Emission rates of beta-myrcene, alpha-pinene, and beta-phellandrene were plotted against limonene emissions and shown in Fig. 4. Limonene was used as the basis for comparison because it was the dominant constitutively emitted compound (Fig. 1). Measurements from the first 36 h while the plants were acclimat-

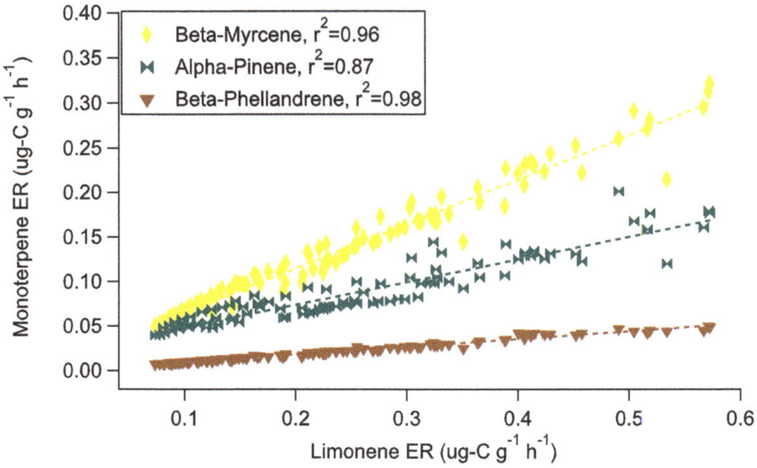

Figure 4. Covariance of constitutively emitted monoterpenes during the *Picea pungens* negative control experiment performed in July (PP-N).

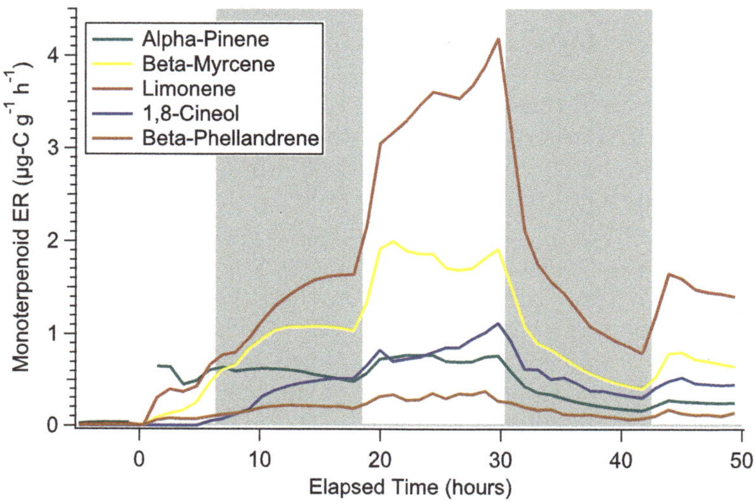

Figure 5. Post-treatment emission rates for five monoterpenoid species during the PP-E1 experiment. The x axis denotes the elapsed time since treatment application in hours. Alternating shaded and unshaded regions demonstrate when the light above the plant enclosure was turned off and on, respectively.

ing to the plant chamber were excluded from the analysis. Correlations between these three constitutively emitted compounds and limonene were high with r^2 values ranging from 0.87 to 0.98. This was also true for the other compounds' emissions, with emission rate correlation coefficients with limonene ranging between 0.85 and 0.96. Camphor was the exception; the correlation between camphor and limonene emissions was 0.35.

In the May MeJA experiment (PP-E1), the dominant pre-treatment constitutive emission was alpha-pinene but after treatment, the major emissions were limonene, beta-myrcene, and 1,8-cineol (Fig. 3). For this experiment, it was informative to look at both the time series of emission rates as well as the covariance between emission rates of difference compounds. A time series of the emission rates after treatment for a subset of the compounds is shown in Fig. 5. Immediately after treatment on 15 May, 2013 at 11:40 LT, alpha-pinene was still the dominant terpene emitted. However, emissions of limonene and beta-myrcene began to increase quickly and had exceeded alpha-pinene emissions by later that evening. Emissions of 1,8-cineol did not begin to increase until 17:00 LT. After that, they continued to increase and surpassed alpha-pinene emissions early the following morning. Beta-phellandrene is also shown on the figure to provide an example of a less dominant emission trend. It immediately began to increase after treatment but never exceeded alpha-pinene emissions. The emission trends of beta-myrcene, limonene, 1,8-cineol, and beta-phellandrene are in contrast to the trend in alpha-pinene emission rates. Alpha-pinene was not impacted by the treatment and maintained a stable emission rate throughout the evening while emission rates of other compounds steadily increased.

The covariance of emission rates after treatment was analyzed by investigating correlations with alpha-pinene (the dominant pre-treatment constitutive emission) and limonene (the dominant post-treatment emission). The correlation between post-treatment emissions of limonene, beta-myrcene, 1,8-cineol, and alpha-pinene were low with r^2 values ranging from 0.13 to 0.45. Emission rates of alpha-pinene were only well-correlated with two compounds, camphene ($r^2 = 0.77$) and beta-pinene ($r^2 = 0.97$). For all other compounds the r^2 ranged between 0.04 and 0.61. Post-treatment correlations between beta-myrcene, 1,8-cineol, and beta-phellandrene and the most stress-enhanced compound, limonene, ranged from 0.85 to 0.90. Limonene emission were also well correlated with ocimene ($r^2 = 0.89$), *p*-cymene ($r^2 = 0.83$), and terpinolene ($r^2 = 0.90$). This could suggest that the MeJA treatment induced de novo emissions of limonene, beta-myrcene, beta-phellandrene, 1,8-cineol, ocimene, *p*-cymene, and terpinolene. After the emissions were induced, these compounds exhibited similar emission patterns because of similar enzymatic controls on de novo production. 3-Carene and *m*-cymene emissions were not well correlated with either alpha-pinene or limonene emissions.

3.3 Western redcedar (*Thuja plicata*)

The VOC daily profiles for the *Thuja plicata* MeJA experiment are summarized in Fig. 6. For this experiment, nine small saplings were kept in the plant chamber for 6 days before applying treatment, and were removed from the chamber the day after treatment. However, for this group of plants there was an exceptionally strong emission response that continued to increase throughout the night following treatment. Consequently, day +1/2 has been included on the chart to capture peak emission response, and refers to the nighttime period that occurred half a day after treatment application. The pre-treatment and post-treatment profiles were plotted separately due to the drastic increase in the emission rate; monoterpene BER increased from an average value of $0.28 \pm 0.02\,\mu$g-C g^{-1} h^{-1} on days −6 to −4 to a maximum average value of $11.88 \pm 0.18\,\mu$g-C g^{-1} h^{-1} during the evening after treatment. This is a 42-fold increase in monoterpenoid BER. Terpinolene, beta-myrcene, and the cymene isomers increased most substantially and dominated the monoterpene profile after treatment.

The post-treatment temporal emissions trends for the *Thuja plicata* experiment exhibited a pattern that was not observed for other trees species. Figure 7 shows the monoterpenoid BER time series immediately following treatment. In Fig. 7, the treatment was applied on 22 September at 08:30 LT, and emissions of all compounds began to increase by 13:00 LT the same day. The emissions of nearly all compounds continued to rise or stabilized at an elevated emission rate for the remainder of the measurement period until 23 September at 05:00 LT when measurements were stopped. However, beta-pinene did not follow this trend; instead, beta-pinene emissions immediately increased after treatment, but began to decrease a few hours later, starting at 15:00 LT on the treatment day. It was the only compound to exhibit this emission pattern.

Terpinolene also demonstrated a slightly different emission pattern from most other monoterpenes. This is evident from the linear regression results presented in Table 5. Terpinolene reached a maximum emission rate on the evening of the treatment day at 20:30 LT (Fig. 7). Afterwards it began to decrease slowly. The only other compound to exhibit

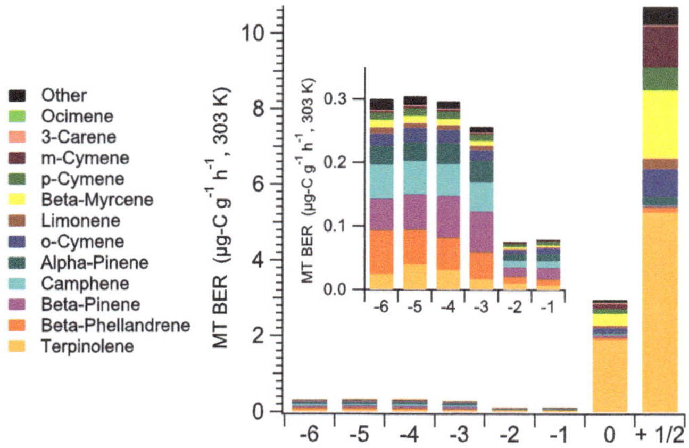

Figure 6. Emission profile of emissions from *Thuja plicata* during methyl jasmonate (MeJA) experiment TP-E. The *x* axis denotes the day relative to treatment application. The *y* axis shows the monoterpenoid BER normalized to 303 K. Note the drastic scale change between the pre- and post-treatment *y* axes. The insert shows an enlarged view of the first 6 days to allow better visualization of the pre-treatment period.

Table 5. Results of linear regression correlation analysis (r^2) between all monoterpenoid emission rates (ERs) vs. terpinolene emission rates and limonene emission rates during experiment TP-E.

	vs. Terpinolene ERs	vs. Limonene ERs
ocimene	0.86	0.26
beta-myrcene	0.48	0.98
p-cymene	0.79	0.93
m-cymene	0.54	0.99
o-cymene	0.58	0.98
limonene	0.56	–
alpha-thujene	0.45	0.98
alpha-pinene	0.26	0.90
gamma-terpinene	0.80	0.93
alpha-phellandrene	0.42	0.98
camphene	0.37	0.92
3-carene	0.57	0.97
beta-phellandrene	0.88	0.83
beta-pinene	0.08	0.59

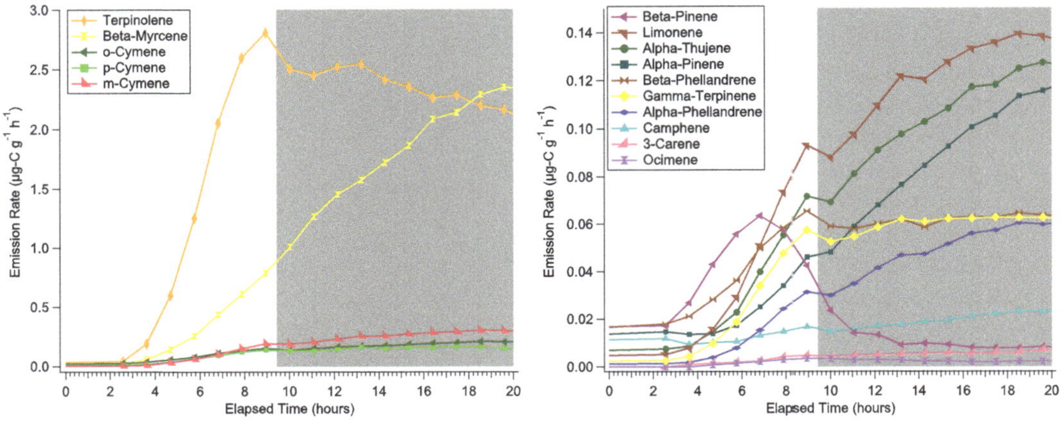

Figure 7. Time series of monoterpene emission rates from *Thuja plicata*. The x axis shows the elapsed time since treatment application in hours. Alternating shaded and unshaded regions demonstrate when the light above the plant enclosure was turned off and on, respectively.

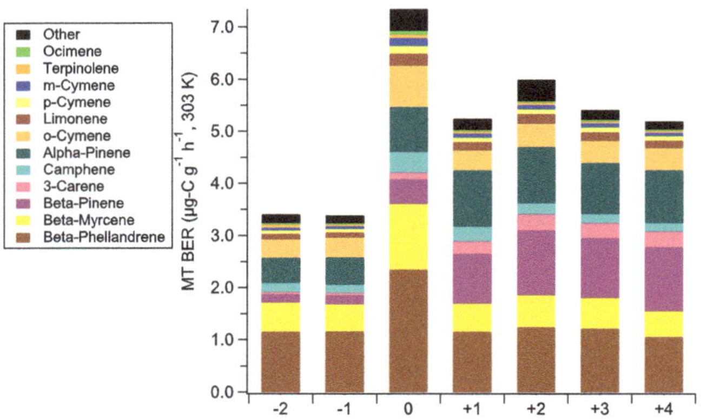

Figure 8. Douglas-fir VOC profile. The x axis denotes the day relative to treatment application. The y axis is the monoterpenoid basal emission rate normalized to 303 K.

this emission trend was ocimene, which had a linear regression correlation with terpinolene emissions of 0.86. Most other compounds continued to increase throughout the night. Thus, most compound emission rates were highly correlated with limonene emissions, which exhibited this continually increasing emission trend. Ten compounds were highly correlated with limonene emissions with $r^2 > 0.90$ (Table 5). Beta-phellandrene and gamma-terpinene were well correlated with both limonene and terpinolene with $r^2 \geq 0.80$. Their emission rates stabilized more quickly than most other compounds during the night. They were best correlated with one another with an $r^2 = 0.96$. This could suggest four different types of emission responses (1) quick increase followed by a slow decrease within 10 h of treatment similar to terpinolene, (2) quick increase followed by a rapid decrease similar to beta-pinene, (3) long-term increase throughout the night similar to limonene, and (4) an increase followed by stabilization within ~12 h of treatment similar to beta-phellandrene.

Monoterpenoid BER values for *Thuja plicata* were the lowest pre-treatment emissions that were measured from all trees in this study. After treatment had been applied, monoterpenoid BERs increased to the third-highest emission rates measured throughout the experiments. This suggests that stress exposure in natural environments could turn normally low-emitting trees into high emitters that could contribute substantially to the net ecosystem BVOC flux. This should be considered in future experimental designs, where it may be tempting to limit tree species representation to only the known highest BVOC emitters in a region because there may be some tree species that are only high emitters under stressed conditions.

3.4 Douglas-fir (*Pseudotsuga menziesii*)

The daily average VOC emission profile from *Pseudotsuga menziesii* is shown in Fig. 8. Some of the minor constituents (< 1 % of BER) have been grouped together within the "Other" category to simplify the presentation. For this experiment, 2 days of measurements were collected prior to treatment after plants had acclimated to the chamber. Following treatment, BVOC emission rates were monitored for another 4 days. Absolute monoterpenoid BERs approximately doubled on the day of treatment. They increased from 3.66 ± 0.88 to $7.34 \pm 1.04\,\mu$g-C g^{-1} h^{-1}. Emissions then remained 34 % higher, on average, than baseline emissions for the following 4 days. Aromatics (predominantly *o*-cymene) comprised more than 10 % of the total *Pseudotsuga menziesii* VOC emissions even before treatment, and thus could be significant contributors to SOA formation in natural forest environments. Emissions of alpha-pinene, beta-pinene, and 3-carene increased most after treatment relative to the other constitutive monoterpenes. Alpha-pinene emissions increased by ~100 %, beta-pinene emissions by ~570 %, and 3-carene emissions by ~640 %. This effect was sustained until measurements ceased 4 days after treatment. One of these stress-enhanced compounds, beta-pinene, co-varied with the dominant constitutive emission, beta-phellandrene, prior to treatment ($r^2 = 0.89$), but was de-coupled from beta-phellandrene emissions after treatment ($r^2 = 0.48$). However, nearly all other compounds continued to co-vary with beta-phellandrene emissions from day +1 to day +4 after treatment. Emissions from beta-myrcene, the cymene isomers, alpha-pinene, limonene, ocimene, and terpinolene all had linear regression results of $r^2 > 0.90$ vs. beta-phellandrene. 3-Carene emissions did not co-vary with any other compound emissions.

The overall stress response exhibited by *Pseudotsuga menziesii* was not as dramatic as the 80-fold increase observed during experiment PP-E1 or the 42-fold increase observed during experiment TP-E. There was also no single stress-enhanced compound that completely dominated the post-treatment emission profile as terpinolene did during experiment TP-E. Despite all this, the three most stress-enhanced compounds (alpha-pinene, beta-pinene, and 3-carene) did contribute significantly to the overall BVOC emissions during this experiment, which were substantial. Pre-treatment, the monoterpenoid BERs for *Pseudotsuga menziesii* were the second-highest pre-treatment values measured in this study (Fig. 1), with a daytime average pre-treatment monoterpenoid BER of $3.39 \pm 0.01\,\mu$g-C g^{-1} h^{-1}. The daytime average post-treatment BER was $5.46 \pm 0.37\,\mu$g-C g^{-1} h^{-1}. This is only a modest increase in overall emission rates relative to some of the other experiments. However, of the $2.06\,\mu$g-C g^{-1} h^{-1} total increase in BER, $1.75\,\mu$g-C g^{-1} h^{-1} was due to the increase in just the three most stress-enhanced compounds: alpha-pinene, beta-pinene, and 3-carene (85 % of the total increase). The post-treatment average BER of these three compounds was $2.48 \pm 0.15\,\mu$g-C g^{-1} h^{-1}, 73 % of the total monoterpenoid pre-treatment BER. Thus, these stimulated monoterpenes can significantly contribute to total BVOC emissions. This is important because different monoterpenes have widely varying chemical reactivity and SOA formation potential (Atkinson and Arey, 1998; Griffin et al., 1999).

3.5 Grand fir (*Abies grandis*)

As shown in Fig. 1, the pre-treatment monoterpene BER for the grand fir experiment was greater than for any other experiment, and was much greater than what had been previously reported elsewhere. This suggests that these trees had been exposed to some unknown external stress while being stored outdoors prior to use. To investigate this, we examined the entire BVOC profile throughout the measurement period (Fig. 9). All monoterpenoid emissions steadily decreased from day −2 to day 0. It is possible that the trees were still acclimating to the plant chamber on day −2, but they should have been well acclimated by day −1 because trees take 12–36 h to acclimate to the plant chamber (having been transported to the chamber on day −3). The observed steady decrease from day to day could be indicative of the hypothesized unknown stress effect waning once the trees were brought into the laboratory. Laboratory notes on tree appearance for this experiment indicate that the trees had a number of dry, orange-red needles when they were transported on 23 June 2013. Another note from 28 June 2013 described large clumps of needles dropping from the trees at the slightest touch during watering. The trees were kept well watered at the greenhouse and in the laboratory chamber and outdoor temperatures were normal for the area, so we do not believe that the needle damage was the consequence of drought or temperature stress. However, this possibility cannot be ruled out completely. Alternatively, the observed effects may have been the result of an unseen herbivore or pathogen that was not detected prior to the experiment.

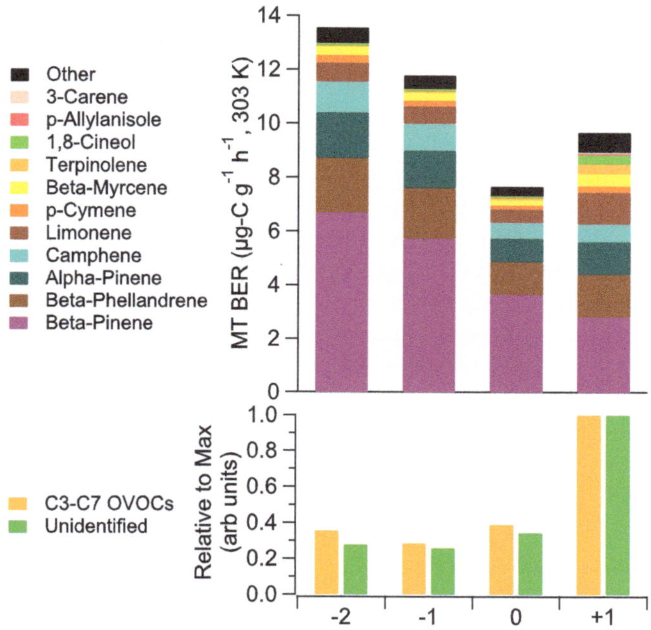

Figure 9. Grand fir BVOC profile. The x axis denotes the day relative to treatment application. The top panel summarizes the monoterpenoid emissions where the y axis is the monoterpenoid basal emission rate normalized to 303 K. The bottom panel summarizes the emissions of small oxy-VOCs and other unidentified compounds where the y axis is the fraction of the emission rate relative to the maximum measured value.

Despite the possible presence of an uncontrolled stressor, the experimental MeJA stress treatment did still have a small effect on BVOC emission rates and profile (Fig. 9). This effect was not immediate; emissions continued their decreasing trend on day 0, but then increased slightly on day +1. The BVOC profile was altered both by the induction of emissions of new compounds and by the alteration of the distribution of constitutive emissions. 1,8-Cineol and, to a much lesser extent, *p*-allylanisole were induced. The former is an oxygenated monoterpene and the latter is a phenylpropanoid produced from the shikhimic acid pathway (Dudareva et al., 2006). These emissions were not observed until 6 h after treatment for 1,8-cineol and 22 h after treatment for *p*-allylanisole. Small OVOCs and unidentified compounds exhibited maximum emissions the day following stress treatment and may also have been induced by the stress treatment. Similar to the other stress-induced and stress-enhanced compounds, they exhibited a delayed response in emissions. These small OVOCs include alcohols, ketones, and aldehydes that have less than eight carbon atoms including small 5-carbon to 6-carbon OVOCs produced from the lipoxygenase biochemical pathway (Connor et al., 2008; Maffei, 2010).

The constitutive monoterpene emission profile also changed. For the first 3 days, the terpene profile was dominated by beta-pinene, beta-phellandrene, and alpha-pinene, and their relative contribution to total emissions did not vary significantly. After the MeJA treatment, beta-pinene emissions continued to decrease as they had been for the previous 3 days, but limonene, beta-myrcene, beta-phellandrene, terpinolene, and alpha-pinene all increased. Increases in these compounds were observed 6 h after treatment, similar to when the induced compound, 1,8-cineol, was first observed. Prior to treatment, constitutive emissions of alpha-pinene, limonene, and terpinolene all co-varied with the dominant constitutive emission, beta-pinene, with all r^2 values greater than 0.90 (Fig. 10, left). Two separate bursts in emissions occurred 24 h apart from one another that produced the three-highest points on the plots (two measurements during one burst and one measurement during the other burst). With those points removed, alpha-pinene and limonene were still well correlated with beta-pinene with r^2 values of 0.97 and 0.89, respectively. The terpinolene r^2 reduced to 0.52 when the two emission bursts were excluded. Other major constitutive emissions also co-varied with beta-pinene prior to treatment but were not shown in the figure; camphene, beta-phellandrene, *p*-cymene, and beta-myrcene also co-varied with beta-pinene prior to treatment with r^2 values ranging from 0.94 to 0.99. However, after treatment, beta-pinene no longer co-varied with alpha-pinene, limonene, or terpinolene with r^2 values of 0.53, 0.25, and 0.12, respectively (Fig. 10, right). Thus, even with the emission bursts removed pre-treatment, all r^2 values decreased relative to the post-treatment correlations. Furthermore, all of the other most highly enhanced constitutive compounds except for beta-phellandrene were well correlated with limonene after treatment with r^2 values > 0.80 (not shown). The MeJA stress treatment de-coupled the dominant constitutive emissions from beta-pinene, which was not enhanced by the stress, while most of the compounds enhanced by the treatment continued to co-vary. 1,8-Cineol, the induced emission, was not well correlated with the most enhanced constitutive emission, limonene ($r^2 = 0.18$).

3.6 Bristlecone pine (*Pinus aristata*)

A time series of the summed monoterpenoid BERs are presented in Fig. 11. There was a large spike in emissions immediately following the MeJA treatment where monoterpenoid emissions increased from 0.54 to 12.52 μg-C g^{-1} h^{-1}. The negative control experiment also demonstrated a slight increase in emissions, but to a much lesser extent than the MeJA experiment; monoterpenoid emissions increased from 0.81 to 2.68 μg-C g^{-1} h^{-1}. The emissions increase was short lived for both experiments and the emissions trend started to reverse within just a few hours following treatment.

The monoterpene profiles for the days before (day −1) and after (day +1) treatment are shown in Fig. 12. The total emissions were slightly reduced for the MeJA experiment on the day following treatment, but not substantially so, and the monoterpenoid profile did not change. The negative control BER and emission profile were similar before and after spraying the trees with water.

Major monoterpene emissions were plotted against the emission rates of the dominant monoterpene throughout these experiments, 3-carene, in Fig. 13. Both the negative control and MeJA experiment demonstrated high correlations ($r^2 > 0.9$) for all monoterpene emissions relative to 3-carene. Beta-pinene, beta-phellandrene, and terpinolene are shown in the figure for illustration, and this was also true for alpha-pinene, *o*-cymene, *p*-cymene, limonene, camphene, beta-myrcene, and *m*-cymene. This indicates that the monoterpene profile did not change substantially during either experiment.

3.7 Summary of emission rate changes

A summary of the change in emission rates after stress treatment for some of the key compounds is summarized for each experiment where a plant stress response was observed (Fig. 14). Note the difference in the *y* axis scale for each experiment because the overall change in emission rates varied between plant types. For the *Thuja plicata* experiment, the delta value was calculated from the day +1/2 post-treatment value minus the baseline daily average from day −4 to day −6. This is a conservative estimate of emissions changes because all emissions decreased during the 2 days prior to treatment (days −1 and −2) but these lower emission values were not used in the calculation. For the *Picea pungens* experiment, the delta BER was calculated by subtracting the average daily value on day −1 from day +1. The maximum response was observed on day +1 and day −2 was excluded because the plants may have still been acclimating to the chamber. For the *Pseudotsuga menziesii* experiments, the delta BER was calculated by subtracting the average daily values on day −2 and day −1 from the average daily values on days +1 to +4. For the *Abies grandis* experiment, the delta BER was calculated as the difference between day 0 and day +1.

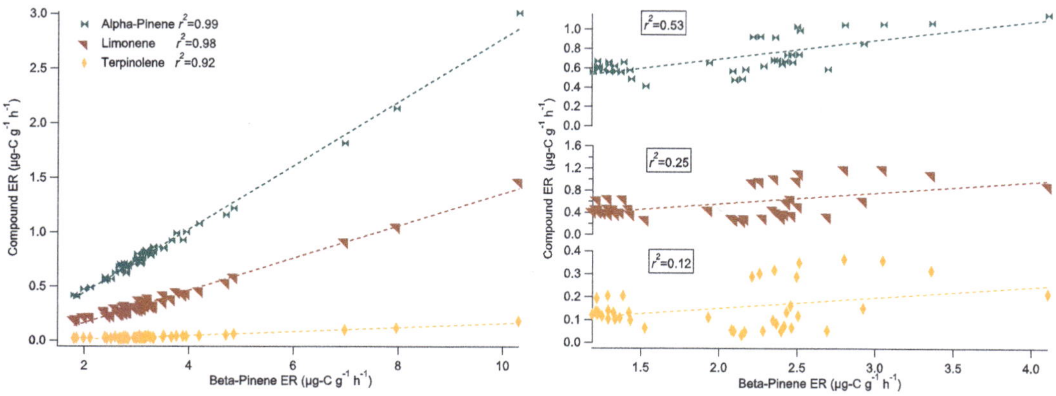

Figure 10. Scatterplots of the constitutive emissions alpha-pinene, limonene, and terpinolene vs. beta-pinene (the dominant constitutively emitted compound during the pre-treatment period) during experiment AG-E. Pre-treatment values are plotted on the left and post-treatment values are plotted on the right. Results of the linear regression analysis are included on the graphs.

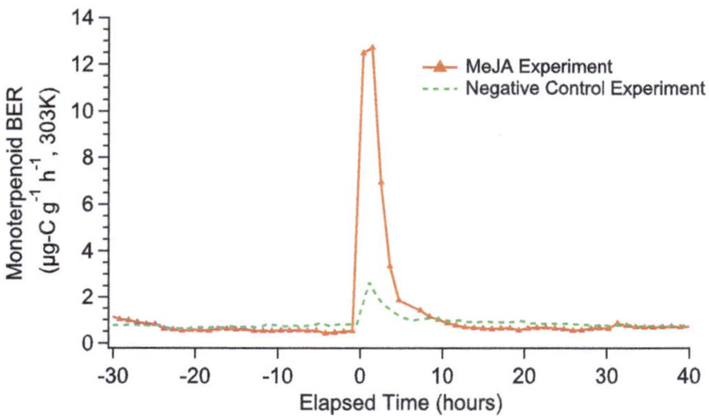

Figure 11. Results from two *Pinus aristata* experiments. Shown above is the time series of the sum monoterpenoid basal emission rates normalized to 303 K as a function of elapsed time since treatment application for the methyl jasmonate (MeJA) experiment (PA-E) and the negative control experiment (PA-C).

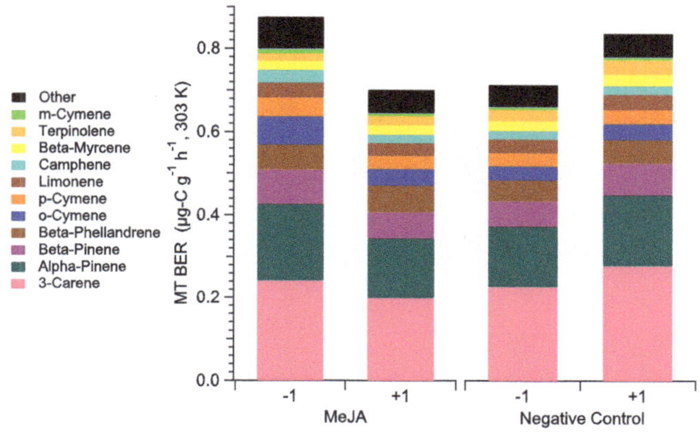

Figure 12. The *Pinus aristata* BVOC profile the day before treatment and the day after treatment for both the methyl jasmonate (MeJA) experiment (PA-E) and the negative control experiment (PA-C). The x axis denotes the day relative to treatment application. The y axis shows the monoterpenoid basal emission rate normalized to 303 K. The left two bars illustrate the BVOC profiles from the methyl jasmonate (MeJA) experiment and the right two bars illustrate the BVOC profiles from the negative control experiment.

The compounds that were most impacted by the stress treatment were highly variable between tree types. In the *Thuja plicata* experiment, the two monoterpenes that increased most were terpinolene and beta-myrcene. The emissions of these compounds increased by a combined 7.04 μg-C g^{-1} h^{-1}. This represents just over 80 % of the total increase in monoterpene BER with terpinolene alone contributing to just over 60 % of the total increase. The cymene isomers also exhibited a significant emission increase. The only other experiment where all three cymene isomers were measured was in the *Pseudotsuga menziesii* experiment. In this case, all cymene isomers increased, but to a lesser extent than during the *Thuja plicata* experiment. The most stress-enhanced compounds in the *Pseudotsuga menziesii* experiment were alpha-pinene, beta-pinene and 3-carene. 1,8-Cineol was identified as an important stress-enhanced or stress-stimulated compound in the *Picea pungens* and *Abies grandis* experiments but was never emitted from the other two plant types. Beta-myrcene was an important stress-enhanced compound for all plant types shown in the figure except for *Pseudotsuga menziesii*. Emissions of other compounds in our experiments generally either increased or stayed the same after treatment. An exception to this was in the *Abies grandis* experiment, where beta-pinene emissions significantly decreased after treatment.

Even though each experiment yielded fundamentally different results, several of the observed behaviors could be more broadly applicable. The differing results that were observed between the two *Picea pungens* MeJA experiments could indicate that plant stress susceptibility changes seasonally. Alternatively, if the *Picea pungens* plants had been exposed to an external unknown stressor for weeks prior to the

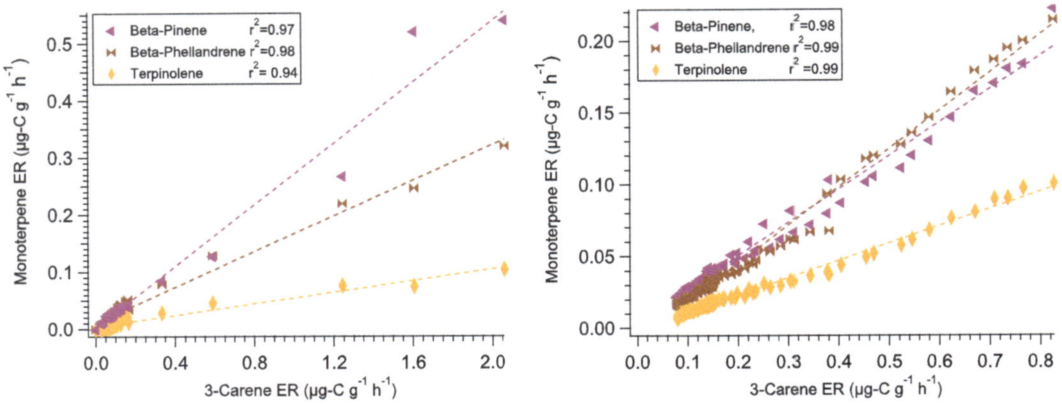

Figure 13. Scatterplots investigating the covariance between major constitutive emissions from *Pinus aristata* vs. 3-carene (the dominant constitutively emitted compound). Results from the linear regression fits of the data are summarized in the legends. The methyl jasmonate (MeJA) experiment is shown on the left and the negative control experiment is shown on the right.

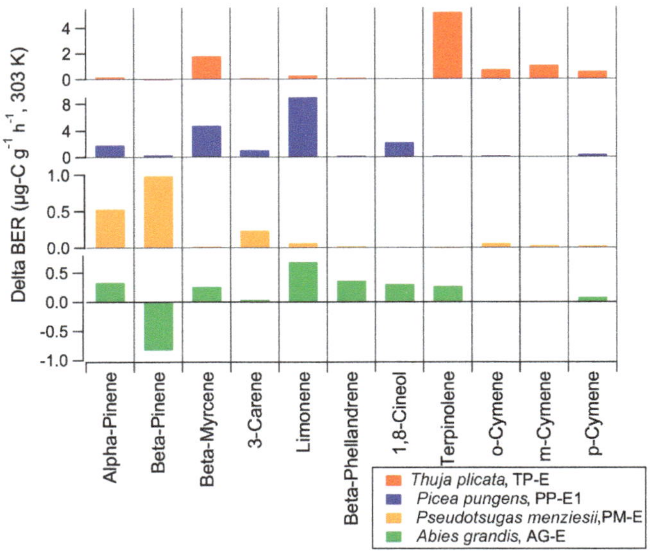

Figure 14. A summary of the change in basal emission rates after stress treatment application for some key compounds for each experiment where a stress response was observed.

second experiment (PP-E2), the results could indicate there is some breaking point where the plants simply do not respond to an additional stressor. These results would be in stark contrast to the *Abies grandis* stress response. The *Abies grandis* results suggest that despite the possible presence of an unknown stress prior to treatment, the simulated herbivory stress still caused additional changes to the emission profile. Thus, the presence of one stressor does not necessarily prevent a tree from responding to another stressor at the same time, and it is possible the effects of the two stressors could be additive. The response of the *Thuja plicata* emissions to the stress treatment can also provide valuable insight. Even though the pre-treatment emissions from the *Thuja plicata* plants were the lowest we measured from all the experiments, the post-treatment emission rates were substantial. This suggests that even naturally low-emitting species that would not contribute significantly to total forest BVOC flux under baseline conditions could be major sources of BVOC emissions under stressed conditions in a changing climate. Consequently, future surveys of BVOC emitters should not be limited to only the highest BVOC emitters in a region because this could change as global change stressors intensify. Finally, the near lack of any long-term response from *Pinus aristata* could indicate that some trees are more resistant to certain types of stress exposure than others. On the other hand, it is possible that, like *Picea pungens*, the *Pinus aristata* could demonstrate a completely different stress response depending on the season. The *Pinus aristata* experiments were conducted in May when pre-treatment emissions were low and the plants may have still been coming out of winter dormancy. This could have contributed to their apparent resistance to the treatment.

3.8 Implications for BVOC atmospheric reactivity

The MeJA stress treatment significantly changed the BVOC profile in many of the experiments. As discussed in the previous section, the specific compounds that were impacted by the treatment were highly variable between the different plant types. Consequently, the overall implications for atmospheric reactivity for the different plant types was also highly variable because different monoterpenoids have widely varying atmospheric reactivity (see Table 3). The pre- and post-treatment BVOC profile for each experiment was used to calculate the concentration-normalized OH and O_3 reactivity by normalizing the relative contribution of each monoterpenoid to a sum monoterpenoid mixing ratio of 1 ppbV. The goal was to isolate the impact on reactivity due to changes in the BVOC profile only. Thus, the focus of this analysis was to investigate the change to the concentration-normalized oxidant reactivity value rather than the absolute pre- and post-

Table 6. Summary of the BVOC pre-treatment (PreT) and post-treatment (PostT) concentration-normalized OH reactivity (rOH) and concentration-normalized O_3 reactivity (rO_3) at 298 ± 2 K. Reactivity values are presented in units of s^{-1}. The σ is the standard deviation of the averaged measurements. The percent difference between the pre-treatment and post-treatment values is also shown.

Exp ID	PreT rOH	σ	PostT rOH	σ	% Diff	PreT rO_3 ($\times 10^{-6}$)	σ ($\times 10^{-6}$)	PostT rO_3 ($\times 10^{-6}$)	σ ($\times 10^{-6}$)	% Diff
PP-E1	2.43	0.13	3.50	0.09	44.0	2.99	0.31	10.7	0.61	257.9
PP-C	3.45	0.06	3.32	0.13	−3.8	6.92	0.69	5.65	1.16	−18.3
PP-E2	3.32	0.12	3.20	0.21	−3.6	5.34	1.03	5.84	1.06	9.4
PA-E	2.16	0.08	2.35	0.12	8.8	5.17	2.61	8.77	0.38	69.6
PA-C	2.37	0.02	2.37	0.04	0.0	7.83	0.66	7.86	0.78	0.4
AG-E	2.43	0.04	2.74	0.12	12.8	3.46	0.50	7.40	1.90	113.9
TP-E	2.21	0.30	4.57	0.13	106.8	3.53	2.59	30.3	2.6	758.4
PM-E	2.75	0.37	2.44	0.29	−11.3	3.37	0.89	2.49	0.75	−26.1

treatment values; the reactivity results are presented in Table 6.

For all experiments where a change in concentration-normalized reactivity was observed, the O_3 reactivity was more significantly affected than the OH reactivity. The three experiments that demonstrated the largest changes were TP-E, PP-E1, and AG-E. For each of these experiments, the stress-induced changes to the BVOC profile increased both the OH and O_3 concentration-normalized reactivity. The normalized OH reactivity of the *Thuja plicata* emission profile (TP-E) approximately doubled with an increase from 2.21 to $4.57\,s^{-1}$ (106.8 % increase). This corresponds to a decrease in OH lifetime from 0.45 to 0.22 s. The normalized O_3 reactivity increased by nearly an order of magnitude from $3.53 \times 10^{-6}\,s^{-1}$ to $30.3 \times 10^{-6}\,s^{-1}$ (758.4 % increase). This corresponds to a decrease in O_3 lifetime from 3.3 days to 9.2 h. This is primarily due to the large increase in the relative amount of terpinolene, which has a high ozone reaction rate constant relative to most other monoterpenoids (Table 3). The normalized OH reactivity of the *Picea pungens* emission profile during the first experiment (PP-E1) increased from 2.43 to $3.50\,s^{-1}$ (44 % increase). This corresponds to a decrease in the OH lifetime from 0.41 to 0.29 s. The normalized O_3 reactivity increased from $2.99 \times 10^{-6}\,s^{-1}$ to $10.7 \times 10^{-6}\,s^{-1}$ (257.9 % increase) corresponding to a decrease in O_3 lifetime from 3.9 days to 1.1 days. The normalized OH reactivity of the *Abies grandis* emissions increased by a small amount from 2.43 to $2.74\,s^{-1}$ (12.8 % increase) corresponding to a decrease in OH lifetime from 0.41 s to 0.36 s. However, the normalized O_3 reactivity significantly increased from $3.46 \times 10^{-6}\,s^{-1}$ to $7.40 \times 10^{-6}\,s^{-1}$ (113.9 % increase) corresponding to a decrease in O_3 lifetime from 3.3 days to 1.6 days.

The *Pinus aristata* experiments (PA-C and PA-E) demonstrated very little change to the BVOC profile (see Sect. 3.6). For the negative control experiment (PA-C), the concentration-normalized reactivity results were consistent with no BVOC profile change; a 0 % change was observed for OH reactivity and a 0.4 % change was observed for O_3 reactivity. The normalized OH reactivity increased slightly after treatment during the PA-E experiment with an increase of 8.8 %. However, the PA-E normalized O_3 reactivity increased significantly by 69.6 % after MeJA treatment despite only minor changes to the BVOC profile (see Fig. 12). These results demonstrate that even small changes to the BVOC profile can have significant impacts on the overall atmospheric reactivity of the BVOC emissions.

Concentration-normalized reactivity of emissions from *Pseudotsuga menziesii* decreased slightly after treatment. The normalized OH reactivity decreased from 2.75 to $2.44\,s^{-1}$ (decrease of 11.3 %) corresponding to a small increase in OH lifetime from 0.36 to 0.40 s. The normalized O_3 reactivity decreased from $3.37 \times 10^{-6}\,s^{-1}$ to $2.49 \times 10^{-6}\,s^{-1}$ (decrease of 26.1 %) corresponding to an increase in O_3 lifetime from 3.4 days to 4.6 days. This was due to an increase in the relative amount of beta-pinene and 3-carene emissions. Both of these compounds have reduced oxidant reactivity relative to other monoterpenoid compounds emitted in higher amounts prior to treatment (Table 3).

4 Conclusions

While many uncertainties remain regarding the impacts of herbivory stress on plant BVOC emissions, it is clear that plant responses are highly variable. Emissions of different compounds were impacted by the stress treatment for different tree types. The compounds that tended to be most affected by the stress treatment were alpha-pinene, beta-pinene, beta-myrcene, 3-carene, limonene, 1,8-cineol, terpinolene, and the cymene isomers. Aromatic cymenes sometimes contributed significantly to the emission profile pre-treatment (i.e. *Pseudotsuga menziesii*), and often increased significantly post-treatment. These aromatic compounds are often not considered to be major precursors of biogenic SOA, but the emission rates observed in these experiments suggest they could be significant contributors to SOA formation in forests.

Four possible plant herbivory response patterns were observed in these experiments: (1) plant susceptibility to herbivory stress changes seasonally; (2) after long-term exposure to one stressor, plant emissions decrease overall and do not respond to additional stressors; (3) alternatively, multiple stressors can be additive, perhaps if the second stressor is applied before the first stressor depletes terpene pools and initiates metabolic shutdown; and (4) herbivory stress could turn naturally low-emitting plants in a region to high emitters that would need to be considered in future climate scenarios with increased herbivory.

Stress-induced changes to the BVOC emission profile can result in significant changes to the concentration-normalized oxidant reactivity of plant emissions in the atmosphere. Increases in reactivity as high as 758.4 % with O_3 and 106.8 % with OH were observed during the *Thuja plicata* experiment (TP-E). Furthermore, even small changes to the BVOC profile during the *Pinus aristata* MeJA experiment (PA-E) increased O_3 reactivity by 69.6 %. These results highlight the importance of making quantitative, compound-specific BVOC emission rate measurements to understand the potential impact of stress-induced emissions on atmospheric chemistry. Changes in the oxidant reactivity of BVOC emissions have significant implications for the production of pollutants like ozone and secondary organic aerosol in forest environments.

Many questions still need to be addressed before stress impacts on BVOC emissions can be incorporated into emissions models. Future research needs to address the seasonality influence on plant susceptibility to herbivory stress. Additionally, the interaction between multiple stressors needs to be addressed because in the natural environment it is likely that plants are being exposed to multiple stressors more often than a single stressor in isolation. A broad survey of plant types should be used in these experiments to investigate which plants could become dominant BVOC emitters under future climate scenarios. Finally, all of these questions need to be asked regarding other types of plant stress including drought, thermal stress, ozone stress, and using different types of real herbivores and pathogens.

Acknowledgements. The authors thank Chuck Cody and the greenhouse staff for taking care of the plants used in this study. This work was supported by the US Department of Energy Early Career Research Program (award no. SC0003899).

Edited by: S. M. Noe

References

Achotegui-Castells, A., Llusia, J., Hódar, J. A., and Peñuelas, J.: Needle terpene concentrations and emissions of two coexisting subspecies of Scots pine attacked by the pine processionary moth (Thaumetopoea pityocampa), Acta Physiol. Plant., 35, 3047–3058, 2013.

Amin, H., Atkins, P. T., Russo, R. S., Brown, A. W., Sive, B., Hallar, A. G., and Huff Hartz K. E.: Effect of bark beetle infestation on secondary organic aerosol precursor emissions, Environ. Sci. Technol., 46, 5696–5703, 2012.

Amin, H., Russo, R. S. Sive, B. Hoebeke, E. R., Dodson, C., McCubbin, I. B. Hallar, A. G., and Huff Hartz, K. E.: Monoterpene emissions from bark beetle infested Engelmann spruce trees, Atmos. Environ., 72, 130–133, 2013.

Arneth, A. and Niinemets, Ü.: Induced BVOCs: how to bug our models?, Trends Plant Sci., 15, 118–125, 2010.

Atkinson, R.: Atmospheric chemistry of VOCs and NOx, Atmos. Environ., 34, 2063–2101, 2000.

Atkinson, R. and Arey, J.: Atmospheric Chemistry of Biogenic Organic Compounds, Accounts Chem. Res., 31, 574–583, doi:10.1021/ar970143z, 1998.

Atkinson, R., Hasegawa, D., and Aschmann, S. M.: Rate constants for the gas-phase reactions of O_3 with a series of monoterpenes and related compounds at 296 K, Int. J. Chem. Kinet., 22, 871–887, 1990.

Bale, J. S., Masters, G. J., Hodkinson, I. D., Awmack, C., Bezemer, T. M., Brown, V. K., Butterfield, J., Buse, A., Coulson, J. C., Farrar, J., Good, J. E. G., Harrington, R., Hartley, S., Jones, T. H., Lindroth, R. L., Press, M. C., Symrnioudis, I., Watt, A. D., and Whittaker, J. B.: Herbivory in global climate change research: direct effects of rising temperature on insect herbivores, Glob. Change Biol., 8, 1–16, doi:10.1046/j.1365-2486.2002.00451.x, 2002.

Berg, A. R., Heald, C. L., Huff Hartz, K. E., Hallar, A. G., Meddens, A. J. H., Hicke, J. A., Lamarque, J.-F., and Tilmes, S.: The impact of bark beetle infestations on monoterpene emissions and secondary organic aerosol formation in western North America, Atmos. Chem. Phys., 13, 3149–3161, doi:10.5194/acp-13-3149-2013, 2013.

Blanch, J.-S., Peñuelas, J., Sardans, J., and Llusia, J.: Drought, warming and soil fertilization effects on leaf volatile terpene concentrations in *Pinus halepensis* and *Quercus ilex*, Acta Physiol. Plant., 31, 207–218, 2009.

Blande, J. D., Tiiva, P., Oksanen, E., and Holopainen, J. K.: Emission of herbivore-induced volatile terpenoids from two hybrid aspen (*Populus tremula & tremuloides*) clones under ambient and elevated ozone concentrations in the field, Glob. Change Biol., 13, 2538–2550, doi:10.1111/j.1365-2486.2007.01453.x, 2007.

Brilli, F., Ciccioli, P., Frattoni, M., Prestininzi, M., Spanedda, A., and Loreto, F.: Constitutive and herbivore-induced monoterpenes emitted by *Populus euroamericana* leaves are key volatiles that orient Chrysomela populi beetles, Plant, Cell, & Environment, 32, 542–552, 2009.

Bryan, A. M., Bertman, S. B., Carroll, M. A., Dusanter, S., Edwards, G. D., Forkel, R., Griffith, S., Guenther, A. B., Hansen, R. F., Helmig, D., Jobson, B. T., Keutsch, F. N., Lefer, B. L., Pressley, S. N., Shepson, P. B., Stevens, P. S., and Steiner, A. L.: In-canopy gas-phase chemistry during CABINEX 2009: sensitivity of a 1-D canopy model to vertical mixing and isoprene chemistry, Atmos. Chem. Phys., 12, 8829–8849, doi:10.5194/acp-12-8829-2012, 2012.

Bäck, J., P. Hari, Hakola, H. Juurola, E., and Kulmala, M.: Dynamics of monoterpene emissions in *Pinus sylvestris* during early spring, Boreal Environ. Res., 10, 409–424, 2005.

Calfapietra, C., Fares, S., and Lofeto, F.: Volatile organic compounds from Italian vegetation and their interaction with ozone, Environ. Pollut., 157, 1478–1486, doi:10.1016/j.envpol.2008.09.048, 2009.

Calvert, J. G., Atkinson, R., Kerr, J. A., Madronich, S., Moortgat, G. K., Wallington, T. J., and Yarwood, G.: The mechanisms of atmospheric oxidation of the alkenes, Oxford University Press New York, available at: http://www.zohu.cn/viewarticle.php?id=233404 (last acces: 13 June 2014), 2000.

Carslaw, K. S., Boucher, O., Spracklen, D. V., Mann, G. W., Rae, J. G. L., Woodward, S., and Kulmala, M.: A review of natural aerosol interactions and feedbacks within the Earth system, Atmos. Chem. Phys., 10, 1701–1737, doi:10.5194/acp-10-1701-2010, 2010.

Connor, E. C., Rott, A. S., Zeder, M., Jüttner, F., and Dorn, S.: ^{13}C-labelling patterns of green leaf volatiles indicating different dynamics of precursors in *Brassica* leaves, Phytochemistry, 69, 1304–1312, 2008.

Constable, J. V. H., Litvak, M. E., Greenberg, J. P., and Monson, R. K.: Monoterpene emission from coniferous trees in response to elevated CO_2 concentration and climate warming, Glob. Change Biol., 5, 252–267, doi:10.1046/j.1365-2486.1999.00212.x, 1999.

Copolovici, L., Kaennaste, A., Remmel, T., Vislap, V., and Niinemets, U.: Volatile Emissions from *Alnus glutionosa* Induced by Herbivory are Quantitatively Related to the Extent of Damage, J. Chem. Ecol., 37, 18–28, doi:10.1007/s10886-010-9897-9, 2011.

Corchnoy, S. B. and Atkinson, R.: Kinetics of the gas-phase reactions of hydroxyl and nitrogen oxide (NO3) radicals with 2-carene, 1, 8-cineole, p-cymene, and terpinolene, Environ. Sci. Technol., 24, 1497–1502, 1990.

Dudareva, N., Negre, F., Nagegowda, D. A., and Orlova, I.: Plant volatiles: recent advances and future perspectives, Crit. Rev. Plant Sci., 25, 417–440, 2006.

Ehn, M., Thornton, J. A., Kleist, E., Sipilä, M., Junninen, H., Pullinen, I., Springer, M., Rubach, F., Tillmann, R., and Lee, B.: A large source of low-volatility secondary organic aerosol, Nature, 506, 476–479, 2014.

Engelberth, J., Alborn, H. T., Schmelz, E. A., and Tumlinson, J. H.: Airborne signals prime plants against insect herbivore attack, P. Natl. Acad. Sci. USA, 101, 1781–1785, 2004.

Faiola, C. L., Erickson, M. H., Fricaud, V. L., Jobson, B. T., and VanReken, T. M.: Quantification of biogenic volatile organic compounds with a flame ionization detector using the effective carbon number concept, Atmos. Meas. Tech., 5, 1911–1923, doi:10.5194/amt-5-1911-2012, 2012.

Faiola, C. L., VanderSchelden, G. S., Wen, M., Elloy, F. C., Cobos, D. R., Watts, R. J., Jobson, B. T., and VanReken, T. M.: SOA Formation Potential of Emissions from Soil and Leaf Litter, Environ. Sci. Technol., 48, 938–946, doi:10.1021/es4040045, 2014a.

Faiola, C. L., Wen, M., and VanReken, T. M.: Chemical characterization of biogenic SOA generated from plant emissions under baseline and stressed conditions: inter- and intra-species variability for six coniferous species, Atmos. Chem. Phys. Discuss., 14, 25167–25212, doi:10.5194/acpd-14-25167-2014, 2014b.

Fall, R. and Monson, R. K.: Isoprene emission rate and intercellular isoprene concentration as influenced by stomatal distribution and conductance, Plant Physiol., 100, 987–992, 1992.

Farmer, E. E. and Ryan, C. A.: Interplant communication: Airborne methyl jasmonate induces synthesis of proteinase inhibitors in plant leaves, P. Natl. Acad. Sci. USA, 87, 7713–7716, 1990.

Filella, I., Peñuelas, J., and Llusia, J.: Dynamics of the enhanced emissions of monoterpenes and methyl salicylate, and decreased uptake of formaldehyde, by *Quercus ilex* leaves after application of jasmonic acid, New Phytol., 169, 135–144, doi:10.1111/j.1469-8137.2005.01570.x, 2006.

Filella, I., Wilkinson, M. J., Llusia, J., Hewitt, C. N., and Peñuelas, J.: Volatile organic compounds emissions in Norway spruce (*Picea abies*) in response to temperature changes, Physiol. Plant., 130, 58–66, doi:10.1111/j.1399-3054.2007.00881.x, 2007.

Gai, Y., Wang, W., Ge, M., Kjaergaard, H. G., Jørgensen, S., and Du, L.: Methyl chavicol reactions with ozone, OH and NO_3 radicals: Rate constants and gas-phase products, Atmos. Environ., 77, 696–702, 2013.

Geron, C., Rasmussen, R., R Arnts, R., and Guenther, A.: A review and synthesis of monoterpene speciation from forests in the United States, Atmos. Environ., 34, 1761–1781, 2000.

Graus, M., Eller, A. S. D., Fall, R., Yuan, B., Qian, Y., Westra, P., de Gouw, J., and Warneke, C.: Biosphere-atmosphere exchange of volatile organic compounds over C4 biofuel crops, Atmos. Environ., 66, 161–168, doi:10.1016/j.atmosenv.2011.12.042, 2013.

Greenberg, J. P., Asensio, D., Turnipseed, A., Guenther, A. B., Karl, T., and Gochis, D.: Contribution of leaf and needle litter to whole ecosystem BVOC fluxes, Atmos. Environ., 59, 302–311, doi:10.1016/j.atmosenv.2012.04.038, 2012.

Griffin, R. J., Cocker, D. R., Flagan, R. C., and Seinfeld, J. H.: Organic aerosol formation from the oxidation of biogenic hydrocarbons, J. Geophys. Res.-Atmosp., 104, 3555–3567, doi:10.1029/1998JD100049, 1999.

Guenther, A. B., Zimmerman, P. R., Harley, P. C., Monson, R. K., and Fall, R.: Isoprene and monoterpene emission rate variability: model evaluations and sensitivity analyses, J. Geophys. Res.-Atmospheres (1984–2012), 98, 12609–12617, 1993.

Guenther, A., Geron, C., Pierce, T., Lamb, B., Harley, P., and Fall, R.: Natural emissions of non-methane volatile organic compounds, carbon monoxide, and oxides of nitrogen from North America, Atmos. Environ., 34, 2205–2230, 2000.

Guenther, A. B., Jiang, X., Heald, C. L., Sakulyanontvittaya, T., Duhl, T., Emmons, L. K., and Wang, X.: The Model of Emissions of Gases and Aerosols from Nature version 2.1 (MEGAN2.1): an extended and updated framework for modeling biogenic emissions, Geosci. Model Dev., 5, 1471–1492, doi:10.5194/gmd-5-1471-2012, 2012.

Hamilton, J. F., Lewis, A. C., Carey, T. J., Wenger, J. C., Borrás i Garcia, E., and Muñoz, A.: Reactive oxidation products promote secondary organic aerosol formation from green leaf volatiles, Atmos. Chem. Phys., 9, 3815–3823, doi:10.5194/acp-9-3815-2009, 2009.

Harley, P., Deem, G., Flint, S., and Caldwell, M.: Effects of growth under elevated UV-B on photosynthesis and isoprene emission in *Quercus gambelii* and *Mucuna pruriens*, Glob. Change Biol., 2, 149–154, 1996.

Heiden, A. C., Hoffmann, T., Kahl, J., Kley, D., Klockow, D., Langebartels, C., Mehlhorn, H., Sandermann Jr, H., Schraudner, M., Schuh, G., and others: Emission of volatile signal and defense molecules from ozone-exposed plants, Ecol. Appl., 9, 1160–1167, 1999.

Helmig, D., Daly, R. W., Milford, J., and Guenther, A.: Seasonal trends of biogenic terpene emissions, Chemosphere, 93, 35–46, doi:10.1016/j.chemosphere.2013.04.058, 2013.

Herrmann, K. M. and Weaver, L. M.: The Shikimate Pathway, Annu. Rev. Plant. Phys., 50, 473–503, doi:10.1146/annurev.arplant.50.1.473, 1999.

Holopainen, J. K. and Gershenzon, J.: Multiple stress factors and the emission of plant VOCs, Trends Plant Sci., 15, 176–184, 2010.

Hu, Z., Shen, Y., Luo, Y., Shen, F., Gao, H., and Gao, R.: Aldehyde volatiles emitted in succession from mechanically damaged leaves of poplar cuttings, J. Plant Biol., 51, 269–275, 2008.

Jansen, R. M. C., Hofstee, J. W., Wildt, J., Verstappen, F. W. A., Bouwmeester, H. J., Posthumus, M. A., and van Henten, E. J.: Health monitoring of plants by their emitted volatiles: trichome damage and cell membrane damage are detectable at greenhouse scale, Ann. Appl. Biol., 154, 441–452, doi:10.1111/j.1744-7348.2008.00311.x, 2009a.

Jansen, R. M. C., Miebach, M., Kleist, E., van Henten, E. J., and Wildt, J.: Release of lipoxygenase products and monoterpenes by tomato plants as an indicator of Botrytis cinerea-induced stress, Plant Biol., 11, 859–868, doi:10.1111/j.1438-8677.2008.00183.x, 2009b.

Kesselmeier, J. and Staudt, M.: Biogenic volatile organic compounds (VOC): an overview on emission, physiology and ecology, J. Atmos. Chem., 33, 23–88, 1999.

Kleist, E., Mentel, T. F., Andres, S., Bohne, A., Folkers, A., Kiendler-Scharr, A., Rudich, Y., Springer, M., Tillmann, R., and Wildt, J.: Irreversible impacts of heat on the emissions of monoterpenes, sesquiterpenes, phenolic BVOC and green leaf volatiles from several tree species, Biogeosciences, 9, 5111–5123, doi:10.5194/bg-9-5111-2012, 2012.

Kroll, J. H. and Seinfeld, J. H.: Chemistry of secondary organic aerosol: Formation and evolution of low-volatility organics in the atmosphere, Atmos. Environ., 42, 3593–3624, doi:10.1016/j.atmosenv.2008.01.003, 2008.

Laothawornkitkul, J., Moore, J. P., Taylor, J. E., Possell, M., Gibson, T. D., Hewitt, C. N., and Paul, N. D.: Discrimination of Plant Volatile Signatures by an Electronic Nose: A Potential Technology for Plant Pest and Disease Monitoring, Environ. Sci. Technol., 42, 8433–8439, doi:10.1021/es801738s, 2008.

Loreto, F. and Delfine, S.: Emission of isoprene from salt-stressed *Eucalyptus globulus* leaves, Plant Physiol., 123, 1605–1610, 2000.

Loreto, F. and Schnitzler, J.-P.: Abiotic stresses and induced BVOCs, Trends Plant Sci., 15, 154–166, 2010.

Loreto, F., Ciccioli, P., Cecinato, A., Brancaleoni, E., Frattoni, M., and Tricoli, D.: Influence of environmental factors and air composition on the emission of α-Pinene from *Quercus ilex* leaves, Plant Physiol., 110, 267–275, 1996.

Loreto, F., Nascetti, P., Graverini, A., and Mannozzi, M.: Emission and content of monoterpenes in intact and wounded needles of the Mediterranean pine, *Pinus pinea*, Funct. Ecol., 14, 589–595, 2000.

Maffei, M. E.: Sites of synthesis, biochemistry and functional role of plant volatiles, S. Afr. J. Bot., 76, 612–631, 2010.

Martin, D. M., Gershenzon, J., and Bohlmann, J.: Induction of volatile terpene biosynthesis and diurnal emission by methyl jasmonate in foliage of Norway spruce, Plant Physiol., 132, 1586–1599, 2003.

Mentel, Th. F., Kleist, E., Andres, S., Dal Maso, M., Hohaus, T., Kiendler-Scharr, A., Rudich, Y., Springer, M., Tillmann, R., Uerlings, R., Wahner, A., and Wildt, J.: Secondary aerosol formation from stress-induced biogenic emissions and possible climate feedbacks, Atmos. Chem. Phys., 13, 8755–8770, doi:10.5194/acp-13-8755-2013, 2013.

Niinemets, Ü.: Mild vs. severe stress and BVOCs: thresholds, priming and consequences, Trends Plant Sci., 15, 145–153, doi:10.1016/j.tplants.2009.11.008, 2010.

Niinemets, Ü., Loreto, F., and Reichstein, M.: Physiological and physicochemical controls on foliar volatile organic compound emissions, Trends Plant Sci., 9, 180–186, 2004.

Niinemets, Ü., Arneth, A., Kuhn, U., Monson, R. K., Peñuelas, J., and Staudt, M.: The emission factor of volatile isoprenoids: stress, acclimation, and developmental responses, Biogeosciences, 7, 2203–2223, doi:10.5194/bg-7-2203-2010, 2010.

Nölscher, A. C., Williams, J., Sinha, V., Custer, T., Song, W., Johnson, A. M., Axinte, R., Bozem, H., Fischer, H., Pouvesle, N., Phillips, G., Crowley, J. N., Rantala, P., Rinne, J., Kulmala, M., Gonzales, D., Valverde-Canossa, J., Vogel, A., Hoffmann, T., Ouwersloot, H. G., Vilà-Guerau de Arellano, J., and Lelieveld, J.: Summertime total OH reactivity measurements from boreal forest during HUMPPA-COPEC 2010, Atmos. Chem. Phys., 12, 8257-8270, doi:10.5194/acp-12-8257-2012, 2012.

Ormeno, E., Fernandez, C., and Mévy, J.-P.: Plant coexistence alters terpene emission and content of Mediterranean species, Phytochemistry, 68, 840–852, 2007.

Ortega, J., Helmig, D., Daly, R. W., Tanner, D. M., Guenther, A. B., and Herrick, J. D.: Approaches for quantifying reactive and low-volatility biogenic organic compound emissions by vegetation enclosure techniques – part B: applications, Chemosphere, 72, 365–380, doi:10.1016/j.chemosphere.2008.02.054, 2008.

Peñuelas, J. and Staudt, M.: BVOCs and global change, Trends Plant Sci., 15, 133–144, doi:10.1016/j.tplants.2009.12.005, 2010.

Priemé, A., Knudsen, T. B., Glasius, M., and Christensen, S.: Herbivory by the weevil, *Strophosoma melanogrammum* causes severalfold increase in emission of monoterpenes from young Norway spruce (*Picea abies*) Atmos. Environ., 34, 711–718, 2000.

Reissell, A., Arey, J., and Atkinson, R.: Atmospheric chemistry of camphor, Int. J. Chem. Kinet., 33, 56–63, 2001.

Rodriguez-Saona, C., Crafts-Brandner, S. J., Pare, P. W., and Henneberry, T. J.: Exogenous methyl jasmonate induces volatile emissions in cotton plants, J. Chem. Ecol., 27, 679–695, 2001.

Semiz, G., Blande, J. D., Heijari, J., Işık, K., Niinemets, Ü., and Holopainen, J. K.: Manipulation of VOC emissions with methyl jasmonate and carrageenan in the evergreen conifer Pinus sylvestris and evergreen broadleaf Quercus ilex, Plant Biol., 14, 57–65, 2012.

Simpraga, M., Verbeeck, H., Demarcke, M., Joo, E., Pokorska, O., Amelynck, C., Schoon, N., Dewulf, J., Van Langenhove, H., Heinesch, B., Aubinet, M., Laffineur, Q., Muller, J.-F., and Steppe, K.: Clear link between drought stress, photosynthesis and biogenic volatile organic compounds in *Fagus sylvatica L.*, Atmos. Environ., 45, 5254–5259, doi:10.1016/j.atmosenv.2011.06.075, 2011.

Staudt, M. and Lhoutellier, L.: Volatile organic compound emission from holm oak infested by gypsy moth larvae: evidence for distinct responses in damaged and undamaged leaves, Tree Physiol., 27, 1433–1440, 2007.

Sternberg, J. C., Gallaway, W. S., and Jones, D. T. L.: Chapter XVIII: The Mechanism of Response of Flame Ionization Detectors, in: Gas Chromatography: Third International Symposium Held Under the Auspices of the Analysis Instrumentation Division of the Instrument Society of America, edited by: Brenner, N., Callen, J. E., and Weiss, M. D., Academic Press, New York and London, 231–267, 1962.

Teuber, M., Zimmer, I., Kreuzwieser, J., Ache, P., Polle, A., Rennenberg, H., and Schnitzler, J.-P.: VOC emissions of Grey poplar leaves as affected by salt stress and different N sources, Plant Biol., 10, 86–96, doi:10.1111/j.1438-8677.2007.00015.x, 2008.

Toome, M., Randjarv, P., Copolovici, L., Niinemets, U., Heinsoo, K., Luik, A., and Noe, S. M.: Leaf rust induced volatile organic compounds signalling in willow during the infection, Planta, 232, 235–243, doi:10.1007/s00425-010-1169-y, 2010.

Trowbridge, A. M., Daly, R. W., Helmig, D., Stoy, P. C., and Monson, R. K. : Herbivory and climate interact serially to control monoterpene emissions from pinyon pine forests, Ecology, 95, 1591–1603, 2014.

United States Environmental Protection Agency: Estimation Programs Interface Suite^{TM} for Microsoft$^{®}$ Windows, v 4.11, US EPA, Washington, DC, USA, available at: http://www.epa.gov/opptintr/exposure/pubs/episuite.htm (last access: 24 January 2015), 2014.

Vuorinen, T., Nerg, A. M., and Holopainen, J. K.: Ozone exposure triggers the emission of herbivore-induced plant volatiles, but does not disturb tritrophic signalling, Environ. Pollut., 131, 305–311, doi:10.1016/j.envpol.2004.02.027, 2004.

Vuorinen, T., Nerg, A.-M., Syrjala, L., Peltonen, P., and Holopainen, J. K.: Epirrita autumnata induced VOC emission of silver birch differ from emission induced by leaf fungal pathogen, Arthropod-Plant Interact., 1, 159–165, doi:10.1007/s11829-007-9013-4, 2007.

Warneke, C., Karl, T., Judmaier, H., Hansel, A., Jordan, A., Lindinger, W., and Crutzen, P. J.: Acetone, methanol, and other partially oxidized volatile organic emissions from dead plant matter by abiological processes: Significance for atmospheric HOx chemistry, Global Biogeochem. Cy., 13, 9–17, 1999.

Winter, T. R., Borkowski, L., Zeier, J., and Rostas, M.: Heavy metal stress can prime for herbivore-induced plant volatile emission, Plant Cell Environ., 35, 1287–1298, doi:10.1111/j.1365-3040.2012.02489.x, 2012.

11

North America's net terrestrial CO_2 exchange with the atmosphere 1990–2009

A. W. King[1], R. J. Andres[1], K. J. Davis[2], M. Hafer[3], D. J. Hayes[1], D. N. Huntzinger[4], B. de Jong[5], W. A. Kurz[3], A. D. McGuire[6], R. Vargas[7], Y. Wei[1], T. O. West[8], and C. W. Woodall[9]

[1]Environmental Sciences Division and Climate Change Science Institute, Oak Ridge National Laboratory, Oak Ridge, Tennessee, USA
[2]Department of Meteorology, The Pennsylvania State University, University Park, Pennsylvania, USA
[3]Canadian Forest Service, Natural Resources Canada, Victoria, British Columbia, Canada
[4]School of Earth Sciences and Environmental Sustainability, Northern Arizona University, Arizona, USA
[5]El Colegio de la Frontera Sur, Unidad Campeche, Campeche, Mexico
[6]US Geological Survey, Alaska Cooperative Fish and Wildlife Research Unit, University of Alaska, Fairbanks, Alaska, USA
[7]Department of Plant and Soil Sciences, University of Delaware, Newark, Delaware, USA
[8]Joint Global Change Research Institute, Pacific Northwest National Laboratory, College Park, Maryland, USA
[9]Northern Research Station, USDA Forest Service, Saint Paul, Minnesota, USA

Correspondence to: A. W. King (kingaw@ornl.gov)

Abstract. Scientific understanding of the global carbon cycle is required for developing national and international policy to mitigate fossil fuel CO_2 emissions by managing terrestrial carbon uptake. Toward that understanding and as a contribution to the REgional Carbon Cycle Assessment and Processes (RECCAP) project, this paper provides a synthesis of net land–atmosphere CO_2 exchange for North America (Canada, United States, and Mexico) over the period 1990–2009. Only CO_2 is considered, not methane or other greenhouse gases. This synthesis is based on results from three different methods: atmospheric inversion, inventory-based methods and terrestrial biosphere modeling. All methods indicate that the North American land surface was a sink for atmospheric CO_2, with a net transfer from atmosphere to land. Estimates ranged from −890 to $-280\,\mathrm{Tg\,C\,yr^{-1}}$, where the mean of atmospheric inversion estimates forms the lower bound of that range (a larger land sink) and the inventory-based estimate using the production approach the upper (a smaller land sink). This relatively large range is due in part to differences in how the approaches represent trade, fire and other disturbances and which ecosystems they include. Integrating across estimates, "best" estimates (i.e., measures of central tendency) are $-472 \pm 281\,\mathrm{Tg\,C\,yr^{-1}}$ based on the mean and standard deviation of the distribution and $-360\,\mathrm{Tg\,C\,yr^{-1}}$ (with an interquartile range of −496 to −337) based on the median. Considering both the fossil fuel emissions source and the land sink, our analysis shows that North America was, however, a net contributor to the growth of CO_2 in the atmosphere in the late 20th and early 21st century. With North America's mean annual fossil fuel CO_2 emissions for the period 1990–2009 equal to $1720\,\mathrm{Tg\,C\,yr^{-1}}$ and assuming the estimate of $-472\,\mathrm{Tg\,C\,yr^{-1}}$ as an approximation of the true terrestrial CO_2 sink, the continent's source : sink ratio for this time period was 1720 : 472, or nearly 4 : 1.

1 Introduction

Only about 45 % of the carbon dioxide (CO_2) released to the atmosphere by global human activities since 1959 (including the combustion of fossil fuels, cement manufacturing and deforestation and other changes in land use) has been retained by the atmosphere (calculated from data in Le Quéré et al., 2013). The remainder has been absorbed by the ocean and terrestrial ecosystems. Given observations of the increase in atmospheric CO_2, estimates of anthropogenic emissions and models of oceanic CO_2 uptake, it is possible to estimate CO_2 uptake by the terrestrial biosphere (i.e., the land sink) as the residual in the global carbon budget (Le Quéré et al., 2013). Le Quéré et al. (2013) thus estimated the mean global land sink for 2002–2011 to be 2.6 ± 0.8 Pg C yr^{-1}. Within the uncertainty of the observations, emissions estimates and ocean modeling, this residual calculation is a robust estimate of the global land sink for CO_2. However, both scientific understanding and policy considerations require more detail than is afforded by a global estimate since the magnitude, spatial pattern and temporal dynamics of the land sink vary considerably at continental and regional scales. Considerations of national and international policy to mitigate climate change by managing net terrestrial carbon uptake must account for this spatial and temporal variability. To do so requires more spatially refined estimates along with an improved understanding of the major controlling factors and underlying ecosystem processes.

The REgional Carbon Cycle Assessment and Processes (RECCAP) project is an effort at regional refinement of terrestrial (and ocean) carbon fluxes based on a synthesis of multiple constraints (Canadell et al., 2011). An international activity organized under the auspices of the Global Carbon Project (GCP; Canadell et al., 2003; http://www.globalcarbonproject.org), the objective of RECCAP is "...to establish the mean carbon balance and change over the period 1990–2009 for all subcontinents and ocean basins" (Canadell et al., 2011, p. 81). RECCAP aims to achieve this objective through a series of regional syntheses designed to "...establish carbon budgets in each region by comparing and reconciling multiple bottom-up estimates, which include observations and model outputs, with the results of regional top-down atmospheric carbon dioxide (CO_2) inversions." Beyond the more spatially (regionally) refined estimates of carbon flux and processes, "[t]he consistency check between the sum of regional fluxes and the global budget will be a unique measure of the level of confidence there is in scaling carbon budgets up and down".

The objective of this study is a synthesis of net land–atmosphere CO_2 exchange for North America combining different approaches (i.e., atmospheric inversion, inventory-based methods and terrestrial biosphere modeling) over the period 1990–2009. The North American land area ($21\,748 \times 10^6$ km^2; Canada: 9985×10^6 km^2; USA (including Alaska, excluding Hawaii): 9798×10^6 km^2; Mexico: 1964×10^6 km^2) is approximately 16 % of the global land area (excluding Greenland and Antarctica). North America's net land–atmosphere exchange is thus a potentially important fraction of the global land sink for atmospheric CO_2. In 2013, fossil fuel and cement CO_2 emissions from North America (Canada, United States and Mexico combined) were second only to those from China (Le Quéré et al., 2014). Quantifying North America's net land–atmosphere CO_2 exchange, potentially offsetting at least a portion of North America's CO_2 emissions, is an important element of understanding and quantifying North America's contribution to the accelerating increase in atmospheric CO_2 concentrations (Le Quéré et al., 2014). Our approach was guided by (a) Canadell et al. (2011); (b) RECCAP syntheses for other regions (Dolman et al., 2012; Gloor et al., 2012; Haverd et al., 2013; Luyssaert et al., 2012; Patra et al., 2013; Piao et al., 2012; Valentini et al., 2014); (c) guidelines found at the RECCAP website (http://www.globalcarbonproject.org/reccap/); and (d) personal communications with J. G. Canadell (2013) as Coordinator of the RECCAP Science Steering Committee. This study focuses on estimates of land–atmosphere CO_2 exchange over Canada, the United States and Mexico. Although the inventory approaches included in this study are based on total carbon changes, we do not report flux estimates of other carbon gases, such as methane and carbon monoxide or N_2O, and other greenhouse gases. This study is a synthesis of the net contribution of the North American land surface to atmospheric CO_2 concentrations and is neither a carbon nor greenhouse gas budget for the region.

2 Methods

We estimated the annual net land–atmosphere exchange of CO_2-C (Tg C yr^{-1}) for North America using results from three different approaches to estimating carbon budgets over large areas: atmospheric inversion modeling, empirical modeling using inventory data and terrestrial biosphere modeling. For each method, we provide estimates for the 1990–1999 and 2000–2009 decades and the entire 20-year 1990–2009 period. We follow the convention that negative values of the estimated net land–atmosphere exchange represent net uptake of CO_2-C by the land surface (predominately in vegetation and soils) or a sink for atmospheric CO_2. Positive values thus represent a net release from the land to the atmosphere or a source of atmospheric CO_2. Lateral flows of carbon as they ultimately influence vertical exchange with the atmosphere, including the trade of grain, wood and fiber, are an important consideration in interpreting and comparing results from each of the approaches. The respective treatments of lateral fluxes in each of the approaches are discussed in the corresponding sections below. More generally, the different approaches include and exclude different contributions to the net land–atmosphere exchange (Fig. 1). Those differences are likewise important in interpreting and comparing results and are described in the respective sections. Here we focus on reporting results aggregated for North America; country-level breakdowns of the three approaches can be found in Hayes et al. (2012) for the 2000–2006 time period.

2.1 Atmospheric inversion models (AIMs)

The methods of atmospheric inversion modeling have been described previously in detail by Enting (2002), Gurney et al. (2008, 2003, 2002), Baker et al. (2006), Peters et

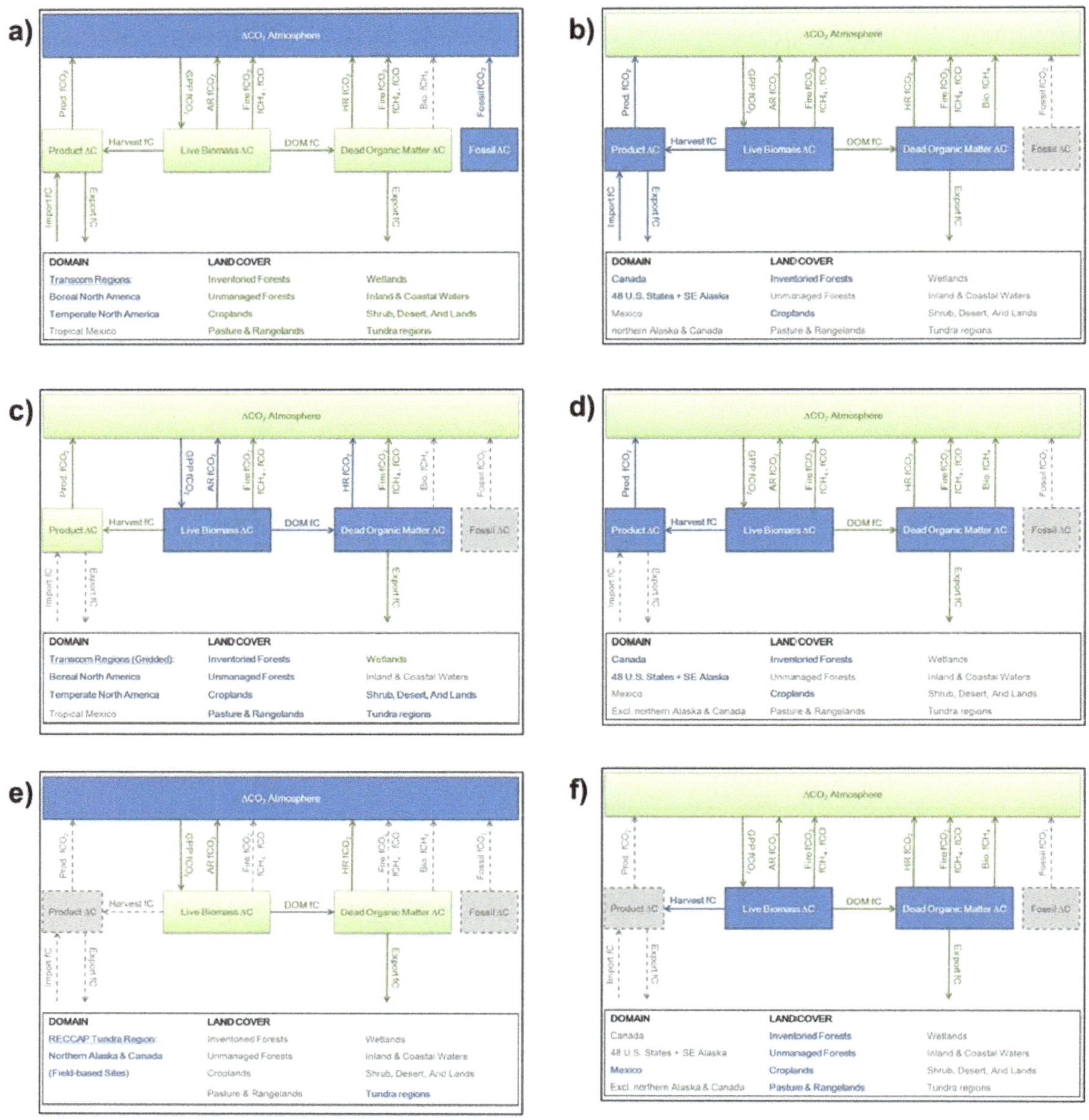

Figure 1. Carbon dioxide budget diagrams illustrating the spatial domains and component fluxes included in each approach and data set synthesized in this study: **(a)** atmospheric inversion models (AIMs), **(b)** atmospheric flow inventory, **(c)** terrestrial biosphere models (TBMs), **(d)** production approach inventory, **(e)** tundra ecosystem flux measurement and **(f)** Mexico land-use change (default approach) inventory. In each diagram, flux components are shown in blue when explicitly estimated (i.e., observed, measured or simulated), in green when implicitly contributing to an aggregated flux but not estimated directly, and in gray when explicitly not included in the estimate. Atmospheric methods **(a, e)** measure the concentration or flux of CO_2 in the atmosphere, which implies all land–atmosphere CO_2 exchange components (and excludes non-CO_2 fluxes). AIMs **(a)** integrate CO_2 concentrations for large regions (boreal and terrestrial North America) and explicitly subtract the contribution of fossil fuel emissions in order to quantify the terrestrial contribution. The eddy covariance flux measurements for the tundra region **(e)** are similar in concept but are site-based and so are not influenced by fire, fossil or harvested-product emissions. Inventory approaches **(b, d, f)** are primarily based on carbon stock change estimates in the major live biomass and dead organic-matter pools. Mostly implicit in the inventories, then, are the fluxes in and out of these pools, with the exception of harvested-carbon (crop and wood) removals that need to be tracked to determine the role of product consumption and decay emissions in the overall budget. The atmospheric flow approach **(b)** considers product imports and exports from international trade in calculating the stock change in the product pool, whereas the production approach **(d)** does not. The default approach **(f)** excludes the harvested-product pools from the accounting. Finally, there is large variation in how TBMs **(c)** explicitly simulate, implicitly include or explicitly exclude the various flux components; here, we represent a "basic case", where all models simulate ecosystem production and respiration and track the major pools. TBMs differ widely, though, as to whether and how they simulate fire, harvest, product emission and dead organic-matter export fluxes (i.e., riverine export). None of the models in this study include estimates of fossil fuel emissions, biogenic methane flux or the lateral transfer of product carbon via international trade.

Table 1. Mean ±1 standard deviation (σ) of annual net land–atmosphere exchange of CO_2-C ($Tg\,C\,yr^{-1}$) for North America by decade and for the 1990–2009 period.

Method	1990–1999	2000–2009	1990–2009
Atmospheric inversion[a]	−929 ± 477	−890 ± 400	−890 ± 409
Inventory: atmospheric flow approach[b]	−159	−348	−356
Terrestrial biosphere modeling[c]	−370 ± 138	−359 ± 111	−364 ± 120
Inventory: production approach[b]	−83	−270	−280
Best estimates			
Mean ± σ	−385 ± 382	−467 ± 285	−472 ± 281
Median (interquartile range)	−264 (−510 to −140)	−354 (−492 to −328)	−360 (−496 to −337)
Mode	> −500 < 0	> −400 < 0	> −400 < 0

[a] The multi-model mean and standard deviation of the time-period means of the RECCAP-selected TransCom3 inversions of Peylin et al. (2013).
[b] See Methods section. Note that there is single inventory estimate and thus no "multi-model" mean or standard deviation.
[c] The multi-model mean and standard deviation of the time-period means of 10 RECCAP-Trendy models' time-averaged annual NBP (see Methods section).

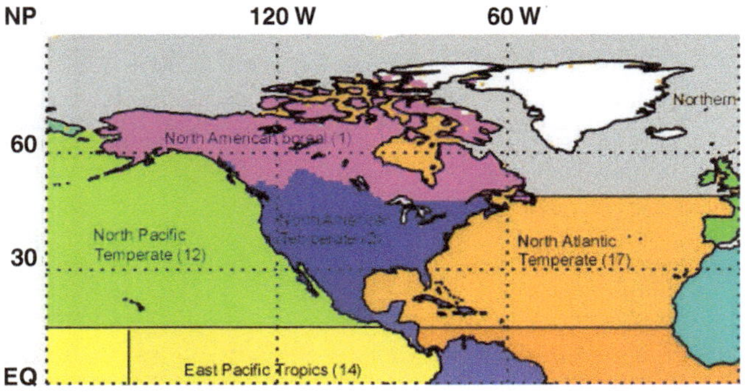

Figure 2. TransCom3 regions of the western Northern Hemisphere (Baker et al., 2006). The combined North American Boreal and North American Temperate regions define North America for the atmospheric inversion model (AIM) and terrestrial biosphere model (TBM) approaches to estimating net land–atmosphere carbon exchange for North America. Adapted from http://transcom.project.asu.edu/transcom03_protocol_basisMap.php.

al. (2007), Butler et al. (2010), Ciais et al. (2011) and others. As summarized by Hayes et al. (2012), AIMs combine data from an observation network of atmospheric CO_2 concentrations with models of surface CO_2 flux and atmospheric transport to infer, from an inversion process, the net land–atmosphere exchange of CO_2-C. Because they provide an integrated estimate of all CO_2 sources and sinks (over a given land area and time period) from the atmospheric perspective, inversions are sometimes referred to as a top-down approach (Canadell et al., 2011; Schulze et al., 2009). In estimating net land–atmosphere exchange, the influence of fossil fuel emissions is assumed to be well-known, and their influence is removed from the problem prior to solving for non-fossil fluxes (Peylin et al., 2013; Schulze et al., 2010). As our primary source, we use the 11-model ensemble of RECCAP-selected TransCom3 (Atmospheric Tracer Transport Model Intercomparison Project Phase 3) inversions (Peylin et al., 2013). The individual models are identified in Table 1 (p. 6703) of Peylin et al. (2013). North America here is defined by the combination of TransCom3 (Baker et al., 2006) regions "North American Boreal" and "North American Temperate" (Fig. 2).

2.2 Terrestrial biosphere models (TBMs)

Terrestrial biosphere modeling employs a model of terrestrial ecosystem carbon dynamics deployed on a geospatial grid to simulate the exchange of carbon with the atmosphere, primarily as CO_2 (Hayes et al., 2012; Huntzinger et al., 2012; Schwalm et al., 2010). The models differ in which ecosystem processes they include and how they conceptually and mathematically represent them. Some, for example, include carbon release to the atmosphere from fire and other disturbances; others do not (see Hayes et al., 2012; Huntzinger et al., 2012). In order to estimate the net land–atmosphere exchange of CO_2 with TBMs, the models must minimally include the processes of CO_2 uptake from the atmosphere in gross primary production (GPP) and the release of CO_2 to the atmosphere in ecosystem respiration (Re), whether separated into autotrophic (Ra) and heterotrophic (Rh) respiration ($Re = Ra + Rh$) or not. Net primary production (NPP) is the balance between GPP and Ra ($NPP = GPP - Ra$). Net ecosystem production (NEP) is the balance between GPP and Re ($NEP = GPP - Re$ or, equivalently, $NEP = NPP - Rh$). Net biome production (NBP) is defined by Schulze et al. (2000) as NEP minus nonrespiratory losses such as fire and harvest. It is defined by Chapin et al. (2006) as the net ecosystem carbon balance (NECB) estimated on large temporal and spatial scales (where NECB is the net rate of organic and inorganic C gain by or loss from an ecosystem) and by RECCAP as NEP plus and/or minus all vertical and horizontal fluxes in and out of an ecosystem. NEP is a subcomponent of net ecosystem exchange (NEE) which is "the net vertical exchange of CO_2 between a specified horizontal surface and the atmosphere above it over a given period of time" (Hayes and

Turner, 2012). NEE is equivalent to the net land–atmosphere exchange of CO_2. However, NEP is often the only net exchange with the atmosphere simulated by TBMs (Hayes et al., 2012; Huntzinger et al., 2012). Thus NEP for these models is, with the sign reversed, a minimal approximation of NEE or the net land–atmosphere exchange of CO_2. When the processes of CO_2 release from fire, land cover change or other disturbances are included in the model (as in NBP), the approximation of net land–atmosphere exchange is even closer. It should be noted, however, that while some TBMs include CO_2-C loss from fire, very few, if any, include the trade and lateral transport of harvested-wood or agricultural products and their subsequent release of CO_2 or the influence of insect outbreaks. These models, as a class, also generally ignore CH_4 emissions from livestock and N_2O emissions from agriculture. However, these absences do not impact our estimate of net land–atmosphere CO_2 exchange from these models

Our source for results from TBMs was Version 2 of the 10-model ensemble of the GCP/RECCAP-trendy activity (http://www-lscedods.cea.fr/invsat/RECCAP/V2/). The models in this ensemble are identified as dynamic global vegetation models (DGVMs), a subset of the larger class of TBMs (Sitch et al., 2008). We used the net biosphere production (NBP) from these models, which includes GPP, Re, and fire emissions, as the near equivalent of NEE approximating the net land–atmosphere exchange of CO_2-C. We extracted the results for North America from these global models, with North America defined by the North American Boreal and North American Temperate regions of Transcom3 (Fig. 2) (Baker et al., 2006).

2.3 Inventory-based methods

Inventory-based methods for estimating net land–atmosphere CO_2 exchange use a combination of field survey, disturbance and land-use and management data, collectively referred to as "activity data", to estimate net carbon emissions over time (IPCC, 2006). In general, repeated measurements and activity data are used to estimate changes in carbon stocks over time, and in this study CO_2 exchange with the atmosphere is inferred from these changes by decomposing them into additions and losses of carbon among the major pools (Hayes et al., 2012; Pan et al., 2011). The inventory-based flux estimates are based on a calculation that includes both the change in ecosystem carbon stocks (from live biomass and dead organic-matter pools) and the change in stocks from product pools and that considers the fate of carbon harvested from the ecosystem as a result of anthropogenic land management and use. Whether, how, where and when carbon stock changes in product pools, including those resulting from trade, are considered as sources or sinks depends on the accounting approach. The different "approaches" represent variations on the conceptual framework for reporting land–atmosphere CO_2 emissions and removals in greenhouse gases inventories. Within each approach, there can be different "methods" based on the underlying data sets and calculations used to estimate these emissions and removals. The inventory-based accounting approaches are conceptually similar and follow common guidelines, though the details of the methods differ by country (i.e., Canada, the USA and Mexico) and sector (e.g., forest lands and crop lands).

For a comparison with estimates from the TBMs and AIMs, here we report net land–atmosphere exchange of CO_2 from inventories using two different accounting approaches: the "production approach" and the "atmospheric flow approach", which differ in where and when the emissions of carbon from harvested products are assigned (IPCC, 2006). The production approach assigns product emissions to the producing country (i.e., the country in which the carbon was harvested). The atmospheric flow approach assigns product emissions to the consuming country, based on stock change in the domestic product pool after adjusting for international imports and exports of harvested products. In both cases, the stock change estimates for harvested-wood product (HWP) pools include "inherited emissions" from products harvested prior to our time period of analysis. In crop lands, the change in harvested-crop product (HCP) pools is zero on an annual basis, so only the adjustment for international imports and exports influences the sink/source estimates (and only when using the atmospheric flow approach). The exception is in our estimates for Mexico, where data on neither carbon stock changes nor the fate of harvested products are currently available to researchers (Vargas et al., 2012). For Mexico we therefore use the "default approach" (IPCC, 2006), which assumes no change in the product pools, and so only carbon stock changes resulting from forest growth, deforestation and reforestation/afforestation are included. As such, we calculate only one inventory-based estimate for Mexico, but we add this same estimate to the continental totals in both the production and atmospheric flow approaches.

The two approaches are complementary in terms of assessing the role of a particular country/sector in the global carbon budget both spatially and temporally. The distinction between the two is important in terms of a comparison with other scaling approaches (Hayes et al., 2012). In general, most TBMs essentially employ the production approach where, if they consider harvested products at all, product carbon is typically assumed to be emitted from within the same grid cell as it was harvested. Thus, stock change estimates using the production approach are more appropriate for comparing inventory-based estimates with those of TBMs. On the other hand, we calculate an inventory-based flux estimate using the atmospheric flow approach as the more appropriate comparison with the AIMs. As they are based on atmospheric CO_2 observations combined with a transport model, AIMs should – in theory – detect a sink where the carbon was originally taken up in vegetation and a source where and when the product carbon is ultimately returned to the atmosphere through consumption or decay. These fluxes may, however, be below detection levels with current AIM technologies.

We used activity data based on national GHG (greenhouse gas) inventories from Canada and the USA to estimate the contribution of forestlands to the net land–atmosphere exchange of CO_2-C for North America. Per IPCC Guidelines (IPCC, 2006), only "managed" forest lands are considered in the inventories, which excludes a large area of forest primarily in the boreal zone (i.e., the northern extent of Canada's forested area as well as interior Alaska). The Canada forest inventory uses the "gain–loss" methodology, which starts with data from a compiled set of inventories of forest carbon pools, which are then modeled forward based on the components of change, including growth, soil C respiration, natural disturbance and forest harvest (Kurz et al., 2009; Stinson et al., 2011). For the USA, forest carbon stock and stock change estimates are based on the "stock change" methodology using repeated measurements in a design-based forest inventory (Bechtold and Patterson, 2005; Smith et al., 2013; USDA Forest Service, 2013). Aboveground standing-tree (both live and dead) carbon pools are directly estimated from allometric equations (Woodall et al., 2011) of individual trees measured across the national plot network, while all other forest pools are estimated from models applied at the plot-level based on specific forest attributes (Smith et al., 2013, 2006; USEPA, 2012).

Both the production and atmospheric flow approaches were used to estimate contributions of HWP to Canadian and USA carbon fluxes. In the atmospheric flow estimate for the USA, the HWP stock change calculations from the production approach (Skog, 2008) were adjusted for both imports and exports from international trade (USEPA, 2012). For Canada, however, the atmospheric flow estimate includes only exports; HWP imports to Canada are known to be very small relative to exports and are not tracked. As noted above, data on changes in HWP are not available for Mexico, and therefore the contribution of HWP is not part of the estimate of carbon fluxes for Mexico.

The estimates of net land–atmosphere CO_2 exchange from cropland in Canada and the USA are based on carbon stock change in agricultural soils and on imports and exports of agricultural commodities. Annual carbon flux from the herbaceous biomass in harvested crops is considered to be net zero because of the fast turnover time (decay and consumption) of this pool, with the exception of the transfer of residue carbon to soils, and the amount of carbon removed in HCP and exported from the region. In the case of agricultural soils, annual soil carbon stock change is estimated directly from activity data since soil carbon stocks are not commonly reported (West et al., 2011). Data on carbon stock change in crop land soils from Canada (Environment Canada, 2013) and the USA (West et al., 2011) were used, and estimates of carbon in HCP imports and exports were available from each country (Canadian Socio-Economic Information Management System, Statistics Canada and Foreign Agricultural Trade of the United States, USDA Economic Research Service).

The contribution of lands in Mexico to the continental estimates of net land–atmosphere CO_2 exchange is derived from that country's Fifth National Communication to the United Nations Framework Convention on Climate Change (SEMARINAT/INECC, 2012). The data represent the carbon accounting for the land use, land-use change and forestry (LULUCF) sector and include estimates of carbon emissions and removals resulting from changes in biomass, the conversion of forests and grasslands to agricultural use, the abandonment of farmland, and carbon stock changes in mineral soils. These estimates use the default accounting approach based on a gain–loss method, where mean carbon stock density by land cover type is distributed according the areal extent of each type at an initial point in time, and stock change is estimated according to the area of land-use change over a subsequent period of time (de Jong et al., 2010).

To these forest land and crop land estimates we also added the estimates of net land–atmosphere CO_2 exchange for the "tundra" region of North America (i.e., Alaska and northern Canada), as reported in the study by McGuire et al. (2012). That study also included modeled estimates, but here we used a synthesis of the observations as analogous to an "inventory" of that region's carbon fluxes. While we add estimates for this large region from an existing study, our continental total estimates do not otherwise include land–atmosphere exchanges from other ecosystem types for which inventories were not available (e.g., arid lands, grasslands, temperate wetlands, shrublands or areas of woody expansion into tundra, and grassland areas previously not forested and not meeting the definition of managed forest). Arid lands generally have low carbon stocks, but in wet years or decades could be an additional sink (Poulter et al., 2014) or source (Thomey et al., 2011) missed by the general exclusion of these lands from inventories. Similarly, a potential contribution to the North American sink is missed by the absence from the national inventories of woody encroachment into previously non-wooded lands (Hayes et al., 2012; King et al., 2012).

2.4 Estimating decadal mean net land–atmosphere exchange

For each of the multi-model approaches (AIMs and TBMs) we first estimated for each decade and the entire 1990–2009 period ($n = 10$ and 20, respectively) the mean and population standard deviation (σ) of each model's time series of annual net exchange for North America. The standard deviation, describing the variability of annual values about the decadal or period mean, is an index of the model's interannual variability for the period. We then averaged the model-specific time averages and standard deviations to estimate the multi-model mean and population standard deviation for each ensemble ($n = 10$ for the AIM ensemble, and $n = 10$ for the

TBM ensemble) for each decade and the entire 1990–2009 period. The resulting multi-model means are the estimate of net land–atmosphere exchange of CO_2-C for each method and time period. There are different opinions of how to best characterize "uncertainty" in CO_2 flux estimates, whether to use, for example, the standard deviation, standard error, 95 % confidence intervals, inter-percentile/quartile ranges, or semiquantitative characterizations such as that used by the IPCC in communicating confidence in scientific findings. For comparison with other RECCAP regional syntheses, we followed Luyssaert et al. (2012) and Ciais et al. (2010) in using the population standard deviation of the multi-model means as a metric of the uncertainty (i.e., variability) in the multi-model estimates.

The two inventory-based estimates (the production approach and the atmospheric flow approach) are both derived from the three regional source data sets (the land carbon stock inventories of Canada, the United States and Mexico). There is no multi-inventory ensemble from which to estimate across-inventory means and standard deviations. The apparent interannual changes in stocks of the USA and Mexico confound inventory uncertainty with actual year-to-year variations in changes in stocks and are unlikely to be a reliable estimate of interannual variability in net exchange with the atmosphere. The Canadian GHG inventory does use annual information on harvest, natural disturbances and land-use change (Stinson et al., 2011), and thus interannual variability resulting from activity data is reflected in those estimates. They do not, however, include changes due to interannual variation (or long-term trends) in atmospheric chemistry and climate. Similarly, the inventories' exclusion of arid lands and range lands means that these approaches also miss interannual variation associated with temporal patterns of precipitation in those regions (Poulter et al., 2014). Accordingly, we estimate net land–atmosphere exchange of CO_2-C from the inventory-based approaches using a single value, the time-averaged mean for each period, and do not report the time-averaged standard deviation either as an index of interannual variability or as a measure of uncertainty.

2.5 Fossil fuel emissions

We also estimated the fossil fuel source for North America to characterize the land sink relative to fossil fuel emissions (King et al., 2007a) or the continent's source-to-sink ratio (King et al., 2012). Estimates were made following Andres et al. (2012) using data from Boden et al. (2013). As with the inventories, we combined emissions data from Canada, the United States and Mexico to estimate North American emissions.

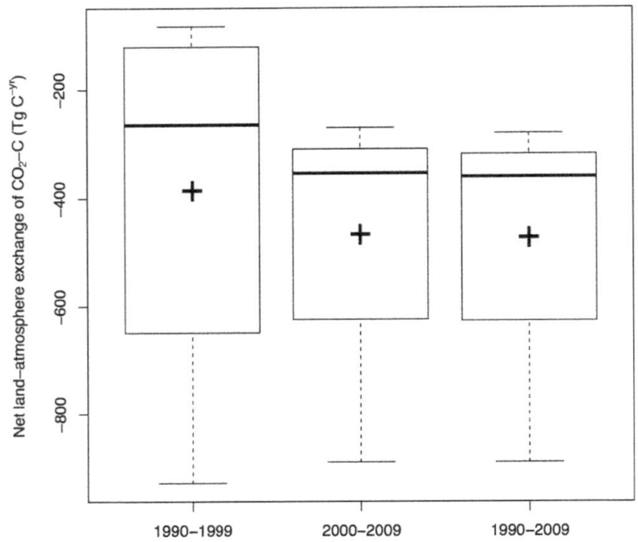

Figure 3. Box-and-whisker diagrams of the estimates from the different methods. The bold horizontal line indicates the median, the + the mean. The upper and lower bounds of the box are the "hinges" of the Tukey box-and-whisker algorithm of R's boxplot and approximate the interquartile range. The whiskers indicate the minimum and maximum values.

3 Results

Table 1 compares the estimates of average annual net land–atmosphere exchange of CO_2-C for North America across the different methods. Table 2 compares the interannual variability. Most notable in Table 1 is the substantially larger estimate for the continental land sink (negative net land–atmosphere CO_2 exchange) from the atmospheric inversions as compared to the estimates from the other methods. The difference is on the order of at least a factor of 2 or more. This pattern has been noted before, most recently in the syntheses of Hayes et al. (2012), Huntzinger et al. (2012) and King et al. (2012).

Because we consider the estimates from the three different methods (Table 1) to all be scientifically credible, the central tendency of the distribution of those estimates can, by synthesizing or integrating across the estimates, provide some indicators of best estimates. Unfortunately the small sample size ($n = 4$) and the asymmetry or skew introduced by the atmospheric inversion estimate (Fig. 3) makes the arithmetic mean and standard deviation across the methods an unreliable estimate of the central tendency and spread in the estimates. However, because the mean is so commonly used to integrate across estimates, we report the across-method mean ± 1 sample standard deviation (σ) in Table 1. The median and interquartile range as measure of central tendency and spread of such a skewed distribution are perhaps a more appropriate best estimate (Table 1 and Fig. 3). The small sample size makes calculation of the mode (i.e., the most frequent/likely value) difficult or a misleading estimate of central tendency.

Table 2. Interannual variability of annual net land–atmosphere exchange of CO_2-C ($Tg\,C\,yr^{-1}$) for North America by decade and for the 1990–2009 period. The population standard deviation (σ) of annual exchange is used as an index of interannual variability.

Method	1990–1999	2000–2009	1990–2009
Atmospheric inversion[a]	316 ± 156	368 ± 115	364 ± 129
Terrestrial biosphere modeling[b]	218 ± 73	250 ± 52	239 ± 58
Best estimates			
Mean $\pm \sigma$	267 ± 69	309 ± 83	302 ± 88
Median (interquartile range)[c]	267 (242 to 292)	309 (280 to 338)	302 (270 to 333)

[a] The multi-model mean ($\pm 1\sigma$) of individual within-model standard deviations from the time-averaged (see Table 1) atmospheric inversion estimates of net land–atmosphere exchange (see Methods section) for each time period for the RECCAP-selected TransCom3 AIMs models (Peylin et al., 2013).
[b] The multi-model mean ($\pm 1\sigma$) of individual within-model standard deviations from the time-averaged annual NBP (Table 1 and Methods section) for each time period for 10 RECCAP-Trendy models.
[c] With only two estimates there is no asymmetry in the distribution as evidenced by the equivalence of mean and median; likewise there is no mode.

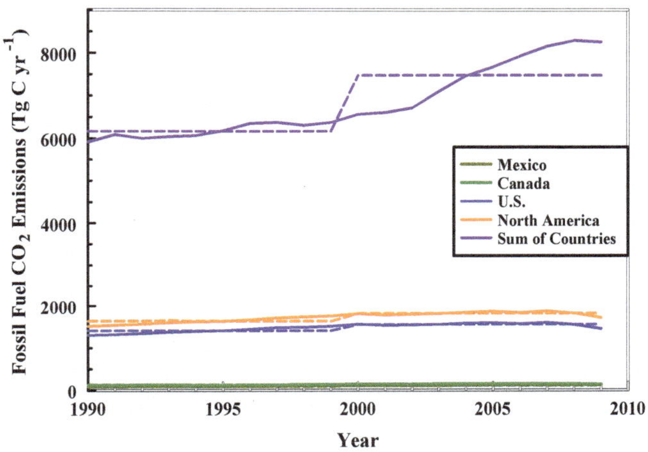

Figure 4. Fossil fuel CO_2 emissions for various political units. Solid lines represent annual emissions, and dashed lines represent the decadal mean of emissions. The sum of countries is used to represent total global emissions in this plot. This allows comparison of emissions on an equal basis as all emissions are based on apparent consumption data and not production data (see Andres et al. (2012) for a fuller discussion of the differences). The global values used here are less than those in the CDIAC (Carbon Dioxide Information Analysis Center) archive (http://cdiac.esd.ornl.gov/trends/emis/tre_glob_2010.html), mainly due to the exclusion of bunker fuels. Data from Boden et al. (2013).

However, inspection and a simple histogram of the estimates suggest a modal estimate of $<400\,Tg\,C\,yr^{-1}$ as an alternative, if imprecise, across-method estimate for 1990–2009.

Results in Table 2 are suggestive of some tendency for an increase in interannual variability in net land–atmosphere exchange in the 2000–2009 decade relative to the preceding 1990–1999 decade. However, given the relatively short 10-year spans and intradecadal variability, any apparent trend should be considered cautiously and the standard deviation for the entire 20-year period a sounder indicator of interannual variability in North America's terrestrial sink. Across approaches, the atmospheric inversions show somewhat greater interannual variability than the TBMs (Table 2). Raczka et al. (2013) similarly showed that TBMs consistently underestimated the amplitude of interannual variability with respect to flux tower records across North America.

Figure 4 displays the fossil fuel CO_2 emissions for the three countries, their sum and the sum of all countries around the world (i.e., global emissions). Solid lines represent annual emissions and dashed lines represent the decadal mean of emissions. For most political units shown, the decadal means well represent the annual emissions on this scale. Only for global emissions, especially in the latter decade, is the decadal mean a poor representation of the annual emissions. Emissions from Mexico and Canada are too similar in magnitude to be easily discernible from each other in this figure.

Table 3 displays the numerical details of Fig. 4 as well as relative percentages of smaller political units to larger political units. In terms of mass emitted globally in the calendar year of 2010, out of 216 countries, the USA is the second largest emitter, Canada is ranked at no. 9 and Mexico is ranked at no. 13. Prior to 2006, USA emissions ranked at no. 1; thereafter China has had the largest emissions (Global Carbon Atlas, 2014; Le Quéré et al., 2014). In 2010, North America as a whole is ranked at no. 2 behind China. For the period 1990–2009, uncertainty (in $Tg\,C\,yr^{-1}$) was higher in Mexico ($\sim 10\,\%$ of mean), lower for Canada ($\sim 2\,\%$ of mean) and substantially lower in the USA ($\sim 0.02\,\%$ of the mean) (Table 3).

Table 4 is as Table 1 but with the entries replaced by the estimates of the terrestrial sink as a percentage of North American fossil fuel emissions. These proportions range across methods and decades from nearly 60 % to as low as 5 %, with a best estimate of perhaps 20–30 %. There is no clear decadal trend in the sink as a proportion of fossil fuel emissions; some methods suggest an increase, others a decrease, and, with the exception of the inventory-based estimates, the changes are small. However, again, as in Table 2, the relatively short record means any apparent change over time in

Table 3. Mean, standard deviation, uncertainty and relative percentage of emissions for various political units and years. The standard deviation of the time-averaged mean is indicated by σ. Uncertainty is our best assessment of how well we know the mean, integrating the variability of the data with knowledge of the quality of the data. North America's percentage of global total does not equal the sum of its components due to rounding. Flux data from Boden et al. (2013); uncertainty estimate from Andres (unpublished data).

	Years	Mean (Tg C)	σ (Tg C)	Uncertainty (Tg C)	Emissions (%) of N. America	Emissions (%) of global total
	1990–1999	129.34	6.42	2.59		
Canada	2000–2009	147.75	4.51	2.95	8	2
	1990–2009	138.54	10.75	2.77		
	1990–1999	93.54	5.75	9.45		
Mexico	2000–2009	115.47	7.92	11.66	6	2
	1990–2009	104.50	12.96	10.55		
	1990–1999	1404.90	69.42	28.10		
United States	2000–2009	1548.94	38.89	30.98	86	22
	1990–2009	1476.92	91.39	29.54		
	1990–1999	1627.78	80.11	34.95		
N. America	2000–2009	1812.16	43.44	39.41	100	25
	1990–2009	1719.97	112.48	37.18		
	1990–1999	6169.80	162.90	203.72		
Global	2000–2009	7471.66	653.98	271.50	–	100
	1990–2009	6820.73	806.73	237.61		

the sink strength relative to fossil fuel emissions should be considered cautiously and should not be considered significant, statistically or otherwise.

Table 5 is as Table 1 but with the entries replaced by the estimates as a percentage of the global land sink estimated by difference to balance the global carbon cycle (Le Quéré et al., 2013). The average global net land–atmosphere exchanges are −2460, −2320 and $-2390\,\mathrm{Tg\,C\,yr^{-1}}$ for the periods 1990–1999, 2000–2009 and 1990–2009, respectively. While this is a crude comparison because the global terrestrial sink is not thought to be uniformly dispersed geographically, the numbers in Table 5 around 15 % are in keeping with the approximately 16 % of the global land surface (minus Greenland and Antarctica) represented by North America (minus Greenland). North America is approximately 21 % of the Northern Hemisphere land surface. While the majority of the global land sink is likely in the Northern Hemisphere (Field et al., 2007), it is unlikely that the entire global sink is in the Northern Hemisphere. Nevertheless, the atmospheric inversion estimates of the North American sink at slightly less than 40 % of the global sink suggest a North American sink disproportional to North America's share of the Northern Hemisphere land surface. However, the across-method mean and mode estimates (Table 5) indicate a sink approximately proportional to North America's relative land area as part of the Northern Hemisphere.

4 Discussion and conclusions

All estimates of North America's net land–atmosphere exchange of CO_2-C synthesized in this study are negative values (Table 1), indicating a net exchange from atmosphere to land (i.e., net land uptake of CO_2-C). We therefore conclude, along with most previous assessments, that the vegetation and soils of North America were a sink for atmospheric CO_2 over the decades of 1990–2009. Our estimates of the net land sink for 1990–2009 range from as large as $-890 \pm 409\,\mathrm{Tg\,C\,yr^{-1}}$ (multi-model mean $\pm\,\sigma$) to as small as $-280\,\mathrm{Tg\,C\,yr^{-1}}$, with the estimates from atmospheric inversions and from the inventory-based production approaching the large and small ends of that range, respectively. The ranges for the decades 1900–1999 and 2000–2009 are -929 ± 477 to $-83\,\mathrm{Tg\,C\,yr^{-1}}$ and -890 ± 400 to $-270\,\mathrm{Tg\,C\,yr^{-1}}$, respectively. The atmospheric inversion and inventory-based production approach are again the high and low ends of those ranges. The State of the Carbon Cycle Report's (SOCCR) (King et al., 2007b) synthesis and assessment of the North American carbon cycle estimate of the North American terrestrial sink circa 2003 based on inventories was $-500\,\mathrm{Tg\,C\,yr^{-1}}$ with an uncertainty of ±50 %[1] (Pacala et al., 2007). Our inventory-based estimates are lower

[1]This is the range relative to the estimate of $-500\,\mathrm{Tg\,C\,yr^{-1}}$ which the authors were highly (95 %) confident included the actual value. This is not a coefficient of variation comparable to the standard deviation used in this paper as a measure of uncertainty (i.e., variability) surrounding a mean estimate. It is also not the 95 % con-

Table 4. Mean annual net land–atmosphere exchange of CO_2-C for North America by decade as a percentage of North American fossil fuel emissions (from Table 3). Note that these are independent proportions and do not add up to 100 %.

Method	1990–1999	2000–2009	1990–2009
Atmospheric inversion	57 %	49 %	52 %
Inventory: atmospheric flow approach	10 %	19 %	21 %
Terrestrial biosphere modeling	23 %	20 %	21 %
Inventory: production approach	5 %	15 %	16 %
Best estimates			
Mean	24 %	26 %	27 %
Median	16 %	20 %	21 %
Mode	<31 %	<28 %	29 %

than that of the SOCCR because, while our estimates include the contribution of tundra, they are based on forest and cropland inventories and exclude additional but highly uncertain sinks such as woody encroachment into previously non-woody ecosystems, wetland sinks and sequestration in rivers and reservoirs included in the SOCCR estimate. The SOCCR found woody encroachment to be a relatively large sink of $-120\,\mathrm{Tg\,C\,yr^{-1}}$, second only to the forest sink but with an uncertainty of >100 %. We feel justified in leaving these additional uncertain sinks out of inventory-based estimates until the uncertainty is reduced by further study. These additional sinks contribute, however, to the estimates from the AIMs and TBMs and may be partially responsible for their larger sink estimates relative to inventory-based estimates. A post-SOCCR assessment for circa 2000–2005 synthesizing atmospheric inversion, TBM and inventory-based approaches estimated a North American land sink of -634 ± 165 [2] $\mathrm{Tg\,C\,yr^{-1}}$ (King et al., 2012). Our best estimate for 2000–2009 based on the average across methods is -472 ± 281 (mean $\pm\sigma$) (Table 1). Our best estimate based on the median of the estimates from different methods is $-360\,\mathrm{Tg\,C\,yr^{-1}}$, with 68 % percent of the estimates (equivalent to the proportion represented by ± 1 standard deviation) in the range -638 to $-316\,\mathrm{Tg\,C\,yr^{-1}}$. Synthesizing across these syntheses, we conclude the North American land sink for the first decade of the 21st century was most likely in the range of -300 to $-600\,\mathrm{Tg\,C\,yr^{-1}}$ but with a relative uncertainty of $\pm$65–78 % to be highly (95 %) confident that the actual value lies within even that large range.

We have made no attempt to resolve temporal trends in the estimates of net land–atmosphere exchange due to the relatively short time frame. However, Kurz et al. (2008) found that Canada's managed forests switched from being a GHG sink to a source in 2002 as a result of large insect outbreaks, and those forests have been a carbon source for all but two (2008–2009) of the subsequent years (through 2012) (Environment Canada, 2014; Stinson et al., 2011). If there had been no change in disturbance in either the United States or Mexico over that period, the North American sink might be expected to decline between the decades of 1990–1999 and 2000–2009. There is perhaps some suggestion of a shift in that direction in the AIM estimates and perhaps the TBM estimates (Table 1), but the uncertainties are very large and any conclusion, as noted above, is tentative at best. Moreover, the inventory-based estimates suggest an increase in the sink (Table 1). Increases in natural disturbances (a declining sink) are offset by simultaneous decreases in harvest rates (an increasing sink), and these two opposing trends in the activity data may make it difficult to identify a clear overall trend in the CO_2 balance using inventory-based methods. Decadal changes in disturbance like those reported by Kasischke et al. (2013) likely influence the North American sink, but a clear definitive signal of that influence in the estimates given their uncertainties is elusive.

The North American land sink is only a fraction of the fossil fuel emissions from the region for that same period (Table 4). The source : sink ratio for the 1990–1999 decadal average ranges across methods from approximately 1628 : 83 (nearly 20 : 1, the estimate from inventories using the production approach) to as low as 1628 : 929 (nearly 2 : 1, the atmospheric inversion estimate). For the 2000–2009 decade that range is from 1812 : 270 (nearly 7 : 1) to 1812 : 890 (approximately 2 : 1), with the inventory-based production approach and atmospheric inversion approach again generating that range. For the entire 1990–2009 period that range is from 1720 : 280 (approximately 6 : 1) to 1720 : 890 (nearly 2 : 1). Based on best estimates of the land sink for that entire period, the ratio is in the range of 1720 : 360 (nearly 5 : 1) based on the median estimate and 1720 : 472 (nearly 4 : 1) based on the average estimate. In the SOCCR the North American source : sink ratio circa 2003 was estimated at approximately 3 : 1 (King et al., 2007a). King et al. (2012) also estimated a source : sink ration of approximately 3 : 1 for the period 2000–2005. The larger potential value of 4 : 1 reported here is attributable to a smaller estimate of the sink based on the median value of the multiple methods (Table 1). Considering both the fossil fuel emissions source and the land sink, North America was a net contributor to the growth of CO_2 in the atmosphere in the late 20th century and early 21st century, with emissions exceeding the land sink by at least a factor of 3.

Both methods (AIMs and TBMs) for which we could calculate the time-average standard deviation as a measure of interannual variability show greater variability in the 2000–2009 decade than in the previous decade. However, as noted in the Results section above, the relatively short record and the averaging by decade make us hesitant to draw any conclusions about changes in interannual variability from decade to decade for any of the approaches. A time series analysis of variability over a longer time period is likely needed to determine whether the North American land sink has been increasing or decreasing, and any such trend may well vary

fidence interval although it is more comparable to that measure of uncertainty than the standard deviation used here.

[2]Multi-method mean ± 1.96 standard error of the mean.

Table 5. Estimates of mean annual net land–atmosphere exchange of CO_2-C for North America by decade and for 1990–2009 as a proportion of the global mean annual net land–atmosphere exchange for those same periods.

Method	1990–1999	2000–2009	1990–2009
Atmospheric inversion	38 %	38 %	37 %
Inventory: atmospheric flow approach	6 %	15 %	15 %
Terrestrial biosphere modeling	15 %	15 %	15 %
Inventory: production approach	3 %	12 %	12 %
Best estimates			
Mean	16 %	20 %	20 %
Median	11 %	15 %	15 %
Mode	< 20 %	< 22 %	< 21 %

with approach. We can say, however, that the AIMs show larger variability than the TBMs (Table 2). Whether this is due to the inversions "seeing" variable net land–atmosphere exchanges not well represented in the TBMs or to some unidentified source of error in the AIMs is unclear. Findings by Poulter et al. (2014) showing the influence of Southern Hemisphere arid grasslands in wet years on interannual variation in the global carbon sink suggest that it may very well be the former. The work of Raczka et al. (2013) showing that TBMs systematically underestimate NEE relative to North American flux towers also points to the conclusion that AIMs are capturing interannual variability in net land–atmosphere CO_2 exchange not well represented by TBMs.

Different methods for estimating the net land–atmosphere exchange of CO_2 of North America continue to generate different estimates of that flux (Hayes et al., 2012; Huntzinger et al., 2012; King et al., 2012) as in this study. Although the different methods all attempt to estimate the same net land–atmosphere flux, the methods account for different components of that exchange (Fig. 1). The atmospheric inversions are influenced by all land–atmosphere exchanges. The TBMs only account for net exchange from those ecosystems and processes that they actually simulate, and the inventory-based estimates are limited to the ecosystems that are actually included in the inventories (e.g., managed forests, as defined by those responsible for the inventory, but not arid lands, grasslands, croplands, wetlands and other non-forest categories). These differences in fluxes captured by the different methods likely contribute to the different estimates.

Disturbance, natural and human, plays an important role in determining North America's net land–atmosphere CO_2 exchange (Kasischke et al., 2013; King et al., 2012). Indeed, much if not most of the early 21st century North American land sink can be attributed to the recovery of forests from earlier disturbance, primarily human clearing and harvesting in the United States (Goodale et al., 2002; Hayes et al., 2012; Huntzinger et al., 2012; King et al., 2012; Myneni et al., 2001; Pacala et al., 2007; Pan et al., 2011). On annual to decadal time scales, the contributions from disturbance are generally greater than those from enhanced GPP with rising atmospheric CO_2 or in response to variations in weather (Luyssaert et al., 2007). The variety of disturbance types, heterogeneity in the spatial and temporal characteristics of disturbance regimes and disturbance intensity, and the many ways in which disturbance can impact terrestrial ecosystem processes in North America (Kasischke et al., 2013) lead to complexity in quantifying the specific contribution of disturbance to net land–atmosphere exchange. The source–sink consequences of disturbance change over time (Amiro et al., 2010; Liu et al., 2011). For example, a forest fire releases CO_2 to the atmosphere during combustion (a source); the reduction in canopy results in an imbalance between GPP and Re, which can reduce the sink represented by a formerly aggrading forest or convert the landscape to a source, while Rh exceeds NPP with lags between Re and Rh (Harmon et al., 2011). Over time, as the forest recovers, NPP exceeds Rh, and the regrowing forest is a sink for atmospheric CO_2 (Kurz et al., 2013).

The three approaches for estimating net land–atmosphere CO_2 exchange differ in how they perceive or represent contributions from disturbance. Atmospheric inversion modeling captures the influence of disturbance contributions to patterns in atmospheric CO_2 concentrations but cannot generally attribute those changes to disturbances or disturbance types without additional effort involving carbon monoxide or other atmospheric gases, carbon isotopes or structured attribution analyses (Keppel-Aleks et al., 2014; Randerson et al., 2005). Inventory-based estimates capture the impact of disturbance on changes in carbon stock, but the carbon accounting might (e.g., the Canadian forest inventory) or might not (e.g., the USA and Mexico forest inventories) explicitly consider disturbances. In the USA, knowledge from other sources about areas burned (and other disturbances) can be used to inform GHG emissions estimates and allow for at least some attribution of specific disturbance to changes in carbon stocks even when disturbances are not explicitly accounted for. Terrestrial biosphere modeling can attribute land–atmosphere CO_2 exchange to specific disturbances but only those which the model explicitly represents, and the models differ considerably in which disturbance types they include and how they represent those disturbances and the consequences for CO_2 exchange with the atmosphere (Hayes et al., 2012; Huntzinger et al., 2012; Liu et al., 2011; Sitch et al., 2013). For example, some models include fire as an internal prognostic variable, others as an external forcing and some not at all (Huntzinger et al., 2012; Sitch et al., 2013). Incomplete representation or misrepresentation of disturbances by the TBMs likely contribute to differences between the TBM estimate and the AIM and inventory-based estimates. Williams et al. (2012) used information on age structure from USA forest inventory data to parameterize the disturbance and recovery processes of a carbon cycle model similar to the TBMs reported on here. They found a much smaller net carbon sink for conterminous USA forests than previous estimates using those inventory data in stock change approaches, like those of the inventory-based estimates here (Williams et al., 2012). The same source of data used in different methods can yield different results. Particulars of how disturbance is represented in inventories are also likely responsible for some portion of the difference between AIM and inventory-based estimates of net atmosphere CO_2 exchange.

Within-method uncertainties also contribute to the differences in estimates and the uncertainty surrounding those estimates (Enting et al., 2012). Each method involves numerous assumptions and myriad sources of uncertainty: transport uncertainty, limited atmospheric data and inversion methodology in the atmospheric inversions; parameter, process and input data uncertainty in the TBMs; and uncertainty in estimating carbon stock from a limited number of observations of tree height and diameter in forest inventories are just a few examples. In principle, the different estimates should agree, but the uncertainty in a method's estimate may cloud that agreement. Multiple and diverse sources of uncertainty within methods make the reconciliation of the estimates by reducing uncertainty more difficult.

The approaches also differ in their coverage of subregional heterogeneity in ecosystem types. Atmospheric inversions estimate the total land–atmosphere CO_2 exchange from a given region, including any fluxes associated with carbon traded across the region's boundaries, while inventory-based approaches estimate only those exchanges from ecosystem types represented in the inventories (most commonly forest and cropland) and may or may not represent trade of products from those ecosystem types. As such, estimates from AIMs may capture fluxes missed by inventory-based estimates, while inventory-based estimates can attribute emissions to specific ecosystems, thereby assisting in the management of carbon sources and sinks. Likewise, the estimates from TBMs only include those ecosystem types and fluxes simulated by the models but can attribute those fluxes to particular processes and ecosystems that might be managed.

Differences in the treatment of trade, fire, insects, land-use change, methane and methane conversions, arid regions, and permafrost and peatland processes are among the many possible contributions to differences in estimated net land–atmosphere exchange among and within the approaches. Years of research have provided information on these various components, but no single comprehensive, integrated, agreed-upon treatment of them in their entirety exists for the attribution of the net flux estimated by the AIMs, to guide national carbon inventories, or for the implementation in TBMs. Efforts to resolve differences among approaches and the specific attribution of the North American sink will likely require a community effort to test specific hypotheses involving, initially at least, one or a very small combination of these components. Recent indications by Poulter et al. (2014) of the influence of arid lands under El Niño conditions combined with the uncertain contribution of woody encroachment to the North American land sink (Hayes et al., 2012; King et al., 2007a) suggest more attention to woody biomass changes in arid and semiarid environments as a promising area of investigation. This attention might include a focus on these lands and dynamics in an inter-model comparison of TBMs or structured synthesis and perhaps additional observations of carbon inventories for these regions.

There is some indication of convergence in the estimates from the different methods across previous syntheses (Hayes et al., 2012; King et al., 2007b, 2012) and the work presented here, suggesting a North American land sink in the first decade of the 21st century in the range of -300 to $-600\,\mathrm{Tg\,C\,yr^{-1}}$. The convergence of inventories with AIMs has been shown for one data-rich region of North America for 1 year (Schuh et al., 2013), but the level of observational and analytic effort put into this study has not yet been replicated on the continental scale. However, with additional synthesis and assessment within continents, the North American Carbon Program's regional and continental interim synthesis activities (Huntzinger et al., 2012; Schuh et al., 2013), for example, and with inter-continental syntheses like RECCAP (Canadell et al., 2011; Ciais et al., 2010), there may be further convergence and an improved understanding of remaining differences. Either or both will improve not only the scientific understanding of the carbon cycle but the input into considerations of national and international carbon policy as well.

Acknowledgements. We thank Devin A. White of the Geographic Information Science and Technology Group, Oak Ridge National Laboratory, for the calculation of internally consistent North American, Northern Hemisphere, and global land areas. Research and preparation of this report was sponsored by the US Department of Energy (DOE), Office of Science, Biological and Environmental Research (BER), Climate & Environmental Sciences Division, and was performed at Oak Ridge National Laboratory (ORNL). ORNL is managed by UT–Battelle, LLC, for the DOE under contract DE-AC05-00OR22725. The manuscript has been co-authored by employees of a contractor of the US government under contract DE-AC05-00OR22725. Accordingly, the US government retains a nonexclusive, royalty-free license to publish or reproduce the published form of this contribution, or allow others to do so, for US government purposes. R. Vargas acknowledges support from NASA under the Carbon Monitoring System (NNX13AQ06G). K. J. Davis acknowledges support from NASA's Terrestrial Ecosystems and Carbon Cycle Program. Any use of trade, firm, or product names is for descriptive purposes only and does not imply endorsement by the USA Government.

Edited by: J. Canadell

References

Amiro, B. D., Barr, A. G., Barr, J. G., Black, T. A., Bracho, R., Brown, M., Chen, J., Clark, K. L., Davis, K. J., Desai, A. R., Dore, S., Engel, V., Fuentes, J. D., Goldstein, A. H., Goulden,

M. L., Kolb, T. E., Lavigne, M. B., Law, B. E., Margolis, H. A., Martin, T., McCaughey, J. H., Misson, L., Montes-Helu, M., Noormets, A., Randerson, J. T., Starr, G., and Xiao, J.: Ecosystem carbon dioxide fluxes after disturbance in forests of North America, J. Geophys. Res.-Biogeo., 115, G00K02, doi:10.1029/2010JG001390, 2010.

Andres, R. J., Boden, T. A., Bréon, F.-M., Ciais, P., Davis, S., Erickson, D., Gregg, J. S., Jacobson, A., Marland, G., Miller, J., Oda, T., Olivier, J. G. J., Raupach, M. R., Rayner, P., and Treanton, K.: A synthesis of carbon dioxide emissions from fossil-fuel combustion, Biogeosciences, 9, 1845–1871, doi:10.5194/bg-9-1845-2012, 2012.

Baker, D. F., Law, R. M., Gurney, K. R., Rayner, P., Peylin, P., Denning, A. S., Bousquet, P., Bruhwiler, L., Chen, Y. H., Ciais, P., Fung, I. Y., Heimann, M., John, J., Maki, T., Maksyutov, S., Masarie, K., Prather, M., Pak, B., Taguchi, S., and Zhu, Z.: TransCom 3 inversion intercomparison: Impact of transport model errors on the interannual variability of regional CO_2 fluxes, 1988–2003, Global Biogeochem. Cy., 20, GB1002, doi:10.1029/2004GB002439, 2006.

Bechtold, W. A. and Patterson, P. J.: The enhanced Forest Inventory and Analysis program – national sampling design and estimation procedures, Southern Research Station, Asheville, NC USDA Forest Service General Technical Report SRS-8, 85 pp., 2005.

Boden, T. A., Andres, R. J., and Marland, G.: Global, regional, and national fossil-fuel CO_2 emissions: 1751–2010, Oak Ridge National Laboratory, US Department of Energy, Oak Ridge, TN, 2013.

Butler, M. P., Davis, K. J., Denning, A. S., and Kawa, S. R.: Using continental observations in global atmospheric inversions of CO_2: North American carbon sources and sinks, Tellus B, 62, 550–572, 2010.

Canadell, J. G., Ciais, P., Gurney, K., Le Quéré, C., Piao, S., Raupach, M. R., and Sabine, C. L.: An International Effort to Quantify Regional Carbon Fluxes, Eos, Transactions American Geophysical Union, 92, 81–82, 2011.

Canadell, J. P., Dickinson, R., Hibbard, K., Raupach, M., and O. Young (Eds.): Global Carbon Project: Science framework and implementation, Rep. 1/Global Carbon Project Rep. 1, 69 pp., Earth Syst. Sci. Partnership, Canberra, ACT, Australia, available at: http://www.globalcarbonproject.org, 2003.

Chapin, F. S., III, Woodwell, G. M., Randerson, J. T., Rastetter, E. B., Lovett, G. M., Baldocchi, D. D., Clark, D. A., Harmon, M. E., Schimel, D. S., Valentini, R., Wirth, C., Aber, J. D., Cole, J. J., Goulden, M. L., Harden, J. W., Heimann, M., Howarth, R. W., Matson, P. A., McGuire, A. D., Melillo, J. M., Mooney, H. A., Neff, J. C., Houghton, R. A., Pace, M. L., Ryan, M. G., Running, S. W., Sala, O. E., Schlesinger, W. H., and Schulze, E. D.: Reconciling Carbon-cycle Concepts, Terminology, and Methods, Ecosystems, 9, 1041–1050, 2006.

Ciais, P., Canadell, J. G., Luyssaert, S., Chevallier, F., Shvidenko, A., Poussi, Z., Jonas, M., Peylin, P., King, A. W., Schulze, E.-D., Piao, S., Rödenbeck, C., Peters, W., and Bréon, F.-M.: Can we reconcile atmospheric estimates of the Northern terrestrial carbon sink with land-based accounting?, Current Opinion in Environmental Sustainability, 2, 225–230, 2010.

Ciais, P., Rayner, P., Chevallier, F., Bousquet, P., Logan, M., Peylin, P., and Ramonet, M.: Atmospheric inversions for estimating CO_2 fluxes: methods and perspectives, in: Greenhouse Gas Inventories, edited by: Jonas, M., Nahorski, Z., Nilsson, S., and Whiter, T., Springer, Netherlands, 69–92, 2011.

de Jong, B., Anaya, C., Masera, O., Olguín, M., Paz, F., Etchevers, J., Martínez, R. D., Guerrero, G., and Balbontín, C.: Greenhouse gas emissions between 1993 and 2002 from land-use change and forestry in Mexico, Forest Ecol. Manag., 260, 1689–1701, 2010.

Dolman, A. J., Shvidenko, A., Schepaschenko, D., Ciais, P., Tchebakova, N., Chen, T., van der Molen, M. K., Belelli Marchesini, L., Maximov, T. C., Maksyutov, S., and Schulze, E.-D.: An estimate of the terrestrial carbon budget of Russia using inventory-based, eddy covariance and inversion methods, Biogeosciences, 9, 5323–5340, doi:10.5194/bg-9-5323-2012, 2012.

Enting, I. G.: Inverse problems in atmospheric constituent transport, Cambridge University Press, 392 pp., 2002.

Enting, I. G., Rayner, P. J., and Ciais, P.: Carbon Cycle Uncertainty in REgional Carbon Cycle Assessment and Processes (RECCAP), Biogeosciences, 9, 2889–2904, doi:10.5194/bg-9-2889-2012, 2012.

Environment Canada: National Inventory Report 1990–2011: Greenhouse Gas Sources and Sinks in Canada, The Government of Canada's Submission to the UN Framework Convention on Climate Change, Environment Canada, Environment Canada, Ottawa, ON, Canada, 538 pp., 2013.

Environment Canada: National Inventory Report: 1990–2012, greenhouse gas sources and sinks in Canada, Environment Canada, Greenhouse Gas Division, Ottowa, Ontario, 507 pp., 2014.

Field, C. B., Sarmiento, J., and Hales, B.: The Carbon Cycle of North America in a Global Context, in: The First State of the Carbon Cycle Report (SOCCR): The North American Carbon Budget and Implications for the Global Carbon Cycle, A Report by the USA Climate Change Science Program and the Subcommittee on Global Change Research, edited by: King, A. W., Dilling, L., Zimmerman, G. P., Fairman, D. M., Houghton, R. A., Marland, G., Rose, A. Z., and Wilbanks, T. J., National Oceanic and Atmospheric Administration, National Climatic Data Center, Asheville, NC, USA, 21–28, 2007.

Global Carbon Atlas: available at: http://www.globalcarbonatlas.org/?q=en/emissions (last access: 21 November 2014), 2014.

Gloor, M., Gatti, L., Brienen, R., Feldpausch, T. R., Phillips, O. L., Miller, J., Ometto, J. P., Rocha, H., Baker, T., de Jong, B., Houghton, R. A., Malhi, Y., Aragão, L. E. O. C., Guyot, J.-L., Zhao, K., Jackson, R., Peylin, P., Sitch, S., Poulter, B., Lomas, M., Zaehle, S., Huntingford, C., Levy, P., and Lloyd, J.: The carbon balance of South America: a review of the status, decadal trends and main determinants, Biogeosciences, 9, 5407–5430, doi:10.5194/bg-9-5407-2012, 2012.

Goodale, C. L., Apps, M. J., Birdsey, R. A., Field, C. B., Heath, L. S., Houghton, R. A., Jenkins, J. C., Kohlmaier, G. H., Kurz, W., Liu, S., Nabuurs, G.-J., Nilsson, S., and Shvidenko, A. Z.: Forest carbon sinks in the Northern Hemisphere, Ecol. Appl., 12, 891–899, 2002.

Gurney, K. R., Law, R. M., Denning, A. S., Rayner, P. J., Baker, D., Bousquet, P., Bruhwiler, L., Chen, Y.-H., Ciais, P., Fan, S., Fung, I. Y., Gloor, M., Heimann, M., Higuchi, K., John, J., Maki, T., Maksyutov, S., Masarie, K., Peylin, P., Prather, M., Pak, B. C., Randerson, J., Sarmiento, J., Taguchi, S., Takahashi, T., and Yuen, C.-W.: Towards robust regional estimates of CO_2 sources and sinks using atmospheric transport models, Nature, 415, 626–630, 2002.

Gurney, K. R., Law, R. M., Denning, A. S., Rayner, P. J., Baker, D., Bousquet, P., Bruhwiler, L., Chen, Y.-H., Ciais, P., Fan, S., Fung, I. Y., Gloor, M., Heimann, M., Higuchi, K., John, J., Kowalczyk, E., Maki, T., Maksyutov, S., Peylin, P., Prather, M., Pak, B. C., Sarmiento, J., Taguchi, S., Takahashi, T., and Yuen, C.-W.: TransCom 3 CO_2 inversion intercomparison: 1. Annual mean control results and sensitivity to transport and prior flux information, Tellus B, 55, 555–579, 2003.

Gurney, K. R., Baker, D., Rayner, P., and Denning, S.: Interannual variations in continental-scale net carbon exchange and sensitivity to observing networks estimated from atmospheric CO_2 inversions for the period 1980 to 2005, Global Biogeochem. Cy., 22, GB3025, doi:10.1029/2007GB003082, 2008.

Harmon, M. E., Bond-Lamberty, B., Tang, J., and Vargas, R.: Heterotrophic respiration in disturbed forests: A review with examples from North America, J. Geophys. Res.-Biogeo., 116, G00K04, doi:10.1029/2010JG001495, 2011.

Haverd, V., Raupach, M. R., Briggs, P. R., J. G. Canadell., Davis, S. J., Law, R. M., Meyer, C. P., Peters, G. P., Pickett-Heaps, C., and Sherman, B.: The Australian terrestrial carbon budget, Biogeosciences, 10, 851–869, doi:10.5194/bg-10-851-2013, 2013.

Hayes, D. and Turner, D.: The need for "apples-to-apples" comparisons of carbon dioxide source and sink estimates, Eos, Transactions American Geophysical Union, 93, 404–405, 2012.

Hayes, D. J., Turner, D. P., Stinson, G., McGuire, A. D., Wei, Y., West, T. O., Heath, L. S., de Jong, B., McConkey, B. G., Birdsey, R. A., Kurz, W. A., Jacobson, A. R., Huntzinger, D. N., Pan, Y., Post, W. M., and Cook, R. B.: Reconciling estimates of the contemporary North American carbon balance among terrestrial biosphere models, atmospheric inversions, and a new approach for estimating net ecosystem exchange from inventory-based data, Glob. Change Biol., 18, 1282–1299, 2012.

Huntzinger, D. N., Post, W. M., Wei, Y., Michalak, A. M., West, T. O., Jacobson, A. R., Baker, I. T., Chen, J. M., Davis, K. J., Hayes, D. J., Hoffman, F. M., Jain, A. K., Liu, S., McGuire, A. D., Neilson, R. P., Potter, C., Poulter, B., Price, D., Raczka, B. M., Tian, H. Q., Thornton, P., Tomelleri, E., Viovy, N., Xiao, J., Yuan, W., Zeng, N., Zhao, M., and Cook, R.: North American Carbon Program (NACP) regional interim synthesis: Terrestrial biospheric model intercomparison, Ecol. Model., 232, 144–157, 2012.

IPCC: 2006 IPCC Guidelines for National Greenhouse Gas Inventories: Volume 4 Agriculture, Forestry and Other Land Use, Prepared by the National Greenhouse Gas Inventories Programme, edited by: Eggleston, H. S., Buendia, L., Miwa, K., Ngara, T., and Tanabe, K., IGES, Japan, 593 pp., 2006.

Kasischke, E. S., Amiro, B. D., Barger, N. N., French, N. H. F., Goetz, S. J., Grosse, G., Harmon, M. E., Hicke, J. A., Liu, S., and Masek, J. G.: Impacts of disturbance on the terrestrial carbon budget of North America, J. Geophys. Res.-Biogeo., 118, 303–316, 2013.

Keppel-Aleks, G., Wolf, A. S., Mu, M., Doney, S. C., Morton, D. C., Kasibhatla, P. S., Miller, J. B., Dlugokencky, E. J., and Randerson, J. T.: Separating the influence of temperature, drought and fire on interannual variability in atmospheric CO_2, Global Biogeochem. Cy., 28, 1295–1310, doi:10.1002/2014GB004890, 2014.

King, A. W., Dilling, L., Zimmerman, G. P., Fairman, D. M., Houghton, R. A., Marland, G., Rose, A. Z., and Wilbanks, T. J.: Executive Summary, in: The First State of the Carbon Cycle Report (SOCCR): The North American Carbon Budget and Implications for the Global Carbon Cycle, A Report by the USA Climate Change Science Program and the Subcommittee on Global Change Research, edited by: King, A. W., Dilling, L., Zimmerman, G. P., Fairman, D. M., Houghton, R. A., Marland, G., Rose, A. Z., and Wilbanks, T. J., National Oceanic and Atmospheric Administration, National Climatic Data Center, Asheville, NC, USA, 1–14, 2007a.

King, A. W., Dilling, L., Zimmerman, G. P., Fairman, D. M., Houghton, R. A., Marland, G., Rose, A. Z., and Wilbanks, T. J.: The First State of the Carbon Cycle Report (SOCCR): The North American Carbon Budget and Implications for the Global Carbon Cycle, A Report by the USA Climate Change Science Program and the Subcommittee on Global Change Research, National Oceanic and Atmospheric Administration, National Climatic Data Center, Asheville, NC, USA, 242 pp., 2007b.

King, A. W., Hayes, D. J., Huntzinger, D. N., West, T. O., and Post, W. M.: North American carbon dioxide sources and sinks: magnitude, attribution, and uncertainty, Front. Ecol. Environ., 10, 512–519, 2012.

Kurz, W. A., Stinson, G., Rampley, G. J., Dymond, C. C., and Neilson, E. T.: Risk of natural disturbances makes future contribution of Canada's forests to the global carbon cycle highly uncertain, P. Natl. Acad. Sci., 105, 1551–1555, 2008.

Kurz, W. A., Dymond, C. C., White, T. M., Stinson, G., Shaw, C. H., Rampley, G. J., Smyth, C., Simpson, B. N., Neilson, E. T., Trofymow, J. A., Metsaranta, J., and Apps, M. J.: CBM-CFS3: A model of carbon-dynamics in forestry and land-use change implementing IPCC standards, Ecol. Model., 220, 480–504, 2009.

Kurz, W. A., Shaw, C. H., Boisvenue, C., Stinson, G., Metsaranta, J., Leckie, D., Dyk, A., Smyth, C., and Neilson, E. T.: Carbon in Canada's boreal forest – A synthesis, Environ. Rev., 21, 260–292, 2013.

Le Quéré, C., Andres, R. J., Boden, T., Conway, T., Houghton, R. A., House, J. I., Marland, G., Peters, G. P., van der Werf, G. R., Ahlström, A., Andrew, R. M., Bopp, L., Canadell, J. G., Ciais, P., Doney, S. C., Enright, C., Friedlingstein, P., Huntingford, C., Jain, A. K., Jourdain, C., Kato, E., Keeling, R. F., Klein Goldewijk, K., Levis, S., Levy, P., Lomas, M., Poulter, B., Raupach, M. R., Schwinger, J., Sitch, S., Stocker, B. D., Viovy, N., Zaehle, S., and Zeng, N.: The global carbon budget 1959–2011, Earth Syst. Sci. Data, 5, 165–185, 2013, http://www.earth-syst-sci-data.net/5/165/2013/.

Le Quéré, C., Moriarty, R., Andrew, R. M., Peters, G. P., Ciais, P., Friedlingstein, P., Jones, S. D., Sitch, S., Tans, P., Arneth, A., Boden, T. A., Bopp, L., Bozec, Y., Canadell, J. G., Chevallier, F., Cosca, C. E., Harris, I., Hoppema, M., Houghton, R. A., House, J. I., Jain, A., Johannessen, T., Kato, E., Keeling, R. F., Kitidis, V., Klein Goldewijk, K., Koven, C., Landa, C. S., Landschützer, P., Lenton, A., Lima, I. D., Marland, G., Mathis, J. T., Metzl, N., Nojiri, Y., Olsen, A., Ono, T., Peters, W., Pfeil, B., Poulter, B., Raupach, M. R., Regnier, P., Rödenbeck, C., Saito, S., Salisbury, J. E., Schuster, U., Schwinger, J., Séférian, R., Segschneider, J., Steinhoff, T., Stocker, B. D., Sutton, A. J., Takahashi, T., Tilbrook, B., van der Werf, G. R., Viovy, N., Wang, Y. P., Wanninkhof, R., Wiltshire, A., and Zeng, N.: Global carbon budget

2014, Earth Syst. Sci. Data Discuss., 7, 521–610, 2014, http://www.earth-syst-sci-data-discuss.net/7/521/2014/.

Liu, S., Bond-Lamberty, B., Hicke, J. A., Vargas, R., Zhao, S., Chen, J., Edburg, S. L., Hu, Y., Liu, J., McGuire, A. D., Xiao, J., Keane, R., Yuan, W., Tang, J., Luo, Y., Potter, C., and Oeding, J.: Simulating the impacts of disturbances on forest carbon cycling in North America: Processes, data, models, and challenges, J. Geophys. Res.-Biogeo., 116, G00K08, doi:10.1029/2009JG001585, 2011.

Luyssaert, S., Inglima, I., Jung, M., Richardson, A. D., Reichstein, M., Papale, D., Piao, S. L., Schulze, E. D., Wingate, L., Matteucci, G., Aragao, L., Aubinet, M., Beer, C., Bernhofer, C., Black, K. G., Bonal, D., Bonnefond, J. M., Chambers, J., Ciais, P., Cook, B., Davis, K. J., Dolman, A. J., Gielen, B., Goulden, M., Grace, J., Granier, A., Grelle, A., Griffis, T., GrÜNwald, T., Guidolotti, G., Hanson, P. J., Harding, R., Hollinger, D. Y., Hutyra, L. R., Kolari, P., Kruijt, B., Kutsch, W., Lagergren, F., Laurila, T., Law, B. E., Le Maire, G., Lindroth, A., Loustau, D., Malhi, Y., Mateus, J., Migliavacca, M., Misson, L., Montagnani, L., Moncrieff, J., Moors, E., Munger, J. W., Nikinmaa, E., Ollinger, S. V., Pita, G., Rebmann, C., Roupsard, O., Saigusa, N., Sanz, M. J., Seufert, G., Sierra, C., Smith, M. L., Tang, J., Valentini, R., Vesala, T., and Janssens, I. A.: CO_2 balance of boreal, temperate, and tropical forests derived from a global database, Glob. Change Biol., 13, 2509–2537, 2007.

Luyssaert, S., Abril, G., Andres, R., Bastviken, D., Bellassen, V., Bergamaschi, P., Bousquet, P., Chevallier, F., Ciais, P., Corazza, M., Dechow, R., Erb, K.-H., Etiope, G., Fortems-Cheiney, A., Grassi, G., Hartmann, J., Jung, M., Lathière, J., Lohila, A., Mayorga, E., Moosdorf, N., Njakou, D. S., Otto, J., Papale, D., Peters, W., Peylin, P., Raymond, P., Rödenbeck, C., Saarnio, S., Schulze, E.-D., Szopa, S., Thompson, R., Verkerk, P. J., Vuichard, N., Wang, R., Wattenbach, M., and Zaehle, S.: The European land and inland water CO_2, CO, CH_4 and N_2O balance between 2001 and 2005, Biogeosciences, 9, 3357–3380, doi:10.5194/bg-9-3357-2012, 2012.

McGuire, A. D., Christensen, T. R., Hayes, D., Heroult, A., Euskirchen, E., Kimball, J. S., Koven, C., Lafleur, P., Miller, P. A., Oechel, W., Peylin, P., Williams, M., and Yi, Y.: An assessment of the carbon balance of Arctic tundra: comparisons among observations, process models, and atmospheric inversions, Biogeosciences, 9, 3185–3204, doi:10.5194/bg-9-3185-2012, 2012.

Myneni, R. B., Dong, J., Tucker, C. J., Kaufmann, R. K., Kauppi, P. E., Liski, J., Zhou, L., Alexeyev, V., and Hughes, M. K.: A large carbon sink in the woody biomass of Northern forests, P. Natl. Acad. Sci., 98, 14784–14789, 2001.

Pacala, S., Birdsey, R. A., Bridgham, S. D., Conant, R. T., Davis, K., Hales, B., Houghton, R. A., Jenkins, J. C., Johnston, M., Marland, G., and Paustian, K.: The North American Carbon Budget Past and Present, in: The First State of the Carbon Cycle Report (SOCCR): The North American Carbon Budget and Implications for the Global Carbon Cycle, A Report by the USA Climate Change Science Program and the Subcommittee on Global Change Research, edited by: King, A. W., Dilling, L., Zimmerman, G. P., Fairman, D. M., Houghton, R. A., Marland, G., Rose, A. Z., and Wilbanks, T. J., National Oceanic and Atmospheric Administration, National Climatic Data Center, Asheville, NC, USA, 29–36, 2007.

Pan, Y., Birdsey, R. A., Fang, J., Houghton, R., Kauppi, P. E., Kurz, W. A., Phillips, O. L., Shvidenko, A., Lewis, S. L., Canadell, J. G., Ciais, P., Jackson, R. B., Pacala, S. W., McGuire, A. D., Piao, S., Rautiainen, A., Sitch, S., and Hayes, D.: A Large and Persistent Carbon Sink in the World's Forests, Science, 333, 988–993, 2011.

Patra, P. K., Canadell, J. G., Houghton, R. A., Piao, S. L., Oh, N.-H., Ciais, P., Manjunath, K. R., Chhabra, A., Wang, T., Bhattacharya, T., Bousquet, P., Hartman, J., Ito, A., Mayorga, E., Niwa, Y., Raymond, P. A., Sarma, V. V. S. S., and Lasco, R.: The carbon budget of South Asia, Biogeosciences, 10, 513–527, doi:10.5194/bg-10-513-2013, 2013.

Peters, W., Jacobson, A. R., Sweeney, C., Andrews, A. E., Conway, T. J., Masarie, K., Miller, J. B., Bruhwiler, L. M. P., Pétron, G., Hirsch, A. I., Worthy, D. E. J., Werf, G. R. v. d., Randerson, J. T., Wennberg, P. O., Krol, M. C., and Tans, P. P.: An Atmospheric Perspective on North American Carbon Dioxide Exchange: CarbonTracker, P. Natl. Acad. Sci. USA, 104, 18925–18930, 2007.

Peylin, P., Law, R. M., Gurney, K. R., Chevallier, F., Jacobson, A. R., Maki, T., Niwa, Y., Patra, P. K., Peters, W., Rayner, P. J., Rödenbeck, C., van der Laan-Luijkx, I. T., and Zhang, X.: Global atmospheric carbon budget: results from an ensemble of atmospheric CO_2 inversions, Biogeosciences, 10, 6699–6720, doi:10.5194/bg-10-6699-2013, 2013.

Piao, S. L., Ito, A., Li, S. G., Huang, Y., Ciais, P., Wang, X. H., Peng, S. S., Nan, H. J., Zhao, C., Ahlström, A., Andres, R. J., Chevallier, F., Fang, J. Y., Hartmann, J., Huntingford, C., Jeong, S., Levis, S., Levy, P. E., Li, J. S., Lomas, M. R., Mao, J. F., Mayorga, E., Mohammat, A., Muraoka, H., Peng, C. H., Peylin, P., Poulter, B., Shen, Z. H., Shi, X., Sitch, S., Tao, S., Tian, H. Q., Wu, X. P., Xu, M., Yu, G. R., Viovy, N., Zaehle, S., Zeng, N., and Zhu, B.: The carbon budget of terrestrial ecosystems in East Asia over the last two decades, Biogeosciences, 9, 3571–3586, doi:10.5194/bg-9-3571-2012, 2012.

Poulter, B., Frank, D., Ciais, P., Myneni, R. B., Andela, N., Bi, J., Broquet, G., Canadell, J. G., Chevallier, F., Liu, Y. Y., Running, S. W., Sitch, S., and van der Werf, G. R.: Contribution of semi-arid ecosystems to interannual variability of the global carbon cycle, Nature, 509, 600–603, 2014.

Raczka, B. M., Davis, K. J., Huntzinger, D., Neilson, R. P., Poulter, B., Richardson, A. D., Xiao, J., Baker, I., Ciais, P., Keenan, T. F., Law, B., Post, W. M., Ricciuto, D., Schaefer, K., Tian, H., Tomelleri, E., Verbeeck, H., and Viovy, N.: Evaluation of continental carbon cycle simulations with North American flux tower observations, Ecol. Monogr., 83, 531–556, 2013.

Randerson, J. T., van der Werf, G. R., Collatz, G. J., Giglio, L., Still, C. J., Kasibhatla, P., Miller, J. B., White, J. W. C., DeFries, R. S., and Kasischke, E. S.: Fire emissions from C_3 and C_4 vegetation and their influence on interannual variability of atmospheric CO_2 and $\delta 13CO_2$, Global Biogeochem. Cy., 19, GB2019, doi:10.1029/2004GB002366, 2005.

Schuh, A. E., Lauvaux, T., West, T. O., Denning, A. S., Davis, K. J., Miles, N., Richardson, S., Uliasz, M., Lokupitiya, E., Cooley, D., Andrews, A., and Ogle, S.: Evaluating atmospheric CO_2 inversions at multiple scales over a highly inventoried agricultural landscape, Glob. Change Biol., 19, 1424–1439, 2013.

Schulze, E. D., Wirth, C., and Heimann, M.: Managing Forests After Kyoto, Science, 289, 2058–2059, 2000.

Schulze, E. D., Luyssaert, S., Ciais, P., Freibauer, A., Janssens, I. A., Soussana, J. F., Smith, P., Grace, J., Levin, I., Thiruchittampalam, B., Heimann, M., Dolman, A. J., Valentini, R., Bousquet, P., Peylin, P., Peters, W., Rödenbeck, C., Etiope, G., Vuichard, N., Wattenbach, M., Nabuurs, G. J., Poussi, Z., Nieschulze, J., Gash, J. H., and the CarboEurope Team: Importance of methane and nitrous oxide for Europe's terrestrial greenhouse-gas balance, Nat. Geosci., 2, 842–850, 2009.

Schulze, E. D., Ciais, P., Luyssaert, S., Schrumpf, M., Janssens, I. A., Thiruchittampalam, B., Theloke, J., Saurat, M., Bringezu, S., Lelieveld, J., Lohila, A., Rebmann, C., Jung, M., Bastviken, D., Abril, G., Grassi, G., Leip, A., Freibauer, A., Kutsch, W., Don, A., Nieschulze, J., BÖRner, A., Gash, J. H., and Dolman, A. J.: The European carbon balance. Part 4: integration of carbon and other trace-gas fluxes, Glob. Change Biol., 16, 1451–1469, 2010.

Schwalm, C. R., Williams, C. A., Schaefer, K., Anderson, R., Arain, M. A., Baker, I., Barr, A., Black, T. A., Chen, G., Chen, J. M., Ciais, P., Davis, K. J., Desai, A., Dietze, M., Dragoni, D., Fischer, M. L., Flanagan, L. B., Grant, R., Gu, L., Hollinger, D., Izaurralde, R. C., Kucharik, C., Lafleur, P., Law, B. E., Li, L., Li, Z., Liu, S., Lokupitiya, E., Luo, Y., Ma, S., Margolis, H., Matamala, R., McCaughey, H., Monson, R. K., Oechel, W. C., Peng, C., Poulter, B., Price, D. T., Riciutto, D. M., Riley, W., Sahoo, A. K., Sprintsin, M., Sun, J., Tian, H., Tonitto, C., Verbeeck, H., and Verma, S. B.: A model-data intercomparison of CO_2 exchange across North America: Results from the North American Carbon Program site synthesis, J. Geophys. Res.-Biogeo., 115, G00H05, doi:10.1029/2009JG001229, 2010.

SEMARINAT/INECC: México: Quinta Comunicación Nacional ante la Convención Marco de las Naciones Unidas sobre el Cambio Climático, 442 pp., 2012.

Sitch, S., Huntingford, C., Gedney, N., Levy, P. E., Lomas, M., Piao, S. L., Betts, R., Ciais, P., Cox, P., Friedlingstein, P., Jones, C. D., Prentice, I. C., and Woodward, F. I.: Evaluation of the terrestrial carbon cycle, future plant geography and climate-carbon cycle feedbacks using five Dynamic Global Vegetation Models (DGVMs), Glob. Change Biol., 14, 2015–2039, 2008.

Sitch, S., Friedlingstein, P., Gruber, N., Jones, S. D., Murray-Tortarolo, G., Ahlström, A., Doney, S. C., Graven, H., Heinze, C., Huntingford, C., Levis, S., Levy, P. E., Lomas, M., Poulter, B., Viovy, N., Zaehle, S., Zeng, N., Arneth, A., Bonan, G., Bopp, L., Canadell, J. G., Chevallier, F., Ciais, P., Ellis, R., Gloor, M., Peylin, P., Piao, S., Le Quéré, C., Smith, B., Zhu, Z., and Myneni, R.: Trends and drivers of regional sources and sinks of carbon dioxide over the past two decades, Biogeosciences Discuss., 10, 20113–20177, doi:10.5194/bgd-10-20113-2013, 2013.

Skog, K. E.: Sequestration of carbon in harvested wood products for the United States, Forest Prod. J., 58, 56–72, 2008.

Smith, J. E., Heath, L. S., Skog, K. E., and Birdsey, R. A.: Methods for calculating forest ecosystem and harvested carbon with standard estimates for forest types of the United States, Northeastern Research Station. Newtown Square, PAUSDA Forest Service General Technical Report NE-343, 216 pp., 2006.

Smith, J. E., Heath, L. S., and Hoover, C. M.: Carbon factors and models for forest carbon estimates for the 2005–2011 National Greenhouse Gas Inventories of the United States, Forest Ecol. Manag., 307, 7–19, 2013.

Stinson, G., Kurz, W. A., Smyth, C. E., Neilson, E. T., Dymond, C. C., Metsaranta, J. M., Boisvenue, C., Rampley, G. J., Li, Q., White, T. M., and Blain, D.: An inventory-based analysis of Canada's managed forest carbon dynamics, 1990 to 2008, Glob. Change Biol., 17, 2227–2244, 2011.

Thomey, M. L., Collins, S. L., Vargas, R., Johnson, J. E., Brown, R. F., Natvig, D. O., and Friggens, M. T.: Effect of precipitation variability on net primary production and soil respiration in a Chihuahuan Desert grassland, Glob. Change Biol., 17, 1505–1515, 2011.

USDA Forest Service: available at: http://apps.fs.fed.us/fiadb-downloads/datamart.html (last access: 23 January 2013), 2013.

USEPA: Forest sections of the Land Use, Land Use Change, and Forestry chapter, and Annex. In: US Environmental Protection Agency, Inventory of US Greenhouse Gas Emissions and Sinks: 1990–2010, United States Environmental Protection Agency,EPA 430-R-12-001, 481 pp., 2012.

Valentini, R., Arneth, A., Bombelli, A., Castaldi, S., Cazzolla Gatti, R., Chevallier, F., Ciais, P., Grieco, E., Hartmann, J., Henry, M., Houghton, R. A., Jung, M., Kutsch, W. L., Malhi, Y., Mayorga, E., Merbold, L., Murray-Tortarolo, G., Papale, D., Peylin, P., Poulter, B., Raymond, P. A., Santini, M., Sitch, S., Vaglio Laurin, G., van der Werf, G. R., Williams, C. A., and Scholes, R. J.: A full greenhouse gases budget of Africa: synthesis, uncertainties, and vulnerabilities, Biogeosciences, 11, 381–407, doi:10.5194/bg-11-381-2014, 2014.

Vargas, R., Loescher, H. W., Arredondo, T., Huber-Sannwald, E., Lara-Lara, R., and Yépez, E. A.: Opportunities for advancing carbon cycle science in Mexico: toward a continental scale understanding, Environ. Sci. Policy, 21, 84–93, 2012.

West, T. O., Bandaru, V., Brandt, C. C., Schuh, A. E., and Ogle, S. M.: Regional uptake and release of crop carbon in the United States, Biogeosciences, 8, 2037–2046, doi:10.5194/bg-8-2037-2011, 2011.

Williams, C. A., Collatz, G. J., Masek, J., and Goward, S. N.: Carbon consequences of forest disturbance and recovery across the conterminous United States, Global Biogeochem. Cycles, 26, GB1005, doi:10.1029/2010GB003947, 2012.

Woodall, C. W., Heath, L. S., Domke, G. M., and Nichols, M. C.: Methods and equations for estimating aboveground volume, biomass, and carbon for trees in the USA forest inventory, 2010, Northern Research Station, Newtown Square, PA USDA Forest Service General Technical Report NRS-88, 30 pp., 2011.

12

Influence of meteorology and anthropogenic pollution on chemical flux divergence of the NO–NO_2–O_3 triad above and within a natural grassland canopy

D. Plake[1,*], M. Sörgel[1], P. Stella[1,], A. Held[2], and I. Trebs[1,***]**

[1]Max Planck Institute for Chemistry, Biogeochemistry Department, P.O. Box 3060, 55020 Mainz, Germany
[2]University of Bayreuth, Atmospheric Chemistry, Bayreuth, Germany
[*]now at: Air Monitoring Department, UCL Umwelt Control Labor GmbH, 44536 Lünen, Germany
[**]now at: AgroParisTech, UMR INRA/AgroParisTech SAD-APT, Paris, France
[***]now at: Luxembourg Institute of Science and Technology, Environmental Research and Innovation (ERIN) Department, 4422 Belvaux, Luxembourg

Correspondence to: I. Trebs (ivonne.trebs@list.lu)

Abstract. The detailed understanding of surface–atmosphere exchange fluxes of reactive trace gases is a crucial precondition for reliable modelling of processes in atmospheric chemistry. Plant canopies significantly impact the atmospheric budget of trace gases. In the past, many studies focused on taller forest canopies or crops, where the bulk plant material is concentrated in the uppermost canopy layer. However, within grasslands, a land-cover class that globally covers vast terrestrial areas, the canopy structure is fundamentally different, as the main biomass is concentrated in the lowest part of the canopy. This has obvious implications for aerodynamic in-canopy transport, and consequently also impacts on global budgets of key species in atmospheric chemistry such as nitric oxide (NO), nitrogen dioxide (NO_2) and ozone (O_3).

This study presents for the first time a comprehensive data set of directly measured in-canopy transport times and aerodynamic resistances, chemical timescales, Damköhler numbers, trace gas and micrometeorological measurements for a natural grassland canopy (canopy height = 0.6 m). Special attention is paid to the impact of contrasting meteorological and air chemical conditions on in-canopy transport and chemical flux divergence. Our results show that the grassland canopy is decoupled throughout the day. In the lowermost canopy layer, the measured transport times are fastest during nighttime, which is due to convection during nighttime and a stable stratification during daytime in this layer. The inverse was found in the layers above. During periods of low wind speed and high NO_x (NO+NO_2) levels, the effect of canopy decoupling on trace gas transport was found to be especially distinct. The aerodynamic resistance in the lowermost canopy layer (0.04–0.2 m) was around 1000 s m^{-1}, which is as high as values determined previously for the lowest metre of an Amazonian rain forest canopy. The aerodynamic resistance representing the bulk canopy was found to be more than 3–4 times higher than in forests. Calculated Damköhler numbers (ratio of transport and chemical timescales) suggest a strong flux divergence for the NO–NO_2–O_3 triad within the canopy during daytime. During that time, the timescale of NO_2 uptake by plants ranged from 90 to 160 s and was the fastest relevant timescale, i.e. faster than the reaction of NO and O_3. Thus, our results reveal that grassland canopies of similar structure exhibit a strong potential to retain soil-emitted NO due to oxidation and subsequent uptake of NO_2 by plants. Furthermore, photo-chemical O_3 production was observed above the canopy, which was attributed to a deviation from the NO–NO_2–O_3 photostationary state by a surplus of NO_2 due to oxidation of NO, by e.g. peroxy radicals. The O_3 production was one order of magnitude higher during high NO_x than during low NO_x periods and resulted in an underestimation of the O_3 deposition flux measured with the EC method.

1 Introduction

Nitric oxide (NO) and nitrogen dioxide (NO_2) play a crucial role in air chemistry as they act as key catalysts for ozone (O_3) production and are therefore involved in the generation of hydroxyl radicals (OH) (Crutzen, 1973). The most significant tropospheric source of O_3 is initiated by photochemical dissociation of NO_2 and subsequent reaction of the oxygen (O) atom with molecular oxygen:

$$NO_2 + h\nu\,(\lambda < 420\,\text{nm}) \rightarrow NO + O(^3P) \tag{R1}$$

$$O(^3P) + O_2 + M \rightarrow O_3 + M. \tag{R2}$$

In the case that O_3 is present, it may oxidize NO and re-form NO_2:

$$O_3 + NO \xrightarrow{k_3} NO_2 + O_2 \tag{R3}$$

In the absence of additional reactions, R1–R3 represent a null cycle. Besides R1–R3, NO is oxidized by peroxy radicals (HO_2+RO_2), which constitutes an important net O_3 production pathway in the troposphere (Warneck, 2000).

Dry deposition to terrestrial surfaces, especially to plant canopies, is an important sink for tropospheric O_3 and NO_2. The uncertainties of dry deposition estimates are substantially higher for NO_2, because its net ecosystem exchange can be bi-directional depending on the ambient NO_2 levels (Lerdau et al., 2000). O_3 instead is exclusively deposited to surfaces. In contrast, NO is known to be mainly net emitted from nearly all soil types. Biogenic NO soil emissions contribute about 20 % to the global NO_x (NO+NO_2) emissions (IPCC, 2013), highlighting the need of detailed investigations on NO_x soil–atmosphere exchange.

A major challenge for studies investigating surface–atmosphere exchange fluxes of these reactive trace gases is the presence of plant canopies. These significantly modify the aerodynamic properties of the surface and, thus, alter trace gas exchange fluxes. Most previous studies focused on taller canopies such as forests. However, grassland canopies represent a highly important land cover class covering globally 41 % and Europe-wide 19 % of the terrestrial land surface (Suttie et al., 2005; Kasanko et al., 2011). In contrast to forests, grasslands feature the main bulk plant area density near the soil (e.g. Ripley and Redman, 1976; Jäggi et al., 2006), accompanied with mean distances between plant elements of only some millimetres (Aylor et al., 1993). Organized coherent structures govern turbulence dynamics within and above plant canopies (Finnigan, 2000). The mean in-canopy transport is slower than above the canopy (e.g. Nemitz et al., 2009). This modification of in-canopy transport has important implications for global atmospheric chemistry. Plant canopies and the soil below are biologically actively emitting and taking up reactive trace gases, and conditions within canopies may provide sufficient time for fast chemical reactions (Nemitz et al., 2009). Subsequently, they modify surface exchange fluxes (e.g. Rinne et al., 2012). For instance, ammonia can be released by a part of the canopy and taken up by another (Nemitz et al., 2000; Denmead et al., 2007). In addition, recapturing of NO_2 originating from biogenic soil NO emissions after reaction with O_3 within plant canopies (Rummel et al., 2002) is accounted for in global models by a so-called canopy reduction factor for NO_x (Yienger and Levy, 1995). However, these estimates are based on only one single experiment in an Amazonian rain forest (Bakwin et al., 1990), and a subsequent model analysis (Jacob and Wofsy, 1990). Canopy reduction for grasslands and other ecosystems has not been experimentally studied up to now. The contrasting canopy structure of grassland and forest ecosystems highlights the need for a detailed analysis and an evaluation of the suggested NO_x canopy reduction factor of e.g. 64 % by Yienger and Levy (1995) for temperate grassland.

Net ecosystem exchange fluxes are typically measured at a certain height above the canopy. They rely on the constant flux layer assumption (e.g. Swinbank, 1968), which however, may be violated for reactive trace gases within or just above the vegetation. To assess the potential chemical divergence of exchange fluxes, the Damköhler number (*DA*) has commonly been applied (e.g. Rinne et al., 2012). *DA* is calculated as the ratio of the transport time (τ_{tr}) and the characteristic chemical timescale (τ_{ch}):

$$DA = \frac{\tau_{tr}}{\tau_{ch}}. \tag{1}$$

Hence, *DA* above unity indicates that chemical reactions occur significantly faster than the transport (flux divergence), whereas *DA* smaller than 0.1 indicates the reverse case. The range in between is commonly addressed as a critical range, where an impact of chemistry cannot be excluded (Stella et al., 2013).

In this paper, we present directly measured transport times, chemical timescales and corresponding Damköhler numbers for three layers above and within a natural grassland canopy under contrasting meteorological and air chemical conditions. For the first time, such a comprehensive analysis involving trace gas and micrometeorological measurements is made for a grassland canopy. Furthermore, the consequences of in-canopy processes for NO_x canopy reduction and simultaneously measured O_3 deposition fluxes will be discussed.

2 Material and methods

2.1 Site description

We performed an intensive field experiment from July to September 2011 at the estate of the Mainz Finthen Airport in Rhineland-Palatinate, Germany (further details given in

Plake and Trebs, 2013; Plake et al., 2014; Moravek et al., 2014, 2015). The vegetation at the site was nutrient-poor grassland with a mean canopy height (h_c) of 0.6 m and a leaf area index (LAI) of 4.8 $m^2\,m^{-2}$. A list of species and an LAI profile are given in Plake et al. (2014), with the latter indicating a high biomass density below 0.2 m corresponding to 85 % of the total LAI. Topographically located on a plateau 150 m above the Rhine valley, the site is located about 9 km south-west of the city centre of Mainz. The site is surrounded by villages and motorways in a distance of 2 to 6 km and 4 to 15 km, respectively. The surrounding area is mainly characterized by agricultural use for vineyards, orchards and crops. The fetch is largest in southwestern direction without significant anthropogenic pollution sources.

2.2 Experimental setup

A vertical Thoron (Tn) profile system was operated at $z_1 = 0.04$, $z_2 = 0.2$ and $z_3 = 0.8$ m for the direct determination of transport times (for details see Plake and Trebs, 2013). Vertical profiles of NO, NO_2, O_3 and CO_2 were measured at z_1, z_2, z_3 and additionally at $z_4 = 4.0$ m by a system described in detail by Plake et al. (2014). Briefly, NO was measured by detection of the chemiluminescence produced during the reaction of NO and O_3 (TEI 42iTL Thermo Scientific, Waltham, USA). NO_2 was photolytically converted to NO by exposure of the sample air to a blue light converter (BLC, Droplet Measurement Technologies, Boulder, USA). O_3 mixing ratios were measured with a UV-absorption analyser (TEI-49i, Thermo Scientific, Waltham, USA). The efficiency of the photolytic conversion of NO_2 to NO was determined by a back-titration procedure involving the reaction of O_3 with NO using a gas phase titration system (SYCOS K-GPT, Ansyco GmbH, Karlsruhe, Germany). Details on the sampling schedule and time resolution of the trace gas profiles system are described in Plake et al. (2014).

This study is based on simultaneous operation of both vertical profile systems at identical heights and, thus, focuses on the period from 19 August to 26 September 2011 when both systems were operational. Vertical profiles of temperature (HMT337, Vaisala, Helsinki, Finland), wind speed and direction (WS425, Vaisala, Helsinki, Finland) were installed at 0.2, 0.8, 1.5, 2.5, 4.0 m. Soil temperature (107L, Campbell Scientific Inc., Logan, USA) was measured at −0.02 m. Global radiation (G) and the NO_2 photolysis frequency (j_{NO_2}) were measured at a height of 2.5 m with a net radiometer (CNR1, Kipp & Zonen, Delft, Netherlands), and a filter radiometer (Meteorology Consult GmbH, Königstein, Germany), respectively. The data of temperature, wind and radiation were recorded by a data logger (CR3000, Campbell Scientific) every 10 s. A 3-D sonic anemometer (CSAT-3, Campbell Scientific) placed at $z_{ref} = 3.0$ m measured 3-D wind and temperature at 20 Hz and the data were recorded by a CR3000 data logger. The friction velocity (u_*) and stability functions (z/L) were computed using the TK3 software (see Mauder and Foken, 2011). Eddy covariance fluxes of O_3 were simultaneously measured and are described in detail by Plake et al. (2014).

2.3 Theory

The data analysis was carried out for three individual layers (L_{1-3}), which are named in ascending order starting at the soil surface. Hence, L_1 is the lowermost canopy layer between the corresponding measurement heights z_{1-2} ($\Delta z(L_1) = 0.16$ m), L_2 the upper canopy layer between z_{2-3} ($\Delta z(L_2) = 0.6$ m), and L_3 the layer above the canopy between z_3 and z_{ref} ($\Delta z(L_3) = 2.2$ m). As shown in Plake et al. (2014) the vertical trace gas gradients between z_{ref} and z_4 were negligible, allowing the use of mixing ratios measured at z_4 for L_3.

2.3.1 Chemical timescales

The overall chemical timescale τ_{ch} (in s) of the NO–NO_2–O_3 triad (Lenschow, 1982) was calculated for each layer (L_i, $i = 1, 2, 3$) as

$$\tau_{ch}(L_i) = 2/\left\{j_{NO_2}(L_i)^2 + k_3(L_i)^2 \times \left(N_{O_3}(L_i) - N_{NO}(L_i)\right)^2 + 2 j_{NO_2}(L_i) \times k_3(L_i) \times \left(N_{O_3}(L_i) + N_{NO}(L_i) + 2N_{NO_2}(L_i)\right)\right\}^{0.5}, \quad (2)$$

where N_{O_3}, N_{NO} and N_{NO_2} are the number densities (in molecules cm^{-3}) of O_3, NO and NO_2 for L_{1-3}, and k_3 the reaction rate constant of Reaction (R3) (in cm^3 $molecule^{-1}\,s^{-1}$) according to Atkinson et al. (2004). Geometric means of the number densities at z_{1-4} were used in Eq. (2) to account for non-linear profiles – e.g. $N_{NO}(L_1) = \sqrt{N_{NO}(z_1) \times N_{NO}(z_2)}$.

Equation (2) gives the chemical timescale of Reactions (R1) and (R3) derived from the O_3 chemical-budget equation, i.e. considering only the reactions between O_3, NO and NO_2 and not taking into account reactions of other compounds (e.g. peroxy radicals and VOCs). It is defined as the time at which the mixing ratio of one of the compounds significantly changes from its initial value when reacting with the other ones. It can also be seen as the time required for reaching a new photostationary state following a change in O_3, NO or NO_2 mixing ratios, or the reaction rate constants j_{NO_2} and k_3 (see Ganzeveld et al., 2012). The underlying assumptions are:

- only source and sink terms of the "triad" are considered, which means other reactions (e.g. RO_2+ NO) are not included;
- covariance terms and other budget terms i.e. horizontal and vertical advection, flux divergence and change in O_3 mixing ratio $d[O_3]/dt$ are neglected.

2.3.2 NO_2 photolysis within the canopy

The data gaps in the measured time series of j_{NO_2} (in s^{-1}) above the canopy were filled using the parameterization of j_{NO_2} as a function of G (in $W\,m^{-2}$) by Trebs et al. (2009). This approach was also used to parameterize in-canopy j_{NO_2} from a vertical in-canopy profile of G. The latter was calculated as a function of the LAI profile using the method of Monsi and Saeki (1953):

$$G\,(\mathrm{LAI}) = G_0 \times \exp^{(-k_{ex} \times \mathrm{LAI})}, \quad (3)$$

where G_0 (in $W\,m^{-2}$) is the above-canopy G and k_{ex} is the dimensionless extinction coefficient of the canopy. In this study, the extinction coefficient of barley ($k_{ex} = 0.69$ by Monteith and Unsworth, 1990) was used. First G (LAI) was deduced and then converted into j_{NO_2}. Finally, geometric means of j_{NO_2} were calculated for $j_{NO_2}(L_{1-3})$.

2.3.3 Transport times

For L_3, height-integrated transport times $\tau_{tr}(L_3)$ (in s) were derived by multiplying the aerodynamic resistance $R_a(L_3)$ (e.g. Hicks et al., 1987; Erisman et al., 1994) with the layer thickness $\Delta z(L_3)$ (see Stella et al., 2013):

$$\tau_{tr}\,(L_3) = R_a(L_3) \times \Delta z(L_3) \quad (4)$$

$$R_a(L_3) = \frac{1}{\kappa \times u_*}\left[\ln\left(\frac{z_{ref}-d}{z_3-d}\right) - \Psi_H\left(\frac{z_{ref}-d}{L}\right) + \Psi_H\left(\frac{z_3-d}{L}\right)\right], \quad (5)$$

where κ is the von Kàrmàn constant (= 0.4), d the displacement height ($d = 0.75 \times h_c$), Ψ_H the stability correction function for heat (Foken, 2008) and L the Obukhov length.

In the canopy, $\tau_{tr}(L_i, i = 1, 2)$ were derived from the vertical Tn profiles (Lehmann et al., 1999; Plake and Trebs, 2013):

$$\tau_{tr}(L_i) = \ln\left[\frac{C_{Tn_{z_l}}(L_i)}{C_{Tn_{z_u}}(L_i)}\right] / \lambda \quad (6)$$

where $C_{Tn_{z_l}}$ and $C_{Tn_{z_u}}$ are the measured Tn concentrations (in $Bq\,m^{-3}$) at the lower (z_l) and upper (z_u) heights of L_i, and λ the radioactive decay rate $\lambda = \ln 2/T_{0.5} = 0.0125\,s^{-1}$ (Hänsel and Neumann, 1995).

3 Results

3.1 Meteorological conditions and mixing ratios

During the field experiment, low and high NO_x periods occurred that were directly coupled to the wind direction and could be attributed to two contrasting synoptic conditions characterized by different wind speeds (see Moravek et al., 2015). Figure 1a displays the dominance of southwesterly winds at the site during 45 % of the field experiment and their relation to relatively low NO_x levels (< 3 ppb). Contrastingly, winds from the northeastern sector were characterized by high NO_x levels often above 13 ppb (Fig. 1a). High NO_x episodes (up to 40 ppb) were accompanied with low wind speed (< $3\,m\,s^{-1}$) and low NO_x (< 5 ppb) with wind speeds above $3\,m\,s^{-1}$ as shown in Fig. 1b. O_3 levels exhibited the opposite dependency on wind speed, while the measured CO_2 levels generally showed a similar pattern for high and low NO_x levels.

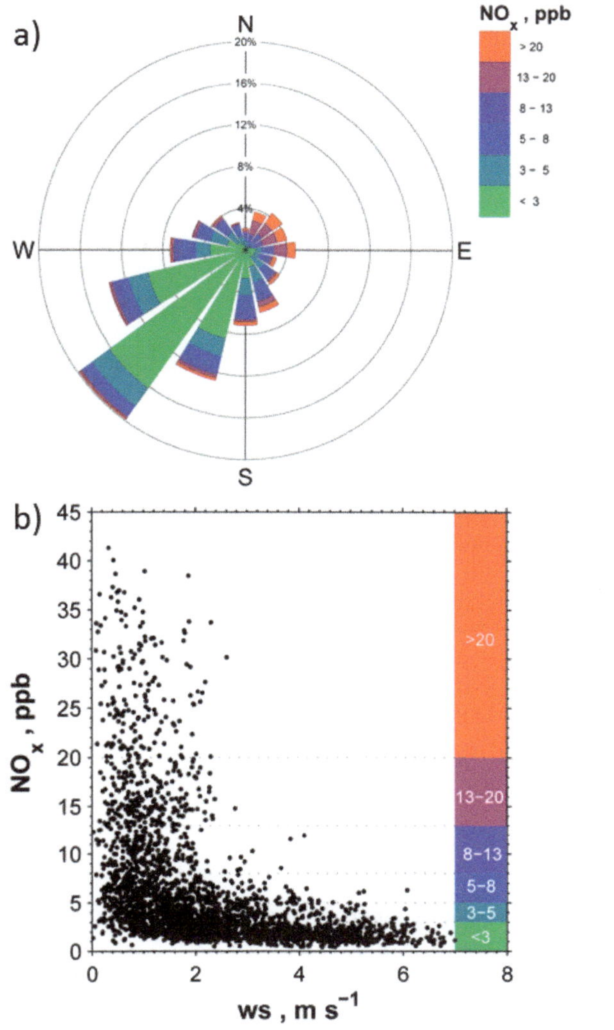

Figure 1. (a) Frequency distribution of wind direction related to NO_x mixing ratios; **(b)** NO_x mixing ratios as function of wind speed at the Mainz Finthen grassland site.

For further data analysis, defined criteria made it possible to account for these specific relationships. In order to clearly separate entire days (24 h) of contrasting conditions from each other, the criteria were defined as low NO_x or high NO_x periods when (i) the mean daytime wind speed was > $3\,m\,s^{-1}$ and the wind direction mainly ranged between 180 and 270°,

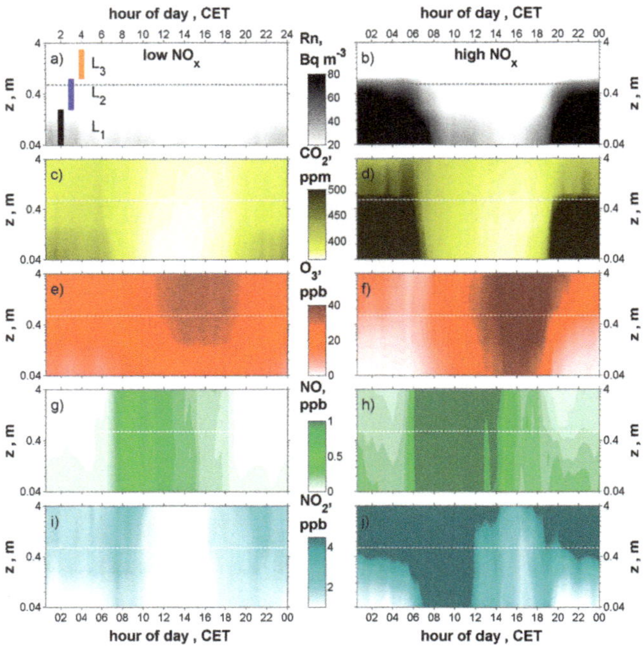

Figure 2. Time height cross-sections indicating the median vertical distribution of **(a, b)** Rn, **(c, d)** CO_2, **(e, f)** O_3, **(g, h)** NO and **(i, j)** NO_2 during low NO_x (left panels) and high NO_x (right panels) conditions at the Mainz Finthen grassland site. The canopy height (dotted line) and L_{1-3} are also shown. The plots were made using the *contourf* function of MATLAB.

or (ii) the mean daytime wind speed was $< 3\,m\,s^{-1}$ and the wind direction was mainly outside 180–270°, respectively. The wind direction definition was fulfilled during 96 % of the low NO_x periods and during 84 % of the high NO_x periods. Following these criteria, we identified 11 and 9 days as low and high NO_x periods, respectively, which were analysed separately.

3.2 Vertical profiles of trace gases

Since the wind field drives vertical exchange of scalars such as trace gases between vegetation and the atmosphere (Finnigan, 2000), it affects their vertical distribution. Passive tracers such as Rn and CO_2 are used especially at nighttime as indicators for vertical exchange processes within plant canopies (e.g. Trumbore et al., 1990; Nemitz et al., 2009). Generally, nighttime wind speeds during the low and high NO_x periods were accordingly higher and lower, respectively. This was reflected by the in-canopy concentrations of both Rn and CO_2 (Fig. 2a–d). During nighttime when both gases are exclusively emitted by soil, a rather weak enrichment within the canopy (Fig. 2a, c) reflected higher wind speeds and, thus, enhanced exchange during the low NO_x periods. In comparison, during the high NO_x periods a strong in-canopy CO_2 and Rn accumulation was observed (Fig. 2b, d). During daytime, photosynthesis prohibits the use of CO_2 as passive tracer, whereas Rn profiles are still useful as no biological processes such as stomatal uptake affect its concentration (Lehmann et al., 1999). The vertical exchange is generally enhanced during daytime causing dilution of the in-canopy Rn concentrations, which was especially pronounced during the low NO_x periods (Fig. 2a) and was less evident during the high NO_x periods (Fig. 2b) with generally lower wind speeds.

The vertical distribution of O_3 (Fig. 2e, f) reflected a typical pattern with lower mixing ratios closer to the ground and higher mixing ratios above. The diurnal O_3 maximum occurred during the afternoon around 16:00 CET (= UTC +1). Nevertheless, in the low NO_x periods the diurnal O_3 maximum was much less pronounced compared to the high NO_x periods with 35 and 50 ppb, respectively. Furthermore, characteristic vertical O_3 distributions were observed during the low and high NO_x periods. Nighttime O_3 gradients were less pronounced during the low NO_x than during the high NO_x periods. Median in-canopy values of O_3 were 10–20 ppb and were 20–25 ppb above the canopy during the low NO_x periods (Fig. 2e). During the high NO_x periods 1–6 ppb of O_3 were measured in the canopy and 10–25 ppb above the canopy (Fig. 2f).

During both the low and the high NO_x periods, significantly enhanced NO mixing ratios prevailed during the morning hours from 06:00 to 14:00 CET (Fig. 2g, h) with median diurnal maxima of 0.6 and 7.2 ppb, respectively, both occurring at 10:00 CET (not visible in Fig. 2h due to scaling). The NO mixing ratios decreased afterwards to approach nighttime minima. These were characterized by small vertical NO gradients during both periods. During low NO_x nights, NO appeared to be mainly present in the in-canopy air layer, with median mixing ratios at z_1 and z_2 of ≤ 0.1 ppb. The median values at z_1 and z_2 during the high NO_x periods were ≤ 0.3 ppb, respectively.

NO_2 mixing ratios were generally found to increase with height (Fig. 2i, j), featuring significantly stronger vertical differences during the high NO_x periods. Similar to NO, also NO_2 mixing ratios were enhanced throughout the profile during the morning hours of both, low and high NO_x periods, with corresponding values of 1–2.5 ppb and 6–14 ppb, respectively. At nighttime, comparable NO_2 mixing ratios of around 1 ppb prevailed during both periods at z_1. NO_2 showed stronger gradients above the canopy during the high NO_x periods. The diurnal NO_2 minima during low and high NO_x periods were observed between 12–16 CET and 14–16 CET, respectively.

3.3 Vertical profiles of chemical timescales

The obtained values for τ_{ch} were generally higher during nighttime than during daytime (Fig. 3a, d, g) and increased from L_3 to L_1. The validity of our applied criteria for separation between low and high NO_x periods is shown by the median values (brown and green lines) that nearly adjoin the interquartile range of the overall data set. Significantly higher

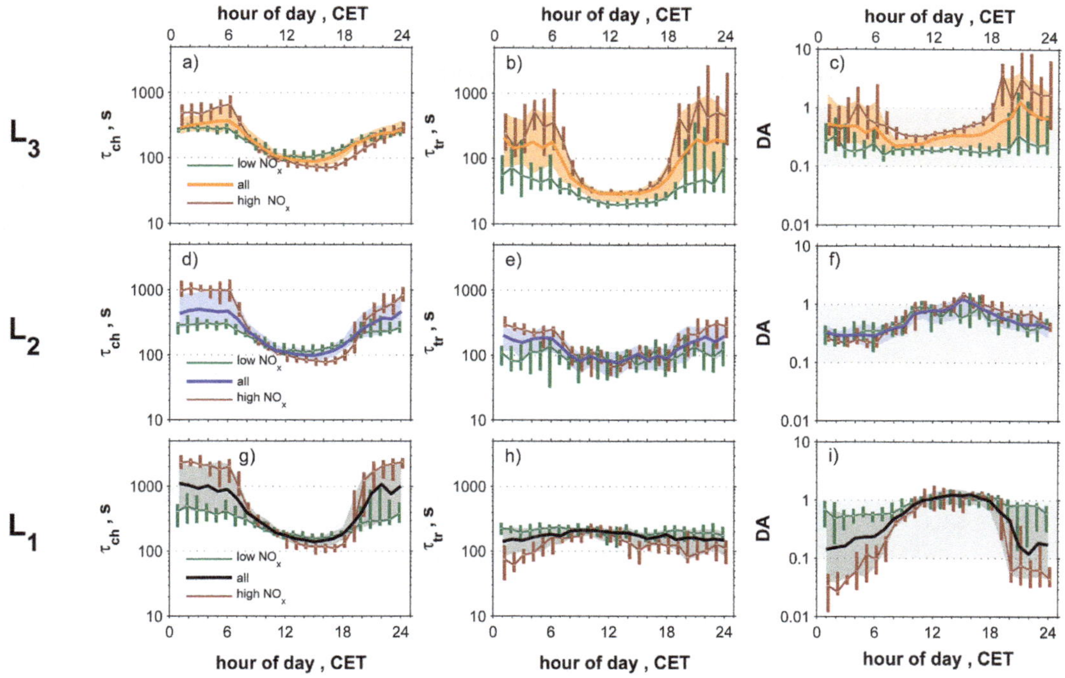

Figure 3. Diurnal courses of **(a, d, g)** $\tau_{ch}(L_{1-3})$, **(b, e, h)** $\tau_{tr}(L_{1-3})$ and **(c, f, i)** $DA(L_{1-3})$ considering all data from 19 August to 26 September 2011 (medians and shaded interquartile ranges) and the low NO_x and high NO_x periods (green and brown medians and interquartile boxes) at the Mainz Finthen grassland site.

τ_{ch} values prevailed during nighttime of the high NO_x periods, ranging from 300 to 2500 s in L_{1-3}. In contrast, low NO_x periods were characterized by τ_{ch} of 250–800 s in L_{1-3}. However, during daytime τ_{ch} was within 100–200 s in L_{1-3} for both periods. During the low NO_x periods τ_{ch} values were slightly higher compared to the high NO_x periods.

3.4 Vertical profiles of transport times

The median $\tau_{tr}(L_3)$ of all data (Fig. 3b) was one order of magnitude smaller during noon than at midnight with 30 and 200 s, respectively. As for τ_{ch} (Sect. 3.3), also in the case of τ_{tr} the medians of the low and high NO_x periods adjoined the interquartile range of the overall data set. For example, $\tau_{tr}(L_3)$ in the low NO_x periods never exceeded $\tau_{tr}(L_3)$ in the high NO_x periods (see Fig. 3b). The difference of $\tau_{tr}(L_3)$ between noon and midnight was largest in the high NO_x and smallest during the low NO_x periods with 470 and 40 s, respectively. Compared to L_{1-2} (Fig. 3e, h), the extreme values of the entire τ_{tr} data set were found above the canopy in L_3. The overall τ_{tr} minimum occurred during daytime of the low NO_x periods, and the maximum during nighttime of the high NO_x periods in L_3.

The diurnal course of $\tau_{tr}(L_2)$ from the entire data set in Fig. 3e exhibited a similar pattern as $\tau_{tr}(L_3)$, with higher $\tau_{tr}(L_2)$ during nighttime than during daytime. Representative nighttime and daytime values were 200 and 100 s, respectively, and a similar nighttime separation between the low and high NO_x periods as in Fig. 3b is observed.

In contrast, both diurnal $\tau_{tr}(L_1)$ medians representing all data and the high NO_x periods (Fig. 3h) were slightly higher during daytime between 08:00 and 13:00 CET than at nighttime with around 200 and 75–175 s, respectively. During the low NO_x periods, the median $\tau_{tr}(L_1)$ was relatively constant throughout the day with about 200 s. The pattern of $\tau_{tr}(L_1)$ was generally opposite to L_{2-3}, with faster $\tau_{tr}(L_1)$ in the high NO_x periods than in the low NO_x periods.

3.5 Vertical profiles of Damköhler numbers

The values for $DA(L_3)$ presented in Fig. 3c were generally smaller during daytime than during nighttime. They exhibited a diurnal minimum of 0.2 and a maximum of 1.3 at 08:00 and 21:00 CET, respectively. During the low NO_x periods, the difference of the $DA(L_3)$ median ($0.2 < DA(L_3) < 0.3$) to a DA of unity was highest, whereas in the high NO_x periods $DA(L_3)$ remained at higher median values ($0.3 < DA(L_3) < 3.9$).

In contrast, the diurnal course of $DA(L_2)$ in Fig. 3f exhibited its maximum of 1.25 at 15:00 CET and featured nighttime minima of about 0.3. The difference in $DA(L_2)$ between the low and high NO_x periods was not as pronounced as for $DA(L_3)$. This was most obvious from 15:00 to 24:00 CET with lower $DA(L_2)$ in the low NO_x periods. Hence, both

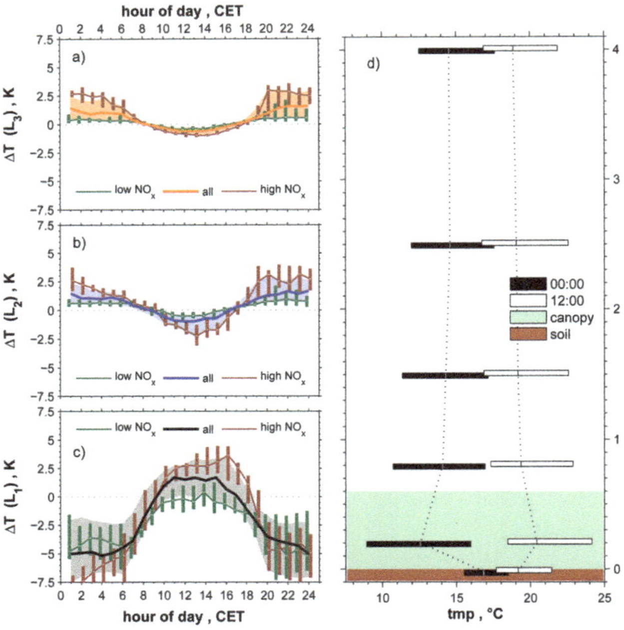

Figure 4. **(a–c)** Diurnal courses of measured $\Delta T(L_{1-3})$ considering all data from 19 August to 26 September 2011 (medians and shaded interquartile ranges) and the low and high NO_x periods (green and brown medians and interquartile ranges); note: $\Delta T(L_1, L_3)$ do not fully cover L_1 and L_3 (Sect. 2.3) due to availability of measurements (Sect. 2.2); **(a)** $\Delta T(L_3)$: 2.5–0.8 m; **(b)** $\Delta T(L_2)$: 0.8–0.2 m; **(c)** $\Delta T(L_1)$: 0.2 to −0.02 m (soil temperature). **(d)** Median vertical temperature profiles and interquartile boxes representing the thermal stratification at 00:00 and 12:00 CET considering all data from 19 August to 26 September 2011 at the Mainz Finthen grassland site.

$DA(L_{2,3})$ values throughout the day were within or above the critical range for DA under all conditions.

Interestingly, the diurnal course of $DA(L_1)$ (Fig. 3i) nearly mirrored that of $DA(L_3)$, with highest and lowest DA during daytime and nighttime, respectively. The diurnal median of $DA(L_1)$ partly exhibited values below 0.1 (transport dominates) during nighttime of the high NO_x periods, values above unity (chemistry dominates) from 12:00 to 17:00 CET under all conditions, and between 0.1 and unity during nighttime in the overall data set and in the low NO_x periods.

4 Discussion

4.1 Transport times and resistances

4.1.1 Thermal stratification

The diurnal courses of the temperature differences $\Delta T(L_{1-3})$ in Fig. 4a–c describe the stability in each layer. They clearly indicate contrasting stability conditions in L_1 and L_{2-3}. During daytime, negative values of $\Delta T(L_{2-3})$ reflected unstable conditions, while positive $\Delta T(L_1)$ reflected stable conditions. In contrast, at nighttime these conditions were reversed. Similar diurnal cycles of stratifications are observed for other canopies (see Jacobs et al., 1994; Kruijt et al., 2000; Nemitz et al., 2000), and are known to decouple the lower canopy from the air layers above (see Fig. 4d). Canopy coupling regimes are typically classified according to the detection of coherent structures in high frequency time series of scalars such as temperature (e.g. Foken et al., 2012; Dupont and Patton, 2012). In our data set $\Delta T(L_1)$ was used to explain why $\tau_{tr}(L_1)$ was generally smaller, i.e. transport was faster during nighttime than during daytime (Fig. 3h). The soil released stored heat as thermal plumes during nighttime that drove an in-canopy nighttime convection, which reached up to the height of the temperature inversion as found by Dupont and Patton (2012) or Jacobs et al. (1994). The $\tau_{tr}(L_1)$ maximum of 200 s from 08:00 to 13:00 CET could accordingly be attributed to positive $\Delta T(L_1)$ values at that time indicating a stable stratification. The thermal stratification was stronger during the high NO_x periods and weaker during the low NO_x periods in all layers (Fig. 4a–c). This was caused by higher wind speeds during the low NO_x periods causing increased turbulence and mixing that yielded smaller vertical temperature differences.

4.1.2 Aerodynamic resistances and transport times

Aerodynamic resistances above (R_a) and within the canopy (R_{ac}) are considered as important input variables for modelling studies on surface–atmosphere exchange fluxes. They can be derived from the transport times through a layer, normalized by the layer thickness ($R_{a(c)} = \tau_{tr}/\Delta z$). In cases when the thicknesses of two layers under consideration differ, the effectiveness of transport can be represented by the corresponding aerodynamic resistances. On the other hand, transport times are required to evaluate the influence of chemical reactions on fluxes (e.g. DA).

Typically, R_{ac} values are parameterized as a function of u_* and LAI (e.g. van Pul and Jacobs, 1994; Personne et al., 2009). These parameterizations are based on experiments above, e.g. crops such as maize (van Pul and Jacobs, 1994), and consider a vertically homogeneous leaf distribution (Personne et al., 2009). However, this approximation may differ substantially within grassland canopies, as their structure is characterized by high biomass density in the lowest layer (see Sect. 2.1).

The importance of our results is underlined by the direct assessment of measured R_{ac} values. From Eq. (6) we can assess R_{ac} for different canopy layers (L_1, L_2 and for the whole canopy ($\tau_{tr}(z_1, z_3)$; $\Delta z = z_3 - z_1$)) of the grassland (see Fig. 5). In the lowermost canopy layer, $R_{ac}(L_1)$ was generally highest with medians of 900 to 1000 s m^{-1} during nighttime and 1000 to 1300 s m^{-1} during daytime (Fig. 5). In comparison, Gut et al. (2002) found the aerodynamic resistance in the lowest metre of an Amazonian rain forest

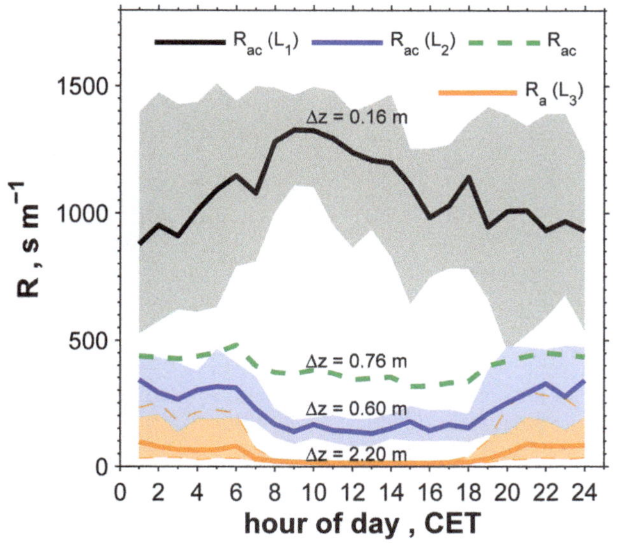

Figure 5. Diurnal courses of in-canopy aerodynamic resistances for each individual canopy layer ($R_{ac}(L_1)$, $R_{ac}(L_2)$) and for the entire grassland canopy ($R_{ac} = \frac{\tau_{tr}(L_1)+\tau_{tr}(L_2)}{z_3-z_1}$) at the Mainz Finthen site (median and shaded interquartile ranges). For comparison, the aerodynamic resistance above the canopy is also displayed ($R_a(L_3)$). The layer thickness (Δz) is indicated. The plot includes all data from 19 August until 26 September 2011.

canopy in a similar range with 600 s m^{-1} during nighttime and 1700 s m^{-1} during daytime, showing the same diurnal pattern.

As found for the transport times, the diurnal course of R_{ac} in the upper layers mirrored that of the lowermost layer (Fig. 5). In the upper canopy, the median of $R_{ac}(L_2)$ typically ranged around 300 s m^{-1} during nighttime and around 150 s m^{-1} during daytime. In comparison, above the canopy the median of $R_a(L_3)$ (Eq. 5) was substantially lower with around 80 and 15 s m^{-1} during nighttime and daytime, respectively. Consequently, the aerodynamic resistances in and above the canopy ($R_{ac}(L_{1,2})$ and $R_a(L_3)$) differed by almost two orders of magnitude during daytime, and by one order of magnitude during nighttime. Accordingly, the efficiency of aerodynamic transport decreased with decreasing height, even if the transport times were occasionally shorter in L_1 compared to L_3. R_{ac} for the entire canopy (Fig. 5) reflects the sum of the measured transport times divided by the entire layer thickness ($\Delta z = z_3 - z_1$) and can be considered as equivalent to a weighted average of $R_{ac}(L_1)$ and $R_{ac}(L_2)$. R_{ac} ranged in-between $R_{ac}(L_1)$ and $R_{ac}(L_2)$ with 440 s m^{-1} during nighttime and 360 s m^{-1} during daytime. The opposite diurnal courses of both $R_{ac}(L_1)$ and $R_{ac}(L_2)$ have an impact on R_{ac}, which in turn showed a smaller diurnal variation. As L_2 contained around 80 % of the layer thickness between z_1 and z_3 (see Fig. 5), R_{ac} was closer to $R_{ac}(L_2)$.

The median transport time through the 0.6 m high natural grassland canopy (also referred to as "canopy flushing time") was presented in the related study of Plake and Trebs (2013) for the same experiment. It was measured using the vertical thoron profile between z_1 and z_3 (Eq. 6). The canopy flushing time is consistent with the sum of $\tau_{tr}(L_1)$ and $\tau_{tr}(L_2)$ in this paper (see Fig. 7) and represents the in-canopy layer down to $0.07 \times h_c$ (z_1/h_c). It was determined to be ≤ 6 min exhibiting only small daytime/nighttime variation. Simon et al. (2005) reported canopy flushing times of around 60 min during any time of the day based on radon measurements within a 40 m high rain forest canopy for the layer between h_c and $0.13 \times h_c$ (canopy top to trunk space). As in the grassland canopy in Mainz Finthen, nighttime in-canopy convection accounted for the small daytime/nighttime variation in their study. Normalization of their canopy flushing time by the layer thickness yielded R_{ac} of the order of 100 s m^{-1}, which is around 3–4 times lower than the corresponding R_{ac} of the grassland site. Other studies (Holzinger et al., 2005; Rummel, 2005) based on surface renewal models reported somewhat lower flushing times. Rummel (2005) found flushing times in a 32 m high rain forest canopy of ≤ 200 s during daytime, which correspond to R_{ac} values ≤ 10 s m^{-1}. In the same way Holzinger et al. (2005) determined flushing times of 90 s during daytime and around 300 s during nighttime within a 6 m high scrubby pine forest. Corresponding R_{ac} values were of the order of 20 and 60 s m^{-1}, respectively.

Thus, it is important to note that even if the canopy height of natural grassland canopies is small compared to forests (around 1–10 %); the corresponding canopy flushing times are within the same order of magnitude as those reported for forest canopies. The typically high biomass density in the lower canopy of grasslands (e.g. Jäggi et al., 2006; Nemitz et al., 2009) is the most obvious explanation. It provides a large aerodynamic resistance (> 900 s m^{-1}) in a small layer adjacent to the ground (here: $R_{ac}(L_1)$). This resistance is large enough to increase the overall aerodynamic resistance of the whole canopy (R_{ac}) by 50 and 140 % during night and daytime, respectively. Consequently, R_{ac} of the grassland canopy was found to be at least 3–4 times higher than R_{ac} values representing corresponding in-canopy layers of forests taken from the literature.

Plake and Trebs (2013) compared their directly measured transport times with the parameterizations of van Pul and Jacobs (1994) and Personne et al. (2009). They found that none of the parameterizations was able to reproduce the entire diurnal course of the in-canopy transport. Agreement with the measured transport times was either found during daytime (Personne et al., 2009) or nighttime (van Pul and Jacobs, 1994), underlining the need for more direct measurements of in-canopy transport.

4.2 Chemical timescales

The non-linear profiles of NO, NO_2 and O_3 might have introduced uncertainties in $\tau_{ch}(L_{1-3})$. The potential uncertainties due to averaging were investigated by determining the

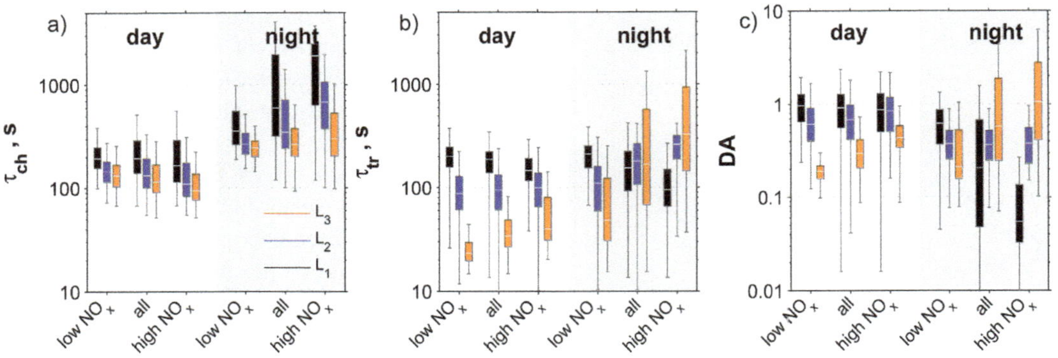

Figure 6. Comparison of box plot statistics for $\tau_{ch}(L_{1-3})$, $\tau_{tr}(L_{1-3})$ and $DA(L_{1-3})$ during daytime and nighttime including all data from 19 August until 26 September 2011 separated for the low and high NO_x periods at the Mainz Finthen grassland site.

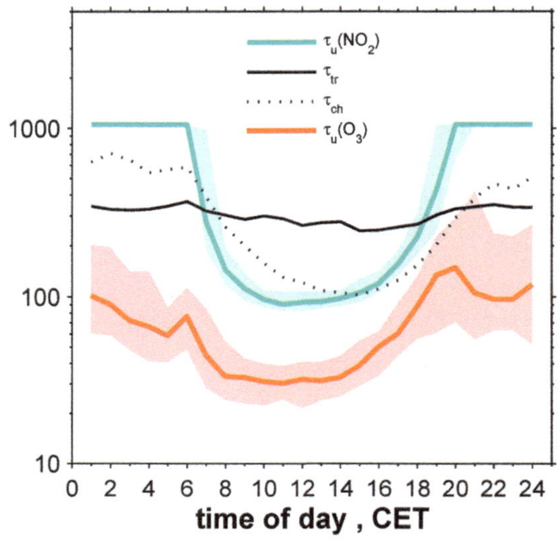

Figure 7. Comparison of median diurnal $\tau_u(NO_2)$, $\tau_u(O_3)$, τ_{tr} and τ_{ch} with interquartile ranges for the whole canopy layer (L_{1+2}) considering all data from 19 August to 26 September 2011 at the Mainz Finthen grassland site.

individual $\tau_{ch}(z_{1-4})$ and their subsequent comparison with $\tau_{ch}(L_{1-3})$. In L_1, L_2 and L_3 they were found to be ≤ 13, ≤ 4 and $\leq 2\,\%$, respectively, during daytime under any condition. During nighttime, the uncertainties in L_2 and L_3 were found to be 6 and 2 % during the low NO_x periods and 57 and 13 % during the high NO_x periods, respectively. In L_1 the uncertainty during nighttime was 30 % for all conditions. Furthermore, the in-canopy parameterization of j_{NO_2} might have introduced additional uncertainties since (i) in reality the attenuation of in-canopy radiation might be more complex than described in Eq. (3), and (ii) the parameterization of j_{NO_2} from G is prone to uncertainties of $> 40\,\%$ for $G < 100\,W\,m^{-2}$, $10 - 40\,\%$ for $G = 100–500\,W\,m^{-2}$ and $\leq 10\%$ for $G > 500\,W\,m^{-2}$ (Trebs et al., 2009). Moreover, omitting the influence of peroxy radical (HO_2+ RO_2) levels for the calculation of the chemical timescales introduces uncertainties. However, measurements of vertical profiles of $HO_2 + RO_2$ inside and above the grassland canopy are challenging and were not made during our experiment. Additionally, no straightforward analytical framework exists to calculate their influence on chemical timescales due to the variety of compounds and reaction rates involved in the complex HO_2 / RO_2 chemistry, which would require numeral modelling (see Heal et al., 2001).

The diurnal maxima and minima of $\tau_{ch}(L_{1-3})$ (Fig. 3a, d, g) were found to coincide with the O_3 minima and maxima (Fig. 2e, f), respectively. The impact of the terms in Eq. (2) on $\tau_{ch}(L_3)$ was examined by a correlation coefficient analysis. It was found to be highest for O_3 followed by NO_2 and NO with $r = -0.57$, $r = 0.46$ and $r = -0.07$, respectively. Consequently, the chemical timescale is dominated by the influence of O_3 as long as O_3 is present in excess compared to the other compounds. As the average air-chemical conditions in Mainz Finthen were characterized by a surplus of O_3 compared to NO_2 or NO (see Sect. 3.2), the magnitude of $\tau_{ch}(L_3)$ was most affected by the mixing ratios of O_3. In contrast, NO was generally less abundant, which explained the low overall impact on $\tau_{ch}(L_3)$. Only during high NO_x periods, when NO levels were above 5 ppb (see Sect. 3.1), was an increased impact of NO on $\tau_{ch}(L_3)$ found.

Figure 6a summarizes the chemical timescales. The temporal variation in τ_{ch}, expressed by higher nighttime and lower daytime values, can be considered as a rather typical pattern based on the diurnal courses of NO, NO_2 and O_3 (Fig. 2e–j) and their strong photochemical link. The vertical variation in $\tau_{ch}(L_{1-3})$ was on the one hand caused by the attenuation of j_{NO_2} in the canopy, and on the other hand by generally increasing mixing ratios of NO, NO_2 and O_3 with height (Fig. 2e–j). It should be noted that the latter finding is exclusively valid for this experimental site. Plake et al. (2014) measured insignificant soil biogenic NO emissions, underlined by weak in-canopy NO gradients (Fig. 2g, h). As already discussed in the previous paragraph, generally low NO, NO_2 and O_3 mixing ratios tend to cause high τ_{ch} values and vice versa. Consequently, at a site with higher NO emissions, such as an intensively managed agricultural field, the vertical τ_{ch} profile would most likely feature smaller vertical differences.

The extremely high $\tau_{ch}(L_{1,})$ during nighttime of the high NO_x periods (Fig. 6a) were a direct consequence of canopy decoupling (see Sect. 4.1.1). Transport of O_3 and NO_2 into the lower canopy was suppressed by the temperature inversion (see Fig. 2f, j). The residual O_3 and NO_2 molecules were convectively circulated within the lower canopy and subsequently deposited efficiently to surfaces until both almost disappeared in the early morning (Fig. 2f, j). Consequently, both the negligible NO emissions and the suppressed supply of O_3 and NO_2 from above yielded very low mixing ratios of all three trace gases, that in turn led to the very high $\tau_{ch}(L_{1,})$ values.

Our results are in line with those of Stella et al. (2013) who reported median diurnal τ_{ch} of 80–300 and 150–600 s above and within the canopy, respectively, for an intensively managed meadow. Their in-canopy τ_{ch} maximum was somewhat lower than in Mainz Finthen, which might be attributed to NO soil emissions or to averaging of different layers.

As mentioned above, the chemistry of HO_2 / RO_2 is not considered in our study. The reaction rate constant of $NO + HO_2$ / RO_2 is about 500 times higher than that of the reaction $NO + O_3$. Assuming relatively high average daytime $HO_2 + RO_2$ mixing ratios of 60 ppt inside the canopy (see Wolfe et al., 2014) the oxidation of NO to NO_2 would be as fast as with 30 ppb of O_3. This implies that these chemical timescales may be comparable to those of the reaction $NO + O_3$, which dominates τ_{ch} derived from Eq. (2). However, it should be noted that peroxides have a high affinity to be lost at surfaces, which may reduce their presence in dense grassland canopies. Additionally, reactions between volatile organic compounds (VOCs) and O_3 (e.g. Atkinson and Arey, 2003) might have influenced chemical timescales. Simultaneously measured biogenic VOC mixing ratios featured very small values at our site (e.g. isoprene < 0.7 ppb, monoterpene < 0.3 ppb, J. Kesselmeier, personal communication, 2013). Due to the absence of measurements the influence of anthropogenic VOCs is not taken into account.

4.3 Influence of meteorology and air pollution on vertical Damköhler number profiles

The summarized daytime $DA(L_{1-3})$ in Fig. 6c exhibited decreasing DA values with increasing layer height. Thus, the likelihood of chemical flux divergence decreased from L_1 to L_3. Throughout L_1 to L_3, the τ_{ch} values (Fig. 6a) showed a lower variation compared to the corresponding τ_{tr} values (Fig. 6b). Hence, the daytime DA profile was mainly caused by the vertical τ_{tr} profile.

Interestingly, the nighttime DA for all data and the high NO_x periods showed opposite vertical profiles, indicating an increasing likelihood of chemical flux divergence with increasing layer height (L_1 to L_3). This was especially pronounced during nighttime of the high NO_x periods, where the only instance without indication for a flux divergence within the entire data set for L_1 was found. The reasons for this were (i) the very high $\tau_{ch}(L_1)$ (Fig. 6a and Sect. 4.2), and (ii) the reversed vertical transport time profiles during nighttime (fastest in L_1) of the high NO_x periods (Fig. 6b). This finding agrees very well with Rummel (2005) who found that the transport timescale in the lowermost layer of an Amazonian rainforest is faster than the chemical timescale of the NO–NO_2–O_3 triad during nighttime.

Above the canopy, the order of magnitude (Fig. 6c) and the median diurnal course (Fig. 3c) of DA compared well with the values of Stella et al. (2013). The in-canopy DA of Stella et al. (2013) was smaller than the DA above the canopy throughout the entire day, which is in contrast to our study. Considering the average canopy flushing time given in Plake and Trebs (2013) (see Sect. 4.1.2) and the $\tau_{ch}(L_2)$ (see Fig. 6a), the average in-canopy DA in Mainz Finthen was in the order of 2 and 1 for daytime and nighttime, respectively. Thus, in our study in-canopy DA values are on average significantly higher than above the canopy throughout the day. Finally, it should be noted that in-canopy DA values may not be fully representative as besides transport and chemistry, additional sources and sinks for trace gases exist within plant canopies. These are specific for each trace gas and will be further discussed below.

4.4 Implications for measured fluxes

4.4.1 Potential NO_x canopy reduction

Within the canopy, $DA(L_{1-2})$ (Fig. 3f, i; Fig. 6c) suggest that chemical reactions exhibited a larger influence on the NO–NO_2–O_3 triad during daytime than during night. However, reactive traces gases in canopies are deposited to soil and vegetation elements. Trace gases can be efficiently taken up by plants due to adsorption/absorption on cuticles and diffusion through stomata (e.g. Breuninger et al., 2012). On the other hand, particularly NO is simultaneously produced by microbial processes and is subsequently released from soil. Although the latter process could be neglected in this study due to insignificant NO soil emissions (Sect. 4.2), the uptake of NO_2 by plants, however, was investigated in order to draw conclusions on potential NO_x canopy reduction within natural grasslands canopies. Hence, the characteristic timescale of plant uptake (τ_u) of NO_2 integrated over the whole canopy (L_{1+2}) was estimated based on a resistance model (Baldocchi, 1988), following an approach of Rummel (2005) as

$$\tau_u(x) = \left(\frac{1}{R_{L_x}} \times \frac{\Delta LAI}{\Delta z}\right)^{-1}, \qquad (7)$$

where x denotes the trace gas of interest (here $x = NO_2$) and R_{L_x} is the leaf resistance of x:

$$R_{L_x} = \left(\frac{1}{R_{bl_x} + R_{s_x} + R_{mes_x}} + \frac{2}{R_{bl_x} + R_{cut_x}}\right)^{-1}, \qquad (8)$$

with R_{bl_x} being the leaf boundary layer resistance of x calculated according to Personne et al. (2009), R_{s_x} the stomatal resistance of x taken from Plake et al. (2014), R_{mes_x} the mesophyll resistance set to $200\,s\,m^{-1}$ for NO_2 and R_{cut_x} the cuticular resistance set to $9999\,s\,m^{-1}$ due to the unimportance of cuticular deposition for NO_2 (both values were taken from Stella et al., 2013).

During daytime, the median of $\tau_u(NO_2)$ calculated from all data was typically found to be the shortest amongst all timescales relevant for NO_2, typically ranging between 90 and 160 s between 09:00 and 17:00 CET (Fig. 7). This timescale was closely followed by $\tau_{ch}(L_{1+2})$ exhibiting values between 100 and 200 s in the same time period, but the minimum was slightly skewed towards the afternoon. In contrast, the canopy flushing time, $\tau_{tr}(L_{1+2})$, ranged from 250 to 290 s (Fig. 7) during this time. For a comparable natural grassland canopy with significant NO soil emissions, this would imply an efficient in-canopy conversion of NO to NO_2 during daytime, followed by an effective NO_2 plant uptake as the transport was found to be 2–3 times slower. Furthermore, the biomass density within the lowest 0.2 m of the canopy (i) strongly inhibits the transport, especially in L_1 during daytime (Figs. 5; 6b), and (ii) attenuates the photolysis of NO_2 at the soil–canopy interface, the location where NO is usually emitted. This indicates that a strong potential for NO_x canopy reduction exists in such grassland ecosystems during daytime, when the precondition of significant NO soil emissions is fulfilled. The presence of peroxy radicals may even amplify this process.

However, during nighttime, $\tau_u(NO_2)$ was found to be very large (Fig. 7) due to plant stomata closure. Hence, the role of turbulence–chemistry interactions ($DA(L_{1-2})$) was dominating over biological uptake processes. In L_1 the transport of soil emitted NO would be slowest under relatively windy nighttime situations (low NO_x periods in Fig. 3h) due to undeveloped in-canopy convection. Thus, a considerably high mixing ratio of O_3 within the canopy (Fig. 2e) would lead to an efficient formation of NO_2 indicated by $DA(L_{1-2})$ close to unity. The uptake of NO_2 by plants would be insignificant (see above), and only soil deposition would lead to a small NO_2 depletion. Most likely, such nighttime conditions would lead to simultaneous NO_2 and NO canopy emission fluxes. During nights with low wind speeds (high NO_x periods), the temperature inversion constitutes a "canopy lid". Within the canopy (L_{1+2}) the reaction of residual O_3 (see Sect. 4.2) and soil-emitted NO would compete with the O_3 deposition to surfaces. Subsequently, a mixture of NO and NO_2 would be trapped inside the canopy. Besides some minor in-canopy NO_2 losses (see above), a distinct NO and NO_2 release may occur in the morning hours, which has been observed for forests (see Foken et al., 2012; Dorsey et al., 2004; Jacob and Wofsy, 1990).

4.4.2 Influence on O3 deposition flux

Similar to NO_2, the application of in-canopy DA values for O_3 remains difficult, since plant uptake and deposition to plant surfaces and the soil are additional O_3 loss pathways. The characteristic timescale of O_3 plant uptake and soil deposition $\tau_u(O_3)$, shown in Fig. 7, was estimated using Eqs. (7) and (8), with $x = O_3$, R_{mes_x} set to $0\,s\,m^{-1}$ (Erisman et al., 1994) and $R_{cut_x} = R_{ns_x} - R_{soil_x}$ (e.g. Lamaud et al., 2009). The R_{ns_x} was taken from Plake et al. (2014) and $R_{soil_x} = 240\,s\,m^{-1}$ according to Lamaud et al. (2009). The $\tau_u(O_3)$ ranged from 30 to 150 s, which clearly illustrates the dominance of in-canopy O_3 plant uptake and soil deposition. The $\tau_u(O_3)$ was significantly faster than both $\tau_{tr}(L_{1-2})$ and $\tau_{ch}(L_{1-2})$ during the entire day (values are given in Sect. 4.4.1).

Consequently, only DA values above the canopy, i.e. $DA(L_3)$, provide an indication for potential O_3 flux divergence. Since the $DA(L_3)$ always exceeded 0.1 (Figs. 3c, 6c), a chemical flux divergence could not be excluded at the Mainz Finthen site. Furthermore, $DA > 1$ (Fig. 3c) during the early evening hours clearly indicated the dominance of chemical reactions over transport. During the low NO_x periods, the probability for flux divergence was lowest. The influence of chemistry on O_3 deposition fluxes determined by Plake et al. (2014) at the Mainz Finthen grassland site will be discussed below. The median O_3 fluxes for the entire measurement period ranged from about -1.5 to $-6\,nmol\,m^2\,s^{-1}$ during night and daytime, respectively.

Due to negligible NO soil emissions, a chemical flux divergence in L_3 resulting from counter-directed fluxes of NO and O_3 was very unlikely. Nevertheless, we use a simplified method proposed by Duyzer et al. (1995) based on R1 and R3 and the law of mass conservation to approximate the flux divergence by the correction factor α_{O_3} as

$$\alpha_{O_3} = \frac{\phi_x}{\kappa \times u_*} \times \left[k_1 \times \left(N_{NC} \times F^*_{O_3} + N_{O_3} \times F^*_{NO}\right) - j_{NO_2} \times F^*_{NO_2}\right], \quad (9)$$

where $\phi_x = \phi_{O_3} = \phi_H$ was the stability correction function for heat (Högström, 1988), $F^*_{O_3}$ the measured O_3 flux at z_{ref} determined by the eddy covariance method (see Plake et al., 2014) and F^*_{NO} and $F^*_{NO_2}$ the corresponding NO and NO_2 fluxes determined by the dynamic chamber technique (see Plake et al., 2014). The estimated O_3 deposition flux at z_3 (F_{z_3}) was then calculated as

$$F_{z_3} = F_{z_{ref}} + \int_{z_3}^{z_{ref}} \left(\frac{\partial F}{\partial z}\right)_z dz = F^*_{O_3} + \alpha_{O_3} \times z_3 \times \left(1 + \ln\frac{z_{ref}}{z_3}\right), \quad (10)$$

where the term $\int_{z_3}^{z_{ref}} \left(\frac{\partial F}{\partial z}\right)_z dz$ is the integrated flux divergence within L_3. The resulting median O_3 flux divergence was quantified to be less than 1 %, confirming our a priori assumption.

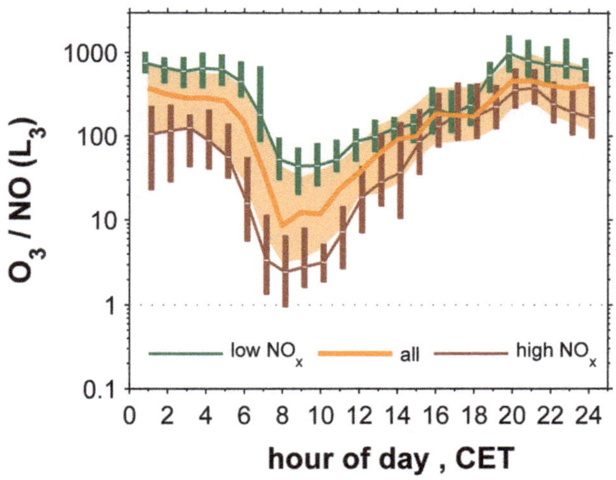

Figure 8. Diurnal course of the O_3 to NO ratio in L_3 considering all data from 19 August to 26 September 2011 (median and shaded interquartile range) and separated for the low NO_x and high NO_x periods (medians and interquartile boxes) at the Mainz Finthen grassland site.

Nevertheless, we examined the influence of the enhanced NO mixing ratios in the morning hours (Sect. 3.2, Fig. 2g, h), accompanied by very low O_3 / NO ratios (Fig. 8) on the measured O_3 fluxes. A chemically induced O_3 flux $F_c(O_3)$ due to production $P(O_3)$ or loss $L(O_3)$ of O_3 by Reactions (R1) and (R3) integrated over the air column of L_3 was quantified according to Rummel et al. (2007) as

$$F_c(O_3) = P(O_3) - L(O_3) = \int_{z_3}^{z_{ref}} \frac{\mu_{NO_2}(z) \times \rho_d(z)}{\tau_{NO_2}(z)} dz - \int_{z_3}^{z_{ref}} \frac{\mu_{O_3}(z) \times \rho_d(z)}{\tau_{O_3}(z)} dz, \quad (11)$$

where ρ_d (in mol m^{-3}) is the molar density of dry air. τ_{NO_2} and τ_{O_3} (in s) are the chemical depletion times of NO_2 and O_3, respectively:

$$\tau_{NO_2} = \frac{1}{j_{NO_2}} \quad (12)$$

$$\tau_{O_3} = \frac{1}{k_1 \times N_{NO}}. \quad (13)$$

Figure 9a displays the diurnal courses of $P(O_3)$ and $L(O_3)$ exhibiting median values of 0 to 1.9 nmol m^{-2} s^{-1} and 0 to −1.4 nmol m^{-2} s^{-1}, respectively. The maximal median values were related to the enhanced NO_x levels in the morning. The resulting median net $F_c(O_3)$ in Fig. 9b ranged between 0.6 and −0.05 nmol m^{-2} s^{-1}, representing a net O_3 production in L_3 during daytime and a net loss during nighttime. Repeatedly, the medians of low and high NO_x periods adjoined the interquartile range of the overall data set, showing a variability of one order of magnitude of net $F_c(O_3)$ during daytime. Considering the median values of all data, the chemical contribution to the measured EC flux of O_3 would be around +10 % during daytime and −3 % during nighttime. Consequently, the actual daytime O_3 deposition to the canopy is higher than measured by the EC method. This finding is interesting, as to our knowledge previous studies only reported chemical O_3 losses above the canopy due to outbalancing of the reactions of O_3 with NO (e.g. Dorsey et al., 2004) or VOCs (e.g. Kurpius and Goldstein, 2003) emitted by soil or plants, respectively. The net O_3 production in our study was attributed to a deviation from the NO–NO_2–O_3 photostationary state by a surplus of NO_2 due to NO oxidation by e.g., peroxy radicals or other oxidants (see Trebs et al., 2012). Unfortunately, we were not able to assess the impact of the reactions involved in the net O_3 production on the calculated chemical timescales as measurements of peroxy radicals were not available. The NO_2 surplus might have originated from simultaneous emissions of non-methane hydrocarbons, carbon monoxide (CO) and NO from motorways surrounding the site in a distance of some kilometres. It is well known that under daytime conditions peroxy radicals are formed that can oxidize NO without consumption of O_3 resulting in net O_3 production (Seinfeld and Pandis, 2006). Although this O_3 production might also prevail at other experimental sites, this effect is most likely balanced or even exceeded by the destruction of O_3 due to biogenic soil NO emissions which were negligible at our site (a nutrient-poor grassland site).

5 Conclusions

For the first time, we simultaneously measured transport times (aerodynamic resistances), vertical profiles of NO–NO_2–O_3 mixing ratios and micrometeorological quantities within and above a natural grassland canopy. The obtained data were analysed to gain insights about the potential NO_x canopy reduction in the grassland canopy, and to analyse the contribution of chemistry on fluxes of purely depositing compounds, such as O_3. We observed two extreme regimes: (a) high wind speed and low NO_x mixing ratios (low NO_x periods) and (b) low wind speed and high NO_x mixing ratios (high NO_x periods). Our study highlights that as a result of in-canopy convection, nighttime transport in the lowermost canopy layer is fastest, while conditions above the canopy are highly stable due to low wind speed during the high NO_x periods.

Interestingly, our results on transport-chemistry interactions within the grassland canopy are partly comparable to those found in the Amazonian rainforest, although the vertical canopy structure differs substantially. Natural grasslands exhibit very high biomass densities in the lowest part of the canopy. Thus, the median aerodynamic resistance in the low-

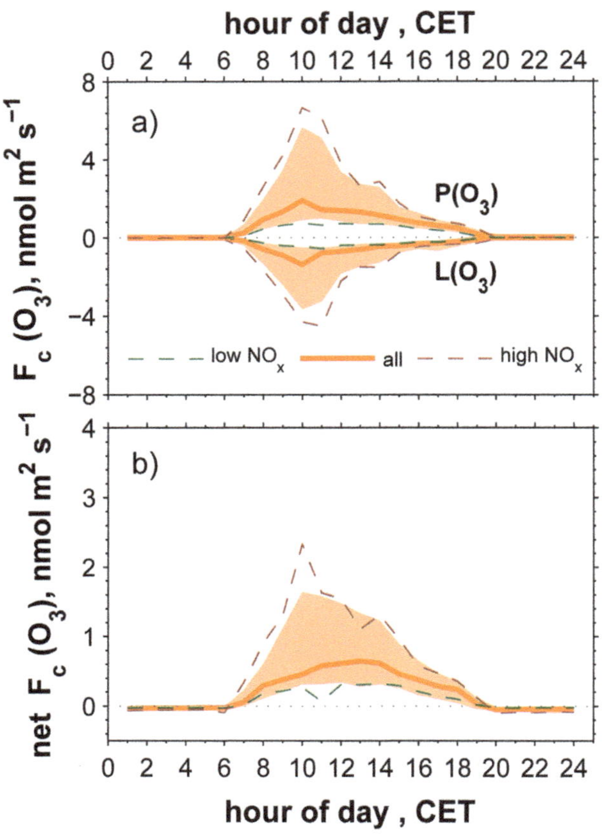

Figure 9. Diurnal courses showing **(a)** $P(O_3)$ and $L(O_3)$ and **(b)** $F_c(O_3)$ (Eq. 11) for L_3 considering all days from 19 August to 26 September 2011 (medians and shaded interquartile ranges) and separated for the low and high NO_x periods (medians) at the Mainz Finthen grassland site.

ermost canopy layer (0.04–0.2 m) was found to be of the same magnitude ($> 900\,s\,m^{-1}$) and to feature the same diurnal pattern (higher during daytime, lower at night) as the aerodynamic resistance determined for the lowest metre of an Amazonian rain forest. The median in-canopy aerodynamic resistance representing the whole grassland canopy was at least 3–4 times higher than in-canopy aerodynamic resistances of forest canopies available from the literature. Our results reveal that even if the canopy height of natural grassland canopies is small compared to forests (around 1–10 %), the corresponding canopy flushing times are of the same order of magnitude as those reported for forest canopies. The median canopy flushing times exhibited only small daytime/nighttime variability, which is in accordance with a detailed study on flushing times within an Amazonian rain forest (Simon et al., 2005). The small daytime/nighttime variability is caused by the compensating transport efficiencies in the lower and upper canopy layers during daytime and nighttime for both canopy types.

The median canopy flushing time of the grassland was found to be ≤ 6 min and the chemical timescale of the NO–NO_2–O_3 triad during daytime ranged between 1–3 min. This has obvious implications, e.g. for soil-emitted reactive compounds such as NO, potentially implying fast chemical conversion of NO to NO_2 within the grassland canopy. During daytime the plant uptake of NO_2 was shown to be 2–3 times faster than the canopy flushing time. Inevitably, this leads to a strong potential NO_x canopy reduction in the presence of biogenic NO soil emissions. This effect may be amplified if substantial levels of peroxy radicals prevail inside the canopy. Due to the extensive global terrestrial coverage with grassland canopies, this finding is highly relevant for the application of global chemistry and transport models. Our results indicate that the daytime NO_x canopy reduction for grasslands may be higher than 64 %. Nevertheless, an improved daily average for the NO_x canopy reduction factor in analogy to Yienger and Levy (1995) cannot be presented here due to the insignificant NO soil emissions at the experimental site.

Moreover, we determined a median net chemical O_3 production of 10 % during daytime within the air column between the EC flux measurement and the canopy, which was due to the absence of soil biogenic NO emission in our study. Hence, in contrast to previous studies our measured O_3 deposition flux by EC is slightly underestimated. The chemical flux divergence for O_3 was one order of magnitude larger during the high NO_x than during the low NO_x periods. In-canopy Damköhler numbers were shown to be relevant for NO_2 only under nighttime conditions, which was due to the minor role of NO_2 uptake by plants at this time. Damköhler numbers above the canopy indicated a potential flux divergence, but did not provide a hint for the observed chemical production of O_3. The only instance without indication of a chemical flux divergence within the entire data set was found during nighttime of the high NO_x periods in the lowermost canopy layer.

Acknowledgements. This project was funded by the Max Planck Society. We are grateful to J.-C. Mayer for the installation and the maintenance of meteorological sensors and for supporting the data evaluation. We are indebted to A. Moravek for his support with the eddy covariance measurements and the corresponding data evaluation and the fruitful discussions on many details of this study.

Edited by: X. Wang

References

Atkinson, R. and Arey, J.: Gas-phase tropospheric chemistry of biogenic volatile organic compounds: a review, Atmos. Environ., 37, S197–S219, 2003.

Atkinson, R., Baulch, D. L., Cox, R. A., Crowley, J. N., Hampson, R. F., Hynes, R. G., Jenkin, M. E., Rossi, M. J., and Troe, J.:

Evaluated kinetic and photochemical data for atmospheric chemistry: Volume I – gas phase reactions of O_x, HO_x, NO_x and SO_x species, Atmos. Chem. Phys., 4, 1461–1738, doi:10.5194/acp-4-1461-2004, 2004.

Aylor, D. E., Wang, Y. S., and Miller, D. R.: Intermittent wind close to the ground within a grass canopy, Bound.-Lay. Meteorol., 66, 427–448, 1993.

Bakwin, P. S., Wofsy, S. C., Fan, S. M., Keller, M., Trumbore, S. E., and Dacosta, J. M.: Emission of nitric-oxide (NO) from tropical forest soils and exchange of NO between the forest canopy and atmospheric boundary-layers, J. Geophys. Res.-Atmos., 95, 16755–16764, 1990.

Baldocchi, D.: A multi-layer model for estimating sulfur-dioxide deposition to a deciduous oak forest canopy, Atmos. Environ., 22, 869–884, 1988.

Breuninger, C., Oswald, R., Kesselmeier, J., and Meixner, F. X.: The dynamic chamber method: trace gas exchange fluxes (NO, NO_2, O_3) between plants and the atmosphere in the laboratory and in the field, Atmos. Meas. Tech., 5, 955–989, doi:10.5194/amt-5-955-2012, 2012.

Crutzen, P.: Discussion of chemistry of some minor constituents in stratosphere and troposphere, Pure Appl. Geophys., 106, 1385–1399, 1973.

Denmead, O. T., Freney, J. R., and Dunin, F. X.: Gas exchange between plant canopies and the atmosphere: Case-studies for ammonia, Atmos. Environ., 42, 3394–3406, 2007.

Dorsey, J. R., Duyzer, J. H., Gallagher, M. W., Coe, H., Pilegaard, K., Weststrate, J. H., Jensen, N. O., and Walton, S.: Oxidized nitrogen and ozone interaction with forests, I: Experimental observations and analysis of exchange with Douglas fir, Q. J. Roy Meteor. Soc., 130, 1941–1955, 2004.

Dupont, S. and Patton, E. G.: Momentum and scalar transport within a vegetation canopy following atmospheric stability and seasonal canopy changes: the CHATS experiment, Atmos. Chem. Phys., 12, 5913–5935, doi:10.5194/acp-12-5913-2012, 2012.

Duyzer, J. H., Deinum, G., and Baak, J.: The interpretation of measurements of surface exchange of nitrogen-oxides – correction for chemical-reactions, Philos. T. R. Soc. A, 351, 231–248, 1995.

Erisman, J. W., Vanpul, A., and Wyers, P.: Parametrization of surface-resistance for the quantification of atmospheric deposition of acidifying pollutants and ozone, Atmos. Environ., 28, 2595–2607, 1994.

Finnigan, J.: Turbulence in plant canopies, Annu. Rev. Fluid Mech., 32, 519–571, 2000.

Foken, T.: Micrometeorology, Springer, Berlin, Heidelberg, 306 pp., 2008.

Foken, T., Meixner, F. X., Falge, E., Zetzsch, C., Serafimovich, A., Bargsten, A., Behrendt, T., Biermann, T., Breuninger, C., Dix, S., Gerken, T., Hunner, M., Lehmann-Pape, L., Hens, K., Jocher, G., Kesselmeier, J., Luers, J., Mayer, J. C., Moravek, A., Plake, D., Riederer, M., Rutz, F., Scheibe, M., Siebicke, L., Sörgel, M., Staudt, K., Trebs, I., Tsokankunku, A., Welling, M., Wolff, V., and Zhu, Z.: Coupling processes and exchange of energy and reactive and non-reactive trace gases at a forest site – results of the EGER experiment, Atmos. Chem. Phys., 12, 1923–1950, doi:10.5194/acp-12-1923-2012, 2012.

Ganzeveld, L., Ammann, C., and Loubet, B.: Review on modelling atmosphere biosphere ex-change of Ozone and Nitrogen oxides. Background document for the joint ÉCLAIRE/COST ES0804 Expert Workshop "From process scale to global scale: integrating our knowledge on biosphere/atmosphere exchange modelling of trace gases and volatile aerosols", 2012.

Gut, A., Scheibe, M., Rottenberger, S., Rummel, U., Welling, M., Ammann, C., Kirkman, G. A., Kuhn, U., Meixner, F. X., Kesselmeier, J., Lehmann, B. E., Schmidt, W., Muller, E., and Piedade, M. T. F.: Exchange fluxes of NO_2 and O_3 at soil and leaf surfaces in an Amazonian rain forest, J. Geophys. Res.-Atmos., 107, 8060, doi:10.1029/2001JD000654, 2002.

Hänsel, H. and Neumann, W.: Physik, Spectrum, Akad. Verl., Heidelberg, Berlin, Oxford, 1995.

Heal, M. R., Booth, B. B. B., Cape, J. N., and Hargreaves, K. J.: The influence of simplified peroxy radical chemistry on the interpretation of NO_2-NO-O_3 surface exchange, Atmos. Environ. 35, 1687–1696, 2001.

Hicks, B. B., Baldocchi, D. D., Meyers, T. P., Hosker, R. P., and Matt, D. R.: A preliminary multiple resistance routine for deriving dry deposition velocities from measured quantities, Water Air Soil Poll., 36, 311–330, 1987.

Högström, U.: Non-dimensional wind and temperature profiles in the atmospheric surface-layer – a re-evaluation, Bound.-Lay. Meteorol., 42, 55–78, 1988.

Holzinger, R., Lee, A., Paw, K. T., and Goldstein, U. A. H.: Observations of oxidation products above a forest imply biogenic emissions of very reactive compounds, Atmos. Chem. Phys., 5, 67–75, doi:10.5194/acp-5-67-2005, 2005.

IPCC: Climate Change 2013: the physical science basis. Working group I contribution to the fifth assessment report of the Intergovernmental Panel on Climate Change, University Press, Cambridge, UK, 2013.

Jacob, D. J. and Wofsy, S. C.: Budgets of reactive nitrogen, hydrocarbons, and ozone over the amazon-forest during the wet season, J. Geophys. Res.-Atmos., 95, 16737–16754, 1990.

Jacobs, A. F. G., Vanboxel, J. H., and Elkilani, R. M. M.: Nighttime free-convection characteristics within a plant canopy, Bound.-Lay. Meteorol., 71, 375–391, 1994.

Jäggi, M., Ammann, C., Neftel, A., and Fuhrer, J.: Environmental control of profiles of ozone concentration in a grassland canopy, Atmos. Environ., 40, 5496–5507,2006.

Kasanko, M., Palmieri, A., and Coyette, C.: Land cover/ land use statistics, in: Agriculture and Fishery Statistics, edited by: Coyette, C., and Schenk, H., Eurostat, Luxembourg, 158, 2011.

Kruijt, B., Malhi, Y., Lloyd, J., Norbre, A. D., Miranda, A. C., Pereira, M. G. P., Culf, A., and Grace, J.: Turbulence statistics above and within two Amazon rain forest canopies, Bound.-Lay. Meteorol., 94, 297–331, 2000.

Kurpius, M. R. and Goldstein, A. H.: Gas-phase chemistry dominates O_3 loss to a forest, implying a source of aerosols and hydroxyl radicals to the atmosphere, Geophys. Res. Lett., 30, 1371, doi:10.1029/2002gl016785, 2003.

Lamaud, E., Loubet, B., Irvine, M., Stella, P., Personne, E., and Cellier, P.: Partitioning of ozone deposition over a developed maize crop between stomatal and non-stomatal uptakes, using eddy-covariance flux measurements and modelling, Agr. Forest Meteorol., 149, 1385–1396, 2009.

Lehmann, B. E., Lehmann, M., Neftel, A., Gut, A., and Tarakanov, S. V.: Radon-220 calibration of near-surface turbulent gas transport, Geophys. Res. Lett., 26, 607–610, 1999.

Lenschow, D. H.: Reactive trace species in the boundary-layer from a micrometeorological perspective, J. Meteorol. Soc. Jpn., 60, 472–480, 1982.

Lerdau, M. T., Munger, L. J., and Jacob, D. J.: Atmospheric chemistry - the NO_2 flux conundrum, Science, 289, 2291, doi:0.1126/science.289.5488.2291, 2000.

Mauder, M. and Foken, T.: Documentation and instruction manual of the eddy-covariance software package TK3, Arbeitsergebnisse Nr. 46, 2011.

Monsi, M. and Saeki, T.: Über den Lichtfaktor in den Pflanzengesellschaften und seine Bedeutung für die Stoffproduktion, Jpn. J. Botany, 14, 22–52, 1953.

Monteith, J. L. and Unsworth, M. H.: Principles of environmental physics, 2nd ed., E. Arnold, London, New York, 291 pp., 1990.

Moravek, A., Foken, T., and Trebs, I.: Application of a GC-ECD for measurements of biosphere–atmosphere exchange fluxes of peroxyacetyl nitrate using the relaxed eddy accumulation and gradient method, Atmos. Meas. Tech., 7, 2097–2119, doi:10.5194/amt-7-2097-2014, 2014.

Moravek, A., Stella, P., Foken, T., and Trebs, I.: Influence of local air pollution on the deposition of peroxyacetyl nitrate to a nutrient-poor natural grassland ecosystem, Atmos. Chem. Phys., 15, 899–911, doi:10.5194/acp-15-899-2015, 2015.

Nemitz, E., Sutton, M. A., Gut, A., San Jose, R., Husted, S., and Schjoerring, J. K.: Sources and sinks of ammonia within an oilseed rape canopy, Agr. Forest Meteorol., 105, 385–404, 2000.

Nemitz, E., Loubet, B., Lehmann, B. E., Cellier, P., Neftel, A., Jones, S. K., Hensen, A., Ihly, B., Tarakanov, S. V., and Sutton, M. A.: Turbulence characteristics in grassland canopies and implications for tracer transport, Biogeosciences, 6, 1519–1537, doi:10.5194/bg-6-1519-2009, 2009.

Personne, E., Loubet, B., Herrmann, B., Mattsson, M., Schjoerring, J. K., Nemitz, E., Sutton, M. A., and Cellier, P.: SURFATM-NH3: a model combining the surface energy balance and bi-directional exchanges of ammonia applied at the field scale, Biogeosciences, 6, 1371–1388, doi:10.5194/bg-6-1371-2009, 2009.

Plake, D. and Trebs, I.: An automated system for selective and continuous measurements of vertical thoron profiles for the determination of transport times near the ground, Atmos. Meas. Tech., 6, 1017–1030, doi:10.5194/amt-6-1017-2013, 2013.

Plake, D., Stella, P., Moravek, A., Mayer, J. C., Ammann, C., Held, A., and Trebs, I.: Comparison of ozone deposition measured with the dynamic chamber and the eddy covariance method, Agr. Forest Meteorol., submitted, 2014.

Rinne, J., Markkanen, T., Ruuskanen, T. M., Petaja, T., Keronen, P., Tang, M. J., Crowley, J. N., Rannik, U., and Vesala, T.: Effect of chemical degradation on fluxes of reactive compounds – a study with a stochastic Lagrangian transport model, Atmos. Chem. Phys., 12, 4843–4854, doi:10.5194/acp-12-4843-2012, 2012.

Ripley, E. A. and Redman, R. E.: Grassland, in: Vegetation and the atmosphere, edited by: Monteith, J. L., Acad. Press, London, 1976.

Rummel, U.: Turbulent exchange of ozone and nitrogen oxides between an Amazonian rain forest and the atmosphere, PhD thesis, Faculty of Biology, Chemistry and Geosciences, University of Bayreuth, Bayreuth, 246 pp., 2005.

Rummel, U., Ammann, C., Gut, A., Meixner, F. X., and Andreae, M. O.: Eddy covariance measurements of nitric oxide flux within an Amazonian rain forest, J. Geophys. Res.-Atmos., 107, 8050, doi:10.1029/2001jd000520, 2002.

Rummel, U., Ammann, C., Kirkman, G. A., Moura, M. A. L., Foken, T., Andreae, M. O., and Meixner, F. X.: Seasonal variation of ozone deposition to a tropical rain forest in southwest Amazonia, Atmos. Chem. Phys., 7, 5415–5435, 2007, http://www.atmos-chem-phys.net/7/5415/2007/.

Seinfeld, J. H. and Pandis, S. N.: Atmospheric chemistry and physics : from air pollution to climate change, 2. Edn., Wiley, Hoboken, NJ, 1203 pp., 2006.

Simon, E., Lehmann, B. E., Ammann, C., Ganzeveld, L., Rummel, U., Meixner, F. X., Nobre, A. D., Araujo, A., and Kesselmeier, J.: Lagrangian dispersion of Rn-222, H_2O and CO_2 within Amazonian rain forest, Agr. Forest Meteorol., 132, 286–304, 2005.

Stella, P., Kortner, M., Ammann, C., Foken, T., Meixner, F. X., and Trebs, I.: Measurements of nitrogen oxides and ozone fluxes by eddy covariance at a meadow: evidence for an internal leaf resistance to NO_2, Biogeosciences, 10, 5997–6017, doi:10.5194/bg-10-5997-2013, 2013.

Suttie, J. M., Reynolds, S. G., and Batello, C.: Introduction, in: Grasslands of the world, edited by: Suttie, J. M., Reynolds, S. G., and Batello, C., FAO, Rome, 2005.

Swinbank, W. C.: A comparison between predictions of dimensional analysis for constant-flux layer and observations in unstable conditions, Q. J. Roy Meteor. Soc., 94, 460–467, 1968.

Trebs, I., Bohn, B., Ammann, C., Rummel, U., Blumthaler, M., Königstedt, R., Meixner, F. X., Fan, S., and Andreae, M. O.: Relationship between the NO_2 photolysis frequency and the solar global irradiance, Atmos. Meas. Tech., 2, 725–739, doi:10.5194/amt-2-725-2009, 2009.

Trebs, I., Mayol-Bracero, O. L., Pauliquevis, T., Kuhn, U., Sander, R., Ganzeveld, L., Meixner, F.X., Kesselmeier, J., Artaxo, P. and Andreae, M. O.: Impact of the Manaus urban plume on trace gas mixing ratios near the surface in the Amazon Basin: Implications for the NO-NO_2-O_3 photostationary state and peroxy radical levels, J. Geophys. Res., 117, D05307, doi:10.1029/2011JD016386, 2012.

Trumbore, S. E., Keller, M., Wofsy, S. C., and Dacosta, J. M.: Measurements of soil and canopy exchange-rates in the Amazon rainforest using Rn-222, J. Geophys. Res.-Atmos., 95, 16865–16873, 1990.

van Pul, W. A. J., and Jacobs, A. F. G.: The conductance of a maize crop and the underlying soil to ozone under various environmental-conditions, Bound.-Lay. Meteorol., 69, 83–99, 1994.

Warneck, P.: Chemistry of the natural atmosphere, 2nd ed., Academic Press, San Diego, California, 927 pp., 2000.

Wolfe, G. M., Cantrell, C., Kim, S., Mauldin III, R. L., Karl, T., Harley, P., Turnipseed, A., Zheng, W., Flocke, F., Apel, E. C., Hornbrook, R. S., Hall, S. R., Ullmann, K., Henry, S. B., DiGangi, J. P., Boyle, E. S., Kaser, L., Schnitzhofer, R., Hansel, A., Graus, M., Nakashima, Y., Kajii, Y., Guenther, A., and Keutsch, F. N.: Missing per-oxy radical sources within a summertime ponderosa pine forest, Atmos. Chem. Phys., 14, 4715–4732, doi:10.5194/acp-14-4715-2014, 2014.

Yienger, J. J. and Levy, H.: Empirical-model of global soil-biogenic NO_x emissions, J. Geophys. Res.-Atmos., 100, 11447–11464, 1995.

13

Simulated anthropogenic CO_2 storage and acidification of the Mediterranean Sea

J. Palmiéri[1,2], J. C. Orr[1], J.-C. Dutay[1], K. Béranger[2], A. Schneider[3], J. Beuvier[4,5], and S. Somot[5]

[1]LSCE/IPSL, Laboratoire des Sciences du Climat et de l'Environnement, CEA-CNRS-UVSQ, Gif-sur-Yvette, France
[2]ENSTA-ParisTech, Palaiseau, France
[3]GEOMAR; Helmholtz-Zentrum für Ozeanforschung Kiel, Germany
[4]Mercator Ocean, Ramonville Saint-Agne, France
[5]CNRM/Météo-France, Toulouse, France

Correspondence to: J. Palmiéri (julien.palmieri@lsce.ipsl.fr)

Abstract. Constraints on the Mediterranean Sea's storage of anthropogenic CO_2 are limited, coming only from data-based approaches that disagree by more than a factor of two. Here we simulate this marginal sea's anthropogenic carbon storage by applying a perturbation approach in a high-resolution regional model. Our model simulates that, between 1800 and 2001, basin-wide CO_2 storage by the Mediterranean Sea has increased by 1.0 Pg C, a lower limit based on the model's weak deep-water ventilation, as revealed by evaluation with CFC-12. Furthermore, by testing a data-based approach (transit time distribution) in our model, comparing simulated anthropogenic CO_2 to values computed from simulated CFC-12 and physical variables, we conclude that the associated basin-wide storage of 1.7 Pg, published previously, must be an upper bound. Out of the total simulated storage of 1.0 Pg C, 75 % comes from the air–sea flux into the Mediterranean Sea and 25 % comes from net transport from the Atlantic across the Strait of Gibraltar. Sensitivity tests indicate that the Mediterranean Sea's higher total alkalinity, relative to the global-ocean mean, enhances the Mediterranean's total inventory of anthropogenic carbon by 10 %. Yet the corresponding average anthropogenic change in surface pH does not differ significantly from the global-ocean average, despite higher total alkalinity. In Mediterranean deep waters, the pH change is estimated to be between −0.005 and −0.06 pH units.

1 Introduction

The Mediterranean region will be particularly affected by climate change (Giorgi, 2006; MerMEX group, 2011; Diffenbaugh and Giorgi, 2012). This region, currently classified as semiarid to arid, is projected to become warmer and drier (Gibelin and Déqué, 2003; Giorgi and Lionello, 2008), amplifying existing water resource problems. At the same time, already heightened anthropogenic pressures are expected to intensify further (Attané and Courbage, 2001, 2004). It has been proposed that the Mediterranean Sea will experience amplified acidification relative to the global average surface ocean (Touratier and Goyet, 2009, 2011). The Mediterranean Sea is able to absorb relatively more anthropogenic CO_2 per unit area for two reasons: (i) its higher total alkalinity gives it greater chemical capacity to take up anthropogenic CO_2 and neutralize acid and (ii) its deep waters are ventilated on relatively short timescales, allowing deeper penetration of this anthropogenic tracer. However the quantity of anthropogenic CO_2 that has been absorbed by the Mediterranean Sea remains uncertain. This quantity cannot be measured directly because the anthropogenic component cannot be distinguished from the much larger natural background. Instead it has been estimated indirectly from observable physical and biogeochemical quantities.

Several indirect methods have been developed, some of which have been compared using the same data sets along basin-wide transects in the Mediterranean Sea. Their first comparison (El Boukary, 2005) revealed large differences between methods. With data from a 1995 transect on the

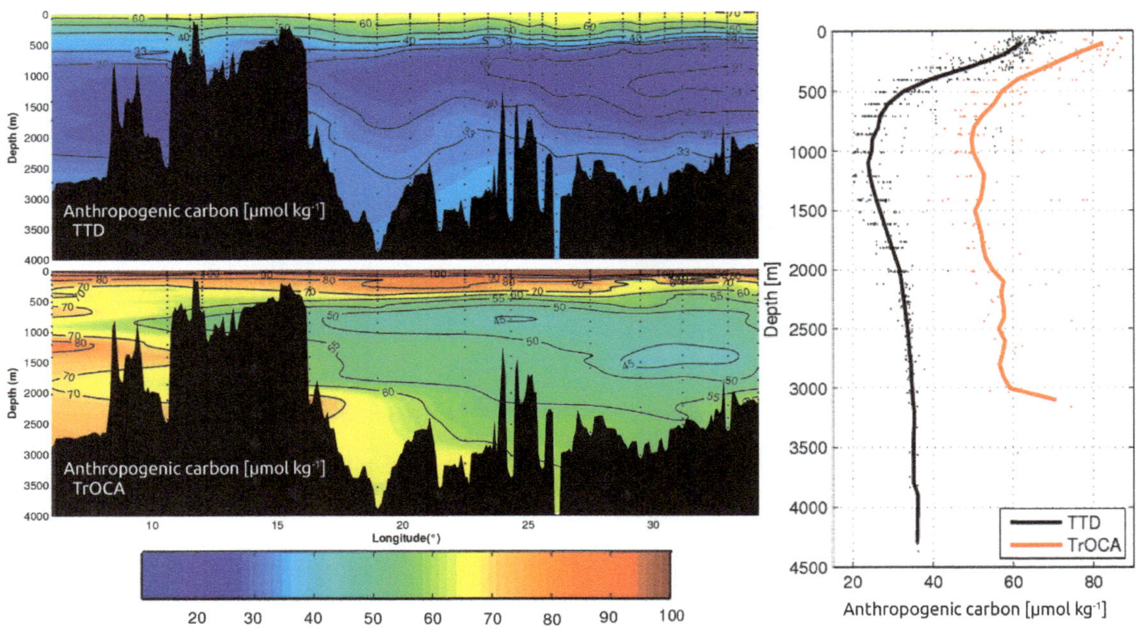

Figure 1. Anthropogenic carbon (μmol kg^{-1}) estimated with two data-based methods – TTD (Schneider et al., 2010) and TrOCA (Touratier and Goyet, 2011) – along the *Meteor* M51/2 section (November 2001). Vertical profiles on the right are for mean anthropogenic C_T along the section estimated by each method. The vertical profile of the TrOCA method does not go below 3500 m depth because it requires measurements, some of which where not available at that depth.

Meteor (M31/1), El Boukary estimated with two methods that the Mediterranean Sea had absorbed 3.1 and 5.6 Pg C, but he concluded that even the lower value was an overestimate. Later, data from a transbasin transect in 2001 (*Meteor* M51/2) was used by two independent studies to estimate anthropogenic CO_2. In the first, Schneider et al. (2010) used those data with the transit time distribution approach (TTD, from Waugh et al., 2006), a back-calculation method based on CFC-12 derived mean ages of water masses. For the second, Touratier and Goyet (2011) used their TrOCA (Touratier et al., 2007) approach, which relies on measured O_2, dissolved inorganic carbon (C_T), and total alkalinity (A_T). Anthropogenic carbon estimated with TrOCA is always greater than that from TTD (Fig. 1), with more than a factor of 2 difference both in the western basin below 500 m depth and in the eastern basin between 500 and 1500 m. These large differences in estimated concentrations further result in opposing estimates for the net transport across the Strait of Gibraltar. With TrOCA, the Mediterranean Sea appears to export anthropogenic carbon to the Atlantic Ocean, whereas with TTD, net calculated exchange is in the opposite direction (Schneider et al., 2010; Aït-Ameur and Goyet, 2006; Huertas et al., 2009; Flecha et al., 2011). These large discrepancies between results from currently used data-based methods fuel a debate about the quantity of anthropogenic carbon that is taken up by the Mediterranean Sea.

Here we take another approach by simulating anthropogenic CO_2 storage of the Mediterranean Sea. Unlike simulations for the global ocean, we cannot rely on coarse-resolution global models because they do not resolve fine-scale bathymetry and circulation features that are critical for the Mediterranean Sea. This semi-enclosed marginal sea is separated into the eastern and the western basins by the Strait of Sicily (Fig. 2). Each of these basins has critical circulation features that are often heavily influenced by bathymetry. For example, Atlantic Water (AW) enters the Mediterranean Sea at the surface via the narrow Strait of Gibraltar and flows counterclockwise along the coast. Surface-water circulation patterns are influenced by deep- and intermediate-water formation driven by strong winds, which are themselves steered and intensified by surrounding mountainous topography. Deep and intermediate waters are formed in four major areas: the Rhodes gyre, where the Levantine Intermediate Water (LIW) originates; the Gulf of Lions and the nearby Ligurian Sea in the Liguro-Provençal sub-basin, which together produce Western Mediterranean Deep Water (WMDW); and two adjacent regions, the Adriatic and the Aegean sub-basins, which together produce Eastern Mediterranean Deep Water (EMDW). Also influencing the deep circulation is the Mediterranean Outflow Water (MOW), a complex mixture of different intermediate and deep waters outflowing at the Strait of Gibraltar underneath the incoming AW.

To capture these and other key features, we used a high-resolution circulation model of the Mediterranean Sea forced by high-resolution air–sea fluxes, interannually varying Atlantic Ocean boundary conditions, and realistic land freshwater inputs. This regional circulation model is combined with

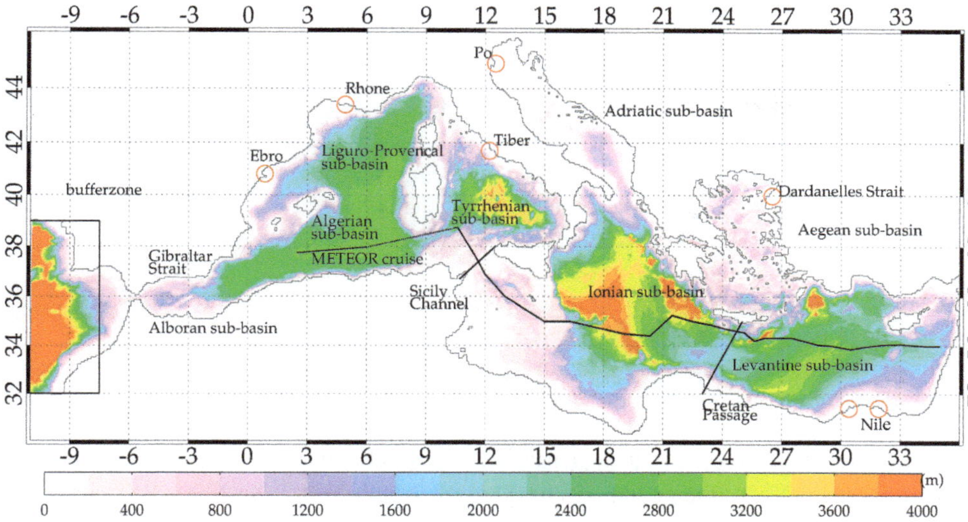

Figure 2. Map of the MED12 model domain and bathymetry with location of the main Mediterranean sub-basins: Adriatic, Aegean, Alboran, Algerian, Liguro-Provençal, Ionian, Levantine, and Tyrrhenian. Red circles indicate the mouths of the main Mediterranean rivers (Ebro, Rhone, Tiber, Po, and Nile) and the input from the Black Sea at the Dardanelles Strait. Black lines indicate the Strait of Sicily, the Crete Passage, and the trans-Mediterranean section from the *Meteor* M51/2 cruise (November 2001). The rectangular area in the western part of the model domain indicates the Atlantic buffer zone (see Sects. 2.1 and 2.2.2). The eastern basin is situated to the east of the Strait of Sicily, while the western basin is situated between the Strait of Gibraltar and the Strait of Sicily. The entire Mediterranean Sea refers to all waters east of the Strait of Gibraltar.

a computationally efficient perturbation approach (Sarmiento et al., 1992) to model anthropogenic CO_2 in the Mediterranean Sea. This geochemical approach simulates only the change in CO_2 uptake due to the anthropogenic perturbation, assuming that the natural carbon cycle is unaffected by the increase in CO_2. For efficiency, it relies on a formulation that relates surface-water changes in the partial pressure of carbon dioxide ($\delta p\text{CO}_2$) to those in dissolved inorganic carbon (δC_T). By focusing only on the C_T perturbation, it needs just one tracer and one simulation that covers only the industrial period. Thus it circumvents the need for the prerequisite simulation of the natural carbon cycle, which requires many tracers and a much longer simulation to allow modeled tracer fields to reach near-steady-state conditions.

Our goal here is to use these simulations to help bracket the Mediterranean Sea's uptake of anthropogenic CO_2 as well as its net transport across the Strait of Gibraltar, while exploring how this marginal sea's heightened total alkalinity affects anthropogenic CO_2 uptake and corresponding changes in pH.

2 Methods

Anthropogenic CO_2 simulations were performed offline with circulation fields from the NEMO circulation model. The same approach was used to perform simulations of CFC-12 in order to evaluate modeled circulation, which heavily influences penetration of both of these passive transient tracers.

2.1 Circulation model

The regional circulation model NEMO-MED12 (Beuvier et al., 2012a) is a Mediterranean configuration of the free-surface ocean general circulation model NEMO (Madec and the NEMO team, 2008). Its domain includes the whole Mediterranean Sea and extends into the Atlantic Ocean to 11°W; it does not include the Black Sea (Fig. 2). The horizontal model resolution is around 7 km, adequate to resolve key mesoscale features. Details of the model and its parametrizations are given by Beuvier et al. (2012a). NEMO-MED12 has been used to study the WMDW formation (Beuvier et al., 2012a), the response of the mixed layer to high-resolution air–sea forcings (Lebeaupin Brossier et al., 2011), and the transport across the Strait of Gibraltar (Soto-Navarro et al., 2014). NEMO-MED12 is descended from a suite of Mediterranean regional versions of OPA and NEMO used by the French modeling community: OPAMED16 (Béranger et al., 2005), OPAMED8 (Somot et al., 2006), and NEMO-MED8 (Beuvier et al., 2010).

The physical simulation used here is very close to that described in Beuvier et al. (2012b). It is initiated in October 1958 with temperature and salinity data representative of the 1955–1965 period using the MEDATLAS data set (MEDAR/MEDATLAS-Group, 2002; Rixen et al., 2005). For the Atlantic buffer, initial conditions are taken from the 2005 World Ocean Atlas for temperature (Locarnini et al., 2006) and salinity (Antonov et al., 2006). Boundary conditions are also needed to specify physical forcing for the atmosphere, freshwater inputs from rivers and the Black Sea, and exchange with the adjacent Atlantic Ocean. For the atmosphere, NEMO-MED12 is forced with daily evaporation, precipitation, radiative and turbulent heat fluxes, and momentum fluxes from the ARPERA data set (Herrmann and Somot, 2008), all over the period 1958–2008. The ARPERA forcing constitutes a 56-year, high-resolution forcing (50 km, daily data) with a good temporal homogeneity (see Herrmann et al., 2010, for more details about the post-2001 period). The SST-relaxation and water-flux correction terms are applied as in Beuvier et al. (2012a). River runoff is derived by Beuvier et al. (2010, 2012a) from the interannual data set of Ludwig et al. (2009) and Vörösmarty et al. (1996). Freshwater input from the Black Sea follows runoff estimates from Stanev and Peneva (2002). Exchange with the Atlantic is modeled through a buffer zone (see Fig. 2) between 11° W and the Strait of Gibraltar, where the model's 3-D temperature and salinity fields are relaxed to the observed climatology (Locarnini et al., 2006; Antonov et al., 2006) while superimposing anomalies of interannual variations from the ENSEMBLES reanalysis performed with a global version of NEMO (Daget et al., 2009). To reproduce the monthly cycle of the Mediterranean Sea's water volume, we restore the total sea-surface height (SSH) in the Atlantic buffer zone toward the monthly climatological values of the GLORYS1 reanalysis (Ferry et al., 2010)

The atmospheric forcing used by Beuvier et al. (2012b) does not include modifications to improve dense water fluxes through the Cretan Arc, which plays a critical role in deep-water formation during the Eastern Mediterranean Transient (EMT). As detailed by Roether et al. (1996, 2007), the EMT was a temporary change in the EMDW formation that occurred when the source of this deep water switched from the Adriatic Sea to the Aegean Sea during 1992–1993. Beuvier et al. (2010) showed that a previous simulation with the circulation model NEMO-MED8 (1/8° horizontal resolution) was able to reproduce a transient in deep-water formation as observed for the EMT, but the simulated transient produced less EMDW. Beuvier et al. (2012b) later performed a simulation with NEMO-MED12 with comparable forcing between October 1958 and December 2012. To improve the characteristics of the simulated EMT, namely the density of newly formed EMDW during 1992–1993, its weak formation rate, and its shallow spreading at 1200 m, we performed a sensitivity test with modified forcing. For that, we modified the ARPERA forcings over the Aegean sub-basin, increasing mean values as done by Herrmann et al. (2008) to study the Gulf of Lions. More specifically, during November to March in the winters of 1991–1992 and 1992–1993, we increased daily surface heat loss by 40 W m^{-2}, daily water loss by 1.5 mm day^{-1}, and the daily wind stress modulus by 0.02 N m^{-2}. That resulted in average wintertime increases in heat loss (+18 %), water loss (+41 %), and wind intensity (+17 %) over the Aegean sub-basin. The increased heat and water losses allow NEMO-MED12 to form denser water masses in the Aegean Sea during the most intense winters of the EMT, while increased wind stress drives more intense mixing via winter convection. Furthermore, enhanced convection accelerates the transfer of surface temperature and salinity perturbations into intermediate and deep layers of the Aegean Sea. In summary for this study, the physical model forcing is identical to that from Beuvier et al. (2012b), except for the enhanced forcing during the two winters mentioned above.

2.2 Passive tracer simulations

2.2.1 CFC-12

The trace atmospheric gas CFC-12 has no natural component. Being purely anthropogenic, its atmospheric concentration has increased since the 1930s and has leveled off in recent decades. Although sparingly soluble, it enters that ocean by gas exchange. There it remains chemically and biologically inert, tracking ocean circulation and mixing. Precise measurements of CFC-12 along several trans-Mediterranean sections make it particularly suited for evaluating these regional model simulations. To model CFC-12, we followed protocols from phase 2 of the Ocean Carbon-Cycle Model Intercomparison Project (OCMIP-2) as described by Dutay et al. (2002). For the air–sea flux of CFC-12 (F_{CFC}), we used the standard formulation for a passive gaseous tracer,

$$F_{CFC} = k_w(C_{eq} - C_{surf}), \quad (1)$$

where k_w is the gas transfer velocity (also known as the piston velocity), C_{surf} is the simulated sea-surface concentration of CFC-12, and C_{eq} is the atmospheric equilibrium concentration. That is,

$$C_{eq} = \alpha p\text{CFC}, \quad (2)$$

where α is the CFC-12 solubility, a function of local seawater temperature and salinity (Warner and Weiss, 1985), and pCFC is the atmospheric partial pressure of CFC-12 computed from the atmospheric mole fraction in dry air. Here we assume atmospheric pressure remains at 1 atm neglecting spatiotemporal variations. The gas transfer velocity is computed from surface-level wind speeds (u) from the ARPERA forcing following the Wanninkhof (1992, Eq. 3) formulation:

$$k_w = au^2\left(\frac{Sc}{660}\right)^{-1/2}, \quad (3)$$

where $a = 0.31$ and Sc is also the CFC-12 Schmidt number computed following Wanninkhof (1992, Table A1).

Regarding lateral boundary conditions, for the Atlantic buffer zone (between 11° W and the Strait of Gibraltar), we assume that net exchange at the boundary may be neglected while relying on atmospheric exchange of this rapidly equilibrating tracer as the dominant factor.

2.2.2 Anthropogenic CO_2

To model anthropogenic CO_2 in the Mediterranean Sea, we use the perturbation approach (Siegenthaler and Joos, 1992; Sarmiento et al., 1992). This classic approach uses a simple relationship between the change in surface-ocean pCO_2, which is needed to compute the air–sea CO_2 flux, and the change in C_T. Such a relationship is necessary for carbon dioxide, unlike for CFC-12, because as CO_2 dissolves in the ocean it does not simply remain as a dissolved gas; it dissociates into two other inorganic species: bicarbonate and carbonate ions. When modeling only the change in the total of the three species (δC_T), the simple relationship that is used allows models to carry only that perturbation tracer.

In the perturbation approach, the geochemical driver is the atmospheric change in carbon dioxide. As written by Sarmiento et al. (1992), that change in terms of the partial pressure of carbon dioxide in moist air is

$$\delta p\mathrm{CO_2}_{\mathrm{a}} = \left(p\mathrm{CO_2}_{\mathrm{a}} - p\mathrm{CO_2}_{\mathrm{a},0}\right)(1 - e_\mathrm{s}/p_\mathrm{a}). \tag{4}$$

For the model simulation, the two pCO_2 terms (in μatm) on the right side of the equation are identical to xCO_2 (in ppm), although units differ, because they both refer to dry air and because the perturbation approach assumes a total atmospheric pressure of 1 atm. Of those two terms, $p\mathrm{CO_2}_{\mathrm{a},0}$ is the preindustrial reference value of 280 μatm (i.e., $xCO_2 = 280$ ppm) and $p\mathrm{CO_2}_{\mathrm{a}}$ is the prescribed atmospheric xCO_2 obtained from a spline fit to observations from the Siple ice core data and atmospheric CO_2 measurements from Mauna Loa, which together span 1800.0 to 1990.0 (Enting et al., 1994), combined with the 12-month-smoothed Mauna Loa atmospheric measurements between 1990.5 and 2009.0 (GLOBALVIEW-CO2, 2010). The final term in Eq. (4) uses the saturation water vapor pressure e_s and the total atmospheric pressure at sea level p_a to convert partial pressure in dry air to that in wet air as needed to compute the air–sea flux.

The modeled air–sea flux of anthropogenic carbon $F_{\mathrm{CO_2}}$ follows the standard formulation

$$F_{\mathrm{CO_2}} = K_{\mathrm{CO_2}}(\delta p\mathrm{CO_2}_{\mathrm{a}} - \delta p\mathrm{CO_2}_{\mathrm{o}}), \tag{5}$$

where $K_{\mathrm{CO_2}}$ is a gas transfer coefficient, $\delta p\mathrm{CO_2}_{\mathrm{a}}$ is described above, and $\delta p\mathrm{CO_2}_{\mathrm{o}}$ is the anthropogenic perturbation in surface-water pCO_2 relative to its reference value in 1800. For the gas transfer coefficient, $K_{\mathrm{CO_2}} = \alpha k_\mathrm{w}$, where α is the CO_2 solubility (Weiss, 1974) and k_w is as in Eq. (3) except that *Sc* is for CO_2 (Wanninkhof, 1992, Table A1).

The $\delta p\mathrm{CO_2}_{\mathrm{o}}$ term is not modeled explicitly but is calculated from the only tracer that is carried in the model, δC_T. The standard formulation from Sarmiento et al. (1992) is based on their finding that the relationship between the ratio $\delta p\mathrm{CO_2}_{\mathrm{o}}/\delta C_T$ and $\delta p\mathrm{CO_2}_{\mathrm{o}}$ is linear, for a given temperature and at constant total alkalinity.

$$\frac{\delta p\mathrm{CO_2}_{\mathrm{o}}}{\delta C_\mathrm{T}} = z_0 + z_1\,\delta p\mathrm{CO_2}_{\mathrm{o}}, \tag{6}$$

where the intercept z_0 and slope z_1 terms are each quadratic functions of temperature. That equation is then rearranged for the model calculation.

$$\delta p\mathrm{CO_2}_{\mathrm{o}} = \frac{z_0\,\delta C_\mathrm{T}}{1 - z_1\,\delta C_\mathrm{T}} \tag{7}$$

To allow for a starting value of $p\mathrm{CO_2}_{\mathrm{a},0}$ that is different than 280 ppm, Lachkar et al. (2007) introduced two corrective terms:

$$\delta p\mathrm{CO_2}_{\mathrm{o}} = \frac{z_0\left[\delta C_\mathrm{T} + \delta C_{T,\mathrm{corr}}\right]}{1 - z_1\left[\delta C_\mathrm{T} + \delta C_{T,\mathrm{corr}}\right]} - p\mathrm{CO_2}_{\mathrm{a,corr}}, \tag{8}$$

where the first correction factor is

$$p\mathrm{CO_2}_{\mathrm{a,corr}} = p\mathrm{CO_2}_{\mathrm{a},0} - p\mathrm{CO_2}_{\mathrm{a,ref}}, \tag{9}$$

determined from the starting xCO_2 in the initial year (1800), i.e., $p\mathrm{CO_2}_{\mathrm{a},0} = 287.78$ ppm, and same reference $p\mathrm{CO_2}_{\mathrm{a,ref}} =$ 280 ppm. With that result, the second correction factor is

$$\delta C_{T,\mathrm{corr}} = \frac{p\mathrm{CO_2}_{\mathrm{a,corr}}}{z_0 + z_1\,p\mathrm{CO_2}_{\mathrm{a,corr}}}. \tag{10}$$

These two minor corrections do not change the way z_0 and z_1 are computed, but they do slightly alter their use in the model simulations, using Eq. (8) instead of Eq. (7).

Equations for the linear regression coefficients z_0 and z_1 are computed in four steps: (i) estimate the initial preindustrial $C_{T,0}$ as a function of temperature from carbonate system thermodynamics, assuming air–sea equilibrium between the atmosphere ($p\mathrm{CO_2}_{\mathrm{a},0} = 280$ ppm) and the surface ocean, constant global-average surface total alkalinity (2300 μmol kg^{-1}), constant salinity (35 psu), and varying temperatures across the observed range; (ii) increase incrementally the $p\mathrm{CO_2}_{\mathrm{a}}$ from 280 to 480 ppm, and recompute the C_T as a function of temperature for each increment; (iii) use those results with Eq. (6) to compute z_0 and z_1 for each temperature; and (iv) fit each of z_0 and z_1 to a quadratic function of temperature. With this approach, Sarmiento et al. (1992) found that Eq. (6) fitted exactly calculated values to within 1 % when $\delta p\mathrm{CO_2}_{\mathrm{o}} \leq 200$ ppm.

The constant value of total alkalinity used in the standard perturbation approach detailed above is the area-weighted mean for the global ocean. That approach with Eq. (8), which we refer to as GLO, will produce biased results for the Mediterranean Sea whose average surface total alkalinity is 10 % greater. Thus we performed a second simulation (MED), where z_0 and z_1 that were used with Eq. (8) were computed following the same four-step procedure as above, except that we replaced the area-weighted surface average total alkalinity for the global ocean (2300 μmol kg^{-1}) with that for the Mediterranean Sea (2530 μmol kg^{-1}).

Finally, to test how variable total alkalinity may affect simulated results, we performed a third simulation (VAR). The perturbation approach was designed for the global, open-ocean waters where total alkalinity varies relatively little,

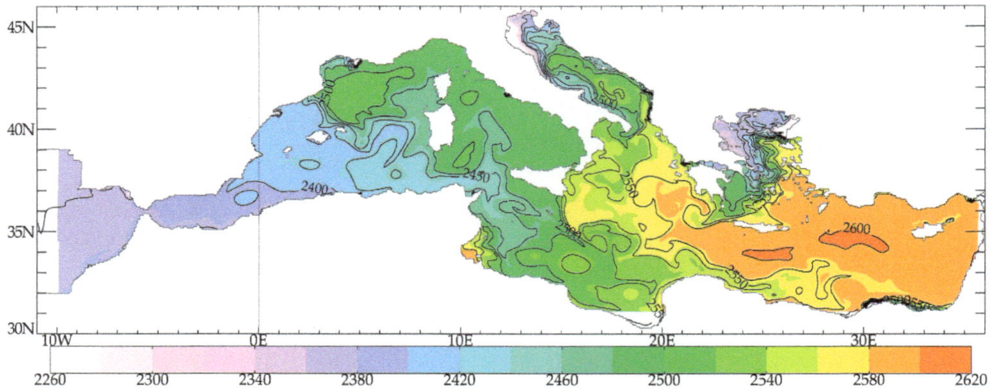

Figure 3. Salinity-derived surface total alkalinity field (μmol kg^{-1}) calculated with the formula of Schneider et al. (2007) (Eq. 11) applied to the model's surface salinity field from November 2001.

Table 1. Coefficients a_i, b_i, and c_i (where the index i varies from 0 to 9) used to compute z_1, z_2, and z_3 with Eq. (13).

i	a_i	b_i	c_i
0	1.177825e+1	9.330105e−2	1.350359e−3
1	−1.614090e−1	−1.857070e−3	−2.422081e−5
2	−5.633789e−1	−5.251668e−3	−8.087972e−5
3	1.102070e−3	1.615968e−5	1.558226e−7
4	1.027733e−2	1.028834e−4	1.655765e−6
5	−4.195387e−6	−5.816404e−8	−3.503140e−10
6	−6.677595e−5	−6.915741e−7	−1.151323e−8
7	5.292828e−3	6.857606e−5	9.547726e−7
8	−1.529681e−5	−2.836387e−7	−3.012886e−9
9	−4.737909e−5	−6.551447e−7	−9.651931e−9

e.g., from 2243 to 2349 μmol kg^{-1} in the zonal mean of the GLODAP gridded data product (Key et al., 2004). Spatial variations in surface total alkalinity in the Mediterranean Sea are more than twice as large, e.g., varying from 2375 to 2625 μmol kg^{-1} between western and eastern margins. To account for variability in Mediterranean total alkalinity, we exploited its tight relationship with salinity derived from the *Meteor* M51/2 transbasin section by Schneider et al. (2007):

$$A_T = 73.7\,S - 285.7, \tag{11}$$

where S is the model's surface salinity and A_T is its computed surface total alkalinity. This equation thus takes much of the A_T spatial variability into account (Fig. 3), although it is expected to be inaccurate near river mouths, where freshwaters with high total alkalinity are delivered to the Mediterranean Sea. This equation also implies that computed A_T varies temporally with simulated salinity.

For VAR to take into account variable salinity (total alkalinity) as well as variable temperature, while maintaining adequate precision, we made two types of modifications to the standard equations. First, we replaced Eq. (6) with a direct relationship between $\delta pCO_{2,o}$ and δC_T but with a cubic formulation instead of a linear formulation, i.e., implying an additional coefficient.

$$\delta pCO_{2,o} = 0 + z_1\,\delta C_T + z_2\,\delta C_T^2 + z_3\,\delta C_T^3 \tag{12}$$

Then for each of three coefficients, we replaced the former two equations, quadratic in temperature T, with three equations, cubic in T and S.

$$\begin{aligned}
z_1 &= a_0 + a_1 T + a_2 S - a_3 T^2 + a_4 S^2 + a_5 T^3 + a_6 S^3 \\
&\quad + a_7 T S + a_8 T^2 S + a_9 T S^2 \\
z_2 &= b_0 + b_1 T + b_2 S + b_3 T^2 + b_4 S^2 + b_5 T^3 + b_6 S^3 \\
&\quad + b_7 T S + b_8 T^2 S + b_9 T S^2 \\
z_3 &= c_0 + c_1 T + c_2 S + c_3 T^2 + c_4 S^2 + c_5 T^3 + c_6 S^3 \\
&\quad + c_7 T S + c_8 T^2 S + c_9 T S^2
\end{aligned} \tag{13}$$

The associated coefficients are listed in Table 1, while the R program used to make these calculations, which exploits the seacarb software package for the carbonate system (Lavigne and Gattuso, 2011), is given in the Supplement. With the VAR approach applied to the range of Mediterranean temperatures (13 to 30 °C), we found that Eq. (12) fitted calculated values to within 0.6 % when $\delta pCO_{2_a} \leq 280$ ppm, i.e., up to a doubling of the preindustrial level of atmospheric CO_2.

Hence the simulated anthropogenic carbon is sensitive to the spatial and temporal changes in T in the GLO and MED experiments and T and S in the VAR experiment (Eq. 12).

For lateral boundary conditions, we restored simulated δC_T throughout the Atlantic buffer zone toward a time-varying, spatially co-located section taken from the global-scale gridded derived product by Khatiwala et al. (2009) for each year between 1765 and 2011, using their values from 1800 (the start year of our anthropogenic CO_2 simulations) as our zero reference. That damping across the entire buffer zone was designed to maintain a reasonable time-varying inflow of δC_T from the Atlantic across the Strait of Gibraltar.

2.3 Looping

For computational efficiency, our geochemical simulations were performed "offline". That is, they were driven by circulation fields that were read in from output from a previous simulation with the NEMO-MED12 circulation model, thereby avoiding the need for us to recalculate them for each passive tracer simulation. Thus those circulation fields were computed by forcing NEMO-MED12 with the ARPERA forcing during 1958–2008. We then repeatedly looped that 51-year sequence of model-circulation fields in order to cover the full 200-year industrial period for anthropogenic carbon simulations. The first 7 years of the circulation model (1958–1964) are considered as a spin-up and are not used in the offline simulations of passive tracers. The next 10 years (ARPERA forcing during 1965–1974) are continuously repeated until 1975 to drive offline simulations of both passive tracers.

For both passive tracers, up until 1975, we began by repeatedly looping the same 10 years of NEMO-MED12 circulation fields, i.e., those forced by the ARPERA atmospheric forcing during 1965–1974. That forcing period was selected because it does not include intense events like the EMT or the Western Mediterranean Transition (Schroeder et al., 2008); we thus considered this period as best suited to produce reasonable circulation fields for the Mediterranean Sea (Beuvier et al., 2010, 2012b; Beuvier, 2011). Thus for the complete CFC-12 simulation, covering 1930 to 2008, the 1965–1974 loop of MED12 circulation fields was repeated 4.5 times to cover offline years 1930–1975. Then to complete the offline CFC-12 simulation, we applied the NEMO-MED12 circulation fields corresponding to the remaining 1975–2008 period forcing. The same 1965–1974 loop of the circulation fields from NEMO-MED12 was likewise repeated for the three anthropogenic CO_2 simulations (GLO, MED, VAR) but instead 17.5 times to cover offline simulation years 1800–1974. Then, as for CFC-12, the last 34 years of the offline anthropogenic CO_2 simulations were driven with the NEMO-MED12 circulation fields from the remaining years (1975–2008) of the ARPERA forcing.

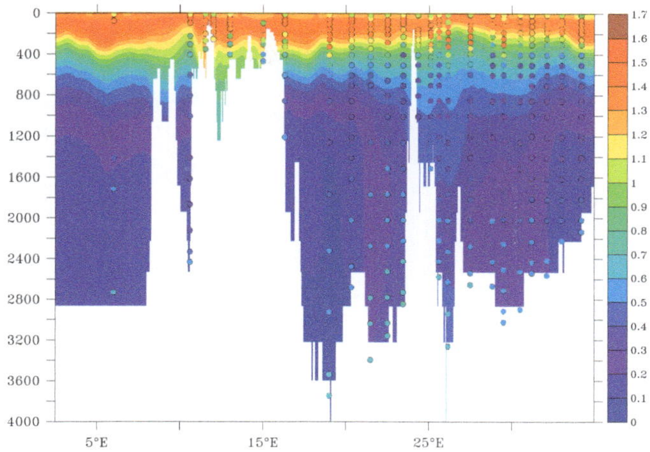

Figure 4. CFC-12 (pmol kg^{-1}) data–model comparison along the *Meteor* M51/2 section. Color-filled contours indicate simulated CFC-12, whereas color-filled dots show in situ observations. Both use the same color scale and are taken at the same time (November 2001).

2.4 δpH

The anthropogenic change in surface in situ pH during 1800 to 2001 was computed from δC_T and prescribed total alkalinity. The preindustrial C_T was computed by assuming it to be in thermodynamic equilibrium with the prescribed total alkalinity and with an atmospheric xCO_2 of 280 ppm at 1 atm total pressure, correcting for humidity. Computations were performed with seacarb, which takes two carbonate system variables and computes all others including pH. Then, to this preindustrial C_T we added our simulated δC_T and recomputed pH. Other input variables, temperature, salinity, and total alkalinity were identical. Concentrations of phosphate and silica were assumed to be zero, a good approximation for the oligotrophic surface waters of the Mediterranean Sea. The anthropogenic change in pH is then just the difference between two computations. This exercise yields a surface map of δpH.

For deep waters, we consider changes only along one transbasin section, Meteor M51/2. Exploiting total alkalinity, C_T, temperature, and salinity measured along from this section in November 2001 (Schneider et al., 2007, 2010), we computed corresponding pH for all data points along the section and throughout the water column. Then we subsampled the simulated δC_T in 2001 at all station locations and sample depths. After removing those simulated results from the measured C_T, we recalculated pH. The difference is the δpH along the same section. For comparison, we repeated this exercise, but instead of simulated δC_T, we used the TTD data-based estimates of anthropogenic C_T from Schneider et al. (2010), already available along the same section.

Table 2. Cumulative air–sea flux between 1800 and 2001 for the three simulations over the eastern and western basins and the entire Mediterranean Sea.

	Average flux (mol C m^{-2})			Total flux (Pg C)		
	East	West	Med. Sea	East	West	Med. Sea
GLO	19.2	24.8	21.0	0.39	0.26	0.65
MED	22.3	31.8	25.5	0.45	0.33	0.78
VAR	23.4	29.4	25.4	0.47	0.31	0.78

Table 3. Budget of anthropogenic carbon accumulated in the Mediterranean Sea (Pg C) between 1800 and 2001. The budget distinguishes Strait of Gibraltar inflow ($G-S_{in}$) via the AW, the corresponding outflow ($G-S_{out}$) via the MOW, and the air–sea flux (Air–sea). Critical combined terms are thus the net inflow–outflow difference (Net $G-S$), the total input ($G-S_{in}+$ air–sea), and the net total (Net $G-S+$ air–sea).

Simulation	$G-S_{in}$	$G-S_{out}$	Net $G-S$	Air–sea	Total Input	Net Total
GLO	0.71	0.42	0.29	0.65	1.36	0.94
MED	0.72	0.46	0.26	0.78	1.5	1.04
VAR	0.72	0.46	0.26	0.78	1.5	1.04

3 Results

3.1 Evaluation

By comparing modeled to observed CFC-12, we evaluated the simulated circulation in regard to ventilation of water masses (Fig. 4). Whereas modeled CFC-12 generally matches observations between 150 m (~ 1.4 pmol kg^{-1}) and 1200 m ($\sim$0.3 pmol kg^{-1}), simulated concentrations do not show the observed mid-depth minimum. For instance, in the Levantine sub-basin, observed CFC-12 concentrations are lowest (~ 0.3 pmol kg^{-1}) between 600 and 1500 m, but below that depth zone concentrations grow with depth, reaching ~ 0.6 pmol kg^{-1} in bottom waters. Conversely, simulated concentrations below 1200 m continue to decline until they bottom out at ~ 0.3 pmol kg^{-1} (Fig. 5).

Generally, the model underestimates the relatively large CFC-12 concentrations observed in deep waters of the eastern and western basins (~ 0.6 pmol kg^{-1}), which are indicative of recently ventilated water masses (Schneider et al., 2010; Roether et al., 2007). Although the model simulates some penetration of CFC-12 south of the Crete Passage with concentrations reaching up to ~ 0.5 pmol kg^{-1}, those remain lower than observed. Ventilation of the model's deep eastern basin is particularly weak in the Adriatic and Ionian sub-basins (Fig. 4). On average below 2000 m, CFC-12 concentrations from the model are only half of those observed. Overall, the CFC-12 evaluation indicates that the model produces an adequate ventilation of intermediate water masses but insufficient ventilation of deep waters.

3.2 Air–sea flux

The invasion of anthropogenic carbon into the Mediterranean Sea is influenced by the δCO_2 flux at the surface and by exchange with the Atlantic Ocean across the Strait of Gibraltar. The simulated air–sea flux of anthropogenic carbon is calculated directly by the model (Eq. 5). When integrated since the beginning of the simulation (cumulative flux), it is found to be similar among the three simulations, all of which exhibit maxima in the same regions (Fig. 6). The highest fluxes occur in the Gulf of Lions and to the east of Crete, both regions of deep and intermediate water formation, and in the Alboran sub-basin, which is highly influenced by the strong Atlantic inflow and by the presence of two standing anticyclonic eddies (Vargas-Yáñez et al., 2002). Along coastlines there are not only local minima but also the maximum uptake at the outflow of the Dardanelles Strait, although that is extremely localized. In the MED simulation, cumulative fluxes over the western basin are on average 25 % larger per unit area than the Mediterranean Sea's mean, whereas they are 13 % lower in the eastern basin (Table 2). Conversely, the larger surface area of the eastern basin means that its total uptake represents 58 % of the total Mediterranean Sea uptake.

The 10 % greater prescribed surface total alkalinity in the MED simulation relative to GLO means that the latter must absorb less anthropogenic carbon (Fig. 6b). Indeed, despite very similar uptake patterns, the basin-wide cumulative uptake is 17 % less in the GLO simulation than in MED, with a greater reduction in the western basin (22 %) than in the eastern basin (14 %). By definition, the salinity-derived total alkalinity in the VAR simulation is more realistic than with MED simulation, varying from 2350 μeq kg^{-1} in the Alboran sub-basin to 2650 μeq kg^{-1} in the eastern basin. That lower western total alkalinity results in an 8 % lower air–sea flux

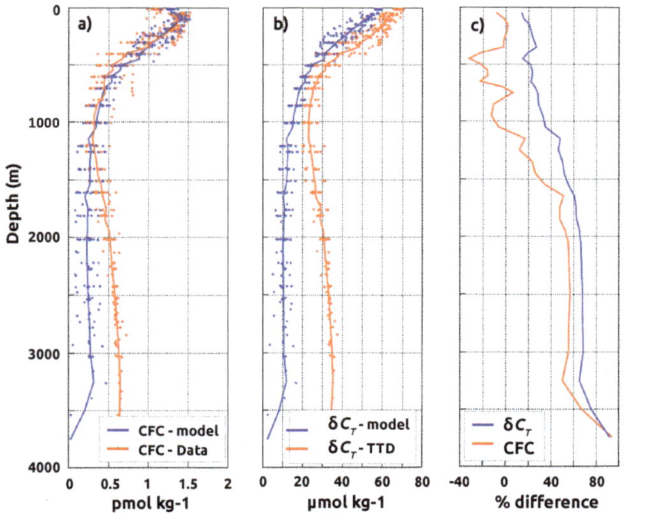

Figure 5. Comparison of average vertical profiles along the *Meteor* M51/2 section for **(a)** CFC-12 (pmol kg^{-1}), **(b)** δC_T (μmol kg^{-1}), and **(c)** the model–data relative difference (in percent). Model results are in blue, while red indicates the CFC-12 data and δC_T data-based estimates; the right panel **(c)** uses blue for δC_T and red for CFC-12. Data-based estimates for δC_T are the TTD results from Schneider et al. (2010).

in the western basin, while the higher eastern total alkalinity drives 5 % greater uptake in the eastern basin (Fig. 6c). Yet despite these east–west differences between VAR and MED, total basin-wide uptake is only 0.3 % less in former than the latter. Overall the eastern basin always dominates, taking up 60 % of the basin-wide integrated flux in VAR and GLO and taking up 57 % in MED. However, it is not only the air–sea flux but also lateral exchange that matters.

3.3 Budget

The Mediterranean Sea's content of anthropogenic carbon is affected not only by the air–sea flux but also by exchange with the Atlantic Ocean through the Strait of Gibraltar. To assess the relative importance of this lateral exchange we constructed a budget of δC_T in the Mediterranean Sea. In that budget, the temporal evolution of the cumulative air–sea flux in the reference simulation MED is compared to the same simulation's total mass of carbon that has entered and left the Mediterranean Sea through the Strait of Gibraltar (Fig. 7). The key terms are thus the flux, the net transfer at the Strait of Gibraltar (inflow – outflow of δC_T), and the actual accumulation of δC_T in the Mediterranean Sea (inventory).

In the MED simulation between 1800 and 2001, there is 1.50 Pg C that enters the Mediterranean Sea via the air–sea flux (0.78 Pg C) and via the Strait of Gibraltar inflow (0.72 Pg C) (Table 3). Yet 64 % of the δC_T inflow (by AW near the surface) is balanced by δC_T outflow at depth (by the Mediterranean outflow). That leaves 1.04 Pg C that remains

Table 4. Average δC_T inventories in the eastern and western basins and for the entire Mediterranean Sea.

	Average inventory (mol C m^{-2})		
Simulation	East	West	Med. Sea
GLO	28.6	33.4	30.2
MED	31.8	36.9	33.5
VAR	32.0	36.6	33.6

in the Mediterranean as the total δC_T inventory. Thus 25 % of the Mediterranean's total δC_T inventory is due to net exchange at the Strait of Gibraltar, while the remaining 75 % is from the air–sea flux. The budget of the VAR simulation is quite similar to that for MED, but both of those differ substantially from the budget for GLO (Table 3). In GLO, the Mediterranean Sea's δC_T inventory in 2001 (0.94 Pg C) is 10 % less, with 69 % of the total input coming from the air–sea flux and 31 % from net exchange across the Strait of Gibraltar. The evolution of the MED simulation's carbon budget (Fig. 8) demonstrates that anthropogenic carbon enters the Mediterranean entirely via the air–sea flux at the beginning of the simulation, but that the fraction entering by lateral exchange across the Strait of Gibraltar grows until stabilizing in the 1960s to one-fourth of the total.

3.4 Simulated δC_T inventory

Having examined how anthropogenic carbon enters the Mediterranean Sea, we now turn to where it is stored, the patterns of which differ from those of the input fluxes due to water mass transport. The vertical integral of the δC_T concentration is termed the inventory. In the Mediterranean Sea, the inventory patterns tend to follow the distribution of bathymetry (Fig. 9). Thus, unlike the global ocean, substantial levels of anthropogenic carbon have already penetrated into deep waters of the Mediterranean Sea, as deduced previously by observational studies (Lee et al., 2011; Schneider et al., 2010; Touratier and Goyet, 2011). Specific inventories (mass per unit area) in the reference simulation (MED) are 10 % higher in the western basin and 6 % lower in the eastern basin relative to the 33.5 mol C m^{-2} average for the Mediterranean Sea (Table 4). For the two other simulations, the basin-wide inventory is 10 % lower in GLO and 0.1 % higher in VAR. Those east–west differences are smaller than those for the air–sea flux (Table 2). There is a strong correlation between latitudinal variations in the inventory and the bathymetry, both along the *Meteor* M51/2 section and in terms of meridional means (Fig. 10). In both cases, the correlation is striking, except in isolated regions such as in the Ionian sub-basin (~ 15–20°E in Fig. 10), where poorly ventilated deep waters have relatively low δC_T concentrations.

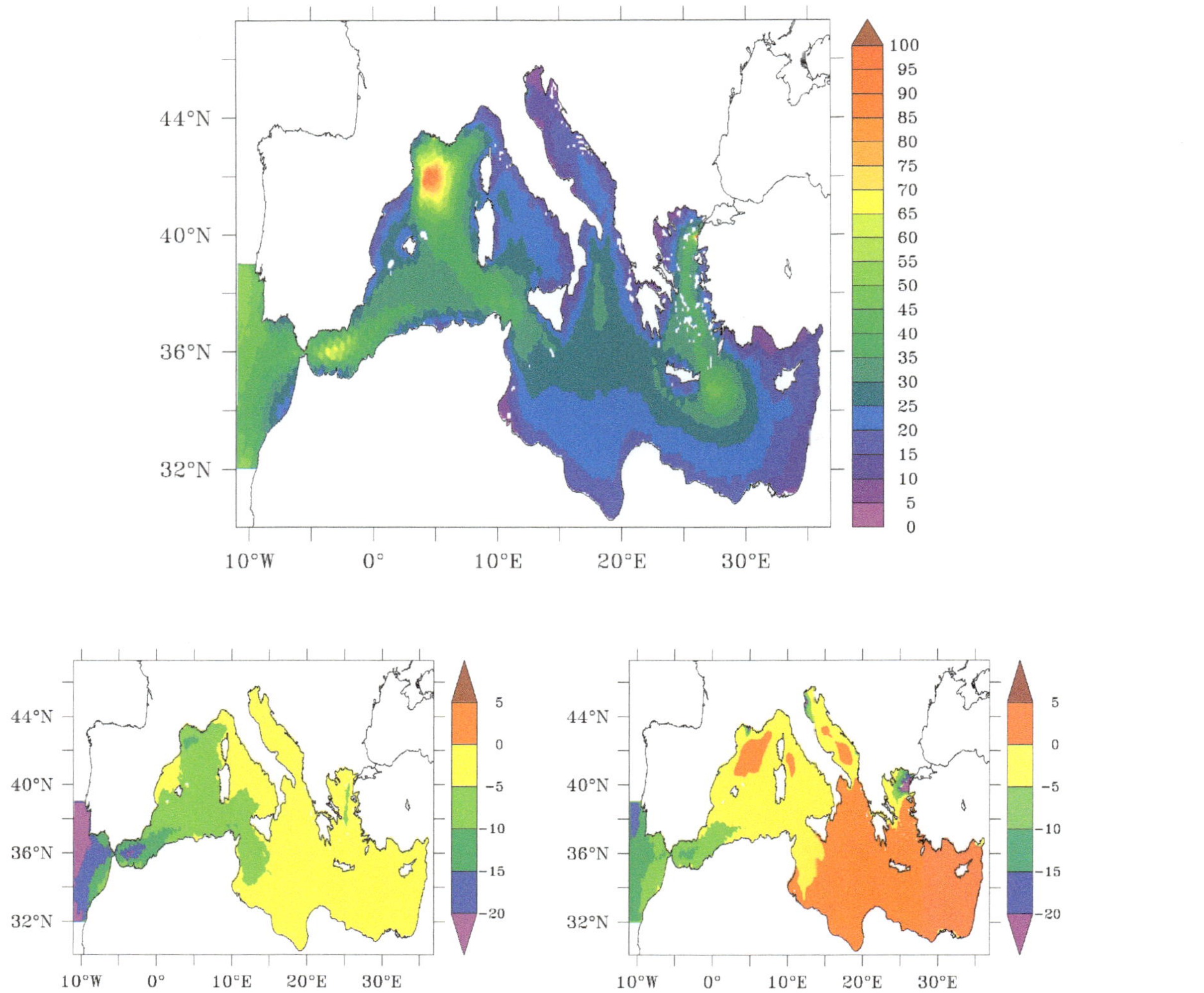

Figure 6. Cumulative air–sea flux of anthropogenic carbon (mol m^{-2}) from 1800 to November 2001 shown as the total flux for the MED reference simulation (top) and for the other two simulations as differences: GLO – MED (bottom left) and VAR – MED (bottom right).

Table 5. Total δC_T inventory (Pg C) for the entire Mediterranean Sea and for the eastern basin as simulated and as estimated by the TTD data-based method (Schneider et al., 2010).

Approach	Med. Sea	East
GLO model	0.93	0.58
MED model	1.03	0.65
VAR model	1.03	0.65
TTD data	1.7 (1.3–2.1)	1.0 (0.7–1.2)

3.5 Comparison with TTD data-based results

To go beyond model comparison of simulated uptake of anthropogenic CO_2, we also compare model results to data-based estimates. In particular, we focused on data-based estimates of anthropogenic carbon deduced with TTD because that method requires only measurements of temperature, salinity, and CFC-12, all of which were simulated, thereby allowing us to test the approach (see Sect. 4.1). For now though, let us simply compare model results to the TTD estimates of δC_T based on observations collected on the *Meteor* M51/2 section in 2001 (Schneider et al., 2010). A first comparison reveals that the modeled δC_T is lower everywhere than the TTD estimates of the inventory along the *Meteor* M51/2 section (Fig. 9). While the TTD inventory along this section averages 83 mol m^{-2} (ranging from 21 to 153 mol m^{-2}), the model average is 50 mol m^{-2}, 40 % less. Expanding the comparison vertically, the model δC_T is seen to underestimate the TTD results throughout the water column, even at the surface (Fig. 11). Surface concentrations are naturally largest, both for the TTD estimates ($\sim$68 μmol kg^{-1}) and for the model (e.g., $\sim$58 μmol kg^{-1} in the MED simulation). Whereas the TTD estimates are lowest (20 to 25 μmol kg^{-1}) in the Levantine sub-basin between 800 and 1500 m and increase below (e.g., reaching up to 35 μmol kg^{-1} in the EMDW), simulated δC_T decreases with depth everywhere, as already seen for simulated

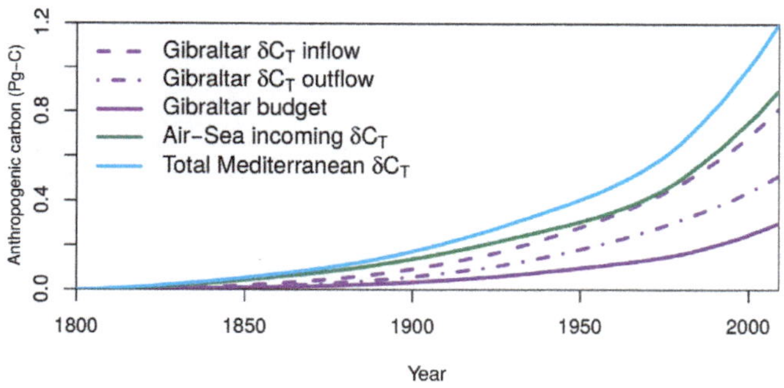

Figure 7. Cumulative increase in anthropogenic carbon (Pg C) in the Mediterranean Sea from 1800 to 2008 due to the Gibraltar inflow (dashed purple line) and outflow (dashed-dotted purple line), i.e., their difference (inflow – outflow, solid purple line) and the air–sea flux (solid green line). Also shown is the total buildup in storage (solid light-blue line). Gibraltar fluxes have been calculated from model monthly mean outputs by multiplying the anthropogenic carbon concentration ($g\,m^{-3}$) of water in a section crossing the Strait of Gibraltar by water fluxes ($m^3\,month^{-1}$) flowing across this section. The sign of the water flux indicates its direction, and hence provides an inflow or an outflow of anthropogenic carbon. The total storage is the sum of the net δC_T flux at the Strait of Gibraltar and of the δC_T air–sea flux, and is consistent with the sum of all the δC_T store in the Mediterranean Sea.

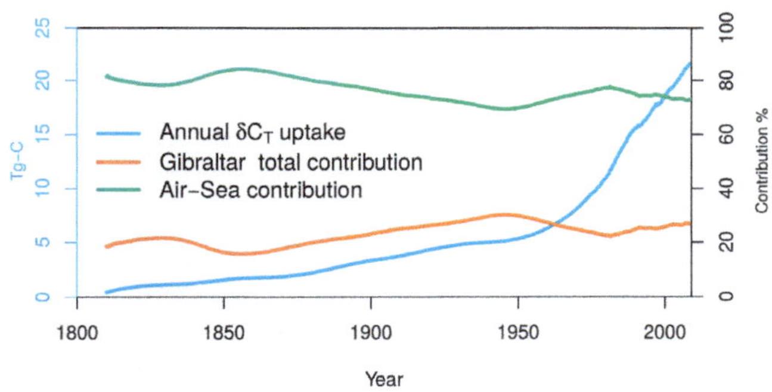

Figure 8. Evolution of the Mediterranean's annual uptake in δC_T (solid light-blue line), shown as a 10-year running average to focus on decadal-scale changes. Also shown is the percentage of that annual δC_T that entered the Mediterranean Sea through the Strait of Gibraltar (solid orange line) and via the air–sea fluxes (solid green line).

CFC-12 (Fig. 4). The lowest simulated δC_T concentrations are found in the bottom waters of the Ionian sub-basin (< $5\,\mu mol\,kg^{-1}$). However, higher deep-water δC_T is simulated in the EMDW near the Crete Passage (up to $15\,\mu mol\,kg^{-1}$), where there is dense-water outflow from the Aegean sub-basin through the Crete Passage during the EMT. In terms of basin totals, Schneider et al. (2010) relied on TTD to help estimate a basin-wide anthropogenic carbon inventory of 1.7 Pg C for the Mediterranean Sea, with 1.0 Pg C of that in the eastern basin (Table 5). Relative to the data-based TTD results, the modeled basin-wide Mediterranean inventory is 40 % less in the MED and VAR and 46 % less in GLO. For the eastern inventory basin, the MED and VAR simulations are 35 % lower than the TTD estimates, whereas GLO is 42 % lower.

3.6 δpH

Anthropogenic changes in surface pH between 1800 and 2001 are remarkably uniform, both between simulations and across the basin. Away from the coast, the change in surface pH between 1800 and 2001 varies between −0.082 and −0.086 in the GLO simulation (Fig. 12). Exceptions include the northern Levantine sub-basin, where the δpH is slightly less (−0.080), and the greater changes seen in the Gulf of Gabes, the Adriatic and Aegean sub-basins, and near the mouths of large rivers such as the Nile and the Rhone. The MED simulation exhibits almost identical patterns and intensities for the change in pH except in Alboran sub-basin and the western portion of the western basin, where pH changes are less intense (−0.076 and −0.074, i.e., a difference of up to ∼0.012 pH units). Conversely, the VAR simulation with its spatially varying total alkalinity produces a more contrasted pattern of pH change. Although VAR's spatial variability in δpH in the western basin is intermediate between that seen for GLO and MED, the eastern basin contrast in VAR is much greater. In particular, VAR 's pH changes are smallest where the salinity derived total alkalinity is highest (Levantine sub-basin), and they are largest where the salinity-derived total alkalinity is smallest (e.g., near the Po, Nile, and Dardanelles outflows). Despite differences in spatial patterns between simulations, their basin-wide average change in surface pH is almost identical: -0.084 ± 0.001 units (total scale).

4 Discussion

4.1 δC_T in the Mediterranean Sea

Our comparison of modeled to measured CFC-12 indicates that the model adequately represents ventilation of near-surface and intermediate waters but underestimates ventilation of deep waters. This CFC-12 evaluation alone implies that our simulated δC_T is likewise too low in Mediterranean Sea deep waters and hence that our simulated total anthropogenic carbon inventory of 1.03 Pg C is a lower limit. Yet even in the top 400 m, where there is tight agreement between simulated and observed CFC-12, the data-based estimates of δC_T from the TTD method are 20 % larger than those simulated (Fig. 5). Hence it is unlikely that the modeled circulation is the primary cause. Simplifications with the perturbation approach, e.g., its steady-state assumption, could be partly to blame, although errors due to circulation-induced changes in biological productivity appear small for the global ocean (Siegenthaler and Sarmiento, 1993; Sarmiento et al., 1998). The treatment of total alkalinity in the perturbation approach also does not appear

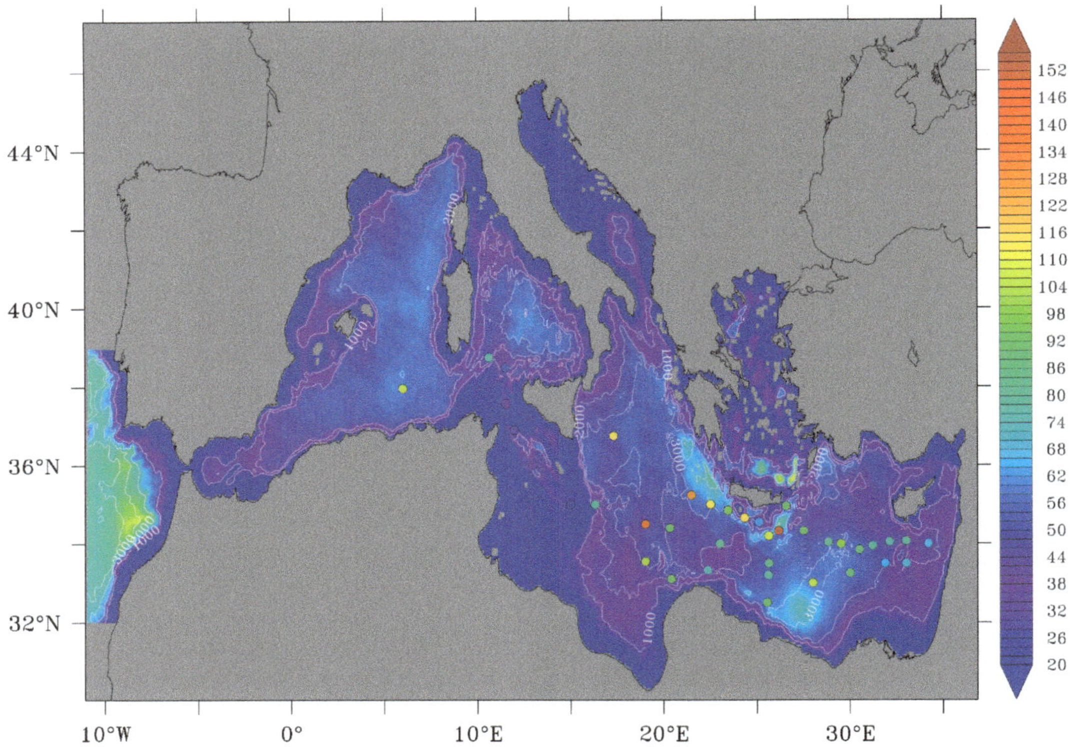

Figure 9. Inventory of δC_T (mol m^{-2}) in November 2001 from the MED simulation (color-filled contours) and from Schneider et al. (2010) data-based estimates (color-filled dots). The Mediterranean bathymetry is shown as white isobaths every 1000 m.

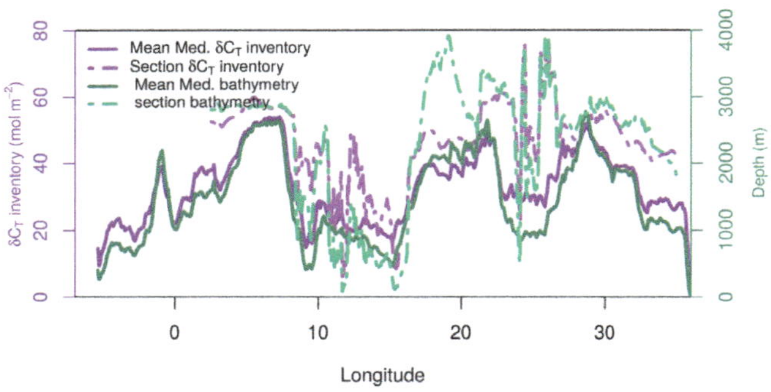

Figure 10. δC_T inventory (mol C m^{-2}) along the *Meteor* M51/2 section (dashed lines) and given as the meridional mean (solid lines) for the MED simulation (purple) along with corresponding model bathymetry (green).

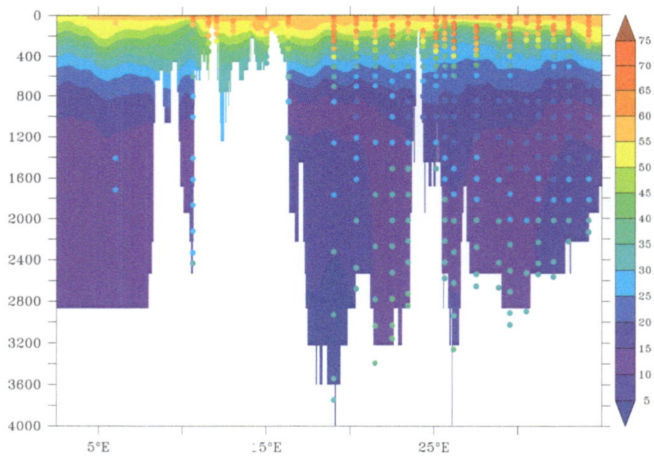

Figure 11. Comparison of δC_T (μmol kg^{-1}) along the *Meteor* M51/2 section for the model (color-filled contours) and the TTD data-based estimates (color-filled dots) in November 2001.

a significant factor, considering that our three treatments with different mean states and spatial variability give results that are quite similar (see Sect. 4.3). Besides these potential simulation biases, it is also possible that the data-based methodology, namely the TTD approach, is biased.

Hence we tested the TTD approach in the model world (MW) by (1) using exactly the same version and parametrizations of the TTD approach described in Schneider et al. (2010) for consistency; (2) using it to estimate δC_T from simulated CFC-12, temperature, and salinity; and (3) comparing those results to the δC_T simulated directly by the model. That comparison reveals that the TTD_{MW} estimates always overestimate the simulated δC_T. Those overestimates start at +10 % in surface waters but reach more than +100 % in Mediterranean Sea bottom waters (Fig. 13). Relative differences are highest where simulated CFC-12 is lowest, i.e., where the ventilation age of water masses is oldest, namely

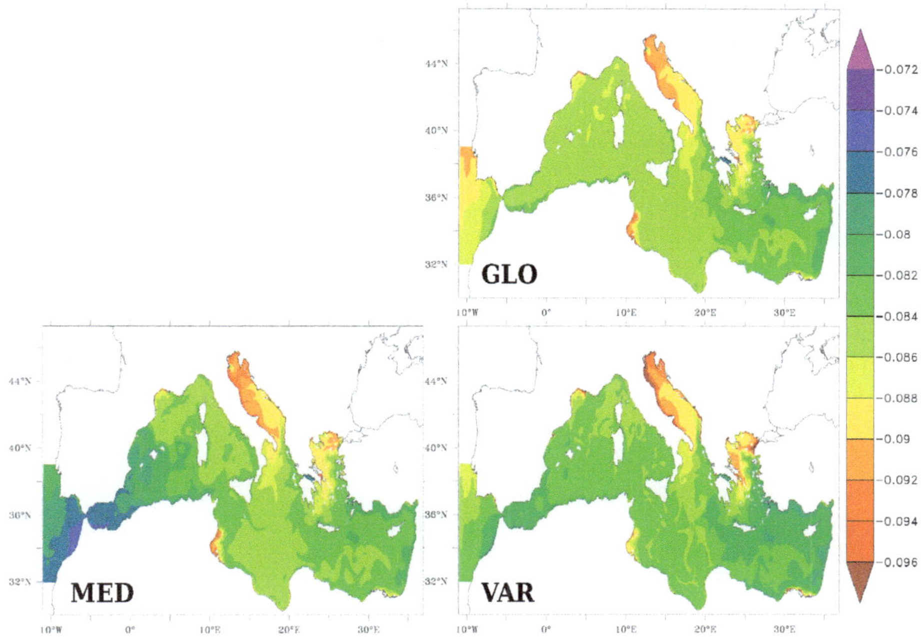

Figure 12. Anthropogenic change in surface pH between 1800 and 2001 for the GLO (top), MED (bottom left), and VAR (bottom right) simulations.

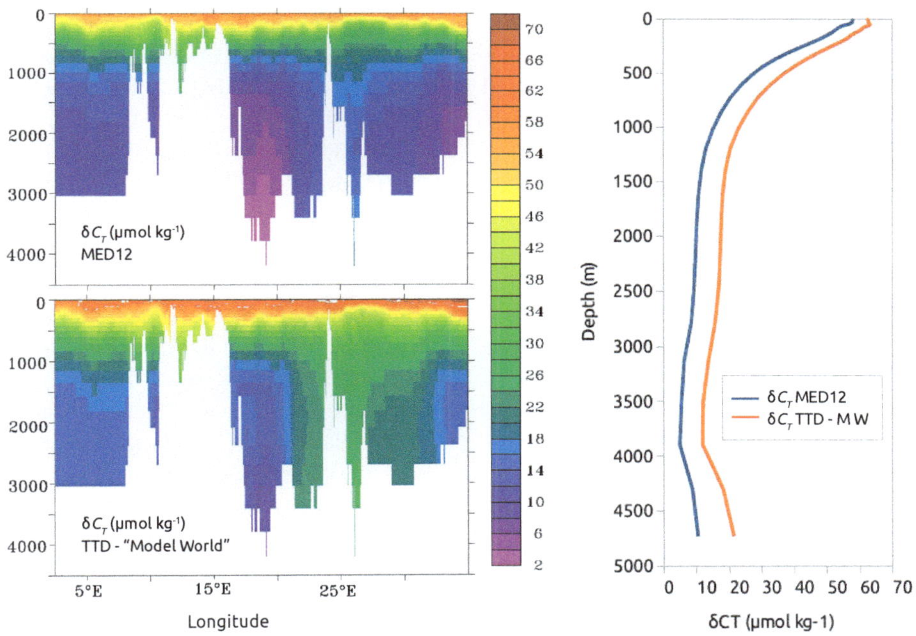

Figure 13. δC_T (μmol kg^{-1}) along the *Meteor* M51/2 section, estimated with the MED reference simulation (top left) and the TTD method in the model world (bottom left). Also shown are the same results but as area-weighted vertical profiles for the whole Mediterranean Sea (right).

in bottom waters, particularly those in the Ionian sub-basin. Whereas the TTD_{MW} estimate of the total anthropogenic carbon inventory in the Mediterranean Sea is 1.4 Pg C, the simulated value in the reference simulation (MED) is 1.0 Pg C. That 40 % overestimate with the TTD approach in the model world could be less in the real Mediterranean Sea, because the model underestimates CFC-12 concentrations in the deep water, which accentuates the discrepancy. Nonetheless, the TTD-based inventory of anthropogenic carbon remains an upper limit.

One reason for this overestimate is that data-based methods such as TTD assume that the change in oceanic pCO_2 is identical to the change in the atmospheric pCO_2.

This convenient assumption, that $\Delta\delta pCO_2 = 0$, has been motivated by measured pCO_2 at three time series (BATS, HOT, and ESTOC) where calculated atmospheric and oceanic trends are not significantly different (Bindoff et al., 2007). However, these stations are all located in subtropical gyres where both the air–sea flux of anthropogenic CO_2 and the corresponding air–sea disequilibrium are the lowest Sarmiento et al. (Fig. 2 in 1992), i.e., where the detection of an air–sea disequilibrium in temporal changes of pCO_2 (i.e., $\Delta\delta pCO_2$) is the most difficult. Our model estimates of this disequilibrium in the Mediterranean Sea (Fig. 14) indicate that it is not negligible. It slowly increases from 14 μatm in 1800 up to 20 μatm in 2001, corresponding to a lag of ocean's δpCO_2 compared to the atmosphere's by 15 to 20 % since ~1850. These $\Delta\delta pCO_2$ values are similar to those simulated by global-ocean models (Sarmiento et al., 1992; Orr et al., 2001; Yool et al., 2010). Assuming that this disequilibrium is zero, as done in TTD, implies a systematic overestimate of anthropogenic carbon uptake.

Other data-based methods that estimate greater anthropogenic carbon inventories than TTD in the Mediterranean Sea, e.g., the TrOCA approach (Fig. 1), must overestimate the true inventory by even more. Although even the upper limit of our range (1.0 to 1.7 Pg C) is small when compared to the global-ocean inventory of anthropogenic carbon of 134 Pg C (Sabine et al., 2004, for year 1994), the Mediterranean Sea contains 2.4 to 4 times as much anthropogenic carbon per unit volume as does the global ocean.

4.2 Transfer across the Strait of Gibraltar

Unlike the global ocean, where outside input of anthropogenic carbon comes only from the atmosphere, in the Mediterranean Sea there is also lateral input and output of anthropogenic carbon via the Strait of Gibraltar. Unfortunately, data-based estimates of that net transport do not agree even in terms of its direction, much less its magnitude. That is, estimates of transport based on data-based estimates of δC_T with the TrOCA method suggest that the Mediterranean Sea is a source of anthropogenic carbon to the Atlantic Ocean (Aït-Ameur and Goyet, 2006; Huertas et al., 2009); conversely, with two other data-based methods – TTD and the ΔC^* approach (Gruber et al., 1996) – there is a net transport of anthropogenic carbon from the Atlantic to the Mediterranean Sea (Huertas et al., 2009; Schneider et al., 2010). The two latter back-calculation data-based methods give similar net fluxes of δC_T: ~4.2 Tg C yr^{-1} with ΔC^* and 3.5 Tg C yr^{-1} with TTD. Both rely on the estimates of water fluxes from Huertas et al. (2009) (Table 6). Both methods also produce similar estimates for the δC_T concentrations in inflowing and outflowing waters: ~60 μmol kg^{-1} in the near-surface inflowing water and ~52 μmol kg^{-1} in the deeper Mediterranean Outflow Water (MOW). However, these δC_T estimates are based on data collected from different periods, i.e., May 2005 to July 2007 for Huertas et al. (2009) and November 2001 for Schneider et al. (2010). Moreover, the transfer deduced from TTD-derived δC_T estimates from (Schneider et al., 2010) are estimated to have a large uncertainty (−1.8 to 9.2 Tg C yr^{-1}). The net δC_T transfer estimated with the TrOCA method is −3 Tg C yr^{-1}. That much stronger net export from the Mediterranean Sea to the Atlantic is due to TrOCA's assessment that the outflowing MOW has higher δC_T (~80 μmol kg^{-1}) than the inflowing AW (~65 μmol kg^{-1}) (Huertas et al., 2009; Schneider et al., 2010; Flecha et al., 2011). Yet that vertical distribution is opposite to that expected from an anthropogenic transient tracer in the ocean with an atmospheric origin.

All three of our model simulations indicate a net transfer of anthropogenic carbon from the Atlantic to the Mediterranean across the Strait of Gibraltar (Sect. 3.3). In the reference simulation (MED), 0.26 Pg C enters the Mediterranean Sea via the Strait of Gibraltar between 1800 and 2001, similar to the TTD- and ΔC^*-based estimates (Table 6). Observational estimates of water transfer across the Strait of Gibraltar are between 0.72 and 1.01 Sv (1 sverdrup (Sv) = 10^6 m^3 s^{-1}) for surface inflow and between 0.68 and 0.97 Sv for deep outflow, resulting in a net transfer of +0.04 to +0.13 Sv (Bryden and Kinder, 1991; Bryden et al., 1994; Tsimplis and Bryden, 2000; Candela, 2001; Baschek et al., 2001; Lafuente et al., 2002; Soto-Navarro et al., 2010). The model falls near the lower limit of these estimates, having an inflow of 0.71 Sv, an outflow of 0.67 Sv, and a net water transfer of +0.04 Sv, when averaged between 1992 and 2008 (Beuvier, 2011). For 2005–2007, the simulated transfer is 0.15 Sv weaker than observational estimates from Huertas et al. (2009) in 2001 both for inflow and outflow, while net transfer is not significantly different: +0.04 vs. 0.05 Sv (Table 6).

Simulated δC_T concentrations in the model's AW are largely determined by damping to data-based estimates from Khatiwala et al. (2009) at the western boundary of the model domain. In the MED simulation, the δC_T in the inflowing AW is 12 to 24 % lower than data-based estimates from Huertas et al. (2009), who used both ΔC^* and TrOCA approaches (Table 6). But the largest discrepancy occurs in the outflowing deeper waters (MOW), for which the simulated δC_T underestimates the data-based ΔC^* for 2005–2007 by 31 %. That underestimate is expected given that simulated CFC-12 in the model's WMDW is only half that observed and that this deep water contributes to the MOW.

The model's underestimate of δC_T in the MOW is the determining factor which results in less outflow and thus more net inflow of anthropogenic carbon to the Mediterranean Sea. It follows that the model must provide an upper limit for the true net inflow of anthropogenic carbon, given that modeled water exchange falls within the observed range and that modeled and data-based estimates of δC_T are more similar in the inflowing water than in the outflowing water. Likewise,

Table 6. Lateral fluxes of water and anthropogenic carbon across the Strait of Gibraltar.

Approach	Year(s)	Net fluxes[7] ($Tg\,C\,yr^{-1}$)	Q_{in} (Sv)	$\delta C_{T;in}$ ($\mu mol\,kg^{-1}$)	Q_{out} (Sv)	$\delta C_{T;out}$ ($\mu mol\,kg^{-1}$)
(ΔC^*)[1,2]	2005–2007	$+4.20 \pm 0.04$	0.89	60	0.85	51
(ΔC^*)[1,3]	2005–2007		0.85	61	0.81	52
(TrOCA)[1,2]	2005–2007	-3.00 ± 0.04	0.89	64	0.85	78
(TrOCA)[1,3]	2005–2007		0.85	69	0.81	81
(TTD)[4]	2001	+3.5 (−1.8 to 9.2)	0.89[5]	62.4[6]	0.85[5]	54.8
Model MED	2001	+4.7	0.70	47.6	0.66	32.1
Model MED	2005–2007	+5.5	0.74	52.6	0.69	35.7

[1] Huertas et al. (2009) estimates based on data near the Strait of Gibraltar during 2005–2007.
[2] Method applied to observations on the Atlantic side of the Strait of Gibraltar.
[3] Method applied to observations on the Mediterranean side of the Strait of Gibraltar during May 2005 to July 2007.
[4] Schneider et al. (2010) estimates using the TTD approach with observations from 2001.
[5] Schneider et al. (2010) water fluxes across the Strait of Gibraltar are from Huertas et al. (2009).
[6] Schneider et al. (2010) δC_T concentration in inflowing AW is from Aït-Ameur and Goyet (2006).
[7] Positive values indicate transfer from the Atlantic to the Mediterranean Sea.

Table 7. Average changes in pH and $[H^+]$ between 1800 and 2001 for the three simulations over western and eastern basins and the entire Mediterranean Sea.

	δpH			$\delta[H^+]$ ($nmol\,kg^{-1}$)		
	West	East	Med. Sea	West	East	Med. Sea
GLO	−0.0851	−0.0849	−0.0850	1.45	1.46	1.46
MED	−0.0823	−0.0848	−0.0840	1.29	1.35	1.33
VAR	−0.0833	−0.0837	−0.0836	1.33	1.32	1.32

a lower limit for net transport of anthropogenic is offered by the computations that use data-based TTD estimates of δC_T. That follows because (1) TTD overestimates deep δC_T by more than surface values and (2) near-surface inflow and deep outflow are similar in magnitude. Hence the TTD-based approach must underestimate net input of anthropogenic carbon to the Mediterranean. Therefore the net input of anthropogenic carbon across the Strait of Gibraltar must be between +3.5 and $+4.7\,Tg\,C\,yr^{-1}$ based on observations collected in 2001. To compare that range to results of Huertas et al. (2009) for 2005–2007, we relied on the simulated δC_T evolution between 2001 and 2005–2007. In that 5 ± 1-year period, simulated δC_T increased by +10.5 % in the inflowing AW and by +11.2 % in the MOW. For the 2005–2007 lower limit, we applied those trends to the lower limit δC_T estimates for 2001 (TTD estimates of Schneider et al., 2010, in the AW and MOW) combined with the 2005–2007 water transfer rates (Huertas et al., 2009); for the upper limit we again used the model result. Hence, for 2005–2007, we consider that the true net input of anthropogenic carbon across the Strait of Gibraltar must fall between +3.7 and $+5.5\,Tg\,C\,yr^{-1}$.

4.3 Sensitivity to total alkalinity

To test the sensitivity of results to total alkalinity, we compared three simulations: GLO with a basin-wide total alkalinity equal to the global-ocean average, MED where the basin-wide inventory is increased by 10 % (equivalent to the Mediterranean Sea's surface average), and VAR where surface total alkalinity varies as a linear function of salinity. The 10 % greater total alkalinity in MED and VAR relative to GLO results in a 10 % greater simulated inventory of anthropogenic carbon (Table 5), but the basin-integrated air–sea flux of anthropogenic in MED and VAR is 20 % greater than in GLO (Table 2). The 10 % difference must be made up by proportionally less input of anthropogenic carbon to the Mediterranean from the Atlantic in MED and VAR relative to GLO.

The MED simulation has greater total alkalinity in the western basin than either GLO or VAR and hence absorbs more anthropogenic carbon there than do the other two simulations (Fig. 6). Yet MED's western basin total alkalinity is too high compared to what actually comes in from the Atlantic and even in terms of the δC_T also coming in with the same water. The latter is determined in all three model runs by restoring to data-based estimates of Khatiwala et al. (2009) in the Atlantic buffer zone. Thus it is less accurate to impose a mean Mediterranean Sea total alkalinity in this area, which artificially increases the surface-water buffer capacity and hence its ability to absorb CO_2. The same artifact results in a lower local change in pH (Fig. 12). Thus the constant Mediterranean surface total alkalinity as used in MED is suboptimal for simulating δC_T near the Strait of Gibraltar.

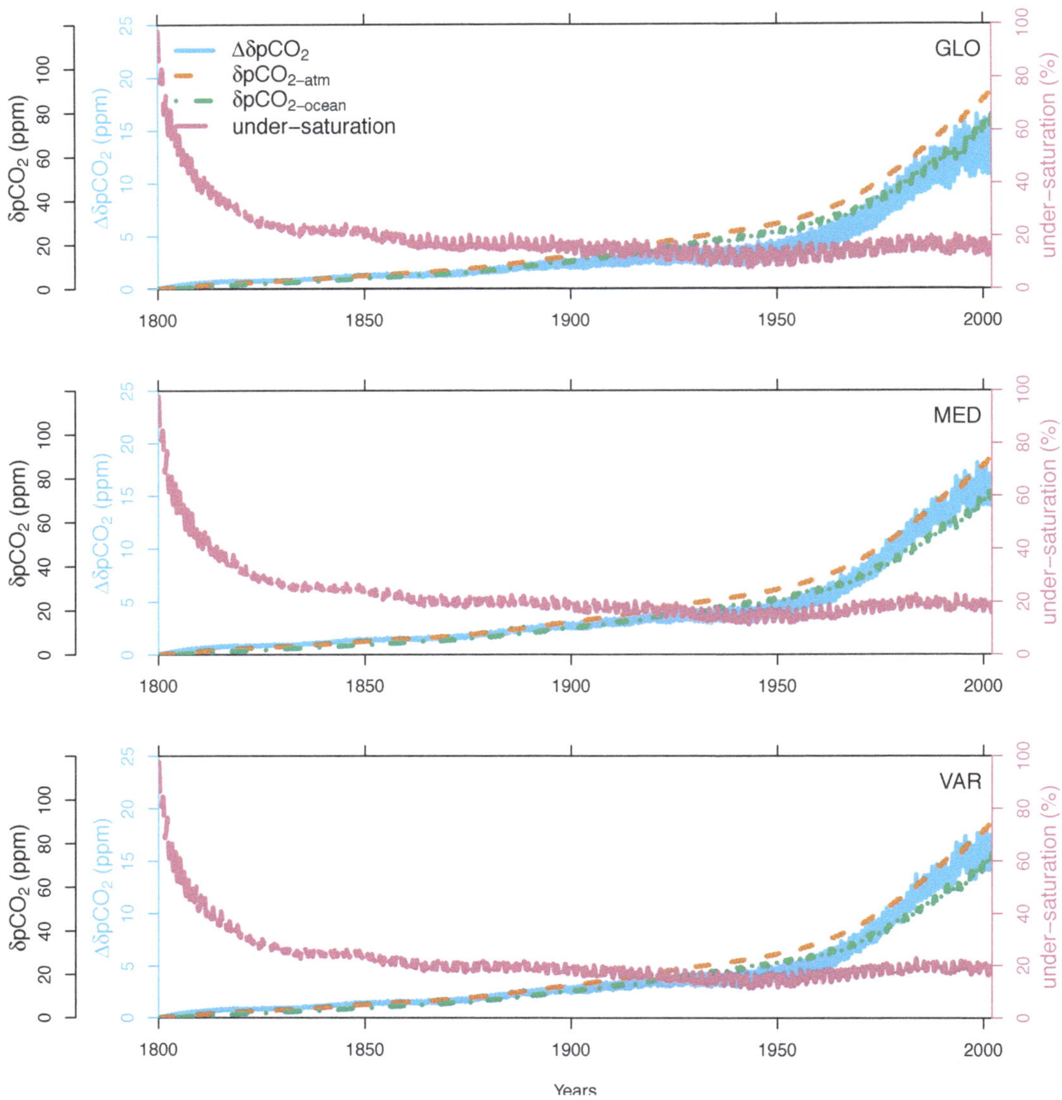

Figure 14. Temporal evolution from 1800 to 2001 of spatially averaged $\delta p\mathrm{CO_2}$ (in ppm) in the atmosphere (dashed orange line), the ocean (dashed-dotted green line), and their difference $\Delta\delta p\mathrm{CO_2}$ (solid light-blue line). Also shown is the corresponding percent undersaturation of oceanic $\delta p\mathrm{CO_2}$ (purple line), defined as $100\left(1-\frac{\delta p\mathrm{CO_{2\,oc}}}{\delta p\mathrm{CO_{2\,atm}}}\right)$.

In contrast, the VAR simulation generally has more realistic total alkalinity that increases from west to east (Fig. 3). That avoids an over-buffered carbonate system near the Strait of Gibraltar, particularly in the Alboran sub-basin, and an under-buffered system in the far eastern Mediterranean. However, VAR is generally less realistic near river mouths than either GLO or MED. By imposing a total alkalinity that is a function of salinity in a model that considers only freshwater riverine input (no total alkalinity delivery), the model-imposed total alkalinity near river mouths is too low. That artifact results in lower air–sea fluxes of anthropogenic carbon when close to river mouths (Fig. 6) and locally more intense reductions in pH near the Nile, Po, and Rhone river mouths and near the outflow of the Dardanelles Strait (Fig. 12); at the latter site, the air–sea flux of anthropogenic carbon even changes sign from ocean uptake to loss, although that is extremely localized.

Despite these local differences, the three approaches yield similar results when integrated across the entire Mediterranean Sea, with spatial variability in total alkalinity leading to differences in global inventory of only 0.1 % and differences between east–west partitioning of less than 1 %.

4.4 Change in pH

Two recent studies have attempted to quantify the decline in the pH of the Mediterranean Sea due to the increase in anthropogenic carbon (Touratier and Goyet, 2009, 2011). Both concluded that the pH reduction in the Mediterranean Sea

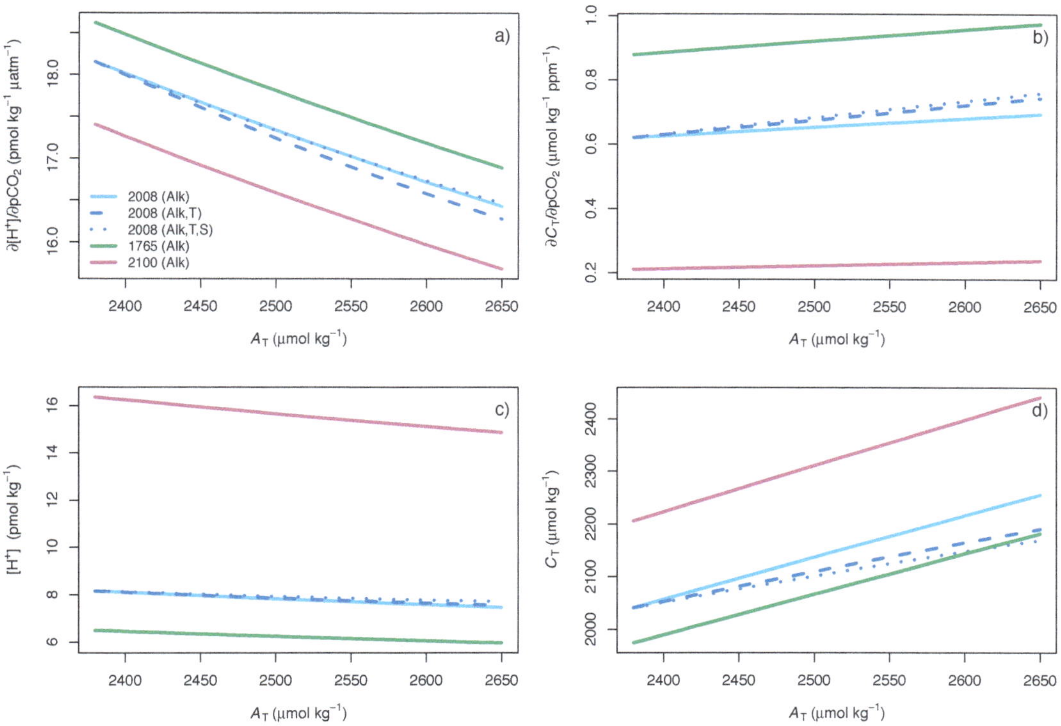

Figure 15. **(a)** Acidification rate ($\partial[H^+] / \partial pCO_2$, in pmol kg^{-1} μatm^{-1}), **(b)** rate of change of total carbon ($\partial C_T / \partial pCO_2$, in μmol kg^{-1} ppm^{-1}), **(c)** surface H^+ ion concentration (pmol kg^{-1}), and **(d)** C_T (μmol kg^{-1}), as a function of the Mediterranean's range of total alkalinity (2380 to 2650 μmol kg^{-1}) for three different atmospheric pCO_2 levels in 1765 (280 ppm, solid green line), 2008 (385 ppm, solid light-blue line), and 2100 (850 ppm, solid purple line). Also shown for year 2008 are effects from varying temperature (dashed blue line) and temperature and salinity (dotted blue line) over the observed west–east range.

(acidification) is larger than that experienced by typical waters of the global ocean. The higher total alkalinity of the Mediterranean Sea was evoked to justify a larger uptake of anthropogenic carbon. Our results support that finding, i.e., with the MED – GLO showing a 10 % increase in anthropogenic carbon inventory that occurs when average surface total alkalinity is increased by 10 % (Mediterranean minus global-ocean average).

The same two studies further suggest that higher levels of δC_T in the Mediterranean Sea also imply greater changes in pH. Yet our sensitivity tests demonstrate that the higher total alkalinity of the Mediterranean Sea does not result in a greater anthropogenic reduction in surface pH. Differences between simulations GLO and MED are negligible (Fig. 12). In both simulations, the simulated decline in surface pH is -0.084 ± 0.001 units when averaged across Mediterranean Sea (Table 7). Hence the decline in pH is quite similar between typical surface waters in the Mediterranean Sea and those in the global ocean. Furthermore, because it is on a log scale, absolute differences in pH actually represent relative changes in $[H^+]$. To avoid such confusion, we prefer to discuss acidification of the Mediterranean Sea in terms of $[H^+]$.

For a quantitative understanding of how total alkalinity affects surface acidification, we made additional equilibrium calculations to assess rates of change in terms of $\partial C_T/\partial pCO_2$ and $\partial[H^+] / \partial pCO_2$ (Fig. 15). Those were computed from analytical expressions for buffer factors (Egleston et al., 2010), corrected by Orr (2011). Both quantities change over the observed west–east gradient of the Mediterranean's surface alkalinity (2380 to 2650 μmol kg^{-1}) when temperature and salinity are each held at their western minimum values.

Our equilibrium thermodynamic equilibrium calculations of the rate of change of the Mediterranean's surface C_T confirm results from our sensitivity tests: the west–east increase in alkalinity, by itself, increases $\partial C_T/\partial pCO_2$ by 12 % (Fig 15b), corresponding to the equivalent west–east change in anthropogenic C_T (Fig 15d).

For the corresponding rate of acidification, the west–east increase in alkalinity, by itself, reduces $\partial[H^+] / \partial pCO_2$ by 8 % (Fig 15a), similar to what was found byAlvarez et al. (2014). Adding in the effect of the west–east temperature increase (6°C during summer) reduces the $\partial[H^+] / \partial pCO_2$ by another 0.5 % (i.e., for the east relative to the west). But the temperature reduction in $\partial[H^+] / \partial pCO_2$ is compensated for by the west–east increase in salinity (3 units on the practical salinity scale) during summer. For comparison, there is a 2 % decrease in $\partial[H^+] / \partial pCO_2$ as atmospheric xCO_2 increases from 280 to 385 ppm and another 4 % decrease when atmospheric xCO_2 increases further to 850 ppm. The

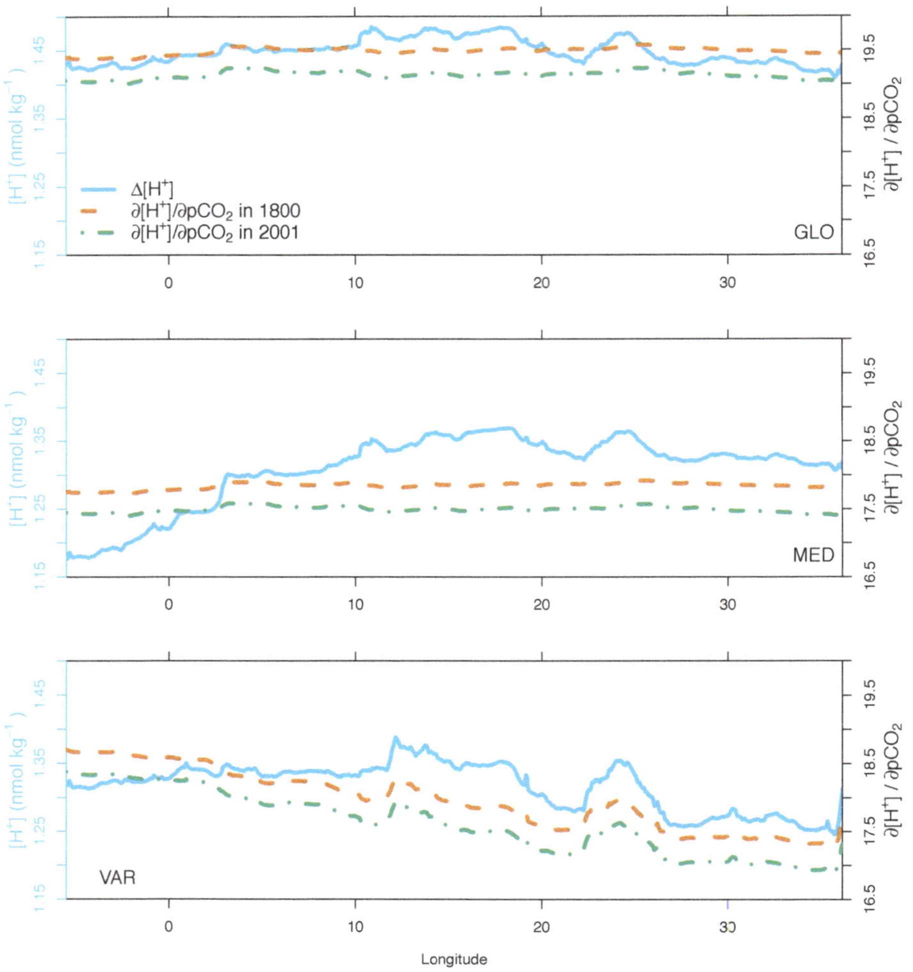

Figure 16. Meridional mean of the simulated surface acidification rate $\partial[H^+] / \partial pCO_2$ (pmol kg^{-1} μatm^{-1}) in 1800 (dashed orange line) and in 2001 (dashed-dotted green line) across the Mediterranean Sea. Also shown is the corresponding simulated $[H^+]$ change between 1800 and 2001 in nmol kg^{-1} (solid light-blue line). Meridional means are computed using grid cells with salinities above 32 to avoid biases near river mouths.

resulting west–east difference in $[H^+]$ decreases by 10.5 % between 1765 and 2008 (Fig 15c). These equilibrium calculations highlight the alkalinity effect on Mediterranean anthropogenic acidification. They confirm the Mediterranean Sea's higher δC_T uptake and its lower $[H^+]$ increase, relative to the global ocean average, both due to its higher alkalinity.

The model's surface acidification rates (Fig. 16) are slightly less intense, because it does not make the simplification that atmospheric and oceanic pCO_2 are identical, as in the equilibrium calculations. The acidification rate is 8 % lower in MED ($\sim$ 17.5 pmol kg$^{-1}\mu$atm^{-1} in 2001) than it is in GLO ($\sim$ 19.1 pmol kg^{-1} μatm^{-1} in 2001). In VAR, the $\partial[H^+] / \partial pCO_2$ decreases by 8 % in 2001 when moving from west to east (from $\sim$18.5 to $\sim$17 pmol kg$^{-1}\mu$atm^{-1}). That modeled west–east gradient is much like that found with the thermodynamic calculations, but curves are displaced downwards by 0.3 units.

Anthropogenic carbon is already present in substantial quantities throughout the deep Mediterranean Sea (Fig. 11). Hence the anthropogenic decline in pH also affects the entire water column. Touratier and Goyet (2011) found that the anthropogenic pH change in some Mediterranean bottom waters has already reached values of up to −0.12, higher even than at the surface. However, they deduce those high values from data-based estimates of δC_T using the TrOCA approach, which overestimates actual values, particularly at depth (see Sects. 1 and 4.1). To estimate subsurface anthropogenic changes in pH, we used a simple three-step method: (1) we used discrete measurements of C_T, total alkalinity, and phosphate and silicate concentrations on the 2001 Meteor M51/2 cruise to compute a modern reference pH using seacarb; (2) we subsampled the MED model at the same time, positions, and depths to get corresponding simulated δC_T; and (3) we subtracted the latter from the modern measurements of C_T to get preindustrial C_T, using that along

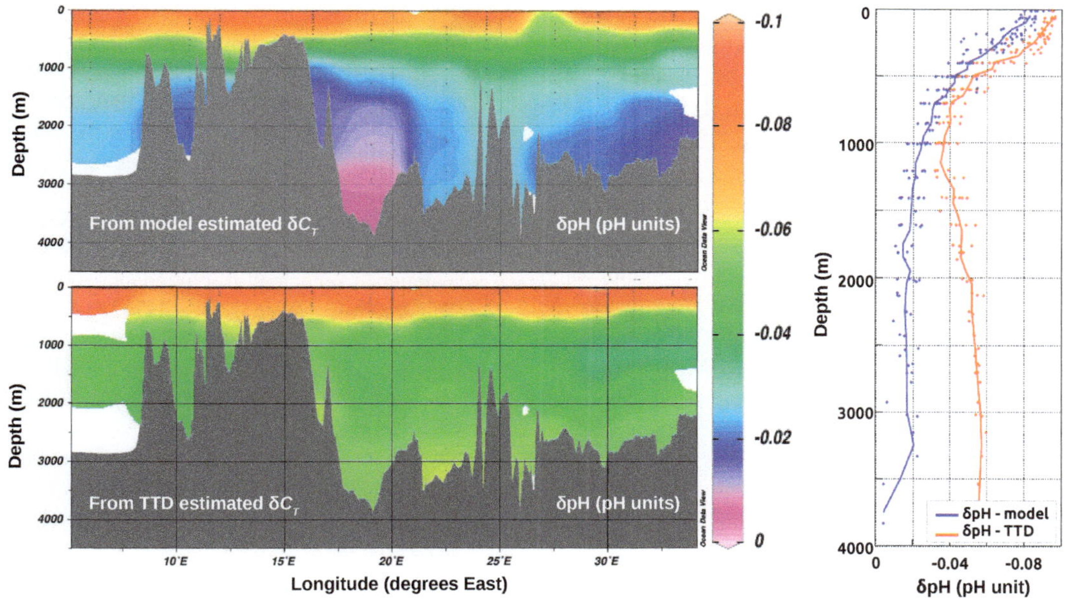

Figure 17. Mediterranean δpH along the *Meteor* M51/2 section calculated with δC_T from the MED simulation (top left) and from the TTD data-based estimates from Schneider et al. (2010) (bottom left). Also shown are the same results but as mean vertical profiles averaged along the section (right).

with the measured values of other input variables (assuming they had not changed) to compute a preindustrial pH. We then compared that model-derived change in pH (δpH_{model}) to the data-based TTD estimates (δpH_{TTD}) calculated in the same fashion, i.e., using TTD δC_T instead of modeled δC_T in the computation sequence (Fig. 17). The resulting anthropogenic change in surface pH ranges from -0.08 to -0.10 units. Below the surface, δpH_{model} gradually becomes less intense until reaching the bottom, where it ranges from -0.005 pH units in the Ionian sub-basin to -0.03 in the Crete Passage. Yet those changes must be underestimates given the model's poor ventilation of deep waters based on the CFC-12 evaluation (Sect. 3.1). The data-based change in δpH_{TTD} exhibits its weakest magnitude (-0.035 pH units) between 1000 and 1500 m in the Levantine sub-basin, where the TTD data-based δC_T is at a minimum. Deeper down, δpH_{TTD} increases in magnitude, reaching up to -0.06 pH units in the bottom waters of the Ionian sub-basin.

As the model results and the TTD data-based approach provide lower and upper limits for the actual changes in deep-water δC_T, it follows that they also provide bounds for the anthropogenic change in pH. The actual change in bottom water pH in the eastern basin thus lies between -0.005 and -0.06 units.

5 Conclusions

A first simulation of anthropogenic carbon in the Mediterranean Sea suggests that it accumulated 1.0 Pg C between 1800 and 2001. That estimate provides a lower limit based on comparison of observed vs. simulated CFC-12 in the same model, which reveals that modeled deep waters are poorly ventilated. Furthermore, we demonstrate that a previous data-based estimate of 1.7 Pg C (Schneider et al., 2010) is an upper limit after testing the associated TTD approach in our model. In 2001 in the reference model, a total of 1.5 Pg C of anthropogenic carbon had entered the Mediterranean Sea with 52 % from the air–sea flux and 48 % from Atlantic Water inflow; however, 31 % of that total had also left via the deep Mediterranean Outflow Water. Out of the net accumulation of 1.0 Pg C, 75 % comes from the air–sea flux and 25 % from net transfer across the Strait of Gibraltar. The rate of net exchange across that strait to the Mediterranean is from 3.5 to 4.7 Tg C yr^{-1} in 2001 and from 3.7 to 5.5 Tg C yr^{-1} in 2005–2007, based on the model and TTD results.

Our estimates of anthropogenic carbon also allow us to assess anthropogenic changes in pH. Although the 10 % higher mean total alkalinity of the Mediterranean Sea is responsible for a 10 % increase in anthropogenic carbon inventory, that does not significantly affect the anthropogenic change in surface pH. The average surface pH change is -0.08 units for both the Mediterranean Sea and the global ocean. Deep waters of the Mediterranean Sea exhibit a larger anthropogenic change in pH than typical global ocean deep waters because ventilation times are faster. In 2001, the δpH in Mediterranean Sea bottom waters is estimated to lie between -0.005 and -0.06 units based on our limits from simulated and TTD data-based δC_T. These findings do not support previous conclusions that the anthropogenic change in the pH of Mediterranean deep waters is as high as -0.12 units, which is more intense even than the surface change (Touratier and Goyet, 2009, 2011). Furthermore, those previous findings rely on the TrOCA data-based estimates of δC_T, which are much larger than the TTD data-based estimates, shown in Sect. 4.1 to be already an upper limit.

Future studies that include the full natural carbon cycle and the effects of climate change are needed to confirm these results and predict future changes while weighing geochemical vs. climate factors. Improved assessment of local changes along coastlines will require improved boundary conditions, particularly for riverine and groundwater discharge of nutrients, carbon, and total alkalinity, combined with developments to improve coastal aspects of the physical and biogeochemical models.

Acknowledgements. We thank Samar Khatiwala for providing his data-based estimates for anthropogenic carbon, which we used as a lateral boundary condition for the Atlantic portion of our model domain. We thank Météo-France/CNRM and in particular Michel Déqué and Florence Sevault for running and providing the ARPERA data set. We also thank Marte Álvarez and an anonymous referee for their insightful reviews, which have helped to improve this manuscript. This work was supported by the French SiMED project (Mercator Ocean), the MORCE and MED-ICCBIO projects (GIS), the EU FP7 project MedSeA (grant 265103), and the French ANR project MACROES (MACRoscope for Oceanic Earth System ANR-09-CEP-003). This work is a contribution to the HyMeX and MERMEX programs, and access was granted to the HPC resources of IDRIS (Institut du Développement et des Ressources en Informatique Scientifique) of the Centre National de la Recherche Scientifique (CNRS) under allocations for years 2010, 2011, and 2012 (project 1010227) made by Grand Equipement National de Calcul Intensif (GENCI).

Edited by: J.-P. Gattuso

References

Aït-Ameur, N. and Goyet, C.: Distribution and transport of natural and anthropogenic CO_2 in the Gulf of Cádiz, Deep-Sea Res. Pt. II, 53, 1329–1343, 2006.

Álvarez, M., Sanleón-Bartolomé, H., Tanhua, T., Mintrop, L., Luchetta, A.,Cantoni, C., Schroeder, K., Civitarese, G.: The CO_2 system in the Mediterranean Sea: a basin wide perspective, Oc. Sci., 10, 69–92, 2014.

Antonov, J. I., Locarnini, R. A., Boyer, T. P., Mishonov, A. V., and Garcia, H. E.: World Ocean Atlas 2005, vol. 2: Salinity, edited by: Levitus, S., NOAA Atlas NESDIS 62, US Government Printing Office, Washington DC, 182 pp., 2006.

Attané, I. and Courbage, Y.: La démographie en Méditerranée, Situation et projections, Economica – Plan Bleu, 2001.

Attané, I. and Courbage, Y.: Demography in the Mediterranean region: situation and projections, Plan Bleu, 2004.

Baschek, B., Send, U., Lafuente, J. G., and Candela, J.: Transport estimates in the Strait of Gibraltar with a tidal inverse model, J. Geophys. Res., 106, 31033–31, 2001.

Béranger, K., Mortier, L., and Crépon, M.: Seasonal variability of water transport through the Straits of Gibraltar, Sicily and Corsica, derived from a high-resolution model of the Mediterranean circulation, Prog. Oceanogr., 66, 341–364, 2005.

Beuvier, J.: Modélisation de la variabilité climatique et des masses d'eau en mer Méditerranée: impact des échanges océan-atmosphère, Ph. D. thesis, Ecole Polytechnique, Palaiseau, France, 2011.

Beuvier, J., Sevault, F., Herrmann, M., Kontoyiannis, H., Ludwig, W., Rixen, M., Stanev, E., Béranger, K., and Somot, S.: Modelling the Mediterranean Sea interannual variability over the last 40 years: focus on the EMT, J. Geophys. Res., 115, C08017, doi:10.1029/2009JC005950, 2010.

Beuvier, J., Béranger, K., Lebeaupin Brossier, C., Somot, S., Sevault, F., Drillet, Y., Bourdallé-Badie, R., Ferry, N., and Lyard, F.: Spreading of the Western Mediterranean Deep Water after winter 2005: Time scales and deep cyclone transport, J. Geophys. Res., 117, C07022 doi:10.1029/2011JC007679, 2012a.

Beuvier, J., Lebeaupin Brossier, C., Béranger, K., Arsouze, T., Bourdallé-Badie, R., Deltel, C., Drillet, Y., Drobinski, P., Lyard, F., Ferry, N., Sevault, F., and Somot, S.: MED12, Oceanic component for the modelling of the regional Mediterranean Earth System, Merc. Oc. Quart. Newsletter, 46, 60–66, 2012b.

Bindoff, N. L., Willebrand, J., Artale, V., Cazenave, A., Gregory, J. M., Gulev, S., Hanawa, K., Le Quéré, C., Levitus, S., Nojiri, Y., Shum, C. K., Talley, L. D., and Unnikrishnan, A. S.: Observations: Oceanic climate change and sea level, in: Climate Change 2007: The Physical Science Basis, Contribution of Working Group I to the Fourth Assessment Report of the Intergovernmental Panel on Climate Change, edited by: Solomon, S., Qin, D., Manning, M., Marquis, M., Averyt, KB., Tignor, M., Miller HL., and Chen Z., chap. 5, Cambridge University Press, Cambridge, United Kingdom and New York, USA, 385–432, 2007.

Bryden, H. L. and Kinder, T. H.: Steady two-layer exchange through the Strait of Gibraltar, Deep-Sea Res. Pt. I, 38, S445–S463, 1991.

Bryden, H. L., Candela, J., and Kinder, T. H.: Exchange through the Strait of Gibraltar, Prog. Oceanogr., 33, 201–248, 1994.

Candela, J.: Mediterranean Water and Global Circulation, vol. 77, Elsevier, 419 pp., 2001.

Daget, N., Weaver, A., and Balmaseda, M.: Ensemble estimation of background-error variances in a three-dimensional variational data assimilation system for the global ocean, Q. J. Roy. Meteor. Soc., 135, 1071–1094, 2009.

Diffenbaugh, N. S. and Giorgi, F.: Climate change hotspots in the CMIP5 global climate model ensemble, Clim. Change, 114, 813–822, 2012.

Dutay, J.-C., Bullister, J. L., Doney, S. C., Orr, J. C., Najjar, R., Caldeira, K., Campin, J.-M., Drange, H., Follows, M., Gao, Y., Gruber, N., Hecht, M. W., Ishida, A., Joos, F., Lindsay, K., Madec, G., Maier-Reimer, E., Marshall, J. C., Matear, R. J., Monfray, P., Mouchet, A., Plattner, G.-K., Sarmiento, J., Schlitzer, R.,

Slater, R., Totterdell, I. J., Weirig, M.-F., Yamanaka, Y., and Yool, A.: Evaluation of ocean model ventilation with CFC-11: comparison of 13 global ocean models, Ocean Model., 4, 89–120, 2002.

Egleston, E. S., Sabine, C. L., and Morel, F. M. M.: Revelle revisited: Buffer factors that quantify the response of ocean chemistry to changes in DIC and alkalinity, Global Biogeochem. Cy., 24, GB1002, doi:10.1029/2008GB003407, 2010.

El Boukary, M. M. S.: Impact des activités humaines sur les cycles biogéochimiques en mer Méditerranée, Ph. D. thesis, Unniversité Paris VI – Pierre et Marie Curie, 2005.

Enting, I. G., Wigley, T. M. L., and Heimann, M.: Future emissions and concentrations of carbon dioxide: key ocean/atmosphere/land analyses, Tech. pap. 31, Div. of Atmos. Res., Commonw. Sci., and Ind. Res. Org., Melbourne, Australia, 1994.

Ferry, N., Parent, L., Garric, G., Barnier, B., Jourdain, N. C., and the Mercator-Ocean-team: Mercator global eddy permitting ocean reanalysis GLORYS1V1: description and results, Merc. Oc. Quart. Newsletter, 36, 15–28, 2010.

Flecha, S., Pérez, F. F., Navarro, G., Ruiz, J., Olivé, I., Rodríguez-Gálvez, S., Costas, E., and Huertas, I. E.: Anthropogenic carbon inventory in the Gulf of Cádiz, J. Marine Syst., 92, 67–75, 2011.

Gibelin, A.-L. and Déqué, M.: Anthropogenic climate change over the Mediterranean region simulated by a global variable resolution model, Clim. Dynam., 20, 327–339, 2003.

Giorgi, F.: Climate change hot-spots, Geophys. Res. Lett., 33, L08707, doi:10.1029/2006GL025734, 2006.

Giorgi, F. and Lionello, P.: Climate change projections for the Mediterranean region, Glob. Planet. Change, 63, 90–104, 2008.

GLOBALVIEW–CO_2: Cooperative Atmospheric Data Integration Project – Carbon Dioxide, NOAA ESRL, Boulder, Colorado, available at: http://www.esrl.noaa.gov/gmd/ccgg/globalview, 2010.

Gruber, N., Sarmiento, J. L., and Stocker, T. F.: An improved method to detect anthropogenic CO_2 in the oceans, Global Biogeochem. Cy., 10, 809–837, 1996.

Herrmann, M. and Somot, S.: Relevance of ERA40 dynamical downscaling for modeling deep convection in the Mediterranean Sea, Geophys. Res. Lett., 35, L04607, doi:10.1029/2007GL032442, 2008.

Herrmann, M., Somot, S., Sevault, F., Estournel, C., and Déqué, M.: Modeling the deep convection in the northwestern Mediterranean Sea using an eddy-permitting and an eddy-resolving model: Case study of winter 1986–1987, J. Geophys. Res., 113, C04011, doi:10.1029/2006JC003991, 2008.

Herrmann, M., Sevault, F., Beuvier, J., and Somot, S.: What induced the exceptional 2005 convection event in the Northwestern Mediterranean basin? Answers from a modeling study., J. Geophys. Res., 115, C12051, doi:10.1029/2010JC006162, 2010.

Huertas, I. E., Ríos, A. F., García-Lafuente, J., Makaoui, A., Rodríguez-Gálvez, S., Sánchez-Román, A., Orbi, A., Ruíz, J., and Pérez, F. F.: Anthropogenic and natural CO_2 exchange through the Strait of Gibraltar, Biogeosciences, 6, 647–662, doi:10.5194/bg-6-647-2009, 2009.

Key, R. M., Sabine., C. L., Lee, K., Wanninkhof, R., Bullister, J., Feely, R. A., Millero, F. J., Mordy, C., and Peng, T.-H.: A global ocean carbon climatology: results from Global Data Analysis Project (GLODAP), Global Biogeochem. Cy., 18, GB4031, doi:10.1029/2004GB002247, 2004.

Khatiwala, S., Primeau, F., and Hall, T.: Reconstruction of the history of anthropogenic CO_2 concentrations in the ocean, Nature, 462, 346–349, 2009.

Lachkar, Z., Orr, J. C., Dutay, J.-C., and Delecluse, P.: Effects of mesoscale eddies on global ocean distributions of CFC-11, CO_2, and $\Delta^{14}C$, Ocean Sci., 3, 461–482, doi:10.5194/os-3-461-2007, 2007.

Lafuente, J. G., Delgado, J., Vargas, J. M., Vargas, M., Plaza, F., and Sarhan, T.: Low-frequency variability of the exchanged flows through the Strait of Gibraltar during CANIGO, Deep-Sea Res. Pt. II, 49, 4051–4067, 2002.

Lavigne, H. and Gattuso, J.-P.: seacarb: seawater carbonate chemistry with R. R package version 2.4, available at: http://CRAN.R-project.org/package=seacarb, 2011.

Lebeaupin Brossier, C., Béranger, K., Deltel, C., and Drobinski, P.: The Mediterranean response to different space-time resolution atmospheric forcings using perpetual mode sensitivity simulations, Ocean Model., 36, 1–25, 2011.

Lee, K., Sabine, C. L., Tanhua, T., Kim, T.-W., Feely, R. A., and Kim, H.-C.: Roles of marginal seas in absorbing and storing fossil fuel CO_2, Energy Environ. Sci., 4, 1133–1146, 2011.

Locarnini, R. A., Mishonov, A. V., Antonov, J. I., Boyer, T. P., and Garcia, H. E.: World Ocean Atlas 2005, Volume 1: Temperature, S. Levitus, Ed., NOAA Atlas NESDIS 61, US Government Printing Office, Washington, D. C., 182 pp., 2006.

Ludwig, W., Dumont, E., Meybeck, M., and Heussner, S.: River discharges of water and nutrients to the Mediterranean and Black Sea: Major drivers for ecosystem changes during past and future decades?, Prog. Oceanogr., 80, 199–217, 2009.

Madec, G. and the NEMO team: NEMO ocean engine, Note du pole de modélisation de l'IPSL no. 27, Institut Pierre Simon Laplace, ISSN 1228–1619, 2008.

MEDAR/MEDATLAS-Group: MEDAR/MEDATLAS 2002 Database, Mediterranean and Black Sea database of temperature, salinity and bio-chemical parameters, Climatological Atlas [4CD], Ifremer, Brest, France, 2002.

Orr, J. C., Monfray, P., Maier-Reimer, E., Mikolajewicz, U., Palmer, J., Taylor, N. K., Toggweiler, J. R., Sarmiento, J. L., Le Quere, C., Gruber, N, Sabine, C. L., Key, R. M., and Boutin, J.: Estimates of anthropogenic carbon uptake from four threedimensionsal global ocean models, Global Biogeochem. Cy., 15, 43–60, 2001.

Orr, J. C.: Recent and future changes in ocean carbonate chemistry, in: Ocean Acidification, edited by: Gattuso, J. and Hansson, L., Oxford University press, Oxford, 41–66, 2011.

Rixen, M., Beckers, J. M., Levitus, S., Antonov, J., Boyer, T., Maillard, C., Fichaut, M., Balopoulos, E., Iona, S., Dooley, H., Garcia, M. J., Manca, B., Giorgetti, A., Manzella, G., Mikhailov, N., Pinardi, N., Zavatarelli, M.: The Western Mediterranean Deep Water: A proxy for climate change, Geophys. Res. Lett., 32, 12, doi:10.1029/2005GL022702, 2005.

Roether, W., Manca, B., Klein, B., Bregant, D., Georgopoulos, D., Beitzel, V., Kovacevic, V., and Luchetta, V.: Recent changes in eastern Mediterranean deep waters, Science, 271, 333–335, 1996.

Roether, W., Klein, B., Manca, B. B., Theocharis, A., and Kioroglou, S.: Transient Eastern Mediterranean deep waters in response to the massive dense-water output of the Aegean Sea in the 1990s, Prog. Oceanogr., 74, 540–571, 2007.

Sabine, C. L., Feely, R. A., Gruber, N., Key, R. M., Lee, K., Bullister, J. L., Wanninkhof, R., Wong, C. S., Wallace, D. W. R., Tilbrook, B., Millero, F. J., Peng, T.-H., Kozyr, A., Ono, T., and Rios, A.: The ocean sink for anthropogenic CO_2, Science, 305, 367–370, 2004.

Sarmiento, J. L., Orr, J. C., and Siegenthaler, U.: A perturbation simulation of uptake in an ocean general circulation model, J. Geophys. Res., 97, 3621–3645, 1992.

Sarmiento, J. L., Hughes, T. M., Stouffer, R. J., and Manabe, S.: Simulated response of the ocean carbon cycle to anthropogenic climate warming, Nature, 393, 245–249, 1998.

Schneider, A., Wallace, D. W. R., and Körtzinger, A.: Alkalinity of the Mediterranean Sea, Geophys. Res. Lett., 34, L15608, doi:10.1029/2006GL028842, 2007.

Schneider, A., Tanhua, T., Körtzinger, A., and Wallace, D. W. R.: High anthropogenic carbon content in the eastern Mediterranean, J. Geophys. Res., 115, C12050, doi:10.1029/2010JC006171, 2010.

Schroeder, K., Ribotti, A., Borghini, M., Sorgente, R., Perilli, A., and Gasparini, G.: An extensive western Mediterranean deep water renewal between 2004 and 2006, Geophys. Res. Lett., 35, L18605, doi:10.1029/2008GL035146, 2008.

Siegenthaler, U. and Joos, F.: Use of a simple model for studying oceanic tracer distributions and the global carbon cycle, Tellus B, 44, 186–207, 1992.

Siegenthaler, U. and Sarmiento, J.: Atmospheric carbon dioxide and the ocean, Nature, 365, 119–125, 1993.

Somot, S., Sevault, F., and Déqué, M.: Transient climate change scenario simulation of the Mediterranean Sea for the 21st century using a high-resolution ocean circulation model., Clim. Dynam., 27, 851–879, 2006.

Soto-Navarro, J., Criado-Aldeanueva, F., García-Lafuente, J., and Sánchez-Román, A.: Estimation of the Atlantic inflow through the Strait of Gibraltar from climatological and in situ data, J. Geophys. Res., 115, C10023, doi:10.1029/2010JC006302, 2010.

Soto-Navarro, J., Somot, S., Sevault, F., Beuvier, J., Criado-Aldeanueva, F., García-Lafuente, J., and Béranger, K.: Evaluation of regional ocean circulation models for the Mediterranean Sea at the strait of Gibraltar: volume transport and thermohaline properties of the outflow, Clim. Dynam., 1–16, doi:10.1007/s00382-014-2179-4, 2014.

Stanev, E. V. and Peneva, L.: Regional sea level response to global climatic change: Black Sea examples, Global Planet. Change, 32, 33–47, 2002.

MerMEX group: Marine ecosystems' responses to climatic and anthropogenic forcings in the Mediterranean, Prog. Oceanogr., 91, 97–166, 2011.

Touratier, F. and Goyet, C.: Decadal evolution of anthropogenic CO_2 in the northwestern Mediterranean Sea from the mid-1990s to the mid-2000s, Deep-Sea Res. Pt. I, 56, 1708–1716, 2009.

Touratier, F. and Goyet, C.: Impact of the Eastern Mediterranean Transient on the distribution of anthropogenic CO_2 and first estimate of acidification for the Mediterranean Sea, Deep-Sea Res. Pt. I, 58, 1–15, 2011.

Touratier, F., Azouzi, L., and Goyet, C.: CFC-11, $\delta^{14}C$ and ^3H tracers as a means to asses anthropogenic CO_2 concentrations in the ocean, Tellus B, 59, 318–325, 2007.

Tsimplis, M. and Bryden, H.: Estimation of the transports through the Strait of Gibraltar, Deep-Sea Res. Pt. I, 47, 2219–2242, 2000.

Vargas-Yáñez, M., Plaza, F., Garcıa-Lafuente, J., Sarhan, T., Vargas, J., and Velez-Belchi, P.: About the seasonal variability of the Alboran Sea circulation, J. Marine Syst., 35, 229–248, 2002.

Vörösmarty, C. J., Fekete, B. M., and Tucker, B. A.: Global River Discharge Database (RivDIS V1.0), International Hydrological Program, Global Hydrological Archive and Analysis Systems, UNESCO, Paris, 1996.

Wanninkhof, R.: Relationship between wind speed and gas exchange over the ocean, J. Geophys. Res., 97, 7373–7382, 1992.

Warner, M. J. and Weiss, R. F.: Solubilities of chlorofluorocarbons 11 and 12 in water and seawater, Deep-Sea Res. Pt. I, 32, 1485–1497, 1985.

Waugh, D. W., Hall, T. M., McNeil, B. I., Key, R., and Matear, R. J.: Anthropogenic CO_2 in the oceans estimated using transit time distributions, Tellus B, 58, 376–389, 2006.

Weiss, R. F.: Carbon dioxide in water and seawater: the solubility of a non-ideal gas, Mar. Chem., 2, 203–215, 1974.

Yool, A., Oschlies, A., Nurser, A. J. G., and Gruber, N.: A model-based assessment of the TrOCA approach for estimating anthropogenic carbon in the ocean, Biogeosciences, 7, 723–751, doi:10.5194/bg-7-723-2010, 2010.

14

Characterization of incubation experiments and development of an enrichment culture capable of ammonium oxidation under iron-reducing conditions

S. Huang and P. R. Jaffé

Department of Civil and Environmental Engineering, Princeton University, Princeton, NJ 08544, USA

Correspondence to: P. R. Jaffé (jaffe@princeton.edu)

Abstract. Incubation experiments were conducted using soil samples from a forested riparian wetland where we have previously observed anaerobic ammonium oxidation coupled to iron reduction. Production of both nitrite and ferrous iron was measured repeatedly during incubations when the soil slurry was supplied with either ferrihydrite or goethite and ammonium chloride. Significant changes in the microbial community were observed after 180 days of incubation as well as in a continuous flow membrane reactor, using 16S rRNA gene PCR-denaturing gradient gel electrophoresis, 454 pyrosequencing, and real-time quantitative PCR analysis. We believe that one of the dominant microbial species in our system (an uncultured *Acidimicrobiaceae* bacterium A6), belonging to the *Acidimicrobiaceae* family, whose closest cultivated relative is *Ferrimicrobium acidiphilum* (with 92 % identity) and *Acidimicrobium ferrooxidans* (with 90 % identity), might play a key role in this anaerobic biological process that uses ferric iron as an electron acceptor while oxidizing ammonium to nitrite. After ammonium was oxidized to nitrite, nitrogen loss proceeded via denitrification and/or anammox.

1 Introduction

The most common removal of nitrogen from soil environments is mineralization (for organic nitrogen), followed by nitrification and then denitrification (Canfield et al., 2010). In water-saturated and organic carbon rich sediments or soils, such as wetland sediments, there is little oxygen for significant nitrification by aerobic ammonium (NH_4^+) oxidation bacteria/archaea (AOB/AOA). A novel anaerobic NH_4^+ oxidation process coupled to iron reduction was first noted in a forested riparian wetland in New Jersey (Clement et al., 2005). In this reaction, NH_4^+ is the electron donor, which is oxidized to nitrite (NO_2^-), and ferric iron – Fe(III) – is the electron acceptor, which is reduced to ferrous iron – Fe(II). The stoichiometry and change in free energy when ferrihydrite is the Fe(III) source is:

$$3Fe_2O_3 \bullet 0.5H_2O + 10H^+ + NH_4^+ \rightarrow 6Fe^{2+} \tag{1}$$

$$+8.5H_2O + NO_2^- \left(\Delta G_r \leq -145.08\ \text{kJ mol}^{-1},\ \text{Supplement 1.1}\right)$$

No proven pathway for the oxidation of NH_4^+ to NO_2^- in anaerobic environments has been described in the literature before this process was reported. Using labeled $^{15}NH_4^+$ in a microcosm experiment resulted in the production of $^{15}N_2$, which conclusively showed that ammonium-N was converted to nitrogen gas (N_2) in these sediments under iron-reducing conditions (Shrestha et al., 2009). Either this same pathway for NH_4^+ oxidation, or a very similar one, was also observed in a biological reactor (Sawayama, 2006) and a tropical rainforest soil (Yang et al., 2012), and coined Feammox (Sawayama, 2006). These pathways have been reported to oxidize NH_4^+ to NO_2^- (Clement et al., 2005; Shrestha et al., 2009), to NO_3^- (Sawayama, 2006), or directly to N_2 (Yang et al., 2012), using Fe(III) as an electron acceptor.

Our understanding of the Feammox process is still incomplete; particularly, information about the microorganism(s) responsible for it is lacking. This makes further study into the mechanism of the Feammox process difficult. Here we focus on a series of incubations and establishing a Feammox enrichment culture to identify the microbial community responsible for the process described previously (Clement et al., 2005; Shrestha et al., 2009). Soil samples were collected from the same location and used for laboratory incubation experiments as well as to set up an enrichment system for Feammox in a continuous flow membrane reactor. Various incubation conditions (Fe(III) sources, inorganic carbon content, NH_4^+ concentration, $^{15}NH_4^+$, and acetylene gas (C_2H_2) as a selected inhibitor) were used to study the Feammox mechanism. Molecular biology methods, such as denaturing gradient gel electrophoresis (DGGE), 454 pyrosequencing, and real-time quantitative PCR (qPCR) analysis were used to investigate the bacterial community change during incubations.

2 Methods

2.1 Sample collection and processing

Soils for all the experiments described in this study were taken from a temperate forested riparian wetland at the Assunpink Wildlife Management Area, New Jersey. Ten soil cores were collected from 10 cm below the surface with polyethylene column containers (8 cm diameter and 30 cm long) and transported to the laboratory within 2 h. The soil pH was between 3.5 and 4.5, and no manganese oxides were detected. The detailed physicochemical characteristics of these wetland soils have been described elsewhere (Clement et al., 2005). Prior to all incubation experiments, soil slurry from the field site was aerated for a month to degrade much of the labile organic carbon. After 30 days of aeration, the dissolved organic carbon (DOC) content was stable at $2.06 \pm 0.20\,mg\,g^{-1}$. Following the aeration treatment, the soil was divided into 400 × 10 g (air-dry equivalent) subsamples, and added to 50 mL serum vials, with 30 mL deionized water. The soil slurries were purged thoroughly with a $CO_2 : N_2$ (80 : 20) mixture, resulting in a final pH of ~4 to 4.5. The vials were sealed tightly with rubber stoppers and were stored in an anaerobic glove box for 30 days at ambient temperature to allow for stabilization before starting the incubations.

2.2 Batch incubation experiments

All incubations, addition of reagents, and sampling were conducted in an anaerobic glove box with a solution of resazurin as the redox indicator. Soil samples were first incubated with different Fe(III) sources to determine which source would yield a more active Feammox process: 6-line ferrihydrite ($Fe_2O_3C \cdot 0.5H_2O$) or goethite (FeO(OH); prepared according to Cornell and Schwertmann, 2003) + NH_4^+ addition; ferric chloride + NH_4^+ addition; ferric citrate + NH_4^+ addition; either only ferrihydrite or NH_4^+ addition; and autoclaved soil with ferrihydrite + NH_4^+ addition ($n = 30$ per treatment). pH was adjusted to 4.5 in the ferrihydrite/goethite augmented samples, and to between 3.5 and ~4.0 in the ferric chloride/citrate augmented samples. Soil-slurry samples, which were prepared to have an initial concentration of $12.0\,mmol\,L^{-1}$ Fe(III) and/or $2.00\,mmol\,L^{-1}$ NH_4^+, were incubated in a series of 50 mL vials with an oxygen-free headspace, created by purging with a $CO_2 : N_2$ (80 : 20) mixture. Triplicate samples were collected destructively every 2 days to analyze iron and nitrogen species.

The second incubation was conducted to extend the anoxic incubation with ferrihydrite to 180 days, with repeated NH_4Cl additions after the NH_4^+ in solution was exhausted. The initial concentration of Fe(III) was 25.0 and $1.00\,mmol\,L^{-1}$ NH_4^+ was added on days 4, 24, and 60; furthermore, $0.20\,mmol\,L^{-1}$ $NaHCO_3$ was added on day 50 and day 90 of the incubation. On day 125, incubation vials were divided into two sets to study the effect of different inorganic carbon contents on Feammox. Either 1.20 or $0.20\,mmol\,L^{-1}$ of $NaHCO_3$ plus $2\,mmol\,L^{-1}$ of NH_4Cl were added to each set. $NaHCO_3$ was then added every 10 days, which increased the soil pH to ~5 in the samples amended with $1.20\,mmol\,L^{-1}$ or $NaHCO_3$. For this incubation, samples were collected every 4 days. Finally, soil samples collected on day 180 of the incubations were used to enrich the Feammox bacteria in a membrane reactor. To study how the organic carbon content affects the Feammox bacteria, $1.00\,mmol\,L^{-1}$ sodium citrate was also supplied on day 125 to four of the $1.20\,mmol\,L^{-1}$ $NaHCO_3$ amended samples.

In the third experiment, inorganic nitrogen species were quantified through incubations in the presence of C_2H_2. Soil slurries were first incubated for 90 days in eighty 50 mL vials. The Fe(III) concentration at the beginning of the incubations was $25\,mmol\,L^{-1}$. One $mmol\,L^{-1}$ NH_4Cl and $0.20\,mmol\,L^{-1}$ $NaHCO_3$ were added on days 24, 60, and 90. After this incubation, 5 mL of pure C_2H_2 gas were added to 40 vials, which resulted in a final C_2H_2 concentration of $100\,\mu mol\,L^{-1}$. Samples with and without C_2H_2 were then incubated anaerobically for 20 days. The headspace gas was sampled every 24 h for N_2O analysis, and soil samples were analyzed every 2 days for Fe and N species.

2.3 Continuous flow membrane Feammox reactor

Soil samples collected on day 180 from the incubation with ferrihydrite, NH_4Cl, and $1.20\,mmol\,L^{-1}$ $NaHCO_3$ additions were inoculated into a continuous flow membrane reactor (Abbassi et al., 2014), which was operated under anaerobic conditions by constantly purging N_2 through the reactor's headspace at room temperature (25°), and with a 48 h hydraulic retention time.

The enrichment medium contained the following components per liter: 177 mg NH_4Cl, 77.9 mg $(NH_4)_2SO_4$, 19.8 mg $NaHCO_3$, 71.0 mg $KHCO_3$, 9.00 mg KH_2PO4, 100 mg $MgSO_4 \times 7H_2O$, and 60.0 mg $CaCl_2 \times 2H_2O$. After autoclaving, 1 mL trace element solution (Van de Graaf et al., 1996) was added to the medium. 50.0 mmol L^{-1} ferrihydrite were added once every 2 weeks directly into the reactor. To aid in maintaining anaerobic conditions, 0.10–0.20 mmol L^{-1} sodium citrate were fed to the reactor about twice per month. pH was controlled at around 4 ~ 5, and dissolved oxygen was less than 0.10 mg L^{-1}. Samples from the outflow were collected every 2 days, and sludge samples from the reactor were collected and kept at −20 °C for molecular biology analysis.

Finally, ^{15}N isotope tracer incubations were conducted using slurries collected form the stable Feammox membrane reactor. Five treatments ($n = 3$ per treatment) were conducted: (1) control with only anoxic deionized (DI) water; (2) $^{15}NH_4Cl$ addition; (3) $^{15}NH_4Cl$ + Fe(III) addition; (4) $^{15}NH_4Cl$ and C_2H_2 addition; and (5) $^{15}NH_4Cl$, C_2H_2, and Fe(III) addition. The headspace gas of each 50 mL incubation vial was sampled every 24 h for $^{15}N_2O$ analysis (Supplement 1.4).

2.4 Chemical analyses

For each sample collection during the incubations, a set of vials was destructively sampled in a glove box under oxygen-free conditions and the pH was measured immediately using a pH electrode. An extraction with 0.5N HCl was conducted for 24 h at room temperature to determine acid-extractable Fe(II) and NH_4^+ concentrations in the soils. Fe(II) was analyzed using the ferrozine assay method (Stookey, 1970; Komlos et al., 2007). Extraction efficiency of Fe(II) was affected by the HCl concentration and the extraction time. About 5–10 % more Fe(II) could be extracted with either 1N HCl extraction over 24 h or with 0.5 N HCl over 36 h as opposed to 0.5 N HCl over 24 h. Furthermore, after more Fe(II) was produced in the system with increasing incubation time, the Fe(II) extraction efficiency improved. Only a 1–2 % difference was observed in the Fe(II) extracted over 24 h using 0.5 N vs. 1 N HCl towards the end of the incubation period. Clays, present in the soil incubations, typically sorb Fe(II) more efficiently when the total Fe(II) is low; furthermore, ferrihydrite is slowly converted to magnetite, resulting in relatively different associations with different phases of the Fe(II) over the duration of the incubation. All of which leads to incomplete Fe(II) extractions, especially when the Fe(II) is low. Here we report Fe(II) data obtained via 0.5 N HCl extractions over 24 h to ensure that the methods and hence data are comparable to those reported by other researchers focusing on iron reduction and iron bioavailability.

NH_4^+ was analyzed using a Dionex™ ion chromatograph (LC3000) with a CS-16 column and a CS-16 guard column (flow rate = 1.00 mL min^{-1}, detection limit = 0.012 ppm). NO_3^- and NO_2^- were extracted with DI water for 1 h anaerobically, and measured via ion chromatography, using an AS-22 column along with an AG-22 guard column (flow rate = 1.20 mL min^{-1}, detection limit = 0.016 ppm). For the total organic carbon (TOC) and total nitrogen (TN) analyses, a Shimadzu TOC-5000(A) was used. N_2O concentrations were determined on a Shimadzu 2014 gas chromatograph equipped with an electron capture detector.

2.5 DNA and RNA isolation

DNA and RNA samples were extracted from soils collected at the wetland prior to any laboratory incubation, from the samples taken at different time points during the incubation experiments, and from the reactor. DNA was extracted from 500 mg soil or sludge samples using the FastDNA® spin kit for soil (MP Biomedicals, USA) as described by the manufacturer, and RNA using the FastRNA® pro Soil Direct Kit. The concentrations were measured using a Nano-drop 2000 spectrophotometer (Thermo Scientific, USA).

2.6 PCR-DGGE and 454 pyrosequencing analysis

Bacterial universal 16S rRNA gene primer sets V3-2/V3-3 (Jensen et al., 1998) were used for PCR amplification (Table 1). DGGE was performed with an 8 % polyacrylamide gel containing a gradient from 40 to 80 % denaturant using the gradient gel electrophoresis system (C.B.S. SCIENTIFIC, USA). The electrophoresis was carried out at 60 V for 15 h. After that the gel was stained with 0.1 μL mL^{-1} SYBR Green I and visualized with an UV transilluminator. All visible bands were excised from the gel and used as templates for re-amplification, using primer set V3-1/V3-2 (Jensen et al., 1998) and followed by cloning (Table 1). PCR products were purified via agarose gel extraction and cloned into a pGEM-T vector (Promega). A total of 10 to 30 positive recombinant clones for each band were identified by colony PCR, and were sent for sequencing to avoid erroneous interpretations. DNA sequencing was then conducted by Genewiz, Inc. Bacteria were classified and the phylogenetic tree of *Acidimicrobiaceae*-related sequences was constructed using the Bayesian inference (Huelsenbeck et al., 2001) (Supplement 1.2). Sequences obtained in this study are available in the GenBank database under accession numbers KC581755–KC581779. To confirm the changes in the bacterial community further, 454 pyrosequencing was performed with samples collected from the incubation on days 0, 30, 90, and 160, and from the membrane reactor after 150 days of reactor operation. Domain-specific primers Bact-338F1/909R, targeting the V3–V5 region of the 16S rDNA of bacteria, were amplified and sequenced following methods suggested by Pinto et al. (2012) (Supplement 1.2).

2.7 Quantitative PCR (qPCR) assay

qPCR experiments were carried out using a StepOnePlus™ Real-Time PCR System (Life Technologies, USA), represented by 16S rRNA genes, using primer sets 1055f/1392r for total bacteria (Harms et al., 2003), Amx368f/Amx820r for anammox bacteria (Schmid et al., 2000; Schmid et al., 2003), acd320f–432r, which we developed for

Table 1. Sequence analysis of bands excised from DGGE gels of soil samples with Feammox activity.

Phylogenetic group	Band	Related sequence	Identity (%)
Actinobacteria	A6, B1, D6	*Ferrimicrobium acidiphilum* strain T23 16S Ribosomal RNA gene (AF251436)	92
		Acidimicrobium ferrooxidans strain TH3 16S Ribosomal RNA gene (EF621760)	90
	A6	Uncultured *Ferrimicrobium* sp. clone D. an-41 16S Ribosomal RNA gene (JX505108)	95
Acidobacteria	A8, D11	Uncultured *Acidobacteria bacterium* clone 3OL11 16S Ribosomal RNA gene (GQ342349)	97
		Geothrix sp. culture clone AP-FeEnrich1 16S Ribosomal RNA gene (JX828409)	94
BetaProteobacteria	A9, B9, C4, D14	Uncultured *Rhodocyclus* sp. clone W4S68 16S Ribosomal RNA gene (AY691423)	97
	A9	*Comamonas* sp. "ARUP UnID 223" 16S Ribosomal RNA gene (JQ259419)	97

Acidobacteriaceae bacteria, and acm342f–439r, which we developed for *Acidimicrobiaceae* bacteria (Supplement 1.3) (Table 1). For the detection of denitrifiers, AOB and AOA, denitrifying functional genes (*nirS and nirk*), ammonia monooxygenase structural genes (*amoA*) were quantified with primer sets NirS3F/NirS5R, NirK1F/NirK5R (Braker et al., 1998) and amoA-1F/amoA-2R (Rotthauwe et al., 1997) for AOB-*amoA* and Arch *amoA*-F/*amoA*-R for AOA-*amoA* (Francis et al., 2005), respectively. Primer pairs Cren-amoA23f/616r were also used to quantify the thaumarchaeal *amoA* genes that represented acidophilic ammonia oxidizers (Tourna et al., 2008). For DNA quantification, each qPCR mixture (20 μL) was composed of 10 μL of SYBR Premix Ex Taq® II (Takara, Japan), 0.8 μL 10 μM of each primer, and $\sim$10 ng DNA template. RNA quantification was conducted through a real-time quantitative reverse transcription PCR (RT-qPCR) analysis, by using the One Step SYBR® PrimeScript® RT-PCR Kit (Takara, Japan) according to the manufacturer's recommendations. Thermal cycling conditions for total 16S rDNA, *nirS* and AOB-*amoA* gene numbers were initiated for 30 s at 94 °C, followed by 40 cycles of 5 s at 94 °C, 30 s at 57 °C, and 30 s at 70 °C. 16S rDNA numbers of anammox, *Acidobacteriaceae* and *Acidimicrobiaceae* bacteria were performed at 56, 55 and 58 °C as annealing temperatures, respectively, with the same program. For the AOB-*amoA* gene and the thaumarchaeal *amoA* genes, the annealing temperatures were 53 and 55 °C, and the annealing time was adjusted to 45 s. For RNA quantification, the cycling conditions were identical to those described for measuring gene numbers, with the exception that an initial incubation was conducted for 5 min at 42 °C to facilitate reverse transcriptase activity. Each assay contained a standard using a serial dilution of plasmids containing specific target genes, independent triplicate templates for each soil sample, and triplicate no template controls (NTC).

3 Results

3.1 Change in Fe and N species under different operational conditions

After incubating the pre-treated soil slurry with the four different Fe(III) sources for 30 days, only samples to which either ferrihydrite or goethite had been added showed measurable NH_4^+ oxidation (Fig. 1 and Supplement Fig. S1). In samples incubated with ferric citrate and NH_4Cl, Fe(III) reduction was much faster than in those supplied with Fe(III) oxides, but the NH_4^+ concentration remained fairly constant (Fig. 1). No detectable Fe(II) reduction or NH_4^+ oxidation was found in the sterilized soils amended with ferrihydrite and NH_4Cl (Fig. S2). Faster iron reduction and NH_4^+ removal was observed in ferrihydrite than in goethite-amended sediments (Fig. S1).

Since samples incubated with ferrihydrite and NH_4Cl resulted in the fastest NH_4^+ oxidation, the anaerobic incubation with ferrihydrite was extended to 180 days. Ferrihydrite as the Fe(III) source results in a larger negative G value than goethite (Clement et al., 2005). The NH_4^+ oxidation rate increased as NH_4Cl was supplied repeatedly, especially after 125 days of incubation, when the $NaHCO_3$ additions were increased from 0.20 to 1.20 mmol L^{-1} in addition to the 2.00 mmol L^{-1} NH_4^+ added. The increased $NaHCO_3$ dosing also increased the generation of Fe(II) (Fig. 2a, b). During the 180-day incubation, the ratio of Fe(II) produced to NH_4^+ removed gradually increased until it reached 5.3 : 1 by day 160, after which it remained stable (Fig. S6).

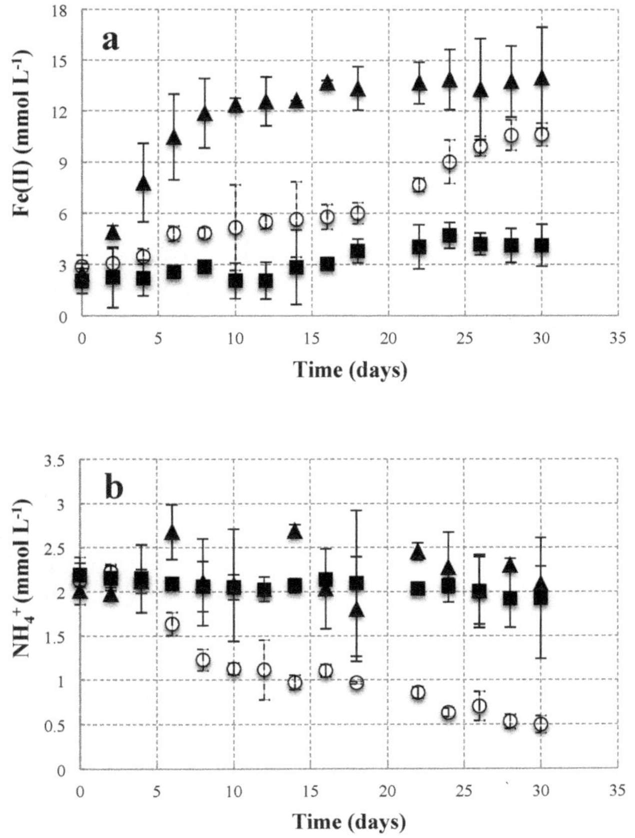

Figure 1. Concentration of Fe(II) and NH_4^+ in incubation with three different Fe(III) sources: ferrihydrite (○), ferric chloride (■), and ferric citrate (▲). The values represent the mean and standard error ($n = 3$).

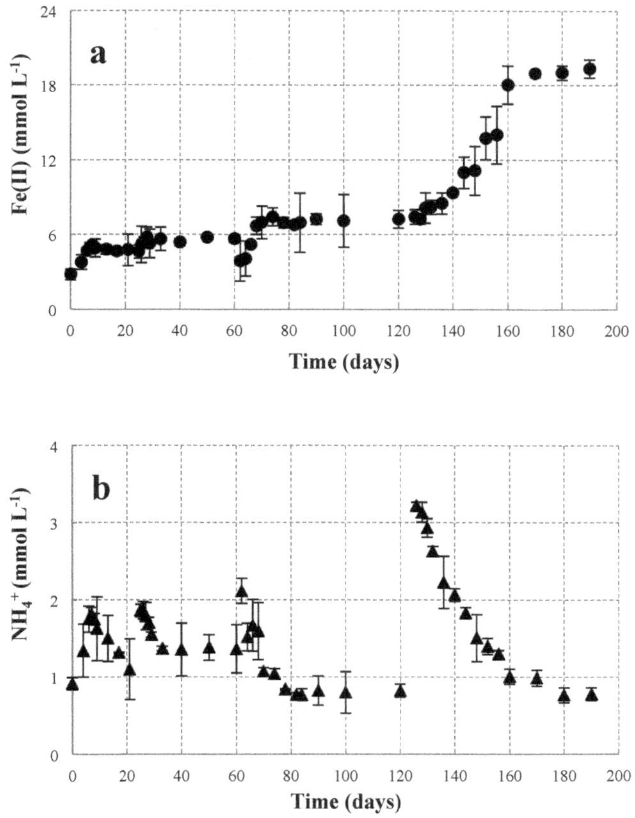

Figure 2. Concentration of **(a)** Fe(II) and **(b)** NH_4^+ during the 180-day incubation. 25 $mmol\,L^{-1}$ Fe(III) was added on day 0. 1.0 $mmol\,L^{-1}$ NH_4^+ was added on days 4, 24, and 60. 0.2 $mmol\,L^{-1}$ $NaHCO_3$ was added on day 50 and day 90. 1.2 $mmol\,L^{-1}$ + 2 $mmol\,L^{-1}$ of NH_4Cl were added on day 125. The values represent the mean and standard error ($n = 3$).

NO_2^- appeared within a few days after the addition of NH_4^+, with a maximum concentration of $0.44 \pm 0.17\,mmol\,L^{-1}$ in the second NH_4^+ oxidation cycle. NO_2^- did not accumulate in the system and was immediately consumed after generation (Fig. S3a). NO_3^- production showed a similar pattern to that of NO_2^- (Fig. S3a), and TN loss similar to the decrease in NH_4^+ (Fig. S3b). During 180 days of incubation, the system experienced a loss of TN of $57.2 \pm 3.13\,mg\,L^{-1}$. The DOC content fluctuated slightly in the early stage of incubation, but overall, the DOC concentration was relatively stable at around $45 \sim 50\,mg\,L^{-1}$ (Fig. S3b).

A 64.5 % NH_4^+ removal, between inflow and outflow was achieved in the membrane reactor after 150 days of operation.

3.2 Phylogenetic analysis of the microbial community based on the 16S rRNA gene

All visible bands observed in the DGGE analysis (significant bands were marked; see Fig. 3) were excised from the gel and sequenced after cloning. Clone libraries from 12 samples resulted in 721 sequences of partial 16S rRNA gene fragments, and six groups of bacteria were classified via a phylogenetic analysis (Table 1 and Supplement Table S1). During this 180-day anaerobic incubation with ferrihydrite and NH4Cl, the microbial communities shifted dramatically and the microbial diversity decreased with time (Fig. 3, lanes 1–4). Some DGGE bands disappeared gradually with time, such as band A5 and band A7. Band A5 represents a dissimilatory iron-reducing bacterium, *Geobacter* sp., which existed in this Fe(III)-rich wetland soil and reappeared for a short time during the initial anaerobic incubation. Band A7 represents an ammonia-oxidizing bacterium, *Nitrosomonas* sp., which showed a strong presence in the samples at 30 days of incubation and was attenuated after longer incubation times. In contrast, DGGE bands A6, A8 and A9 became more significant as the incubation time increased, showing that there were three groups of bacteria dominating in the system after 160 days of incubation. Band A6 represents a group of bacteria belonging to the *Acidimicrobiaceae* family. Bacteria from the *Acidobacteriaceae* family are represented by band

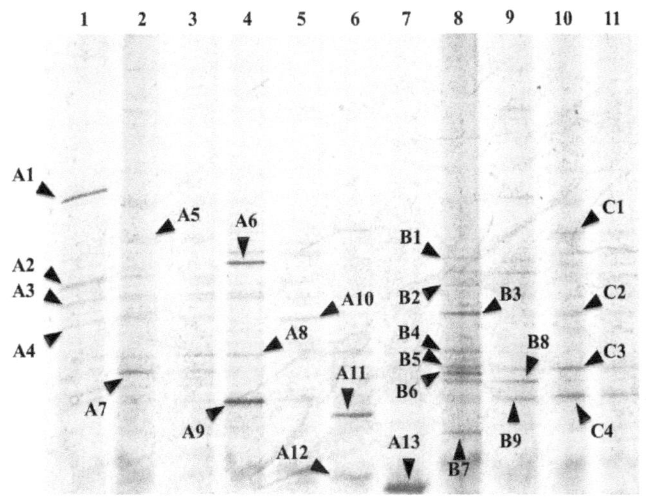

Figure 3. Comparison of DGGE analysis profiles of soil communities during anaerobic incubations. Samples from 0, 30, 90 and 160 days of incubation with ferrihydrite + NH_4Cl + $NaHCO_3$ (lanes 1–4); 160 days of incubation with only ferrihydrite (lane 5); ferric chloride + NH_4Cl (lane 6); ferric citrate + NH_4Cl (lane 7); 120 days incubation with ferrihydrite + NH_4Cl + $NaHCO_3$+ organic carbon (band 8); ferrihydrite + NH_4Cl + organic carbon (lane 9). Samples from 6 and 120 days of incubation without any addition (lanes 10 and 11) were used as controls.

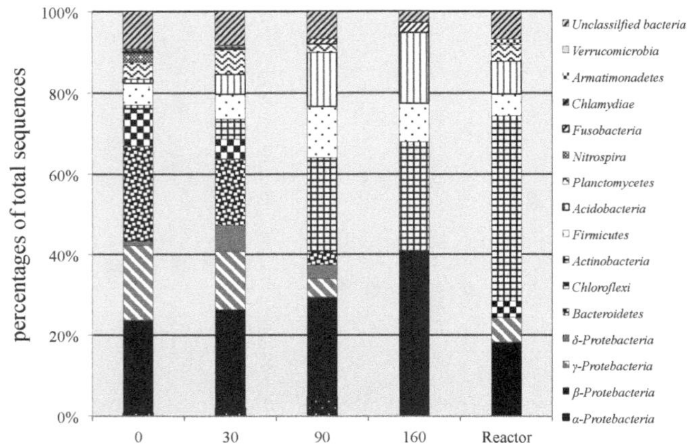

Figure 4. Relative abundance of bacterial phyla for each soil sample during anaerobic incubations (days 0, 30, 90, 160) and enrichment culture from the reactor.

A8. Some species in this family have been described as iron reducers and obligate heterotrophs (Kishimoto et al., 1991; Rowe et al., 2007; Coupland and Johnson, 2008). DGGE band A9 represents bacteria of the *Rhodocyclaceae* family. This family contains mainly denitrifying bacteria, which exhibit very versatile metabolic capabilities (Smith et al., 2005; Huang et al., 2011).

Microbial communities also differed between samples incubated with various Fe(III) sources, and between samples with or without the addition of inorganic carbon. Samples supplied with either ferric chloride or ferric citrate as the Fe(III) source plus NH_4^+, and samples supplied with just ferrihydrite and no NH_4^+, had a decreased diversity in their bacterial communities (Fig. 3, lanes 5–7). Samples supplied with both organic carbon (1.00 mmol L^{-1} sodium citrate) and inorganic carbon (1.20 mmol L^{-1} $NaHCO_3$) had a higher bacterial diversity (DGGE Fig. 3, lane 8).

Changes in the microbial community after 180 days of incubation were also confirmed via 454 pyrosequencing, and the obvious growth of *Actinobacteria*, *Acidobacteria* and *β-Proteobacteria* groups (which bands A6, A8 and A9 belong to) was consistent with the DGGE results, where the *Actinobacteria* cell number increased the most (Fig. 4). The *Planctomycetes* phylum, with which anammox bacteria are affiliated, was detected in the first 90 days of incubation, but disappeared or was below detection on day 160. *Actinobacteria* were also the dominant species in the Feammox enrichment reactor, based on the results of the 16S rDNA library obtained via pyrosequencing (Fig. 4).

The *Acidimicrobiaceae* bacterium, represented by band A6, which belongs to the *Actinobacteria* phylum, was the dominant species in the incubation experiments after 180 days of incubation (14.8 % of total 16S rRNA gene sequences) as well as in the membrane reactor after 150 days of operation (40.2 % of total 16S rRNA gene sequences). Its similarity to other *Acidimicrobiaceae*-related sequences is shown using a phylogenetic tree (Fig. 5). Unlike the bacteria represented by bands A8 and A9, which were also found in samples that did not show Feammox transformations, this *Acidimicrobiaceae* bacterium was only detected in incubations (or reactors) that were augmented simultaneously with ferrihydrite, NH_4Cl, and $NaHCO_3$, and have shown Feammox activity (Fig. 3).

3.3 Changes in bacterial abundance and activity during incubations and in the reactor

The total bacterial abundance determined via the 16S rRNA gene copy number decreased during the 180-day incubation (Fig. 6). Both, 16S rRNA gene and rRNA fragment copies of *Acidimicrobiaceae* bacteria (DGGE band A6), increased during the incubation, particularly after 90 days. The rRNA numbers increased slowly during the first 3 months and doubled between day 130 and day 140 of the incubation period (Fig. 6). rRNA as a biomarker for changes in protein level, even though not as specific as mRNA, is a good indicator of bacterial activity (Poulsen et al., 1993; Park et al., 2010). The abundance of *Acidobacteriaceae* bacteria (DGGE band A8), and that of denitrifiers (represented by the number of copies of the *nirS* gene and the *nirK* gene), increased over the 180-day incubation (Fig. S4a, b, c), although their growth over that period was less than that of the *Acidimicrobiaceae* bacteria (DGGE band A6). An increase in copies of the *nirS* and *nirK* genes showed similar trends, although the number

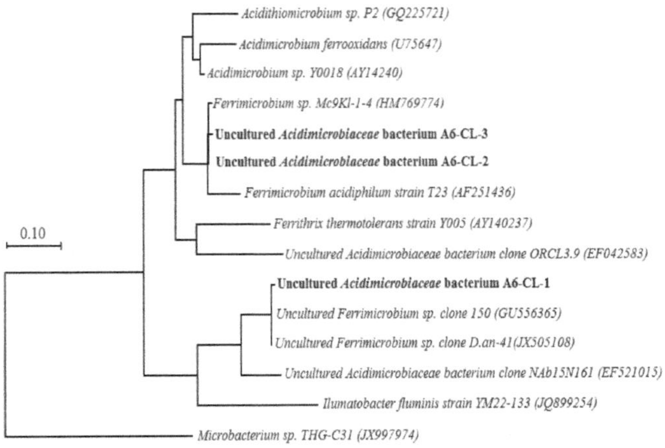

Figure 5. Phylogenetic tree of *Acidimicrobiaceae*-related sequences. The tree was constructed using the Bayesian inference (BI) method with 16S rRNA gene sequences from DGGE band A6 and bacteria from the *Acidimicrobiaceae* family from other studies. Sequences determined in this study are in bold. Bootstrap values were based on 1000 replicates each and are shown at the nodes with less than 50 % bootstrap support. The scale bar represents 10 % sequence divergence.

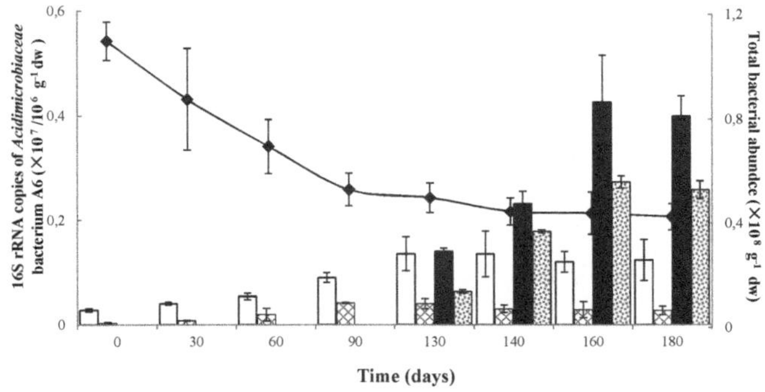

Figure 6. Abundance of total bacteria (♦) during 180 days of anaerobic incubation. 16S rRNA gene (■) and RNA (⊡) copy numbers of *Acidimicrobiaceae* bacterium A6 in soil samples with 1.20 mmol L^{-1} $NaHCO_3$ addition. 16S rRNA gene (□) and RNA (⊠) copy numbers of bacterium A6 with 0.20 mmol L^{-1} $NaHCO_3$ addition.

of *nirK* gene copies was 2 orders of magnitude lower than that of *nirS* (Fig. S4b, c). An increase in the denitrifier activity was most likely stimulated by the NO_2^- generated via Feammox. The number of the *amoA* gene, representing the abundance of ammonia-oxidizing bacteria, decreased sharply with time, and was hardly detected after 90 days of incubation (Fig. S4e). Through quantification of thaumarchaeal *amoA* genes, none of the acidophilic ammonia oxidizers was detected in our system.

In the Feammox reactor, the copy numbers of *Acidimicrobiaceae* bacterium A6, Anammox bacteria and the *nirS* gene were 0.37×10^7, 0.13×10^6, and 0.92×10^6 copies g^{-1}dw, respectively, while the amoA gene was not detected.

3.4 Changes in bacterial abundance and activities with $NaHCO_3$ amendment

Abundance and activity of *Acidimicrobiaceae* bacteria, represented by band A6, were compared between samples incubated under the same conditions, except for the amounts of $NaHCO_3$ added (0.20 mmol L^{-1} vs. 1.20 mmol L^{-1}). From day 125 to day 180 of the incubation, both 16S rRNA gene and rRNA fragment numbers of *Acidimicrobiaceae* bacteria were higher in the soils with the higher inorganic carbon content. The 16S rRNA gene copies of samples augmented with 1.20 mmol L^{-1} $NaHCO_3$ were four times higher than those in samples that had been augmented with only 0.20 mmol L^{-1} $NaHCO_3$ (Fig. 6). The rRNA copies of the *Acidimicrobiaceae* bacteria showed even larger differences in response to the amounts of $NaHCO_3$ added. In the samples augmented with 1.20 mmol L^{-1} $NaHCO_3$, the rRNA copy number increased from $0.04 \pm 0.06 \times 10^6$ to $0.19 \pm 0.09 \times 10^6$ copies g^{-1}dw over 50 days of incubation (day 130 to day 180). However, in the samples to which only 0.2 mmol L^{-1} $NaHCO_3$ were added on day 125, the rRNA number gradually deceased from $0.29 \pm 0.10 \times 10^5$ to $0.19 \pm 0.05 \times 10^5$ copies g^{-1}dw during the same 50-day incubation (Fig. 6).

3.5 Nitrogen species changes in samples incubated in the presence of C_2H_2

C_2H_2 can inhibit the oxidation of NH_4^+ to NO_2^- under aerobic conditions, and the reduction of N_2O to N_2 as well as the anammox pathway under anaerobic conditions (Yoshinari et al., 1977; Jensen et al., 2007; Kartal et al., 2011). To gain further insights into the nitrogen removal process observed, incubations with C_2H_2 were conducted. After 20 days of incubation, less NH_4^+ was oxidized in the samples amended with C_2H_2, compared to those incubated without C_2H_2 (Fig. S5a). NO_2^-, which is postulated to be the direct product of the NH_4^+ oxidation, accumulated slowly in samples incubated with C_2H_2 (Fig. S5b). NO_3^- reached a higher concentration in samples without C_2H_2 than in samples incubated with C_2H_2 (Fig. S5c). N_2O, a product of NO_2^- reduction, accumulated in samples incubated with C_2H_2, which inhibits the reduction of N_2O to N_2 (Fig. S5d). Fe(II) production was not much affected by the presence of C_2H_2, and after 20 days incubation was 4.36 ± 0.72 and 5.71 ± 0.67 mmol L^{-1} in sample incubated with and without C_2H_2, respectively.

In the ^{15}N isotope tracer incubations, detectable $^{15}N\text{–}N_2O$ was only found in samples amended with both, $^{15}NH_4Cl$ and Fe(III), with $^{15}N\text{–}N_2O$ production rates 2.14 ± 0.059 or $0.072 \pm 0.023\,\mu g\,g^{-1}\,d^{-1}$ in samples incubated with or without C_2H_2 treatment (Table S2).

4 Discussion

DGGE band A5 represents dissimilatory iron-reducing bacteria, which appeared for a short time at the beginning of the anaerobic incubation. For longer incubation times, these heterotrophic bacteria decreased rapidly (Fig. 3). Over a 25-day incubation period, more than three times the mass of Fe(II) was produced in samples amended with ferrihydrite and NH_4^+ as compared to the samples amended only with ferrihydrite, indicating that most of the Fe(III) reduction came from the Feammox reaction, and not from dissimilatory Fe(III) reduction (Fig. 1a vs. Fig. S2a).

AOB represented by DGGE band A7 (Fig. 3), as well as the *amoA* gene, decreased after 30 days of incubation (Fig. S4e). Also, no AOA or acidophilic ammonia oxidizers were detected, although the pH condition seems suitable for them in these incubations. Oxygen deficiency was the most likely reason for the decline in AOB in this system over time (Laanbroek et al., 1994). Even though a small amount of AOB would be enough for NH_4^+ oxidation, in our control samples to which no Fe(III) was added, which AOB do not require, no NH_4^+ consumption was detected (Fig. S2). Moreover, the decrease in the *amoA* gene at a time of increasing NH_4^+ oxidation also indicates that neither AOB nor acidophilic ammonia oxidizers were the drivers of the NH_4^+ oxidation in the later incubation times.

An uncultured *Acidimicrobiaceae* bacterium became the dominant species during the 180-day anaerobic incubation period, increasing from 0.92 % on day 0 in terms of cell numbers to 14.8 % on day 160. In the incubation experiments conducted (which included controls with only NH_4^+, only iron, autoclaved, and various Fe(III) sources), this *Acidimicrobiaceae* bacterium was only detected and growing in samples to which NH_4^+ was supplied as an electron donor, ferrihydrite was supplied as electron acceptor, and $NaHCO_3$ was supplied as a carbon source (Figs. 1, 3, S1, S2). The abundance and activity of this *Acidimicrobiaceae* bacterium increased along with the Feammox activity during the incubations. During the incubation period its rRNA changed from $(0.22 \pm 0.01) \times 10^5$ copies g^{-1}dw to $(0.28 \pm 0.07) \times 10^6$ copies g^{-1}dw, indicating a substantial increase in its activity (Fig. 6). In the continuous flow membrane reactor, which had a high NH_4^+ removal and Fe(III) reduction rate, this *Acidimicrobiaceae* bacterium was enriched from an initial 14.8 to 40.2 % after 150 days operation, and no other known NH_4^+ oxidizers (AOB or anammox) were detected (Fig. 4). These results indicated that this *Acidimicrobiaceae* bacterium might play an important role in the Feammox reactions described in this study. According to a phylogenetic analysis, this bacterium has a 92 % identity with *Ferrimicrobium acidiphilum* sp. (Table 1). *F. acidiphilum*, which belongs to the *Acidimicrobiaceae* family, was first isolated from mine environments (Johnson et al., 2009), and *F. acidiphilum* strain T23 is the only pure strain with a comprehensive characterization. Uncultured *Ferrimicrobium* sp. has been detected in mine water, but so far not in wetland soils (Gonzalez-Toril et al., 2003; Johnson et al., 2009; Bruneel et al., 2011). *Ferrimicrobium* sp. is an acidophilic heterotrophic ferrous iron oxidizing bacterium, which can also reduce Fe(III) under anoxic conditions (Johnson et al., 2009). The uncultured *Acidimicrobiaceae* bacterium, also has a 90 % identity with *Acidimicrobium ferrooxidans* (Table 1), a facultative autotroph in the same family, which can reduce Fe(III) in anaerobic environments while oxidizing sulfide to sulfur, and exists widely in soil environments (Clark and Norris, 1996; Bond et al., 2000; Hartmann et al., 2009).

According to a phylogenetic comparison with similar clones from studies reported in the GenBank (Fig. 5), and taking into account its special growth characteristics (stimulated by inorganic carbon, oxidizing NH_4^+ coupled to Fe(III) reduction), also its gradual activity increase with increased Feammox activity, as well as a strong link with a Feammox enrichment reactor, this uncultured *Acidimicrobiaceae* bacterium A6 is probably a previously unreported species in the *Acidimicrobiaceae* family that might be either responsible or play a key role in the Feammox process described here. *Acidimicrobiaceae* bacterium A6 was more active and the Feammox pathway was faster in samples with higher $NaHCO_3$ amendments (Figs. 2 and 6), which, in addition to the fact that ΔG in Eq. (1) is negative, indicates that if this *Acidimicrobiaceae* bacterium is actually responsible for conducting the Feammox reaction as depicted in Eq. (1), it may be an autotroph. Growth of the *nirS* gene suggested that denitrification pathways were also active in the incubations described here. NO_2^- that was being produced during the anaerobic NH_4^+ oxidation was reduced to N_2 by denitrifiers, and NO_2^- did not accumulate in the system.

The Feammox reaction studied here proceeded only when iron oxides (ferrihydrite or goethite) were supplied as electron acceptor, whereas samples incubated with ferric chloride or ferric citrate as the Fe(III) source showed no measurable NH_4^+ oxidation (Figs. 1 and S1). In the incubations to which ferric citrate was added as the Fe(III) source, Fe(III) was reduced rapidly by dissimilatory iron reducers, using organic carbon as electron donor. The DGGE results for incubations with ferric citrate (Fig. 3, lane 7) show that the most dominant species was an *Actinobacterium* (Table S1), known to reduce iron under anaerobic conditions (Lin et al., 2007; Lentini et al., 2012). *Acidimicrobiaceae bacterium* A6 was not detected in these incubations. Since acidic conditions as well as minimal dissolved Fe(II) and NO_2^- concentrations are required to make the Feammox reaction energetically favorable as shown in Eq. (1), the presence of iron oxides as the main Fe(III) source may have helped to maintain the concentrations of Fe(II) in solution below the detection limit through the incubation since iron oxides can sorb Fe(II) and/or incorporate it into their structure.

Various NH_4^+ oxidation products, i.e., NO_3^-, NO_2^- and N_2, generated through the Feammox process, are thermodynamically feasible, and were reported in different Feammox studies (Sawayama, 2006; Shrestha et al., 2009; Yang et al., 2012). Because there was no initial nitrate or nitrite in the system, because all experiments were conducted under strict oxygen-free conditions, and because of the rapid decrease in *amoA* genes, neither NO_3^- reduction nor aerobic NH_4^+ oxidation could be the reason for the formation of NO_2^- during the incubations. In all incubations where NH_4^+ was removed, the production of NO_2^- was observed, although NO_2^- did not accumulate. Most of the NO_2^- produced from the NH_4^+ oxidation was reduced rapidly by denitrifiers, which were present in the incubations.

When C_2H_2 was used to stop the reduction of N_2O to N_2, the total N_2O ($0.72 \pm 0.23\,mmol\,L^{-1}$) plus NO_2^- produced ($0.13 \pm 0.07\,mmol\,L^{-1}$) was equal to the NH_4^+ consumed, showing that NH_4^+ was not oxidized directly to N_2 in our samples. ^{15}N–NH_4^+ incubations, as an extension of C_2H_2 treatment, showed that ^{15}N–N_2O built up when $^{15}NH_4Cl$ was added as the NH_4^+ source (Table S2), demonstrating that NH_4^+ was oxidized during the Feammox process rather than adsorbed or taken up by microorganisms in the system.

Although nitrification might happen in suboxic environments (oxygen $< 5\,\mu M$, Lam et al., 2007), NH_4^+ oxidation in the presence of C_2H_2 has never been reported. C_2H_2 is an inhibitor of ammonia monooxygenase (AMO), and can restrain aerobic NH_4^+ oxidizers from using oxygen by binding covalently to AMO (Hynes and Knowles, 1982; Hyman and Wood, 1985; Gilch et al., 2009). C_2H_2 can also inhibit the NH_4^+-activation step of anammox cells, which use NO_2^- as the oxidant (Kartal et al., 2011). Therefore, these Feammox bacteria might differ from common NH_4^+ oxidizers, by using an alternative NH_4^+ oxidation pathway that is not inhibited by C_2H_2, and AMO might not play a role in Feammox. The fact that NH_4^+ oxidation was not affected by the presence of acetylene is a further indication that AOB are not responsible for this process since they would be affected by acetylene. Furthermore, in the isotope tracer incubations, ^{15}N–N_2O was below the detection limit in samples to which Fe(III) was not supplied, showing again that NH_4^+ oxidation proceeded only when iron was being reduced.

The role of anammox during the incubations was also evaluated. During the incubations the ratio of NH_4^+ oxidized to Fe(III) reduced increased gradually from $1:1.9$ to $1:5.3$ (Fig. S6), which is close to the stoichiometry of $1:6$, shown in Eq. (1). Although the discrepancies in the Feammox stoichiometry between iron and NH_4^+ are attributed in part to incompletely Fe(II) extraction, the influence of anammox activity in the earlier incubations would have contributed to a lower Fe(II) produced to NH_4^+ removed ratio than the theoretical value of $1:6$.

The Feammox reaction became more dominant in terms of NH_4^+ oxidation after 125 days of incubation due to a relative increase in the activity of the Feammox bacteria. A parallel pathway to Feammox, such as anammox, could, as mentioned above, explain the lower stoichiometric ratio, especially at earlier incubation times. In the samples taken before the incubation, $0.17 \pm 0.05 \times 10^6$ copies g^{-1}dw of the anammox rRNA gene were found, which decreased to $0.09 \pm 0.06 \times 10^5$ on day 130 (Fig. S4d). We postulate that anammox was responsible for some initial NH_4^+ and NO_2^- removal, and denitrification became more dominant for NO_2^- removal later during the incubation period (Fig. 5 and Fig. S4b, d). NH_4^+ removal via anammox in the early incubations may also explain why the observed NH_4^+ oxidation rates and the abundance of *Acidimicrobiaceae* A6 did not change proportionally over the full incubation period.

The results and analyses described here have shown that a Feammox enrichment reactor has the capacity to oxidize NH_4^+ coupled to iron reduction under anaerobic conditions, and that an uncultured *Acidimicrobiaceae* bacterium A6, which became the dominant species over time, might be responsible for this Feammox reaction. Without access to samples from other reported biological NH_4^+ oxidation by Fe(III) reducers (Sawayama, 2006; Yang et al., 2012), it is not possible to know if the organisms for the processes reported by these investigators are the same as those identified here. Conclusive linkage between *Acidimicrobiaceae* A6 and the Feammox process requires the isolation of the strain and then the conducting of incubations with the pure strain. At this point, the observations (i) that NH_4^+ removal only occurred in samples when the presence of *Acidimicrobiaceae* bacterium A6 was detected and when iron was being reduced, (ii) that *Acidimicrobiaceae* A6 numbers increased gradually after sequential NH_4^+ and Fe(III) additions, and (iii) the results from the enrichment culture that was operated for an extended time period while only adding NH_4^+ and iron oxide sources and during which *Acidimicrobiaceae* A6 became the dominant bacterial species, while no other known NH_4^+ oxidizer (AOB or anammox) was detected after 150 days of operation, indicate that *Acidimicrobiaceae* A6 is likely to play an important role in the oxidation of NH_4^+ under iron reduction conditions.

Acknowledgements. This research was supported by Project X from Princeton University. We thank L. Young, J. Stock, A. Mumford and S. Zhang for technical assistance.

Edited by: J. Middelburg

References

Abbassi, R., Yadav, A. K., Huang, S., and Jaffé, P. R.: Laboratory Study of Nitrification, Denitrification and Anammox Processes in Membrane Bioreactors Considering Periodic Aeration, J. Environ. Manage., 142, 53–59, 2014.

Bond, P. L., Druschel, G. K., and Banfield, J. F.: Comparison of Acid Mine Drainage Microbial Communities in Physically and Geochemically Distinct Ecosystems, Appl. Environ. Microbiol., 66, 4962–4971, 2000.

Braker, G., Fesefeldt, A., and Witzel, K. P.: Development of PCR primer systems for amplification of nitrite reductase genes (*nirK* and *nirS*) to detect denitrifying bacteria in environmental samples, Appl. Environ. Microbiol., 64, 3769–3775, 1998.

Bruneel, O., Volant, A., Gallien, S., Chaumande, B., Casiot, C., Carapito, C., Bardil, A., Morin, G., Brown, G. E. Jr., Personné, C. J., Le Paslier, D., Schaeffer, C., Van Dorsselaer, A., Bertin, P. N., Elbaz-Poulichet, F., and Arsène-Ploetze, F.: Characterization of the active bacterial community involved in natural attenuation processes in arsenic-rich creek sediments, Microb. Ecol., 61, 793–810, 2011.

Canfield, D. E., Glazer, A. N., and Falkowski, P. G.: The evolution and future of earth's nitrogen cycle, Science, 330, 192–196, 2010.

Clark, D. A. and Norris, P. R.: *Acidimicrobium ferrooxidans* gen. nov., sp. nov.: mixed-culture ferrous iron oxidation with *Sulfobacillus* species, Microbiology, 142, 785–790, 1996.

Clement, J. C., Shrestha, J., Ehrenfeld, J. G., and Jaffé, P. R.: Ammonium oxidation coupled to dissimilatory reduction of iron under anaerobic conditions in wetland soils, Soil Biol. Biochem., 37, 2323–2328, 2005.

Cornell, R. M. and Schwertmann, U.: The Iron Oxides: Structure, Properties, Reactions, Occurrences, and Uses, John Wiley and Sons Ltd., 2003.

Coupland, K. and Johnson, D. B.: Evidence that the potential for dissimilatory ferric iron reduction is widespread among acidophilic heterotrophic bacteria, FEMS Microbiol. Lett., 279, 30–35, 2008.

Francis, C. A., Roberts, K. J., Beman, J. M., Santoro, A. E., and Oakley, B. B.: Ubiquity and diversity of ammonia-oxidizing archaea in water columns and sediments of the ocean, Proc. Natl. Acad. Sci., 102, 14683–14688, 2005.

Gilch, S., Meyer, O., and Schmidt, I.: A soluble form of ammonia monooxygenase from Nitrosomonas europaea, Biol. Chem., 390, 863–873, 2009.

Gonzalez-Toril, E., Llobet-Brossa, E. O., Casamayor, R., Amann, R., and AmilsMicrobial, R.: Microbial ecology of an extreme acidic environment, the Tinto River, Appl. Environ. Microbiol., 69, 4853–4865, 2003.

Harms, G., Layton, A. C., Dionisi, H. M., Gregory, I. R., Garrett, V. M., Hawkins, S. A., Robinson, K. G., and Sayler, G. S.: Real-time PCR quantification of nitrifying bacteria in a municipal wastewater treatment plant, Environ. Sci. Technol., 37, 343–351, 2003.

Hartmann, M., Lee, S., Hallamand, S. J., and Mohn, W. W.: Bacterial, archaeal and eukaryal community structures throughout soil horizons of harvested and naturally disturbed forest stands, Environ. Microbiol., 11, 3045–3062, 2009.

Huang, S., Chen, C., Yang, X., Wu, Q., and Zhang, R.: Distribution of typical denitrifying functional genes and diversity of the *nirS*-encoding bacterial community related to environmental characteristics of river sediments, Biogeosciences, 8, 5251–5280, 2011, http://www.biogeosciences.net/8/5251/2011/.

Huelsenbeck, J. P., Ronquist, F. R., Nielsen, R., and Bollback, J. P.: Bayesian inference of phylogeny and its impact on evolutionary biology, Science, 294, 2310–2314, 2001.

Hyman, M. R. and Wood, P. M.: Suicidal inactivation and labeling of ammonia monooxygenase by acetylene, Biochem. J., 227, 719–725, 1985.

Hynes, R. K. and Knowles, R.: Effect of acetylene on autotrophic and heterotrophic nitrification, Can. J. Microbiol., 28, 334–340, 1982.

Jensen, M. M., Thamdrup, B., and Dalsgaard, T.: Effect of specific inhibition on anammox and denitrification in marine sediments, Appl. Environ. Microbiol., 73, 3151–3158, 2007.

Jensen, S., Øvreås, L., Daae, F. L., and Torsvik, V.: Diversity in methane enrichments from agricultural soil revealed by DGGE separation of PCR amplified 16 s rDNA fragments, FEMS Microbiol. Ecol., 26, 17–26, 1998.

Johnson, D. B., Bacelar-Nicolau, P., Okibe, N., Thomas, A., and Hallberg, K. B.: Characteristics of *Ferrimicrobium acidiphilum* gen. nov., sp. nov., and *Ferrithrix thermotolerans* gen. nov., sp. nov.: heterotrophic iron-oxidizing, extremely acidophilic actinobacteria, Int. J. Syst. Evol. Microbiol., 59, 1082–1089, 2009.

Kartal, B., Maalcke, W. J., de Almeida, N. M., Cirpus, I., Gloerich, J., Geerts, W., Op den Camp, H. J. M., Harhangi, H. R., Janssen-Megens, E. M., Francoijs, K., Stunnenberg, H. G., Keltjens, J. T., Jetten, M. S. M., and Strous, M.: Molecular mechanism of anaerobic ammonium oxidation, Nature, 479, 127–130, 2011.

Kishimoto, N., Kosako, Y., and Tano, T.: *Acidobacterium capsulatum* gen. nov., sp. nov.: an acidophilic chemoorganotrophic bacterium containing menaquinone from acidic mineral environment, Curr. Microbiol., 22, 1–7, 1991.

Komlos, J., Kukkadapu, R. K., Zachara, J. M., and Jaffé, P. R.: Biostimulation of Iron Reduction and Subsequent Oxidation of Sediment Containing Fe-silicates and Fe-oxides: Effect of Redox Cycling on Fe(III) Bioreduction, Water Res., 41, 2996–3004, 2007.

Laanbroek, H. J., Bodelier, P. L. E., and Gerards, S.: Oxygen consumption kinetics of *Nitrosomonas europaea* and *Nitrobacter hamburgenis* grown in mixed continuous cultures at different oxygen concentrations, Arch. Microbiol., 161, 156–162, 1994.

Lam, P., Jensen, M. M., Lavik, G., McGinnis, D. F., Müller, B., Schubert, C. J., Amann, R., Thamdrup, B., and Kuypers, M. M. M.: Linking crenarchaeal and bacterial nitrification to anammox in the Black Sea, Proc. Natl. Acad. Sci. USA, 104, 7104–7109, 2007.

Park, H., Rosenthal, A., Ramalingam, K., Fillos, J., and Chandran, K.: Linking community profiles, gene expression and N-removal in anammox bioreactors treating municipal anaerobic digestion reject water, Environ. Sci. Technol., 44, 6110–6116, 2010.

Pinto, A. J. and Raskin, L.: PCR biases distort bacterial and archaeal community structure in pyrosequencing datasets, PLoS One, 7, 43093, doi:10.1371/journal.pone.0043093, 2012.

Poulsen, L. K., Ballard, G., and Stahl, D. A.: Use of rRNA fluorescence in situ hybridization for measuring the activity of single cells in young and established biofilms, Appl. Environ. Microbiol., 59, 1354–1360, 1993.

Rosenzweig, B. R., Smith, J. A., Baeck, M. L., and Jaffé, P. R.: Monitoring Nitrogen Loading and Retention in an Urban Stormwater Detention Pond, J. Environ. Qual., 40, 598–609, 2011.

Rotthauwe, J. H., Witzel, K. P., and Liesack, W.: The ammonia monooxygenase structural gene amoA as a functional marker: molecular fine-scale analysis of natural ammonia-oxidizing populations, Appl. Environ. Microbiol., 63, 4704–4712, 1997.

Rowe, O. F., Sánchez-España, J., Hallberg, K. B., and Johnson, D. B.: Microbial communities and geochemical dynamics in an extremely acidic, metal-rich stream at an abandoned massive sulfide mine (Huelva, Spain) underpinned by two primary production systems, Environ. Microbiol., 9, 1761–1771, 2007.

Sawayama, S.: Possibility of anoxic ferric ammonium oxidation, J. Biosci. Bioeng, 101, 70–72, 2006.

Schmid, M., Twachtmann, U., Klein, M., Strous, M., Juretschko, S., Jetten, M., Metzger, J. W., Schleifer, K., and Wagner, M.: Molecular evidence for genus level diversity of bacteria capable of catalyzing anaerobic ammonium oxidation, Syst. Appl. Microbiol., 23, 93–106, 2000.

Schmid, M., Walsh, K., Webb, R., Rijpstra, W. I., van de Pas-Schoonen, K., Verbruggen, M. J., Hill, T., Moffett, B., Fuerst, J., Schouten, S., Damsté, J. S., Harris, J., Shaw, P., Jetten, M., and Strous, M.: Candidatus "Scalindua brodae," sp. nov., Candidatus "Scalindua wagneri", sp. nov., two new species of anaerobic ammonium oxidizing bacteria, Syst. Appl. Microbiol., 26, 529–538, 2003.

Shrestha, J., Rich, J., Ehrenfeld, J., and Jaffé, P. R.: Oxidation of ammonium to nitrite under iron-reducing conditions in wetland soils: Laboratory, field demonstrations, and push-pull rate determination, Soil Sci., 174, 156–164, 2009.

Smith, R. L., Buckwalter, S. P., Repert, D. A., and Miller, D. N.: Small-scale, hydrogen-oxidizing-denitrifying bioreactor for treatment of nitrate-contaminated drinking water, Water Res., 39, 2014–2023, 2005.

Stookey, L. L.: Ferrozine-a new spectrophotometric reagent for iron, Anal. Chem., 42, 779–781, 1970.

Tourna, M., Freitag, T. E., Nicol, G. W., and Prosser, J. I.: Growth, activity and temperature responses of ammonia oxidising archaea and bacteria in soil microcosms, Environ. Microbiol., 10, 1357–1364, 2008.

Van de Graaf, A. A., de Bruijn, P., Robertson, L. A., Jetten, M. S. M., and Kuenen, J. G.: Autotrophic growth of anaerobic ammonium-oxidizing micro-organisms in a fluidized bed reactor, J. Microbiol., 142, 2187–2196, 1996.

Yang, W. H., Weber, K. A., and Silver, W. L.: Nitrogen loss from soil through anaerobic ammonium oxidation coupled to iron reduction, Nat. Geosc., 5, 538–541, 2012.

Yoshinari, T., Hynes, R., and Knowles, R.: Acetylene inhibition of nitrous oxide reduction and measurement of denitrification and nitrogen fixation in soil, Soil Biol. Biochem., 9, 177–183, 1977.

Biogeochemical variations at the Porcupine Abyssal Plain sustained Observatory in the northeast Atlantic Ocean, from weekly to inter-annual timescales

S. E. Hartman[1], Z.-P. Jiang[2], D. Turk[3,4], R. S. Lampitt[1], H. Frigstad[5], C. Ostle[6,7], and U. Schuster[8]

[1]National Oceanography Centre, Southampton, SO14 3ZH, UK
[2]Ocean Collage, Zhejiang University, 310012 Hangzhou, China
[3]Dalhousie University, Halifax, B3H 4R2, Canada
[4]Lamont-Doherty Earth Observatory, Palisades, NY 10964, USA
[5]Norwegian Environment Agency, 0663 Oslo, Norway
[6]Sir Alister Hardy Foundation for Ocean Science, Plymouth, PL1 2PB, UK
[7]School of Environmental Sciences, University of East Anglia, Norwich, NR4 7TJ, UK
[8]University of Exeter, Exeter, EX4 4PS, UK

Correspondence to: S. E. Hartman (suh@noc.ac.uk)

Abstract. We present high-resolution autonomous measurements of carbon dioxide partial pressure $p(CO_2)$ taken in situ at the Porcupine Abyssal Plain sustained Observatory (PAP-SO) in the northeast Atlantic (49° N, 16.5° W; water depth of 4850 m) for the period 2010–2012. Measurements of $p(CO_2)$ made at 30 m depth on a sensor frame are compared with other autonomous biogeochemical measurements at that depth (including chlorophyll *a* fluorescence and nitrate concentration data) to analyse weekly to seasonal controls on $p(CO_2)$ flux in the inter-gyre region of the North Atlantic. Comparisons are also made with in situ regional time series data from a ship of opportunity and mixed layer depth (MLD) measurements from profiling Argo floats. There is a persistent under-saturation of CO_2 in surface waters throughout the year which gives rise to a perennial CO_2 sink. Comparison with an earlier data set collected at the site (2003–2005) confirms seasonal and inter-annual changes in surface seawater chemistry. There is year-to-year variability in the timing of deep winter mixing and the intensity of the spring bloom.

The 2010–2012 period shows an overall increase in $p(CO_2)$ values when compared to the 2003–2005 period as would be expected from increases due to anthropogenic CO_2 emissions. The surface temperature, wind speed and MLD measurements are similar for both periods of time. Future work should incorporate daily CO_2 flux measurements made using CO_2 sensors at 1 m depth and the in situ wind speed data now available from the UK Met Office Buoy.

1 Introduction

A persistent feature of the subpolar North Atlantic is under-saturation of carbon dioxide (CO_2) in surface waters throughout the year, which gives rise to a perennial CO_2 sink (Körtzinger et al., 2008). This makes the north east Atlantic a region of great importance in the global carbon cycle. There is evidence for inter-annual variation in the CO_2 sink (1–3 $\mathrm{mol\,m^{-2}\,a^{-1}}$) due to changes in wintertime mixing and stratification (Schuster and Watson, 2007). Changes in the amount of CO_2 absorbed by the ocean may have implications for the global carbon cycle now and for the role of the ocean as a carbon sink in the future. Studies of the physical and biological processes regulating surface water $p(CO_2)$ (partial pressure of CO_2) are required to estimate future trends in the ability of the ocean to act as a sink for increasing CO_2 in the atmosphere. Frequent observations from fixed positions are critical to make these calculations (McGillicuddy et al., 1998).

Accurate, high-resolution, long-term data sets are offered by time series studies such as the Porcupine Abyssal Plain sustained Observatory (PAP-SO) in the northeast Atlantic at 49° N, 16.5° W (4850 m water depth) where a fixed-point mooring has been in place since 2002 (Hartman et al., 2012). The PAP-SO is in the North Atlantic Drift Region, a biogeographical province defined by deep winter convective mixing (Longhurst, 2006; Monterey and Levitus, 1997). The surface mixed layer depth can change from 25 m in the summer to over 400 m in winter. A 2-fold decrease in winter nitrate concentration over a 3-year period from 2003 has been attributed to a combination of shallower winter convective mixing and changes in surface circulation (Hartman et al., 2010). The PAP-SO is in an area with relatively high wind speeds, frequently greater than $10\,m\,s^{-1}$. High wind speeds have a significant effect on CO_2 flux (Takahashi et al., 2002). The CO_2 flux at the PAP-SO was calculated from $p(CO_2)$, between 2003 and 2005 as a net flux into the ocean of over $3\,mol\,m^{-2}\,a^{-1}$ (Körtzinger et al., 2008). This is a significant sink compared with subtropical time series sites such as ESTOC (near the Canary Islands, 29.17° N, 15.5° W), which is an overall annual CO_2 source region ($0.05\,mol\,m^{-2}\,a^{-1}$, González-Dávila et al., 2003).

Recently, the decline in North Atlantic CO_2 uptake from 1994–1995 to 2002–2005 has been linked to a variation in the North Atlantic Oscillation (Schuster and Watson, 2007; Padin et al., 2011). The decreased uptake may be a consequence of declining rates of wintertime mixing and ventilation between surface and subsurface waters due to increasing stratification. Enhanced stratification forms a barrier to nutrient exchange, which may result in a progressive decline in primary production (Field et al., 1998), as was seen in the North Atlantic between 1999 and 2004 (Behrenfeld et al., 2006). The observed decrease in nitrate concentration and productivity in this region (Behrenfeld et al., 2006), may in turn affect the oceanic uptake of $p(CO_2)$.

In this paper, we present recent year round time series data of temperature, salinity, nitrate concentration, chlorophyll *a* fluorescence and $p(CO_2)$ collected at 30 m depth from May 2010 to June 2012. The data are compared with an earlier published data set (from July 2003 to July 2005) and additional $p(CO_2)$ measurements made from a ship of opportunity. The in situ data set is considered in relation to convective mixing processes using mixed layer depth (MLD) estimates calculated from profiling Argo floats. The weekly air–sea CO_2 flux at the PAP-SO site was calculated from in situ $p(CO_2)$ measurements and ancillary satellite wind speed data sets. The objective of this study is to examine the biogeochemical variations at the PAP-SO in the northeast Atlantic over different periods from weekly, seasonal to annual.

2 Materials and methods

2.1 Study site

The position of the PAP-SO at 49° N, 16.5° W is shown in Fig. 1. Lampitt et al. (2001) has summarised the hydrography, meteorology and upper mixed layer dynamics in the region.

2.2 In situ data

The instrumentation of the PAP-SO has been described in detail by Hartman et al., 2012 (see Table 1 and Fig. 1 therein) and is briefly summarised here. Since 2002 instruments on a mooring at the PAP-SO (49° N, 16.5° W) have recorded a suite of parameters in the mixed layer. Temperature and salinity measurements were made on a frame at a nominal depth of 30 m, using Seabird SBE 37-IM MicroCAT recorders (Sea-Bird Electronics Inc., Bellevue, Washington, USA). Measurements of nitrate concentration, chlorophyll *a* fluorescence and $p(CO_2)$, were also made using biogeochemical sensors on the frame, often within the deep chlorophyll maximum. Between 2002 and 2007 the sensor frame depth varied from 20 to 225 m, deflecting in response to local currents. A surface buoy was added in 2007 so that measurements were consistently made at 30 m depth. In 2010, collaboration with the UK Met Office led to a redesigned infrastructure, providing simultaneous surface physical and biogeochemical measurements with surface meteorological data.

$p(CO_2)$ data during the two periods of time examined here were collected using different instrumentation. From 2003 to 2005 it was measured using a SAMI (Sunburst Sensors LLC, USA) sensor, which is based on equilibration of a pH indicator solution, contained in a gas-permeable membrane, with ambient $p(CO_2)$ and subsequent spectrophotometric determination in the equilibrated solution (DeGrandpre et al., 1995). Twice daily $p(CO_2)$ measurements, from 2010 to 2012, were made using a membrane-based PRO-CO2 sensor (Pro-Oceanus, Canada), which uses an infrared detector and is internally calibrated through an auto-zero calibration function (Jiang et al., 2014). Note that a measurement error of an early version of the PRO-CO2 sensor during the deployment, induced by the fluctuation of detector cell temperature, was identified and corrected (see Jiang et al., 2014 for further details). A pump was used (Seabird Inc.) to improve water flow across the sensor membrane to accelerate attaining equilibrium. The surface in situ $p(CO_2)$ time series ceased between 2006 and 2009 due to funding issues.

Although measured by different instruments, the two $p(CO_2)$ data sets were calibrated in a similar way to make them comparable: the sensor outputs were calibrated against $p(CO_2)$ values calculated from dissolved inorganic carbon (DIC) and total alkalinity (TA) from discrete samples taken at the mooring site during deployment/recovery cruises; and

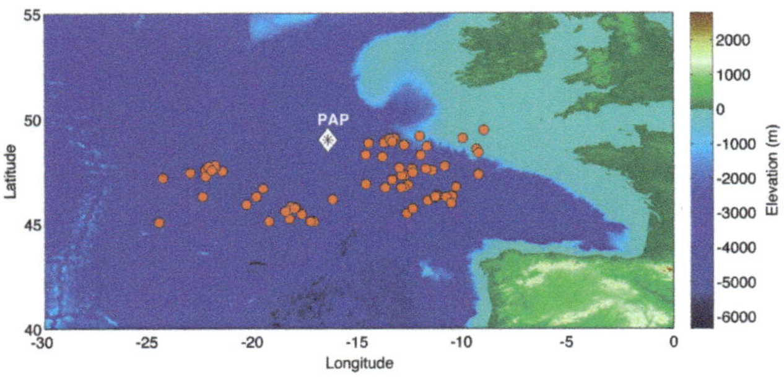

Figure 1. Map of the inter-gyre region of the northeast Atlantic showing the bathymetry around the PAP observatory (white diamond) and the ship of opportunity (SOO) sampling positions (red circles) from 2010 to 2012.

plausibility checks were made with underway $p(CO_2)$ measurements around the PAP site. The 2003–2005 data were previously published (see Körtzinger et al., 2008 for details) with a precision of 1 μatm and accuracy estimated as 6–10 μatm. The 2010–2012 data have a similar precision (1 μatm) and accuracy (6 μatm).

A Hobi Labs Inc., HS-2 fluorometer (Arizona, USA) was used on the PAP-SO mooring to estimate chlorophyll *a* concentration until 2005 when an alternative ECO FLNTU (WETlab, USA) fluorometer came into use. The quoted precision for fluorescence measured by these fluorometers is 0.04 %, however as described by Hartman et al. (2010), fluorescence output can only provide an approximation of the chlorophyll *a* concentration. The fluorescence/chlorophyll *a* concentration ratio changes throughout the year, due to variations in the phytoplankton species composition. On the mooring, chlorophyll *a* fluorescence measurements were taken every 2 h over the 1 year deployments and biofouling was controlled using motorised copper shutters on each of the fluorometers.

Nitrate concentration measurements were initially made using wet chemical NAS Nitrate Analysers (EnviroTech LLC, USA), precision 0.2 μmol L^{-1}, as described in Hydes et al. (2000), with twice daily sampling frequency and internal calibration as described by Hartman et al. (2010). From 2010 additional higher frequency inorganic nitrate measurements were made using UV detection methods (ISUS, Satlantic), with a precision of 1 μmol L^{-1}.

For each instrument, the manufacturer's calibration was checked at the start of each deployment and a correction for instrument drift was made using a second calibration check on recovery of the instruments. Biogeochemical data from the PAP-SO are available from EuroSITES (2014) and the British Oceanographic Data Centre (BODC). Data presented here cover the period when $p(CO_2)$ measurements are available, July 2003 to the end of June 2005 (with deployments in July 2003, November 2003, June 2004) and the period from May 2010 to June 2012 (with sensor deployment in May 2010, September 2010, July 2011, May 2012). All of the measurements are within the mixed layer although the depth of measurements is closer to the 30 m nominal depth after mooring redesign to incorporate a surface float in 2007.

2.3 Other observational data sources

Temperature and salinity data were taken from Argo floats (http://www.coriolis.eu.org), extracting (30 ± 5) m depth data. To obtain a continuous seasonal description, a large region around the PAP site was selected (45–52° N and 26.08–8.92° W, excluding the shelf area). The Argo data have a potentially lower accuracy (0.005 °C for temperature and 0.1 for salinity) than the in situ MicroCAT data (0.002 for salinity and 0.002 °C for temperature). However the Argo data were chosen over the in situ data for all calculations as they have a larger temporal coverage and are internally consistent.

The $p(CO_2)$ time series was compared with surface data from a ship of opportunity (SOO) running from Portsmouth, UK, to the Caribbean (Schuster and Watson, 2007). Onboard the SOO continuous $p(CO_2)$ measurements are made using a calibrated system with a showerhead equilibrator (Schuster et al., 2009). Data are available from the Surface Ocean CO_2 Atlas (SOCAT; http://www.socat.info/). Discrete nutrient samples were collected at 4 h intervals along the same route and were analysed ashore (Hartman et al., 2008). This provides an approximately monthly nutrient sample and $p(CO_2)$ data points close to the PAP-SO on the return route of the ship. The nominal depth of these samples is 5 m, which is shallower than the 30 m samples from the PAP-SO. We selected SOO data between 52 and 45° N and 8.92 and 26.08° W, and then took the average $p(CO_2)$ values that were within that area on the same day as the sample from the PAP-SO site.

Through collaboration with the UK Met Office in situ wind speed data are available since 2010. However for consistency in calculations of CO_2 flux between the two time periods (2003–2005 and 2010–2012) considered here we took wind speed data from weekly satellite data: Fleet Numerical Meteorology and Oceanography Center (FNMOC) 1° by 1°. We calculated a weekly mean from the 6 hourly, 10 m height data; available from http://las.pfeg.noaa.gov/.

The air–sea CO_2 flux (in mmol m^{-2} d^{-1}) was calculated from the air–sea $p(CO_2)$ difference, temperature and salinity (30 m) and wind speed at 10 m height, using the following equation:

$$F(CO_2) = k \cdot K_0 \left[p(CO_2, \text{sea}) - p(CO_2, \text{air}) \right], \qquad (1)$$

where k is the transfer coefficient based on the wind speed-dependent formulation of Nightingale et al. (2000), scaled to the temperature-dependent Schmidt number according to Wanninkhof (1992), K_0 is the CO_2 solubility at the in situ temperature and salinity after Weiss (1974). While $p(CO_2$, sea) and $p(CO_2$, air) are the CO_2 partial pressures of seawater and average CO_2 dry mole fraction measured in air, respectively. As $p(CO_2)$ was reported throughout this manuscript, we used $p(CO_2)$ for the air–sea flux calculation. Using fugacity for the calculation would generate the same results of flux estimates. The atmospheric $p(CO_2)$ is calculated from monthly averaged $p(CO_2)$ measured at Mace Head (53.33° N, 9.90° W) assuming 100 % water vapour saturation under 1 atm air pressure. Please note that 1 atm = 1.01325 bar. This is an appropriate pressure to use at the PAP-SO as the average (and standard deviation) of the air pressure, measured on the buoy at the PAP-SO between September 2010 and July 2011, was (1.01354 ± 13.14) bar.

Total alkalinity (TA) was calculated from Argo temperature and salinity (30 m), following the relationship for the North Atlantic developed by Lee et al. (2006) with an uncertainty of $\pm 6.4\,\mu$mol kg^{-1} (Lee et al., 2006). The DIC concentration was then calculated from TA and $p(CO_2)$ using the "seacarb" package (Lavigne and Gattuso, 2011), with Argo temperature and salinity (30 m) and nutrient concentrations set to zero. The chosen constants were Lueker et al. (2000) for K_2 and K_2, Perez and Fraga (1987) for K_f and the Dickson (1990) constant for K_s, as recommended by Dickson et al. (2007). We followed Körtzinger et al. (2008)'s method to correct the DIC changes driven by air–sea exchange:

$$\Delta \mathrm{DIC}_{\mathrm{gas}} = F(\mathrm{CO_2})/\mathrm{MLD}.$$

Using TA and $p(CO_2)$ to calculate DIC, and taking the various uncertainties in the calculation into account, introduces an error in the order of $7.0\,\mu$mol kg^{-1}.

The MLD was calculated from density profiles using global gridded fields of temperature and salinity collected by Argo floats, XBTs, CTDs and moorings. These data are collected and made freely available by the Coriolis project and programmes that contribute to it (http://www.coriolis.eu.org). We used the near real-time mode data as these data sets have been quality control checked. Before deciding on a MLD definition an inter-comparison of many definitions commonly used in the literature was done such as density differences, temperature differences and density gradients (Kara et al., 2000; Thomson and Fine, 2003; Montegut et al., 2004). A subset of the global density profiles calculated from the gridded temperature and salinity fields was used to compare the different methods. The depth of the mixed layer was estimated through visual inspection of over 3000 profiles, following a similar approach used by Fiedler (2010). The Holte and Talley (2009) density difference algorithm gave the closest match with the visually estimated MLD (RMSD 29.38 m). The depth of the mixed layer was defined by a density difference of 0.03 kg m^{-3} from the density at a reference depth (in this case 10 m to avoid diurnal changes in temperature and salinity at the surface). This Holte and Talley (2009) density difference algorithm incorporates linear interpolation to estimate the depth at which the density difference is crossed.

The North Atlantic Oscillation (NAO) index (after Hurrell, 1995) was obtained from the University of East Anglia web site http://www.cru.uea.ac.uk/cru/data/nao/.

3 Results

Figure 2a–c show the in situ observations from the PAP-SO at 30 m depth, including $p(CO_2)$, chlorophyll *a* fluorescence and nitrate concentration. Figure 2a shows the range of $p(CO_2)$ from 2003 to 2005, which was also shown in Körtzinger et al. (2008). The range was $74\,\mu$atm (300–$374\,\mu$atm) and the mean was $339\,\mu$atm. In comparison, $p(CO_2)$ between 2010 and 2012 had a $57\,\mu$atm range (327 to $384\,\mu$atm) with a higher mean of $353\,\mu$atm. The $p(CO_2)$ data for the 2010–2012 period are confirmed by SOO data from the Portsmouth to Caribbean route in Fig. 2a (see Fig. 1 for positions of the SOO samples). Körtzinger et al. (2008) also reported a good comparison with a SOO route from Kiel, to the north of the Portsmouth to Caribbean route, for the 2003–2005 data. The SOO data fill in the gap in the time series when PAP-SO $p(CO_2)$ data were not available due to failure of the instrument logger. The higher $p(CO_2)$ values in the 2010–2012 period are confirmed by the SOO data.

In situ chlorophyll data in Fig. 2b shows the characteristic chlorophyll *a* fluorescence increase for this area during the spring bloom. There is large inter-annual variability in both the timing and magnitude of the spring bloom for the two time periods shown. For example the spring bloom in 2004 started in late May compared with an earlier bloom in 2011 (that started in April). The increase in chlorophyll *a* fluorescence during the 2011 spring bloom was also larger compared with the other years shown.

Nitrate concentration data in Fig. 2c shows the characteristic seasonality, with increased winter nitrate concentrations and depletion following the spring bloom (seen in Fig. 2b). The seasonality in the nitrate concentration is similar for the two periods shown (2003–2005 and 2010–2012). SOO nitrate concentration data show a good agreement with the PAP-SO data throughout 2010–2012 and fill in the gaps in early 2011 when nutrient measurements at the PAP-SO are not available. Overall, the in situ data show a characteristic increase in inorganic nitrate concentrations, and $p(CO_2)$, through the winter as fluorescence decreases. However, winter nitrate concentrations are significantly lower in the 2004/05 winter compared with other years as has been discussed in Hartman et al. (2010).

Figure 3a shows the Argo temperature data extracted at 30 m depth and the in situ MicroCAT temperature data at the PAP-SO. Temperature shows opposite seasonal variations to the $p(CO_2)$ and nitrate concentration from in situ data. A comparison of Argo temperature with in situ 30 m MicroCAT data ($n = 112$, comparison not shown) suggests errors of up to 1 % for temperature in the Argo data compared with the in situ data (when available). Both data sets show that the temperature variations in these years are very similar, showing a summer–winter difference of 6 °C (Fig. 3a).

The seasonality of the in situ data can be put in context when looking at the MLD in Fig. 3b. The increase in $p(CO_2)$ and nitrate concentration corresponds to deeper convective

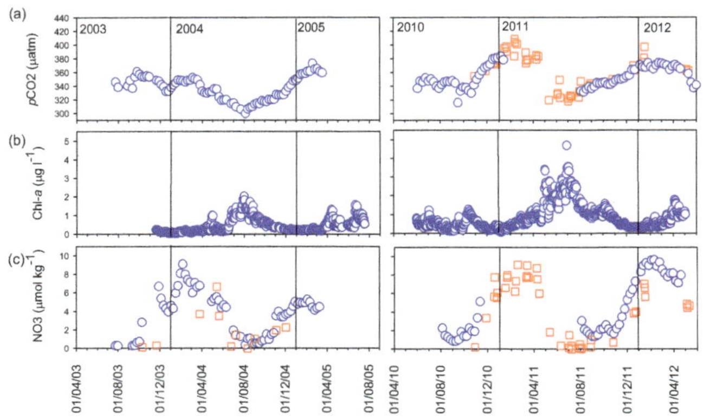

Figure 2. In situ 30 m PAP-SO data (blue circles) from 2003 to 2005 and 2010 to 2012 and 5 m SOO data (red squares) with vertical lines to represent the start of each year showing: **(a)** $p(CO_2)$; **(b)** chlorophyll *a* concentration; **(c)** weekly-averaged nitrate concentration.

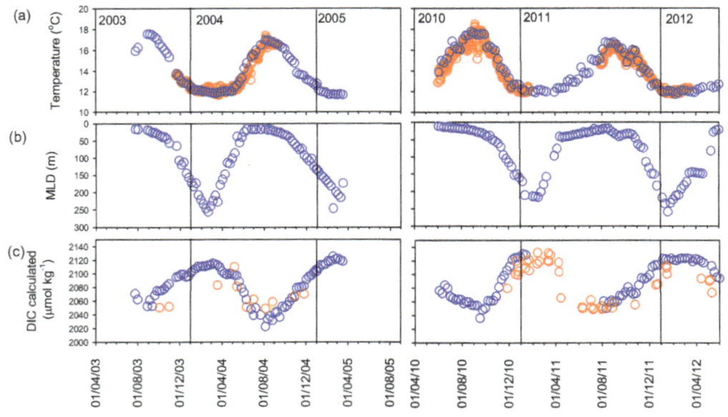

Figure 3. Data from 2003 to 2005 and 2010 to 2012 (blue circles) with vertical lines to represent the start of each year showing: **(a)** Argo temperature data from 30 m depth around the PAP-SO and in situ MicroCAT temperature data at 30 m (red circles); **(b)** calculated mixed layer depth (MLD) data; **(c)** calculations of weekly dissolved inorganic carbon (DIC) concentrations based on in situ PAP-SO $p(CO_2)$ and salinity-based TA parameterisations (see text for details) with additional DIC calculations based on SOO data (red circles).

mixing in winter. The MLD range varies little over the winters considered here (Fig. 3b) and the maximum MLD does not exceed 260 m. However the timing of the maximum winter mixed layer depth at PAP-SO varies from year to year. For example the maximum MLD (Fig. 3b) for the 2010/11 winter reached 215 m in February 2011 compared with earlier and deeper mixing (to 257 m) in the following 2011/12 winter (December 2011).

The calculated DIC concentrations (Fig. 3c) show a closer relationship to the MLD seasonality than nitrate concentration data. Seasonal variation in the concentration of both DIC and nitrate is similar apart from the 2004/05 winter; when low DIC concentrations were not seen at the same time as the low nitrate concentrations (Fig. 3c).

The interrelation between DIC and nitrate concentrations can be considered by comparing the C : N ratios to the Redfield ratio (Redfield, 1958). The 2003–2005 time period has already been considered in Körtzinger et al. (2008) so is not reproduced here. Following Körtzinger et al. (2008) we calculated DIC, corrected for gas exchange. DIC concentrations were plotted against the in situ nitrate concentrations in different seasons for 2010–2012 (Fig. 4). The C : N ratio differed from the Redfield ratio of 6.6 with especially high values in spring (14.3).

Figure 5a shows weekly satellite wind speed data used to calculate the CO_2 flux. The wind speeds were similar in the two periods. There is an earlier period of days with high wind speeds towards the end of 2011 that can be compared with the CO_2 data presented. The annual average wind speed was 8.2 m s^{-1} for both time periods. The maximum was 14 m s^{-1}, although in situ winds of up to 20 m s^{-1} were seen from the Met Office data (eurosites.info/pap), this is not seen in the weekly averaged satellite wind speed data presented.

Figure 5b shows the sea-to-air CO_2 flux (where a positive flux is defined as from sea to the atmosphere). This was calculated from in situ $p(CO_2)$ data and satellite wind speed

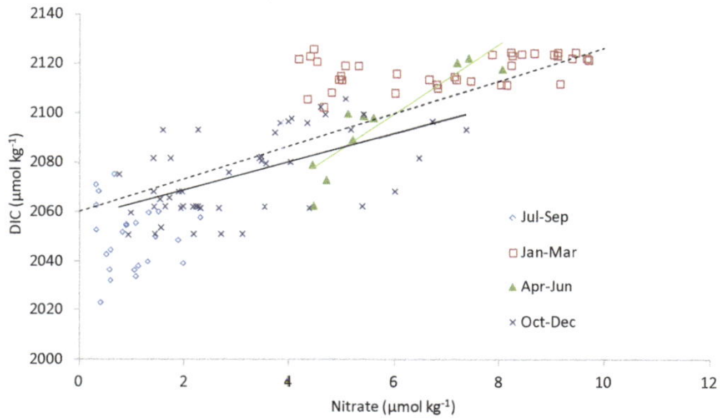

Figure 4. The relationship between concentrations of gas exchange-corrected DIC and nitrate (2010–2012) at the PAP-SO showing 4 different seasons: winter (January–March, red squares); spring (April–June, green triangles); summer (July–September, blue diamonds); autumn (October–December, dark blue crosses). The green line shows the ratio in spring (14.3) and the blue line is the ratio in autumn (6.4), with the Redfield ratio of 6.6 shown for reference as a dashed line.

data (Fig. 5b). The week by week variation in CO_2 flux is shown and an overall average for the two periods of time has been calculated as (-5.7 ± 2.8) mmol m^{-2} d^{-1} for the 2003–2005 period and (-5.0 ± 2.2) mmol m^{-2} d^{-1} for the 2010–2012 period. SOO data have been used at the start of 2011 when in situ $p(CO_2)$ data were unavailable. The start and end months of the two periods of time differs, which will contribute to the errors in the flux measurements. However the errors are comparable for the two periods of time considered and overall the average for the two time periods is similar.

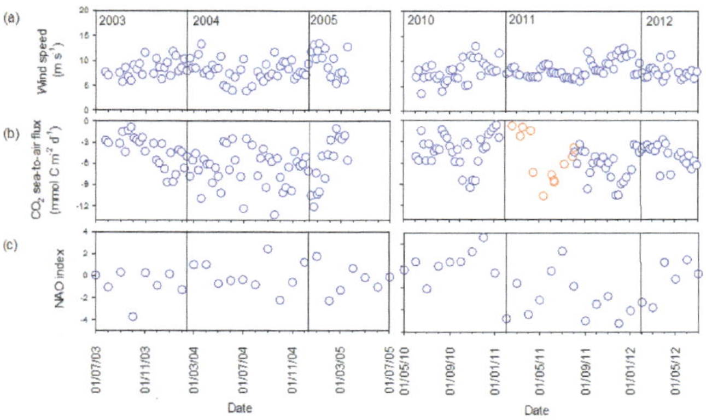

Figure 5. Data from 2003 to 2005 and 2010 to 2012 (blue circles) for **(a)** weekly satellite wind data in the region of the PAP-SO; **(b)** calculations of weekly sea-to-air CO_2 flux (negative: into the ocean) from in situ PAP-SO $p(CO_2)$ data and satellite wind (see text for details) with additional flux calculations from SOO data (red circles); **(c)** the monthly NAO index.

There is little variation in CO_2 flux and MLD between the years shown but for completeness the NAO index is shown in Fig. 5c. The 2003/04 winter NAO was near zero and the 2004/05 winter NAO was also low, between -2 to $+1$. In contrast there is a large range in the winter NAO in the 2010/11 winter when the NAO changed from -4 to $+3$. Overall the range in the NAO values was larger for the 2010–2012 time period shown.

4 Discussion

4.1 PAP-SO seasonal variation

The 2003–2005 and 2010–2013 data sets show very similar seasonal patterns between the years. Concentrations of nitrate and DIC exhibit seasonal variations opposite to temperature. The seasonal variation in nitrate and DIC concentrations is controlled by convective mixing (resulting in the winter maximum) and biological uptake during the spring bloom period (resulting in the summer minimum), which is similar to elsewhere in the North Atlantic (Jiang et al., 2013).

The $p(CO_2)$ distribution pattern at the PAP-SO site is characterised by a single annual peak (high in winter and low in summer), which is similar to that of nutrient and DIC concentrations, but in antiphase to the temperature signal. Jiang et al. (2013) compared seasonal carbon variability between different sites in the North Atlantic and suggested a latitudinal change in $p(CO_2)$ seasonality from the temperature-dominated oligotrophic subtropical gyre to the subpolar region where $p(CO_2)$ is dominated by changing concentrations of DIC. Our $p(CO_2)$ observations at the PAP-SO site show the subpolar-like seasonal pattern, which is similar to that of the ocean weather station M (Skjelvan et al., 2008). The surface $p(CO_2)$ is mainly governed by the varying DIC concentration while the seasonal cooling and warming have a contrasting effect.

The time integrated uptake of DIC and nitrate during the spring bloom is reflected by the slope of the linear regression between them (Fig. 4). The ratio of DIC and nitrate concentrations from 2010 to 2012 shows higher values than the Redfield C : N ratio of 6.6. For example the spring-time ratio of 14.3 (± 5) was considerably higher than the Redfield ratio, in agreement with similar "carbon overconsumption" ratios seen for the North Atlantic (e.g. 14.2, Sambrotto et al., 1993). This value is in agreement with the single C : N ratio reported previously at the PAP-SO of 11.0 (Körtzinger et al., 2008). In both cases the DIC concentrations were calculated and therefore associated with errors in the order of $7.0\,\mu mol\,kg^{-1}$. We have demonstrated seasonal variation in the C : N ratio at the PAP-SO, with an autumn C : N value that is closer to the Redfield ratio and large deviations from the Redfield ratio in winter.

4.2 Air–sea CO_2 flux

Wind speeds have an indirect impact on the biogeochemistry, in particular $p(CO_2)$. In the North Atlantic the strength and frequency of wintertime storms is significantly increasing (Donat et al., 2011). Wind speeds are similar for the two time periods considered here. However there is some suggestion of an earlier increase in winds at the start of the 2011/12 winter (Fig. 5a) coinciding with an earlier increase in mixing (Fig. 3b). Although the CO_2 flux is not linked linearly to the wind speed there is a corresponding decrease in CO_2 flux into the ocean at this time.

It is well known that the northeast Atlantic is a strong CO_2 sink with large variability. The observations at the PAP-SO provide high frequency data to follow the variability in CO_2 exchange. The largest CO_2 flux shown here was in September 2004, as a combined result of low seawater $p(CO_2)$ (Fig. 2a) and high wind speed (Fig. 5b). Larger CO_2 flux into the ocean may have occurred in 2011 considering the large, early spring bloom seen in that year but we do not have in situ PAP-SO $p(CO_2)$ data to calculate the flux at that time. However flux calculations from SOO data in early 2011 do not suggest an increase in CO_2 flux. Increases in productivity do not necessarily result in enhanced oceanic CO_2 uptake as the gas exchange is also affected by other factors such as temperature and wind speed (Jiang et al., 2013). The average is similar for the years presented with values of $-5.7\,mmol\,m^{-2}\,d^{-1}$ in 2003–2005 and $-5.0\,mmol\,m^{-2}\,d^{-1}$ from 2010 to 2012.

4.3 PAP-SO inter-annual variations

It is suggested that NAO plays an important role in modulating the inter-annual variability in the northeast Atlantic region by affecting the intensity of winter convection (Bennington et al., 2009; Jiang et al., 2013). The Gibraltar minus Iceland version of the NAO index is really most applicable to the winter half of the year. During positive NAO periods, the PAP-SO region experiences subpolar-like conditions, with strong wind stress and deep mixed layers (Henson et al., 2012). However the MLD did not vary significantly at the PAP-SO between the 2003–2005 and 2010–2012 time periods shown here (with a range of only 215–257 m for deepest winter MLD between the years. In previous years such as 2009/10 deep winter mixing of 390 m has been seen with an NAO reaching −3, (not shown). NAO is unlikely to have a large role at the PAP-SO as winter sea surface temperature and MLD were similar in the time periods 2003–2005 and 2010–2012. Data from a winter with deeper mixing would need to be put into the comparison to resolve this.

There was a 2-fold decrease in nitrate concentrations in the 2004–2005 winter despite sea surface temperature and MLD values being close to other years. The low values were confirmed by SOO data, also shown in Hartman et al. (2010). As discussed in Hartman et al. (2010) the lower winter nitrate concentration seen in 2004/05 did not correlate with a decrease in the MLD and this showed the influence of horizontal mixing at the PAP-SO. It was suggested that lateral advection to the site at that time introduced a subtropical water mass with a lower nitrate concentration. Earlier time series studies largely ignored circulation at the PAP-SO site, assuming convective mixing is a dominant process influencing mixed layer temperature and nitrate concentrations in the region (Williams et al., 2000; Körtzinger et al., 2008). However, fixed-point time series observations are influenced by spatial variability passing the point of observation (McGillicuddy et al., 1998; Painter et al., 2010). It is clear from Hartman et al. (2010) that lateral advection may significantly influence the surface temperature and nitrate concentrations in the region of the PAP-SO site.

The observed seawater $p(CO_2)$ increased from $(339 \pm 17)\,\mu$atm in 2003–2005 to $(353 \pm 15)\,\mu$atm in 2010–2012, which largely agrees with the increasing rate of surface seawater $p(CO_2)$ observed in the North Atlantic basin of $(1.84 \pm 0.4)\,\mu\text{atm}\,\text{a}^{-1}$ (Takahashi et al., 2009). Despite similar maximum winter MLD in 2003–2005 and 2010–2012, the timing and intensity of the spring bloom is quite different and the cause of this requires further investigation.

5 Conclusions and further work

We have presented recent year-round surface time series biogeochemical data at the PAP-SO and compared it with previous observations. The surface $p(CO_2)$, and concentrations of DIC and nitrate, at the PAP-SO all show a clear seasonal cycle, which is mainly controlled by winter convective mixing and biological activity in the spring bloom. However the suggestion that inter-annual variability is dominated by convection (Bennington et al., 2009) is not clear as the MLD did not vary significantly between the winter periods shown. An especially low winter nitrate concentration in 2005 was observed, thought to be due to surface advection and this highlights the need to consider advection when dealing with time series data in the future. Despite the similar winter physical conditions (temperature and MLD), there is a year to year difference in the timing and intensity of the spring blooms, which requires further investigation. At PAP-SO, increasing mean seawater $p(CO_2)$ from $(339 \pm 17)\,\mu$atm in 2003 to $(353 \pm 15)\,\mu$atm in 2011 was observed. However the mean air–sea CO_2 flux did not show a significant change. It varied from $(-5.7 \pm 2.8)\,\text{mmol}\,\text{m}^{-2}\,\text{d}^{-1}$ in 2003–2005 to $(-5.0 \pm 2.2)\,\text{mmol}\,\text{m}^{-2}\,\text{d}^{-1}$ in 2010–2012.

In 2010, collaboration between the UK's Natural Environment Research Council (NERC) and Meteorological Office led to the first simultaneous monitoring of in situ meteorological and ocean variables at the PAP-SO (Hartman et al., 2012). From 2013 additional measurements of $p(CO_2)$ will be made at the site, at the shallower depth of 1 m, and should further improve the SOO comparison. The site could be used to investigate the effect of different parameterisations (Prytherch et al., 2010) and wind products on calculations of CO_2 flux, in particular during the high wind conditions seen. Using the contemporaneous atmospheric and ocean data sets we will be able to investigate the effect of storms on CO_2 flux and resolve daily variability.

Acknowledgements. We would like to acknowledge the various ship crew, engineers and scientists involved in preparation, deployment and recovery of the PAP-SO moorings. We would especially like to thank J. Campbell, M. Hartman and M. Pagnani for the PAP time series and H. Cole who calculated MLD from the Argo float data. The research leading to these results was supported through the EU FP7 project CARBOCHANGE "Changes in carbon uptake and emissions by oceans in a changing climate", which received funding from the European Commission's Seventh Framework Programme under grant agreement no. 264879. Funding that supports the running of the SOO network used in this project also includes EU grant 212196 (CO-COS), and UK NERC grant NE/H017046/1 (UKOARP); Funding for part of this study was provided by NERC CASE studentship grant reference number NE/J500069/1 in collaboration with SAHFOS. Argo data were made freely available by the Coriolis project and programmes that contribute to it http://www.coriolis.eu.org. FNMOC wind speed data were available from http://las.pfeg.noaa.gov. Mooring data and support for this research was provided by the European research projects ANIMATE (Atlantic Network of Interdisciplinary Moorings and Time-for Europe), MERSEA (Marine Environment and Security for the European Sea), EUR-OCEANS (European Network of Excellence for Ocean Ecosystems Analysis) and EuroSITES grant agreement EU 202955. The work was also supported through the Natural Environment Research Council (NERC), UK, project Oceans 2025 and National Capability. The PAP-SO also contributes to the EU funded FixO3 project EU 312463 and the NERC Greenhouse Gas TAP NE/k00249x/1. D. Turk was supported by the Canada Excellence Research Chair (CERC) in Oceans Science and Technology. LDEO contribution number 7768.

Edited by: J. Kaiser

References

Behrenfeld, M. J., O'Malley, R. T., Siegel, D. A., McClain, C. R., Sarmiento, J. L., Feldman, G. C., and Milligan, A. J.: Climate-driven trends in contemporary ocean productivity, Nature, 444, 752–755, 2006.

Bennington, V., McKinley, G. A., Dutkiewicz, S., and Ullman, D.: What does chlorophyll variability tell us about export and air-sea CO_2 flux variability in the North Atlantic?, Global Biogeochem. Cy., 23, GB3002, doi:10.1029/2008GB003241, 2009.

DeGrandpre, M. D., Hammar, T. R., Smith, S. P., and Sayles, F. L.: In situ measurements of seawater pCO_2, Limnol. Oceanogr., 40, 969–975, 1995.

Dickson, A. G.: Standard potential of the (AgCl(s) + $1/2H_2$ (g) = Ag(s) + HCl(aq)) cell and thedissociation constant of bisulfate ion in synthetic sea water from 273.15 to 318.15 K, J. Chem. Thermodyn., 22, 113–127, doi:10.1016/0021-9614(90)90074-Z, 1990.

Dickson, A. G., Sabine, C. L., and Christian, J. R.: Guide to Best Practices for Ocean CO_2 Measurements, PICES Special Publication 3, Sidney, British Columbia, 191 pp., 2007.

Donat, M. G., Renggli, D., Wild, S., Alexander, L. V., Leckebusch, G. C., and Ulbrich, U.: Reanalysis suggests long-term upward trends in European storminess since 1871, Geophys. Res. Lett., 38, L14703, doi:10.1029/2011GL047995, 2011.

EuroSITES: European Ocean Observatory Network, http://www.eurosites.info/pap, last access: August 2014.

Fiedler, P. C.: Comparison of objective descriptions of the thermocline, Limnol. Oceanogr.-Meth., 8, 313–325, 2010.

Field, C. B., Behrenfeld, M. J., and Randerson, J. T.: Primary production of the biosphere: integrating terrestrial and oceanic components, Science, 281, 237–240, 1998.

González-Dávila, M., Santana-Casiano, J. M., Rueda, M. J., Llinas, O., and González-Dávila, E. F.: Seasonal and interannual variability of sea-surface carbon dioxide species at the European Station for Time Series in the Ocean at the Canary Islands (ESTOC) between 1996 and 2000, Global Biogeochem. Cy., 17, 1076, doi:10.1029/2002gb001993, 2003.

Hartman, S. E., Larkin, K. E., Lampitt, R. S., Lankhorst, M., and Hydes, D. J.: Seasonal and inter-annual biogeochemical variations in the Porcupine Abyssal Plain 2003–2005 associated with winter mixing and surface circulation, Deep-Sea Res. Pt. II, 57, 1303–1312, 2010.

Hartman, S. E., Lampitt, R. S., Larkin, K. E., Pagnani, M., Campbell, J., Gkritzalis, A., and Jiang, Z. P.: The Porcupine Abyssal Plain fixed-point sustained observatory (PAP-SO): variations and trends from the Northeast Atlantic fixed-point time-series, ICES J. Mar. Sci., 57, 776–783, 2012.

Henson, S., Lampitt, R., and Johns, D.: Variability in phytoplankton community structure in response to the North Atlantic Oscillation and implications for organic carbon flux, Limnol. Oceanogr., 57, 1591–1601, doi:10.4319/lo.2012.57.6.1591, 2012.

Holte, J. and Talley, L.: A New Algorithm for Finding Mixed Layer Depths with Applications to Argo Data and Subantarctic Mode Water Formation, J. Atmos. Ocean. Tech., 26, 1920–1939, 2009.

Hurrell, J. W.: Decadal trends in the North Atlantic oscillation: regional temperatures and precipitation, Science, 269, 676–679, 1995.

Hydes, D. J., Wright, P. N., and Rawlinson, M. B.: Use of a wet chemical analyser for the in situ monitoring of nitrate, Chemical sensors in Oceanography, in: Chemical Sensors in Oceanography, edited by: Varney, M., Gordon and Breach, Amsterdam, 95–105, 2000.

Jiang, Z.-P., Hydes, D. J., Tyrrell, T., Hartman, S. E., Hartman, M. C., Dumousseaud, C., Padin, X. A., Skjelvan, I., and González-Pola, C.: Key controls on the seasonal and inter- annual variations of the carbonate system and air-sea CO_2 flux in the Northeast Atlantic (Bay of Biscay), J. Geophys. Res.-Oceans, 118, 785–800, 2013.

Jiang, Z.-P., Hydes, D. J., Tyrrell, T., Hartman, S. E., Hartman, M. C., Campbell, J. M., Johnson, B. D., Schofield, B., Turk, D., Wallace, D., Burt, W., Thomas, H., Cosca, C., and Feely, R.: Application and assessment of a membrane-based pCO_2 sensor under field and laboratory conditions, Limnol. Oceanogr.-Meth., 12, 264–280, 2014.

Kara, A. B., Rochford, P. A., and Hurlburt, H. E.: An optimal definition for ocean mixed layer depth, J. Geophys. Res., 105, 16803–16821, 2000.

Körtzinger, A., Send, U., Lampitt, R. S., Hartman, S., Wallace, D. W. R., Karstensen, J., Villagarcia, M. G., Llinás, O., and DeGrandpre, M. D.: The seasonal pCO_2 cycle at 49° N/16.5° W in the northeastern Atlantic Ocean and what it tells us about biological productivity, J. Geophys. Res., 113, C04020, doi:10.1029/2007jc004347, 2008.

Lampitt, R. S., Bett, B. J., Kiriakoulis, K., Popova, E., Ragueneau, O., Vangriesheim, A., and Wolff, G. A.: Material supply to the abyssal seafloor in the Northeast Atlantic, Prog. Oceanogr., 50, 27–63, 2001.

Lavigne, H. and Gattuso, J.-P.: Seacarb: seawater carbonate chemistry with R, R package version 2.4., available at: http://CRAN.R-project.org/package=seacarb, last access: January 2014, edited 2011.

Lee, K., Tong, L. T., Millero, F. J., Sabine, C. L., Dickson, A. G., Goyet, C., Park, G. H., Wanninkhof, R., Feely, R. A., and Key, R. M.: Global relationships of total alkalinity with salinity and temperature in surface waters of the world's oceans, Geophys. Res. Lett., 33, L19605, doi:10.1029/2006gl027207, 2006.

Longhurst, A.: Ecological geography of the sea, Ecological Geography of the Sea Series, 2nd Edn., Academic Press, San Diego, 560 pp., 2006.

Lueker, T. J., Dickson, A. G., and Keeling, C. D.: Ocean pCO_2 calculated from dissolved inorganic carbon, alkalinity, and equations for K-1 and K-2: validation based on laboratory measurements of CO_2 in gas and seawater at equilibrium, Mar. Chem., 70, 105–119, 2000.

McGillicuddy, D. J., Robinson, A. R., Siegel, D. A., Jannasch, H. W., Johnson, R., Dickey, T. D., McNeil, J., Michaels, A. F., and Knap, A. H.: Influence of mesoscale eddies on new production in the Sargasso Sea, Nature, 394, 263–266, 1998.

McKinley, G. A., Fay, A. R., Takahashi, T., and Metzl, N.: Convergence of atmospheric and North Atlantic carbon dioxide trends on multidecadal timescales, Nat. Geosci., 4, 606–610, 2011.

Montegut, C. D., Madec, G., Fischer, A. S., Lazar, A., and Iudicone, D.: Mixed layer depth over the global ocean: An examination of profile data and a profile-based climatology, J. Geophys. Res., 109, C12003, doi:10.1029/2004JC002378, 2004.

Monterey, G. and Levitus, S.: Seasonal variability of mixed layer depth for the World Ocean, NOAA NESDIS Atlas, US Government Printing Office, Washington, DC, 5 pp., 1997.

Nightingale, P. D., Malin, G., Law, C. S., Watson, A. J., Liss, P. S., Liddicoat, M. I., Boutin, J., and Upstill-Goddard R. C.: In situ evaluation of air-sea gas exchange parameterizations using novel conservative and volatile tracers, Global Biogeochem. Cy., 14, 373–387, doi:10.1029/1999GB900091, 2000.

Padin, X. A., Castro, C. G., Ríos, A. F., and Pérez, F. F.: Oceanic CO_2 uptake and biogeochemical variability during the formation of the Eastern North Atlantic Central water under two contrasting NAO scenarios, J. Marine. Syst., 84, 96–105, 2011.

Painter, S. C., Pidcock, R. E., and Allen, J. T.: A mesoscale eddy driving spatial and temporal heterogeneity in the productivity of the euphotic zone of the Northeast Atlantic, Deep-Sea Res. Pt. II, 57, 1281–1292, 2010.

Perez, F. F. and Fraga, F.: The pH measurements in seawater on the NBS scale, Mar. Chem., 21, 315–327, 1987.

Prytherch, J., Yelland, M. J., Pascal, R. W., Moat, B. I., Skjelvan, I., and Neill, C. C.: Direct measurements of the CO_2 flux over the ocean: development of a novel method, Geophys. Res. Lett., 37, L03607, doi:10.1029/2009GL041482, 2010.

Redfield, A. C.: The biological control of chemical factors in the environment, Am. Sci., 64, 205–221, 1958, 1958.

Sambrotto, R. N., Savidge, G., Robinson, C., Boyd, P., Takahashi, T., Karl, D. M., and Codispoti, L.: Elevated consumption of carbon relative to nitrogen in the surface ocean, Nature, 363, 248–250, 1993.

Schuster, U. and Watson, A. J.: A variable and decreasing sink for anthropogenic CO_2 in the North Atlantic, J. Geophys. Res., 112, C1106, doi:10.1029/2006JC003941, 2007.

Schuster, U., Watson, A. J., Bates, N. R., Corbiere, A., Gonzalez-Davila, M., Metzl, N., Pierrot, D., and Santana-Casiano, M.: Trends in North Atlantic sea-surface fCO_2 from 1990 to 2006, Deep-Sea Res. Pt. II, 56, 620–629, doi:10.1016/j.dsr2.2008.12.011, 2009.

Skjelvan, I., Falck, E., Rey, F., and Kringstad, S. B.: Inorganic carbon time series at OceanWeather Station M in the Norwegian Sea, Biogeosciences, 5, 549–560, doi:10.5194/bg-5-549-2008, 2008.

Takahashi, T., Sutherland, S. C., Sweeney, C., Poisson, A., Metzl, N., Tilbrook, B., and Bates, N.: Global sea–air CO_2 flux based on climatological surface ocean pCO_2, and seasonal biological and temperature effects, Deep-Sea Res. Pt. II, 49, 1601–1622, 2002.

Takahashi, T., Sutherland, S. C., Wanninkhof, R., Sweeney, C., Feely, R. A., Chipman, D. W., Hales, B., Friederich, G., Chavez, F., Sabine, C., Watson, A., Bakker, D. C. E., Schuster, U., Metzl, N., Yoshikawa-Inoue, H., Ishii, M., Midorikawa, T., Nojiri, Y., Kortzinger, A., Steinhoff, T., Hoppema, M., Olafsson, J., Arnarson, T. S., Tilbrook, B., Johannessen, T., Olsen, A., Bellerby, R., Wong, C. S., Delille, B., Bates, N. R., and de Baar, H. J. W.: Climatological mean and decadal change in surface ocean pCO_2, and net sea-air CO_2 flux over the global oceans, Deep-Sea Res. Pt. II, 56, 554–577, doi:10.1016/j.dsr2.2008.12.009, 2009.

Thomson, R. E. and Fine, I. V.: Estimating Mixed Layer Depth from Oceanic Profile Data, J. Atmos. Ocean. Tech., 20, 319–329, 2003.

Wanninkhof, R.: Relationship between wind-speed and gas-exchange over the ocean, J. Geophys. Res., 97, 7373–7382, doi:10.1029/92JC00188, 1992.

Weiss, R. F.: Carbon dioxide in water and seawater: the solubility of a non-ideal gas, Mar. Chem., 2, 203–215, doi:10.1016/0304-4203(74)90015-2, 1974.

Williams, R. G., McLaren, A. J., and Follows, M. J.: Estimating the convective supply of nitrate and implied variability in export production over the North Atlantic, Global Biogeochem. Cy., 14, 1299–1313, 2000.

16

Evaluating the ocean biogeochemical components of Earth system models using atmospheric potential oxygen and ocean color data

C. D. Nevison[1], M. Manizza[2], R. F. Keeling[2], M. Kahru[2], L. Bopp[3], J. Dunne[4], J. Tiputra[5], T. Ilyina[6], and B. G. Mitchell[2]

[1]University of Colorado, Boulder, Institute for Arctic and Alpine Research, Boulder, Colorado, USA
[2]Scripps Institution of Oceanography, La Jolla, California, USA
[3]IPSL/LSCE, UMR8212, CNRS-CEA-UVSQ, Gif sur Yvette, France
[4]National Oceanic and Atmospheric Administration/Geophysical Fluid Dynamics Laboratory, Princeton, New Jersey, USA
[5]Uni Climate, Uni Research and Bjerknes Centre for Climate Research, Bergen, Norway
[6]Max Planck Institute for Meteorology, Hamburg, Germany

Correspondence to: C. D. Nevison (cynthia.nevison@colorado.edu)

Abstract. The observed seasonal cycles in atmospheric potential oxygen (APO) at a range of mid- to high-latitude surface monitoring sites are compared to those inferred from the output of six Earth system models (ESMs) participating in the fifth phase of the Coupled Model Intercomparison Project phase 5 (CMIP5). The simulated air–sea O_2 fluxes are translated into APO seasonal cycles using a matrix method that takes into account atmospheric transport model (ATM) uncertainty among 13 different ATMs. Three of the ocean biogeochemistry models tested are able to reproduce the observed APO cycles at most sites, to within the large TransCom3-era ATM uncertainty used here, while the other three generally are not. Net primary production (NPP) and net community production (NCP), as estimated from satellite ocean color data, provide additional constraints, albeit more with respect to the seasonal phasing of ocean model productivity than overall magnitude. The present analysis suggests that, of the tested ocean biogeochemistry models, the community ecosystem model (CESM) and the Geophysical Fluid Dynamics Laboratory (GFDL) ESM2M are best able to capture the observed APO seasonal cycle at both northern and southern hemispheric sites. In most models, discrepancies with observed APO can be attributed to the underestimation of NPP, deep ventilation or both in the northern oceans.

1 Introduction

Ocean physical and biogeochemical processes have profound influences on Earth's climate. Phytoplankton in the sunlit part of the ocean convert carbon from inorganic to organic form via photosynthesis, thereby establishing the base of the ocean food chain. Primary production and subsequent export of organic carbon from the mixed layer (export production, EP) and remineralization at depth are key components of the so-called biological pump, which regulates the partition of carbon between the ocean and atmosphere (Gruber and Sarmiento, 2002; Boyd and Doney, 2003).

Net community production (NCP) and the related process of EP are also important for understanding the distribution of dissolved O_2 within the ocean and the flux of O_2 (F_{O2}) at the air–sea interface. NCP is defined here as the net amount of organic carbon fixed through photosynthesis over the depth of the mixed layer after accounting for grazing and both autotrophic and heterotrophic respiration. NCP is closely linked to F_{O2}, since each mole of photosynthetically fixed carbon that persists beyond the timescale of air–sea exchange (2–3 weeks) leaves a stoichiometric amount of O_2 available for release to the atmosphere. This release of O_2 to the atmosphere in association with NCP occurs mainly in the spring and summer at extratropical latitudes (Keeling et al., 1993). EP more or less balances NCP when averaged over a full year or if the upper ocean is in a long-term steady state and advective fluxes are zero (Laws et al., 2000). The exported carbon subsequently is respired in the subsurface ocean, leading to O_2 depletion at depth. O_2 is replenished by absorption from the atmosphere when deep waters mix back to the surface in fall and winter. Deep ventilation and NCP thus are distinct processes that are largely separate in time and space but are both closely linked to the biological pump that draws carbon out of surface waters and is critical for ocean uptake of atmospheric CO_2.

To explore the impacts of future climate change on Earth's climate and ecosystems, the Coupled Model Intercomparison Project phase 5 (CMIP5) relies on three-dimensional numerical Earth system models (ESMs), which incorporate descriptions of biogeochemical impacts of land and marine biota. Projections of future atmospheric CO_2 levels and associated climate warming in the CMIP5 depend not only on fossil fuel use projections but also on assumptions about uptake and storage of carbon by the land and ocean. The oceans have absorbed approximately one-third of the anthropogenic carbon released to the atmosphere since the beginning of the industrial era (Khatiwala et al., 2009), but this fractional rate of uptake is unlikely to continue in the future as the buffering capacity of surface waters declines and the export of carbon from the surface to the deep ocean fails to keep pace with anthropogenic fossil fuel combustion (Arora et al., 2013). Changes in ventilation of abyssal deepwater are an additional possible consequence of future climate forcing that current models may or may not be able to predict accurately (Sigman et al., 2010).

Recent studies have tested the present-day skill of the ocean components of ESMs and some have also examined future projections (Schneider et al., 2008; Steinacher et al., 2010, Bopp et al., 2013; Anav et al., 2013). These evaluations have compared model output to both hydrographic measurements and remotely sensed ocean color products, most commonly net primary production (NPP). The models predict spatial–annual patterns in NPP that reproduce some of the main features seen in satellite data, but differ over a factor of 2 in NPP magnitude. Some evaluations have examined seasonal variability and have found that ocean models tend to underestimate observed NPP at high latitudes (poleward of 44°) in the Northern Hemisphere and overestimate it in the Southern Hemisphere. The models also fail to capture the timing of the observed high-latitude peak in NPP in both hemispheres, with predictions that are often 1–2 months earlier than observations (Anav et al., 2013; Henson et al., 2013). However, ocean color-derived NPP values are uncertain, especially in the Southern Ocean, reducing confidence in the observed constraints.

Many biogeochemical processes that are expected to occur in the future, such as responses to warming and stratification, are also highly relevant on seasonal timescales (Keeling et al., 2010; Anav et al., 2013). Thus, challenging models against known seasonal variations can aid in the development of credible predictions of future changes. Here, we evaluate six ESMs used in the CMIP5 against two cross-cutting metrics, which test the models' ability to account for changes in ocean biogeochemistry on seasonal time frames. This work is intended primarily as a demonstration of method using an available subset of the CMIP5 ESMs rather than as a comprehensive evaluation of all the CMIP5 models. The first metric is based on satellite-derived estimates of ocean color, focusing on NPP and NCP. The second metric is based on the seasonal cycles in atmospheric potential oxygen (APO), an atmospheric tracer that varies seasonally mainly due to air–sea exchanges of O_2 (Stephens et al., 1998; Manning and Keeling, 2006).

EP is the ocean color-derived flux most relevant to the biological pump, but cannot be directly observed by remote sensing. It is derived by a combination of remote measurements and poorly constrained models, which inherently increases its uncertainty (Schneider et al., 2008; Nevison et al., 2012a). The quantity actually observed from space is spectral top-of-the-atmosphere radiance, which is used to estimate chlorophyll (or another proxy of phytoplankton biomass); chlorophyll and other variables such as photosynthetic radiation are used to estimate NPP and, finally, NPP is used to estimate EP. The first step, estimation of chlorophyll, is known to have significant bias (underestimation by $\sim$2–3 times) in the Southern Ocean which is transferred to higher level products. We correct for that bias by using algorithms tuned to Southern Ocean data sets blended with more or less standard products elsewhere (Mitchell and Kahru, 2009; Kahru and Mitchell, 2010). While our satellite estimates of EP are improved, they are still subject to high uncertainty.

Observed seasonal cycles in APO provide a new benchmark for the ocean biogeochemistry model components of ESMs. They offer evaluation metrics complementary to ocean color products by providing additional information on deep ventilation processes unavailable from satellite data alone. The main drawback of APO seasonal cycles is that atmospheric transport models (ATMs) are needed to translate ocean model air–sea O_2 fluxes into a seasonal APO signal, which inevitably introduces uncertainty (Stephens et al., 1998; Nevison et al., 2012a). A first attempt has been made to use APO seasonal cycles to evaluate ocean-only marine biogeochemistry models (Naegler et al., 2007), but the models in that study implemented a simplified parameterization of the biological processes affecting O_2 and CO_2 air–sea fluxes and were considerably less advanced than the current ecosystem dynamics and biogeochemical components used in state-of-the-art ESMs. Further, while Naegler et al. (2007) asserted that the uncertainty introduced by ATMs was too large to provide a strong constraint on ocean model fluxes, their study relied on only two ATMs. Here, we translate the model air–sea fluxes into APO signals using a wider range of ATMs and show that, in many cases, the discrepancies between modeled and observed APO seasonal cycles transcend ATM uncertainty.

2 Methods

2.1 Ocean biogeochemistry models

The CMIP5 models analyzed in this study include the Geophysical Fluid Dynamics Laboratory (GFDL) ESMs (depth-based ESM2M and density-based ESM2G vertical oceans; Dunne et al., 2012) from Princeton, New Jersey; the Institut Pierre Simon Laplace Coupled Model 5 in its low-resolution version (IPSL-CM5A-LR, hereafter referred to as IPSL) model from Paris, France; the community ecosystem model (CESM) from the National Center for Atmospheric Research in Boulder, CO; the Max Planck Institute for Meteorology (MPIM) ESM, version MPI-ESM-LR, from Hamburg, Germany; and the Norwegian ESM (NorESM1-ME, hereafter referred to as NorESM1). The ocean biogeochemical models embedded in the respective ESMs are represented by the Tracers of Phytoplankton with Allometric Zooplankton (TOPAZ) (GFDL) (Dunne et al., 2013), Pelagic Interaction Scheme for Carbon and Ecosystem Studies (PISCES) (IPSL) (Aumont and Bopp, 2006), biogeochemical elemental cycling (BEC) (CESM) (Moore et al., 2002, 2004, 2013), and Hamburg Ocean Carbon Cycle (HAMOCC) (MPIM) (Ilyina et al., 2013). NorESM1 uses a variant of HAMOCC, adapted to a sigma-coordinate ocean circulation model (Tjiputra et al., 2013).

The six ESMs differ in their physical components and implement ocean biogeochemical schemes that vary in their specifics, but have many common features. All include explicit representations of upper ecosystem dynamics that distinguish at least one phytoplankton group and one size class of zooplankton. Four of the models (CESM, both GFDL variants and IPSL) divide phytoplankton further into at least two size classes: large (micro) and small (nano + pico). GFDL and CESM also explicitly model diazotrophs. Phytoplankton growth rates in all models are co-limited by light, temperature and nutrient (N, P, Si, Fe) availability. Carbon export flux is closely linked to ecosystem structure and dynamics, with higher sinking rates assumed for large phytoplankton, representing, e.g., diatoms.

For each model, the following output fields were obtained for the CMIP5 standard historical simulation, which is driven by prescribed atmospheric CO_2 from 1850 to 2005: carbon export flux at 100 m depth (EP_{100}), vertically integrated NPP, net air–sea O_2 and CO_2 fluxes, net surface heat flux (Q), and sea surface salinity and temperature (SST). Many of these fields were available through public web interfaces, but some variables, particularly Q, required assistance from the individual modeling groups, which effectively limited the study to the six models listed above. The EP_{100} and NPP fields were compared directly to the corresponding satellite ocean color products. The remaining five output fields were used in the estimation of APO time series, with the final three fields used to estimate air–sea N_2 fluxes based on the $Q(\mathrm{d}S/\mathrm{d}T)_{N_2}/C_p$ equation (Keeling et al., 1993; Manizza et al., 2012) with modifications from Jin et al. (2007). In this equation, Q is heat flux, $(\mathrm{d}S/\mathrm{d}T)_{N_2}$ is the temperature derivative of the N_2 solubility coefficient and C_p is the heat capacity of sea water. The resulting N_2 fluxes, together with the prognostic O_2 and CO_2 air–sea fluxes, were used as described below to force atmospheric transport model simulations to compute atmospheric time series of APO (Naegler et al., 2007; Nevison et al., 2008; 2012a). Since all the ocean models operated on an irregular, off-polar grid with two-dimensional latitude and longitude coordinates, these were first interpolated to a regular $1° \times 1°$ latitude–longitude grid using climate data operator (CDO) freeware (https://code.zmaw.de/projects/cdo). The climate data operator interpolation was not mass conservative, but resulted in global O_2 flux differences generally of less than 1 %. An exception was the CESM, whose output was converted conservatively to a regular grid using a CESM-specific mapping file.

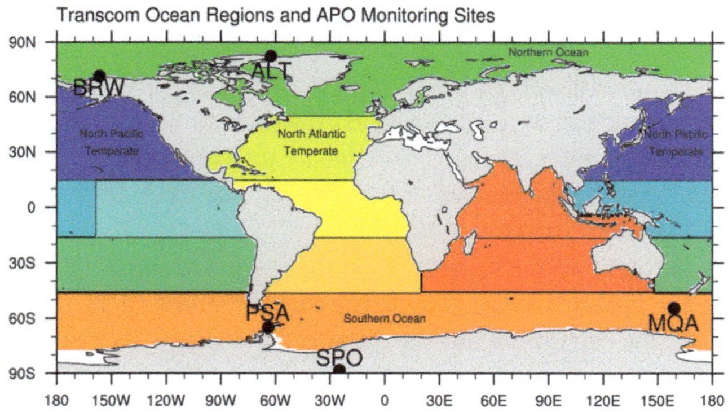

Figure 1. TransCom3 Level 2 ocean regions used in the matrix-based atmospheric transport method. Locations of the five APO monitoring sites featured in Fig. 2 are superimposed.

2.2 Atmospheric transport model simulations

2.2.1 Matrix method

A matrix method was used to translate the ocean model air–sea O_2, N_2 and CO_2 fluxes into corresponding annual mean cycles in atmospheric potential oxygen (APO). The method was based on the pulse-response functions from the TransCom3 Level 2 (T3L2) ATM intercomparison. Each of the 13 ATMs that participated in T3L2 conducted forward simulations in which a uniformly distributed CO_2 flux, normalized to $1\,\mathrm{PgC\,yr^{-1}}$, was released from each of 11 ocean regions (Fig. 1) for each of 12 emission months, i.e., January–December, allowed to decay for 35 months, using an annually repeating cycle of meteorology that was model specific for each ATM, and sampled every month at a range of surface monitoring sites (Gurney et al., 2003; 2004). The APO code was developed from an earlier pulse-response matrix code, which has been described in detail in Nevison et al. (2012b), that translates terrestrial net ecosystem exchange (NEE) fluxes of carbon into the corresponding annual mean cycles in atmospheric CO_2. The matrix method is considerably faster than a full forward ATM simulation, allowing annual mean cycles in APO from 13 different ATMs to be computed in seconds, rather than the days or weeks required for a single forward simulation.

The pulse-response matrix code was applied separately to the O_2, N_2 and oceanic CO_2 fluxes from the last 12 years of the historical simulations, spanning 1994–2005, converting from carbon to oxygen or nitrogen units where appropriate, to create three separate time series of atmospheric O_2, N_2 and CO_2 as mole fraction anomalies (μmol mol^{-1}) on a H_2O-free basis, where the O_2 and N_2 anomalies are computed as though O_2 and N_2 were trace gases, similar to CO_2. These were combined to calculate a 9-year time series in APO in per meg units, spanning 1997 to 2005, according to Eq. (1) (Stephens et al., 1998):

$$\mathrm{APO} = \frac{1}{X_{\mathrm{O_2}}}(\mathrm{O_2}) - \frac{1}{X_{\mathrm{N_2}}}(\mathrm{N_2}) + \frac{1.1}{X_{\mathrm{O_2}}}(\mathrm{CO_2}), \tag{1}$$

where X_{O_2} and X_{N_2} are the dry air mole fractions of O_2 and N_2 in H_2O-free air, treated here as constants (0.2094 and 0.7808, respectively). The mean seasonal cycle was computed by detrending the time series with a third-order polynomial and then taking the average of the detrended data for all Januaries, Februaries, etc. The matrix method involves calculating separately the components of APO at each measurement site arising from fluxes from each ocean region. These components are then summed to compute the net APO signal. The model definition of APO in Eq. (1) ignores contributions to APO from land biospheric exchanges at ratios other than 1 : 1 and fossil fuel burning, but these are very small in comparison to oceanic contributions on seasonal timescales (Manning and Keeling, 2006; Nevison et al., 2008).

2.2.2 Evaluation of matrix method based on APO TransCom

An evaluation exercise was conducted in which the APO pulse-response matrix code was forced by climatological O_2 and N_2 fluxes from Garcia and Keeling (2001) and used to compute the mean seasonal cycle in APO as described above using Eq. (1) (minus the oceanic CO_2 term). The matrix-based results were evaluated against the mean seasonal cycles from archived station output from the forward ATM simulations of the APO TransCom Experiment, which also used the Garcia and Keeling O_2 and N_2 forcing fluxes (Blaine, 2005; Nevison et al., 2012b). This evaluation was conducted using a subset of 9 of the original 13 T3L2 ATMs that also participated in APO TransCom. For this subset, the matrix method performed well in relatively homogeneous regions like the Southern Ocean and at northern high-latitude sites like Barrow, Alaska (BRW) and Alert, Canada (ALT). It was less reliable in capturing the forward simulation cycle at sites located within northern mid-latitude ocean regions, including Cold Bay, Alaska, and La Jolla, California, where the uniform distribution of fluxes assumed by T3L2 did not accurately capture the impact of strong heterogeneity in air–sea fluxes from these regions (Tables S1, S2 and Figs. S1, S2 in the Supplement). These same North Pacific stations are subject to large uncertainty in full forward ATM simulations due to uncertainty in vertical mixing (Stephens et al., 1998; Battle et al., 2006; Tohjima et al., 2012). We therefore focus in Sect. 3 on ALT, BRW and three Southern Ocean sites, including Macquarie Island (MQA), Palmer Station, Antarctica (PSA) and the South Pole (SPO) in our use of APO to evaluate the ESM-simulated air–sea O_2, N_2 and CO_2 fluxes. The locations of these five sites with respect to the T3L2 ocean regions are shown in Fig. 1.

While the evaluation exercise indicates that the matrix method reproduces the shape and phase of the seasonal cycles with high reliability at the above sites, it tends to underestimate the seasonal amplitude by about 4–5 % at ALT and BRW and by 11–12 % at MQA and SPO and to slightly overestimate the amplitude at PSA. In applying the matrix code to the ESM oceanic fluxes, we therefore scaled up the estimated cycles by site and ATM-specific scaling factors obtained from the evaluation exercise (Tables S1, S2, Fig. S2). Since these scaling factors were only available for the subset of 9 of the 13 T3L2 ATMs that also participated in APO TransCom, we subsequently (Sect. 3.1) compare observations alternatively to the scaled nine-model subset, or to all 13 unscaled models.

2.2.3 Component O_2 fluxes

The net air–sea O_2 flux for each ESM can be divided into three components, associated with NCP, deep ventilation and thermal processes (Nevison et al., 2012a):

$$F_{\mathrm{O_2,total}} = F_{\mathrm{O_2,NCP}} + F_{\mathrm{O_2,vent}} + F_{\mathrm{O_2,therm}}. \tag{2}$$

These in turn can be used to force the matrix model and the resulting total APO cycle can be presented as the sum of component signals according to Eq. (3).

$$\mathrm{APO} = \mathrm{APO_{NCP}} + \mathrm{APO_{vent}} + \mathrm{APO_{therm}} \tag{3}$$

Here, the APO_{therm} term also includes the effects of N_2 fluxes, as per the second right-hand term in Eq. (1). The atmospheric signal due to oceanic CO_2 (last term in Eq. 1) is not easily included in any of the component terms in Eq. (3) based on available ESM output, but in principle all three component processes may lead to changes in CO_2 fluxes as well as O_2 fluxes. In practice, CO_2 has only a small influence on the amplitude and phasing of APO in most of the ESMs and thus is ignored in the component analysis. An exception is the MPIM, in which the oceanic CO_2 signal has a peak–trough seasonal amplitude of up to 5 ppm in the Southern Ocean that opposes the O_2 cycle, as noted previously (Anav et al., 2013) and discussed further below.

Table 1. Vertically integrated NPP, EP at 100 m (both in $Pg\,C\,yr^{-1}$) and EP/PP for six CMIP5 models and two satellite products.

Model	CESM	ESM2M	ESM2G	IPSL	NorESM1	MPIM	VGPM	SPGANT[a]
Global								
EP	7.97	7.78	5.27	7.02	8.00	8.26	8.20	N/A
NPP	56.3	82.2	66.5	33.6	41.0	57.9	45.42	N/A
EP/NPP	0.14	0.095	0.08	0.21	0.20	0.14	0.18	N/A
40–60° N								
EP	0.71	0.83	0.53	0.75	0.66	0.51	1.47	N/A
NPP	3.85	4.71	3.92	2.42	3.45	3.77	4.97	N/A
EP/NPP	0.19	0.18	0.14	0.31	0.19	0.13	0.30	N/A
60–90° N								
EP	0.34	0.33	0.21	0.19	0.15	0.08	0.46	N/A
NPP	1.48	1.35	0.95	0.58	0.74	0.75	1.29	N/A
EP/NPP	0.23	0.24	0.22	0.33	0.20	0.11	0.36	N/A
40–60° S								
EP	1.25	1.18	0.82	1.42	1.93	1.77	1.60	2.85
NPP	6.77	9.36	8.53	5.24	10.3	12.5	6.01	8.81
EP/NPP	0.18	0.13	0.10	0.27	0.19	0.14	0.27	0.32

[a] SPGANT totals are only shown for the 40–60° S band because the algorithm is optimized for the Southern Ocean but not well validated in the Northern Hemisphere.

Among the terms in Eq. (2), $F_{O_2,total}$ was provided outright by the ESMs and the thermal component $F_{O_2,therm}$ can be derived easily from standard ESM output following the approach described above for N_2. The remaining terms, $F_{O_2,NCP}$ and $F_{O_2,vent}$ are more challenging to estimate from available ESM output. In Nevison et al. (2012a), $F_{O_2,NCP}$ was estimated from EP multiplied by a molar ratio of 1.4 mol O_2 per mol C exported. The assumption that $F_{O_2,NCP} = 1.4\,EP$ was shown in Nevison et al. (2012a) to yield reasonable results for EP derived from satellite data (and indeed was applied to the satellite data described below in Sect. 2.3), but this approach proved unsatisfactory for EP_{100} from the ESMs, especially in the Southern Ocean as discussed further below, since it yielded an atmospheric signal that was unreasonably small. The assumption also led to phasing uncertainties for some models (IPSL, NorESM1 and MPIM) that use finite sinking velocities for particulate organic carbon (as opposed to instantaneous vertical redistribution, as assumed, e.g., by CESM) with a resulting delay in EP_{100} relative to NPP. Since the timing of $F_{O_2,NCP}$ is likely to be more closely related to NPP than EP_{100} (Nevison et al., 2012a), we estimated $F_{O_2,NCP}$ from the ESMs alternatively as $1.4\,EP_{100}$ and $1.4\,ef^*$ NPP, where NPP is the standard, vertically integrated ESM output variable and ef is the model-specific annual mean EP_{100}/NPP ratio, integrated over the 40–60° N or 40–60° S latitude band for northern and southern stations, respectively (Table 1).

Finally, $F_{O_2,vent}$ in principle can be estimated as a residual of the other three terms in Eq. (2). $F_{O_2,vent}$ was estimated with reasonable success at the northern hemispheric sites, but generally looked unreasonable in the Southern Ocean for most models, with the exception of the IPSL. The signals were judged to be unreasonable on the basis of whether the APO_{vent} term, if estimated as a residual from Eq. (3), dominated the APO_{NCP} term in driving the springtime rise in APO. In reality, the APO_{NCP} term must be primarily responsible for this rise (Keeling et al., 1993; Bender et al., 1996; Nevison et al., 2012a). We therefore do not attempt to explicitly resolve or present APO_{vent} signals in the Southern Hemisphere. While the problems with APO_{vent} necessarily imply a corresponding problem in one or both of the other component terms APO_{NCP} and APO_{therm}, as discussed below, the shape of these latter terms is still informative and is less sensitive to the uncertainties inherent in the residually estimated APO_{vent} term.

2.3 Satellite ocean color data

The primary output product of satellite ocean color measurements historically has been the concentration of chlorophyll *a* (Chl), which is also the main input to most satellite-based ocean primary productivity models (Behrenfeld and Falkowski, 1997). However, the standard Chl product based on empirical band ratios of reflectances represents primarily the coefficient of total absorption of blue light and is inherently biased if the distributions of the optically active components deviate from the global mean (Lee et al., 2011; Siegel et al., 2005; Sauer et al., 2012). In the Southern Ocean

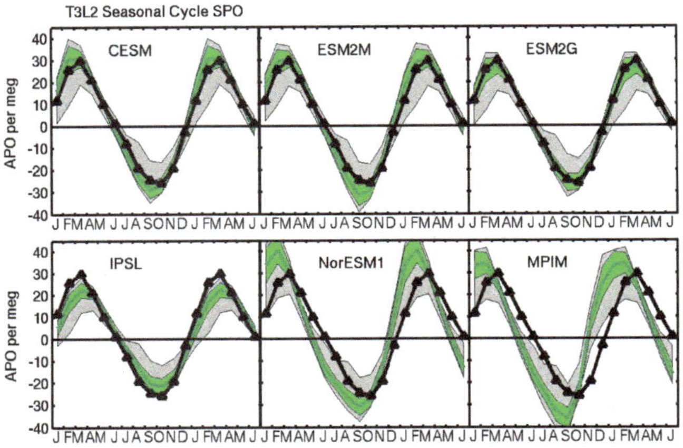

Figure 2a. Results of the pulse-response code forced by O_2, N_2 and CO_2 air–sea fluxes from six ESM ocean biogeochemistry model components. The dark green line and light green window show the mean and standard deviation, respectively, of the nine ATMs participating in both T3L2 and APO TransCom. The amplitudes are scaled for each ATM and monitoring site based on the validation exercise described in Sect. 2.2.2 and illustrated in the Supplementary material. The gray window shows the full range of responses from all 13 T3L2 ATMs, uncorrected based on the APO TransCom validation exercise. The heavy black line shows the observed APO mean annual cycle. **(a)** Results at South Pole, compared to SIO observations.

the standard Chl algorithms underestimate in situ Chl by 2–3 times (Mitchell and Kahru, 2009), whereas in the Arctic they overestimate it (Mitchell, 1992). These errors are directly transferred into errors in estimates of NPP and EP.

For the Southern Hemisphere we used an empirical Chl algorithm (Scripps Photobiology) that was tuned to in situ Chl in the Southern Ocean and spatially blended with the standard SeaWiFS OC4 algorithm (Kahru and Mitchell, 2010). The same blending scheme was applied when blending NPP between two versions of the Vertically Generalized Productivity Model (VGPM) algorithm (Behrenfeld and Falkowski, 1997): the Southern Ocean version and the low-latitude version of Kahru et al. (2009). EP was calculated using a modified version of the Laws (2004) model according to Nevison et al. (2012a). The mean annual cycles for Chl, NPP and EP were calculated for 1997–2009 using data derived from SeaWiFS.

For the Northern Hemisphere we used NPP data calculated according to the standard VGPM using MODIS-Aqua Chl. NPP was downloaded from http://science.oregonstate.edu/ocean.productivity. EP was calculated according to Dunne et al. (2005). The mean annual cycles for NPP and EP were calculated for 2002–2011 using monthly composites derived from MODIS-Aqua. While the Laws (2004) and Dunne et al. (2005) methods of deriving EP are not identical, they both estimate export efficiency as a function of sea surface temperature and NPP, are fitted to in situ data, and generally produce similar estimates. In Nevison et al. (2012a) the Southern Ocean EP derived with the Laws (2004) model was modified by constraining to the bulk nutrient budget estimated in the ocean inversion of Schlitzer (2002). That reduced the unrealistically high export efficiency of the Laws (2004) model at cold temperatures and brought it into closer agreement with the Dunne et al. (2005) export efficiency.

Both the SPGANT and VGPM/OSU (Oregon State University) satellite algorithms for NCP were converted to air–sea O_2 fluxes using $F_{O_2,NCP} = 1.4\,NCP$, where 1.4 refers to the molar ratio between O_2 produced and carbon fixed in photosynthesis. $F_{O_2,NCP}$ was used to force the pulse-response code to estimate the corresponding APO signal associated with NCP as per Nevison et al. (2012a).

2.4 APO data

APO is a unique atmospheric tracer of ocean biogeochemistry that is calculated by combining high precision O_2 and CO_2 data according to $APO = O_2 + 1.1CO_2$ (Stephens et al., 1998). By design, APO is mostly insensitive to exchanges with the land biosphere, which have a nearly fixed stoichiometry that produces compensating changes in O_2 and CO_2. In contrast, the exchanges of O_2 and CO_2 across the air–sea interface are not strongly correlated, largely because variability in dissolved CO_2 is strongly damped by carbonate chemistry in seawater on seasonal timescales. As a result, seasonal variability in APO reflects changes in atmospheric oxygen occurring almost solely due to oceanic processes (Manning and Keeling, 2006).

Atmospheric O_2 data, reported in terms of deviations in the O_2 / N_2 ratio, were obtained from the Scripps Institution of Oceanography (SIO) and Princeton University (PU) networks. Data are available from the early to mid-1990s, depending on the station (Bender et al., 2005; Manning and Keeling, 2006). In Fig. 2, we use SIO data from SPO, PSA and ALT and PU data from MQA and BRW. Details of the station locations and time spans of data used to calculate the mean seasonal cycle are listed in Table S2 and shown in Fig. 1. For MQA (1997–2007) and BRW (1993–2008), the time spans overlapped mostly but not perfectly with the CMIP5 model output (1994–2005) and the satellite data (1997–2009 for SPGANT, 2002–2011 for VGPM).

APO was calculated according to

$$APO = \delta(O_2 / N_2) + \frac{1.1}{X_{O_2}} CO_2, \qquad (4)$$

where $\delta(O_2 / N_2)$ is the relative deviation in the O_2 / N_2 ratio from a reference ratio in per meg units, $X_{O_2} = 0.2094$ is the O_2 mole fraction of dry air (Tohjima et al., 2005), CO_2 is the mole fraction of carbon dioxide in parts per million (μmol mol^{-1}), and 1.1 is a qualitative estimate of the $-O_2 : C$ ratio of terrestrial respiration and photosynthesis. Mean seasonal cycles for observed APO were obtained using the same detrending and averaging methodology described in Sect. 2.2.1. The uncertainty in the observed mean seasonal

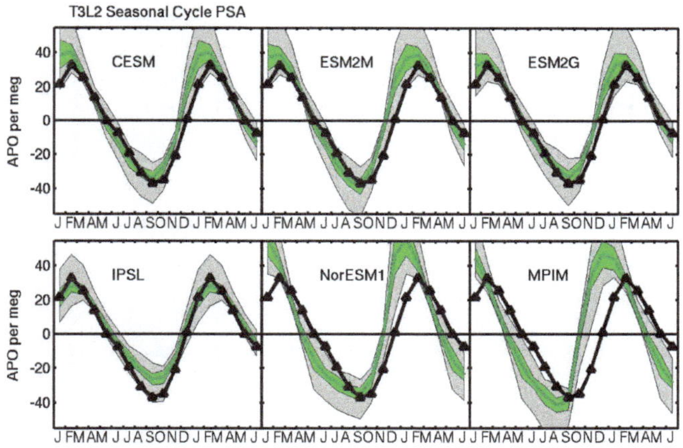

Figure 2b. Results at Palmer Station (64.9° S, 64° W), compared to SIO observations.

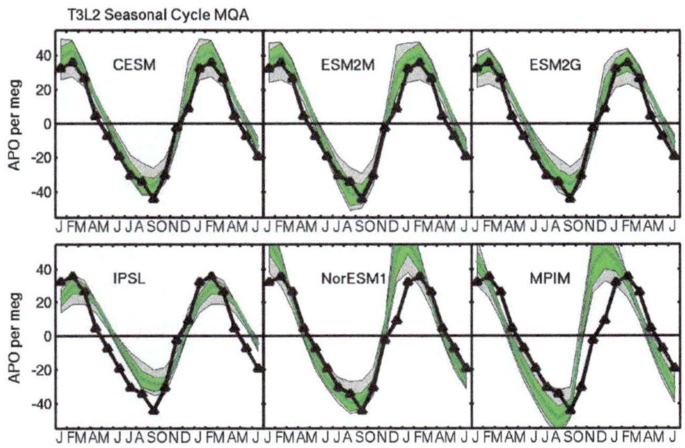

Figure 2c. Results at Macquarie Island (54.5° S, 159° E), compared to PU observations.

cycles over the timespan of available data is less than 6 % at extratropical latitudes, reflecting a combination of instrumental precision, synoptic variability and interannual variability (IAV) in the seasonal cycle. The current study is focused on the mean seasonal cycle in APO as a first-order challenge for the CMIP5 ocean models. Here, model, APO and satellite seasonal cycles are evaluated over roughly comparable periods that are dictated by data availability. The examination of IAV is deferred to future research, which will require ATM simulations of APO driven by interannually varying meteorology.

2.5 Phase metrics

The time of year of the seasonal maximum in APO and NPP was used as a phase metric. For APO, monthly mean, station-specific time series, both modeled and observed, were fit to a third-order polynomial plus first two-harmonics function. The harmonic components of the fit were used to construct a mean seasonal cycle with daily resolution and the day of the

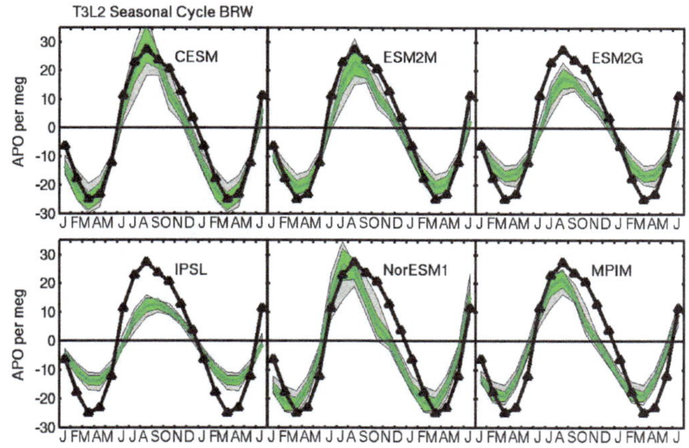

Figure 2d. Results at Barrow, Alaska (71.3° N, 156.6° W), compared to PU observations.

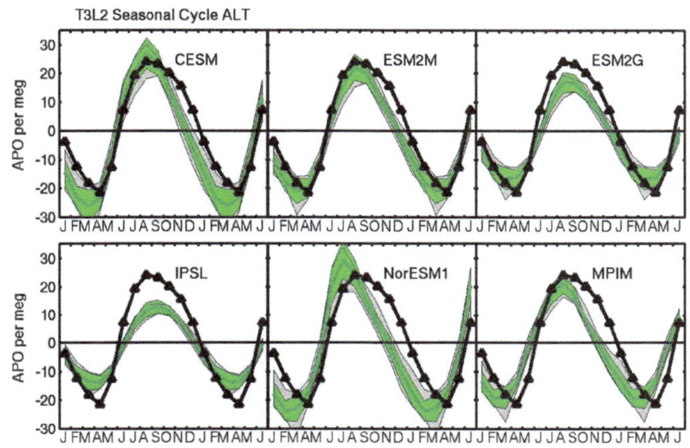

Figure 2e. Results at Alert, Canada (82.5° N, 62.5° W), compared to SIO observations.

seasonal maximum was identified. The same approach was used to derive the day of the seasonal NPP maximum, except that the fit was applied to monthly mean satellite-derived and ESM NPP integrals summed from 40 to 60° S and 40 to 60° N, which were compared to the APO phase metric at southern and northern stations, respectively.

3 Results

3.1 APO comparison to ESMs

The APO cycles estimated from the six sets of ESM air–sea fluxes were compared to observations at three Southern Ocean and two northern monitoring sites (Fig. 2). In these plots, the green envelope reflects our best estimate of the ATM uncertainty in the ocean model APO signal based on the nine scaled ATM results, while the gray window reflects the more complete range of uncertainty using all 13 unscaled ATM results. In general, the distinction between the green and gray windows is only moderately important, as the observed APO cycle in most cases either falls within both envelopes or lies outside of both envelopes.

The MPIM and related NorESM1 ocean biogeochemistry models are examples in which the observed APO cycle lies outside both ranges of uncertainty at all five evaluation sites (Fig. 2, lower middle and right panels). For these models, the rise in the APO cycle occurs too early in the springtime in both hemispheres, while the overall amplitude of the cycle is too large at all the southern stations. Here, it is notable that the MPIM APO amplitude would be even larger in the Southern Ocean if it were not offset by the unrealistically large seasonal cycle in oceanic CO_2 described above. The large CO_2 cycle, however, does not substantially alter the phase of APO, which is determined mainly by the timing of the O_2 fluxes.

The IPSL is another ocean biogeochemistry model for which the observed APO cycle lies outside of both the best guess and full range of uncertainty at all monitoring sites, with the exception of Palmer Station (64.9° S), where observed APO falls within the wider gray window of uncertainty (Fig. 2b, lower left panels). Unlike the MPIM and NorESM1, the rise in the IPSL APO cycle occurs somewhat later in the springtime than observed, while the overall amplitude of the cycle tends to be underestimated. The underestimate is mild at all the southern stations, and even falls within the broader range of uncertainty at PSA, but is more pronounced at the northern monitoring sites, where the IPSL amplitude is too small by nearly a factor of 2.

The CESM is the top performer among the six ESMs evaluated, consistently yielding green (gray) windows that encompass the observed APO cycle at most (all) of the five monitoring sites (Fig. 2, upper left panels). The GFDL ESM2M (depth-based coordinates) is the second most consistent performer, yielding cycles that generally agree with observations, with exceptions at BRW, where ESM2M tends to mildly underestimate the depth of the APO trough, and at PSA, where the rise in the APO cycle may be up to 1 month too early. The sigma-coordinate GFDL ESM2G model is the third best performer, capturing the observed APO cycle relatively well at most southern stations, but underestimating the seasonal amplitude at the northern stations.

3.1.1 Regional analysis of APO cycle

The matrix method can partition the ocean model APO cycles into regional contributions from the 11 ocean regions used in T3L2. At the southern stations of SPO, PSA and MQA, this partitioning reveals, not surprisingly, that the Southern Ocean (defined as all ocean regions south of 44° S) dominates the APO cycle (not shown). However, at BRW and ALT at least three regions make important contributions, including the temperate North Pacific (extending from 15° N to the Bering Strait around 65° N and thus including the subpolar region), the temperate North Atlantic (extending from 15 to 48° N) and the Northern Ocean (including the Arctic

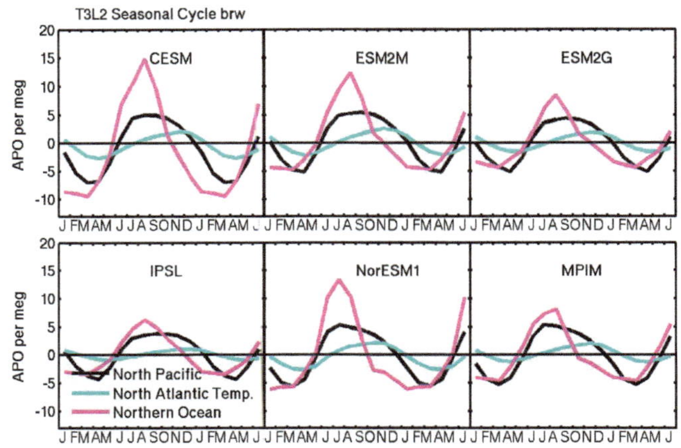

Figure 3. Partitioning the APO cycle at Barrow, Alaska, into its main regional contributions, North Pacific (black), temperate North Atlantic (cyan) and Northern Ocean (magenta), which includes the North Atlantic north of 48° N and the Arctic Ocean. All curves reflect the unscaled model mean of the 13 ATMs used in the matrix method.

Ocean and the North Atlantic north of 48° N). The Northern Ocean is the most important contributor to the APO seasonal cycle at both BRW (Fig. 3) and ALT and is by far the most variable component among the six ESMs. The largest APO amplitudes in the Northern Ocean are produced by the CESM and NorESM1, which are the only two models that capture the total observed APO amplitude at BRW (Fig. 2d).

3.1.2 Partitioning of APO cycle among component signals

To probe further into the underestimate of the APO amplitude at BRW by most of the ESMs, we partitioned APO into thermal and NCP-related components, as described in Sect. 2.2.3 (Fig. 4). A comparison of the CESM and ESM2M in Fig. 4 indicates that both have similar APO_{therm} and APO_{NCP} signals, but that the CESM captures total APO more or less correctly while ESM2M underestimates the total APO amplitude. By inference, the missing APO_{vent} term accounts for the difference. However, as discussed in Sect. 2.2.3, APO_{vent} can be estimated only as a residual of three other terms using standard CMIP5 output and thus its shape and phasing are sensitive to even small uncertainties in those other terms. Thus, the residual ventilation curves in Fig. 4 should be interpreted with caution (e.g., the MPIM curve is clearly unreasonable in phasing). The four remaining ESMs have APO_{NCP} cycles of similar or smaller amplitude than the CESM, which in the case of the ESM2G and MPIM is due primarily to their relatively low export fraction ratios (ef ratios), and all these models substantially underestimate the total APO amplitude at BRW. This suggests that these models probably also underestimate some combination of deep ventilation and NCP.

A similar partitioning of APO was attempted in the Southern Ocean, but the estimation of APO_{NCP} from model EP_{100}

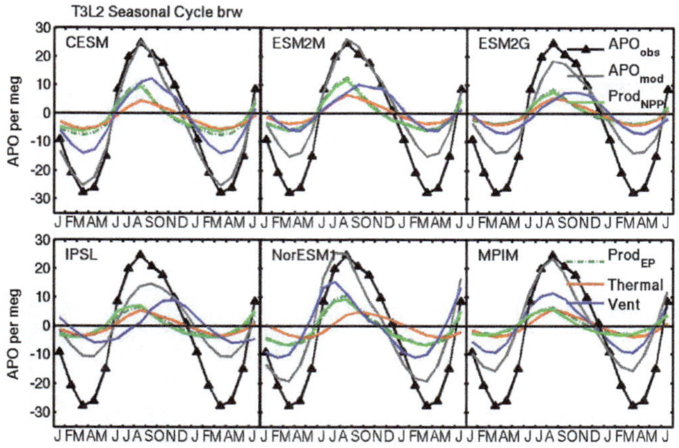

Figure 4. Partitioning of the model mean APO cycle into NCP, thermal and residual ventilation components at Barrow, Alaska. The APO_{NCP} components are estimated alternatively based on ocean model EP at 100m ($Prod_{EP}$ light green, solid curve) and vertically integrated NPP ($Prod_{NPP}$) scaled by the mean ratio of EP_{100} / NPP (f ratio) between 40 and 60° N of the given ocean model (dark green, dashed curve). All components were translated into atmospheric signals as described in Sect. 2.2.3. Also shown is APO_{vent} (blue), calculated as a residual of APO − APO_{NCP} − APO_{therm}. With the exception of observed APO, all curves reflect the unscaled mean of the 13 ATMs used in the matrix method.

generally did not give plausible results in this region. This problem is discussed in more detail in Sect. 4.

3.2 Satellite data compared to ESMs

Estimates of NPP display a wide variety of spatial patterns among models and satellite data (Fig. 5). Global totals range over more than a factor of 2 (34–82 $Pg\,C\,yr^{-1}$) among the ESMs, with most models tending to exceed the VGPM satellite-based estimate of 45 $Pg\,C\,yr^{-1}$ (Table 1). Global EP is more consistent among the models, with a value around 8 $Pg\,C\,yr^{-1}$ in most cases, in good agreement with the satellite-based estimate. Global EP converges among the ESMs because the model with the highest global NPP (ESM2M) has the smallest ef ratio of <0.1 and the models with the lowest global NPP (IPSL, NorESM1) have the largest ef ratios of about 0.2 (Table 1).

The high global NPP totals in the ESMs are driven in large part by high tropical NPP values, which generally are not reflected in the satellite data except along coastlines (Fig. 5). In this paper, we focus on the 40–60° latitude bands, which are more important than the tropics in driving the seasonal cycles in NPP, EP (NCP) and APO (Garcia and Keeling, 2001; Anav et al., 2013). In the Southern Ocean 40–60° S band, global NPP ranges among ESMs from 5.2 to 12.5 $Pg\,C\,yr^{-1}$, encompassing the satellite-based estimates (Table 1, Fig. 6). However, the ESMs tend to underestimate EP relative to the satellite-derived values, particularly the SPGANT Laws (2004) model, due largely to the small model ef ratios. In the 40–60° N band, the ESMs generally underestimate both NPP and ef ratios relative to the satellite-derived values. This combination leads to model EP values that are smaller than satellite EP by a factor of 2 on average (Table 1). In both hemispheres, the model NPP maximum tends to occur earlier than the satellite-derived maximum, with some models (IPSL, MPIM) predicting a maximum that is up to 1–2 months early (Fig. 6).

3.3 Combining APO and satellite data

In the previous sections we considered APO and satellite data as separate evaluation metrics for ESMs. Below we consider the two as combined metrics. While this analysis is limited by uncertainties in the absolute magnitude of satellite NPP and EP/NCP and our imperfect ability to partition the ESM total APO signal into its NCP and other components, it nevertheless provides some additional insight into the behavior of the ESMs.

3.3.1 Phase metrics

The phase metrics defining the timing of the observed and model seasonal maximum in APO reveal characteristic patterns for each ESM, which are relatively consistent across APO monitoring sites (Fig. 7). The APO seasonal maxima of the MPIM and NorESM1 are earlier than observed by about 1 month and 3 weeks, respectively, on average, while the IPSL APO maximum (with the exception of PSA) tends to be later than observed by 2–3 weeks. The remaining models, CESM, ESM2M and ESM2G, have seasonal APO maxima that are relatively consistent with observations, although with some variation among different stations.

The observed seasonal maximum of NPP occurs about 30–40 days earlier than the observed APO maximum in the Southern Ocean stations and about 50 days earlier at BRW and ALT. Of the models, ESM2G, CESM and ESM2M capture the phase of the NPP maximum to within about 1–3 weeks, although as noted above in Fig. 6 the model NPP maxima tend to occur earlier than the satellite-based maxima. In the MPIM, the NPP maximum is about 1 to 1.5 months earlier than observed, and the APO maximum is also corresponding early (Fig. 7). The IPSL is an outlier from the general slope of the APO vs. NPP phase relationship, as defined by the rest of the ESMs. The IPSL NPP maximum occurs about 40 days earlier than observed in the Southern Hemisphere and nearly 2 months earlier than observed in the Northern Hemisphere, but the IPSL, curiously, also has the latest APO seasonal maximum of any of the models. NorESM1 is another outlier in the opposite direction off the general APO vs. NPP phase slope, at least in the Northern Hemisphere. There, NorESM1's seasonal maximum in NPP has a relatively small lag from the APO maximum compared to the other models. NorESM1 is also unusual in that

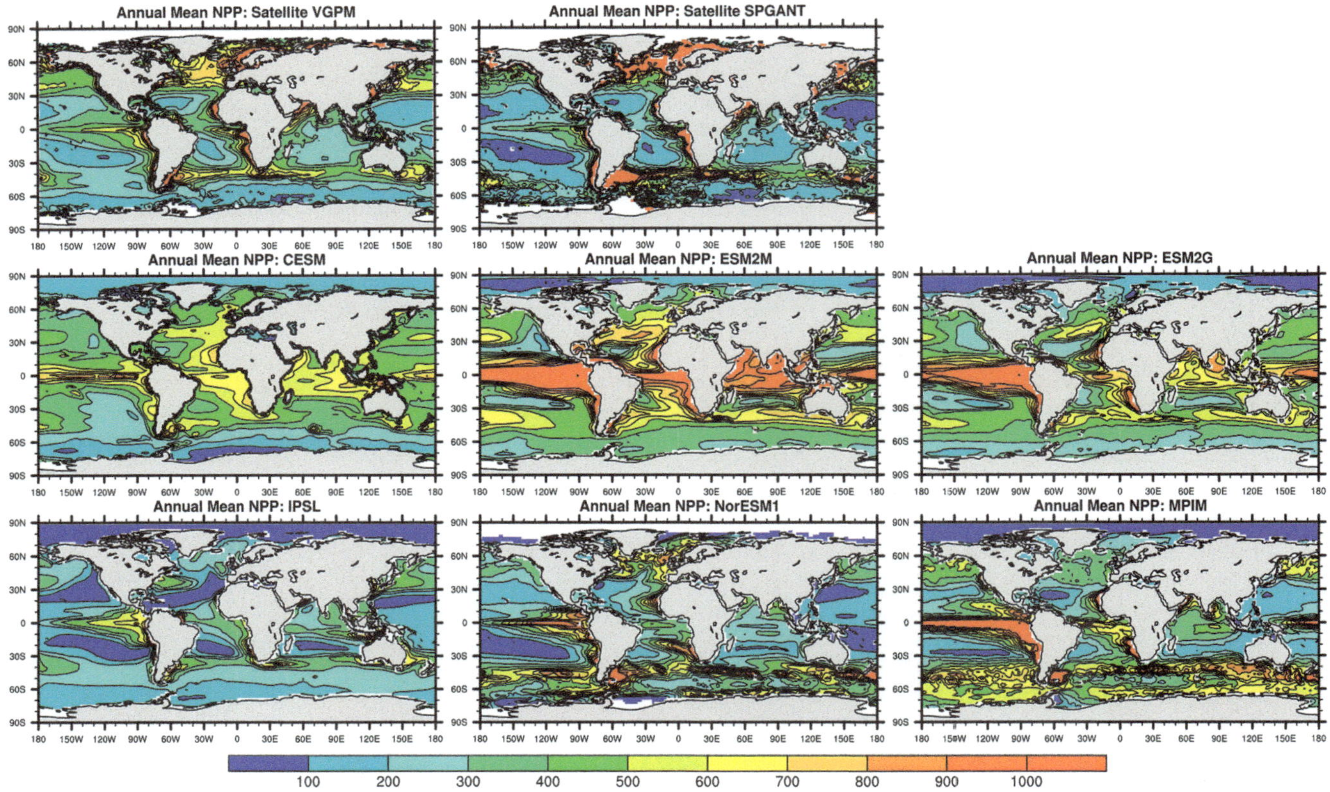

Figure 5. Annual mean NPP (in mg C m^{-2} day^{-1}). Top row: MODIS-Aqua data input to the VGPM NPP model and **(b)** SeaWiFS data input to the SPGANT algorithm as described in Nevison et al. (2012a). Rows 2 and 3 show the corresponding NPP fields from six ESMs for the mean of 1997–2005.

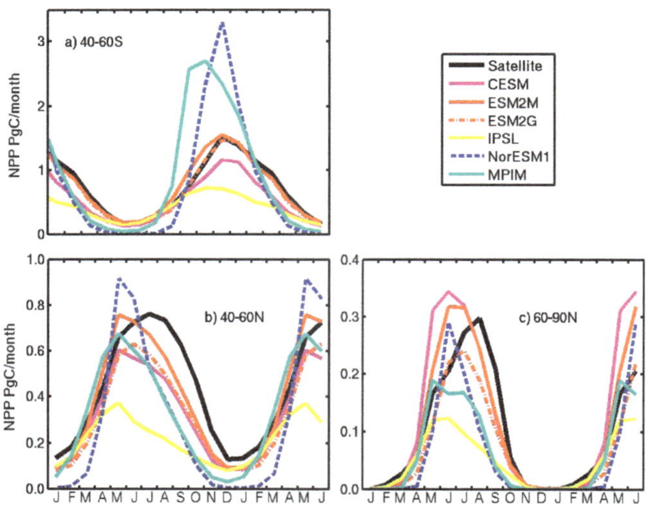

Figure 6. Comparison of the NPP (PgC month) mean annual cycle as simulated by ESMs and satellite-derived observations integrated over **(a)** 40–60° S, **(b)** 40–60° N, **(c)** 60–90° N. The satellite data are from SPGANT Laws (2004) model in panel **(a)** and VGPM Dunne model in panels **(b–c)**.

the APO_{therm} seasonal maximum at Barrow occurs about 1 month later than in any of the other ESMs (Fig. 4).

3.3.2 Seasonal amplitudes

In addition to evaluating the phasing of the ocean model APO and NPP cycles, we examined the amplitude of the cycles, with the caveat that the absolute magnitude of satellite-based NPP is not well determined and at present provides a relatively weak constraint on the models. Furthermore, the APO seasonal amplitude in principle is more closely related to NCP (or EP) than NPP. However, we chose NPP for the seasonal amplitude analysis due to the strong discrepancies in ef ratio among models and satellite data indicated in Table 1, which may unduly bias the results.

A cross plot of the seasonal amplitude in APO against the seasonal amplitude of NPP integrated between 40 and 60° S suggests a strong correlation between the amplitudes of APO and NPP among the ocean biogeochemistry models, with larger NPP amplitudes associated with larger APO cycles. The strong correlation holds at all Southern Ocean stations and is illustrated in Fig. 8a at MQA. The cluster of top-performing ESMs (CESM, ESM2M, ESM2G) agrees relatively well with the observed APO and SPGANT amplitudes. Meanwhile both amplitudes are underestimated by the IPSL and overestimated by the NorESM1 and MPIM.

Cross plots of the seasonal amplitudes of APO and NPP in the Northern Hemisphere reveal that these amplitudes

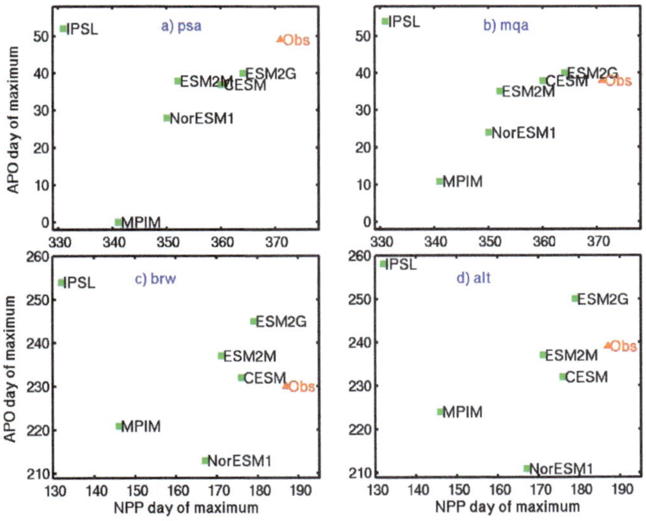

Figure 7. Day of APO maximum plotted against day of NPP maximum. The observed data point is derived from APO data at **(a)** Palmer Station, **(b)** Macquarie Island, **(c)** Barrow and **(d)** Alert plotted against satellite NPP data integrated over the 40–60° latitude band of the appropriate hemisphere.

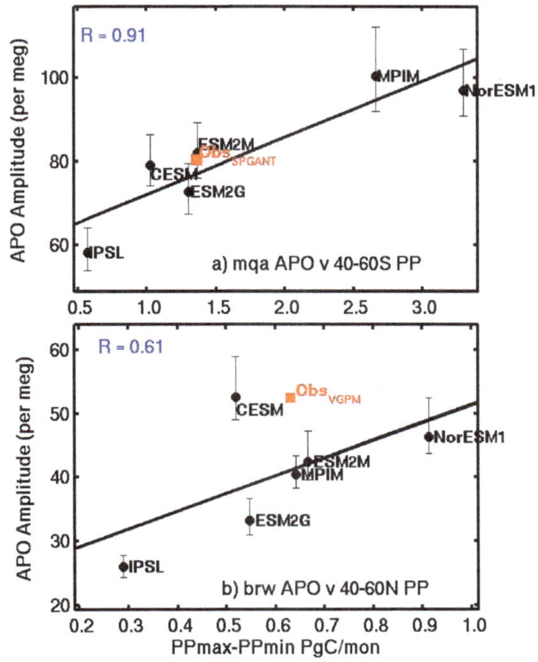

Figure 8. **(a)** Seasonal amplitude in APO at Macquarie Island (MQA), located at 54.5° S, 159° E, as estimated from the air–sea O_2, CO_2 and heat fluxes from six ESMs, plotted against the seasonal amplitude of NPP integrated from 40 to 60° S. Error bars represent the ATM uncertainty in model APO as estimated with the matrix method. The observed (Obs) data points (in red) are based on APO data from the PU network at MQA and NPP from the SPGANT satellite ocean color algorithms, as described in the text. The correlation coefficient R refers to regression through ESM points only, **(b)** Same as 8a, but plotting seasonal amplitude in APO at Barrow, Alaska, against the seasonal amplitude of NPP integrated from 40 to 60° N. The observed data point is based on APO data from the PU network and the VGPM algorithm with MODIS-Aqua input.

are positively correlated at BRW (Fig. 8b) and ALT (not shown), although the correlation is weaker than in the Southern Hemisphere. The CESM, ESM2G, ESM2M and MPIM all capture the satellite-based NPP seasonal amplitude relatively well, while both the CESM and NorESM1 capture the observed APO amplitude accurately. However, the CESM is the only model that reproduces both the NPP and APO seasonal amplitudes well relative to the observations.

4 Discussion

4.1 Northern oceans

Most ESMs tend to underestimate substantially the observed seasonal amplitude of APO at Barrow, Alaska. A combination of region-specific results (Fig. 3) and O_2 component analysis (Fig. 4) suggests that some combination of fall–winter deep ventilation and spring–summer EP in the northern oceans (defined to include the North Atlantic north of 48° N) in particular may be underestimated in many models. The combined analysis of the APO vs. NPP seasonal amplitudes (Fig. 8b) supports these conclusions and suggests that, while several models may be capturing primary production well in the northern oceans, accurate representation of EP and deep ventilation is also important for reproducing the observed APO cycle. The inference from the APO component analysis in Fig. 4 that the GFDL models may have weak ventilation in the North Atlantic appears to contradict the analysis of Dunne et al. (2012), who found robust North Atlantic Deep Water (NADW) formation in both the ESM2M and ESM2G versions, but possibly could be reconciled if the biogeochemical gradients across which deep water formation acts are too weak.

We investigated several mechanisms that might explain the differences among models in the APO cycle at high northern latitudes, including subpolar heat transport and Arctic sea ice cover. Here, stronger northward heat transport should lead to more deep ventilation, while lower sea ice cover will permit more production and ventilation in the Arctic Ocean. Subdividing the northern oceans region into the Arctic Ocean and North Atlantic components revealed that some models (IPSL and ESM2G) have a very small component (<2 per meg) of APO seasonal amplitude coming from the Arctic Ocean alone (Fig. 9). In ESM2G this may be related to the extensive winter sea ice cover, which exceeds the observed covered area reported by the National Snow and Ice Data Center (http://nsidc.org/data/seaice_index/archives.html) by about 2×10^6 km^2. However, sea ice cover is lower than observed in the IPSL, suggesting the small Arctic APO

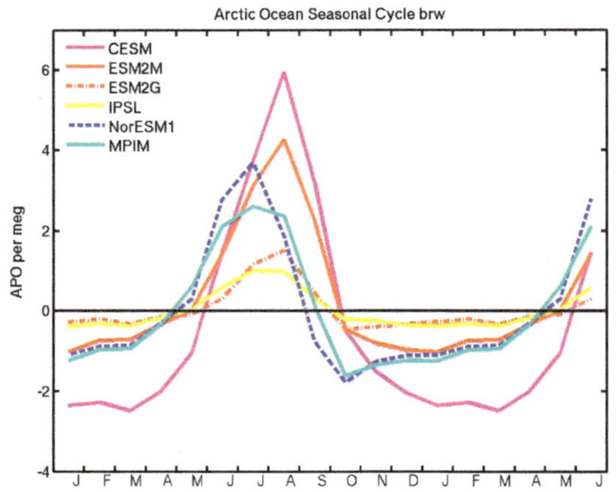

Figure 9. APO cycle at Barrow, Alaska, from the TransCom northern oceans region, restricted to latitudes north of 65° N to estimate the contribution of the Arctic Ocean. All curves reflect the unscaled model mean of 13 ATMs used in the matrix method.

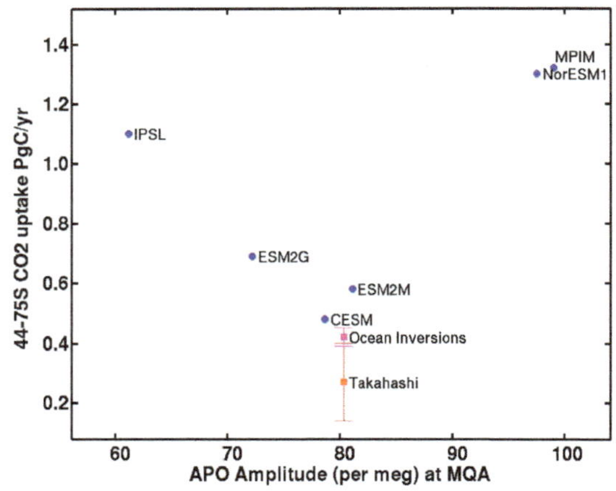

Figure 10. Annual mean CO_2 uptake in the Southern Ocean for 1997–2005 integrated from 44–75° S and plotted vs. mean APO amplitude at Macquarie Island (MQA) over the same period, as predicted by six ESMs. Independent estimates of carbon uptake from ocean inversions and observed pCO_2 databases (Lenton et al., 2013), plotted against the observed APO amplitude at MQA are shown for reference.

component in that model is more related to general underestimate of primary and EP (e.g., as shown in Figs. 6b and 8b). While it seems clear that the strong APO seasonality in the CESM can be attributed in part to its high productivity and EP in the northern subpolar and polar regions (Fig. 6 and Table 1), a full explanation for the underlying mechanisms of the CESM fidelity on APO compared to the other models is not readily apparent from surface-only data. This suggests the need for a more detailed exploration of ocean interior ventilation and biological response interactions outside the scope of the present work.

4.2 Southern Ocean

Compared to the Northern Hemisphere stations, the ESMs generally are more successful in the Southern Ocean in capturing the observed APO cycle (Fig. 2). Within the range of ATM uncertainty, at least three models, CESM, ESM2M, ESM2G (and IPSL at Palmer Station), predict seasonal APO amplitudes in agreement with observations. Although the Southern Ocean APO amplitude in these models varies over as much as 20 per meg, we currently are not able to distinguish which of the underlying air–sea O_2 flux fields is the most realistic, due to the uncertainty associated with translating these fluxes into an atmospheric signal using TransCom3 era model responses to uniformly distributed regional fluxes. However, even with our current matrix method, the APO constraint is sufficiently robust to indicate that the NorESM1 and MPIM substantially overestimate some combination of production and deep ventilation in the Southern Ocean, while the IPSL probably tends to underestimate these fluxes (Table 1, Fig. 8a). Notably, the ESMs that reproduce APO the best in the Southern Ocean tend to predict a smaller net carbon uptake between 44 and 75°, and are in better agreement with independent estimates (Lenton et al., 2013) of carbon uptake from ocean inversions and observed pCO_2 databases (Fig. 10).

Reducing ATM uncertainty is a challenge that potentially can be addressed by using column-integrated APO signals from aircraft data (Wofsy, 2011), or conversely, by using vertical profiles to identify top-performing ATMs (Stephens et al., 2007). In addition, the spread in ATM results has been reduced substantially for CO_2 inversions using post-TransCom3-era ATMs (Peylin et al., 2013), suggesting that ATM uncertainty also may be reduced for forward simulations of APO. If this is the case, then new forward simulations with several different modern-era ATMs may be sufficient to characterize ATM uncertainty. Alternatively, it may be valuable to continue with a matrix-based approach, using basis functions from many ATMs, but with redefined regional boundaries that are not defined based simply on latitude, as in T3L2 (Fig. 1), but rather that correspond to the biogeography of major ocean regions (Fay and McKinley, 2014). The definition of such basis functions could help extend the utility of the matrix approach to lower latitude APO monitoring sites and allow for the partitioning of the Southern Ocean into multiple regions defined around biogeochemical function, while still retaining the advantages of the matrix method, i.e., the ability to quickly and easily compare multiple ATMs forced with the same air–sea fluxes.

A second complication in the Southern Ocean analysis is that the EP_{100} values reported by the ESMs clearly are not directly comparable to satellite NCP(EP) data, particularly our SPGANT product, and thus can not be translated with confidence into air–sea O_2 fluxes associated with NCP. A likely problem is that the 100 m depth horizon used to compute EP may not be comparable across satellite algorithms and ocean biogeochemistry models. EP_{100} will underestimate the model's true NCP-related O_2 outgassing flux if organic matter is respired as it sinks from the actual model mixed layer depth to 100 m depth (Najjar et al., 2007). It is also puzzling that the ef ratios predicted by the ESMs (Table 1) appear to have decreased considerably in some cases relative to those reported for earlier versions of the same models (Steinacher et al., 2010). For example, the Southern Ocean ef ratios for the MPIM and IPSL in that earlier study were about 0.2 and 0.4, respectively, compared to 0.14 and 0.27, respectively, in the current study. The mean global ef ratio for the six ESMs in the current study is only 0.14 and, even in the Southern Ocean, is only 0.17 on average, compared to satellite-based estimates of 0.18 globally and about 0.3 at high latitudes.

The small ef ratios in the GFDL models (of less than 0.1 globally and only 0.10 to 0.13 in the Southern Ocean) appear consistent with the relatively deep summer mixed layer depths (MLDs) in the Southern Ocean, which even at their minimum are often deeper than 100 m in both ESM2M and ESM2G (Dunne et al., 2012). In the CESM the Southern Ocean summer MLDs are generally shallower than 100 m and in many regions are only around 10–40 m deep (Moore et al., 2013). The shallower summer MLDs may contribute to the CESM's larger ef ratio of 0.18, although this ratio is still small compared to the satellite-based estimates. The small GFDL ef ratios may also be related to an overvigorous picophytoplankton component wherein a prochlorococcus-like form is capable of competing relatively well even in cold polar waters. Small picophytoplankton are more likely to be reoxidized and remineralized within the mixed layer, whereas larger, heavier microphytoplankton (e.g., diatoms) are more likely to be exported out of the oceanic mixed layer (Uitz et al., 2010).

4.3 Phase relationships

While much of our analysis focuses on the seasonal amplitude of APO and NPP at mid- to high latitudes, both of these metrics involve relatively large uncertainty. This derives from TransCom3-era uniform flux ATM uncertainty in the case of APO, while for NPP the uncertainty results from the lack of strong constraints on the absolute magnitude of the satellite fluxes. In contrast, we have relatively high confidence in the phasing of model APO, as represented by the matrix method (see the Supplement) and in NPP observationally derived from satellite data, based on the close correspondence in phasing between the SPGANT and VGPM algorithms. For these reasons, we used a phase metric, i.e., the timing of the seasonal maximum, to examine relationships between observed and model APO and NPP. As in the seasonal amplitude analysis, MPIM, NorESM1 and IPSL displayed phasing patterns that tended to deviate from observations and the other three top-performing models, albeit in diverging ways. A complete diagnosis of the model physics responsible for the phasing anomalies (e.g., IPSL's early NPP maxima and late APO maxima) described in Sect. 3.3.1 is beyond the scope of this paper. Here we note mainly that the phase metrics are a robust and relatively good indicator of overall model performance with respect to APO.

5 Summary

We have used measurements of the seasonal cycles in APO to challenge and test the ocean model components of six ESMs. The model/data comparison reveal that three of the ESMs tested reproduce the observed cycles reasonably well, within the range of ATM uncertainty, while three do not. ESM performance in general is more favorable in the Southern Hemisphere than in the Northern Hemisphere, where most models appear to underestimate the wintertime ventilation of O_2-depleted deepwater that drives the declining branch of the APO seasonal cycle and many may also underestimate both primary and EP, particularly at high northern latitudes. We used NPP and NCP(EP) products derived from satellite ocean color data as complementary constraints on the models in an effort to tighten the APO constraint, which reflects a combination of production and ventilation processes. However, while the satellite data provide relatively strong constraints with respect to phasing, they are more uncertain with respect to the absolute magnitudes of NPP and NCP(EP).

At least two primary uncertainties limit our ability to place stronger constraints on ocean model biogeochemistry based on currently available information from APO and satellite data: (1) the relatively large ATM uncertainty involved in translating air–sea O_2 fluxes into APO signals. (2) The uncertainty in how model EP_{100} relates to the true model $F_{O_2,NCP}$ flux and how this relationship varies across models and satellite algorithms. The first of these, ATM uncertainty, is large, as quantified using our TransCom3-based matrix method. However, it probably has been overstated in previous analyses, which in some cases went so far as to suggest that APO does not provide a useful constraint on ocean model fluxes (e.g., Naegler et al., 2007). Further, ATM uncertainty could be reduced substantially in future work with modern ATMs and O_2-specific flux patterns, or with new regional boundaries defined based on ocean biogeography rather than simple latitude. Even within the limits of our current approach, we have shown that half of the six ESMs tested here produce APO cycles whose mismatch with observed APO clearly transcends ATM uncertainty, suggesting underlying deficiencies in those models' physics and biogeochemistry.

Improving the understanding of the relationship between model air–sea O_2 fluxes and quantities like NPP, NCP and EP is a more tractable problem that can be dissected with appropriate model diagnostics, e.g., as per Manizza et al. (2012). In the current analysis, using standard CMIP5 output from six ocean biogeochemistry models, we encountered difficulties in relating F_{O2} to EP and NCP, which hindered our ability to diagnose the mechanisms responsible for model performance and to compare ESM-derived APO_{NCP} directly to satellite-based APO_{NCP} signals. Extending model-derived insights to satellite products likely will require a shift in emphasis from EP at an arbitrary reference depth to near-surface processes like NCP, which are more relevant for exchanges of O_2 and CO_2 at the air–sea interface and more directly related to upward radiances detected by satellites.

Acknowledgements. The authors gratefully acknowledge the CMIP5 ocean modelers for providing the output that made this project possible. In particular, we thank Keith Lindsay, Laure Resplandy and Cristoph Heinze for their assistance. We also thank Britt Stephens and an anonymous reviewer for their helpful comments, which much improved the manuscript. Lastly, we thank Michael Bender and Robert Mika for providing APO data and acknowledge support from NASA Ocean Biology and Biogeochemistry grant NNX11AL73G.

Edited by: M. Heimann

References

Anav, A., Friedlingstein, P., Kidston, M., Bopp, L., Ciais, P., Cox, P., Jones, C., Jung, M., Myneni, R., and Zhu, Z. : Evaluating the Land and Ocean Components of the Global Carbon Cycle in the CMIP5 Earth System Models, J. Climate, 26, 6801–6843, 2013.

Arora, V. K., Boer, G. J., Friedlingstein, P., Eby, M., Jones, C., Christian, J., Bonan, G., Bopp, L., Brovkin, V., Cadule, P., Hajima, T., Ilyina, T. Lindsay, K, Tjiptura, J., and Wu, T.: : Carbon–Concentration and Carbon–Climate Feedbacks in CMIP5 Earth System Models. J. Climate, 26, 5289–5314, 2013.

Aumont, O. and L. Bopp: Globalizing results from ocean in situ iron fertilization studies, Glob. Biogeochem. Cy., 20, GB2017, doi:10.1029/2005GB002591, 2006.

Battle, M., Mikaloff Fletcher, S., Bender, M. L., Keeling, R. F., Manning, A. C., Gruber, N., Tans, P. P., Hendricks, M. B., Ho, D. T., Simonds, C., Mika, R., and Paplawsky, B.: Atmospheric potential oxygen: New observations and their implications for some atmospheric and oceanic models, Glob. Biogeochem. Cy., 20, GB1010, doi:10.1029/2005GB002534, 2006.

Behrenfeld, M. J. and Falkowski, P. G.: Photosynthetic rates derived from satellite based chlorophyll concentration, Limnol. Oceanogr., 42, 1–20, 1997.

Bender, M., Ellis, J. T., Tans, P. P., Francey, R. J., and Lowe, D.: Variability in the O_2 / N_2 ratio of southern hemisphere air, 1991–1994: implications for the carbon cycle, Global Biogeochem. Cy., 10, 9–21, 1996.

Bender, M., Ho, D., Hendricks, M. B., Mika, R., Battle, M. O., Tans, P., Conway, T. J., Sturtevant, P. B., and Cassar, N.: Atmospheric O_2 / N_2 changes, 1993-02002: Implications for the partitioning of fossil fuel CO_2 sequestration, Global Biogeochem. Cy., 19, GB4017, doi:10.1029/2004GB002410, 2005.

Blaine, T. W.: Continuous Measurements of Atmospheric Ar/N_2 as a Tracer of Air-Sea Heat Flux: Models, Methods, and Data, University of California, San Diego, La Jolla, 22–66, 2005.

Bopp, L., Resplandy, L., Orr, J. C., Doney, S. C., Dunne, J. P., Gehlen, M., Halloran, P., Heinze, C., Ilyina, T., Séférian, R., Tjiputra, J., and Vichi, M.: Multiple stressors of ocean ecosystems in the 21st century: projections with CMIP5 models, Biogeosciences, 10, 6225–6245, doi:10.5194/bg-10-6225-2013, 2013.

Boyd, P. W. and Doney, S. C.: The impact of Climate Change and Feedback Processes on the Ocean Carbon Cycle, Chapter 7 ,in: The role of the Ocean Carbon Cycle in Global Change, edited by: Fasham, M. J. R., Ocean Biogeochemistry, Springer-Verlag, 157–198, 2003.

Dunne, J. P., Armstrong, R. A., Gnanadesikan, A., and Sarmiento J. L.: Empirical and mechanistic models for the particle export ratio, Global Biogeochem. Cy., 19, GB4026, doi:10.1029/2004GB002390, 2005.

Dunne, J. P., John, J., Adcroft, A., Griffies, S., Hallberg, R., Shevliakova, E., Stouffer, R., Cooke, W., Dunne, K., Harrison, M., Krasting, J., Malyshev, S., Milly, P., Phillips, P., Sentman, L., Samuels, B., Spelman, M., Winton, M., Wittenberg, A., and Zadeh, N.: : GFDL's ESM2 global coupled climate-carbon Earth System Models Part I: Physical formulation and baseline simulation characteristics, J. Clim., 25, 6646–6665, doi:10.1175/JCLI-D-11-00560.1, 2012

Dunne, J. P., John, J., Shevliakova, E., Stouffer, R., Krasting, J., Malyshev, S., Milly, P., Sentman, L., Adcroft, A., Cooke, W., Dunne, K., Griffies, S., Hallberg, R., Harrison, M., Levy, H., Wittenberg, A., Phillips, P., and Zadeh, N.: GFDL's ESM2 global coupled climate-carbon Earth System Models Part II: Carbon system formulation and baseline simulation characteristics, J. Clim., 26, 2247–2267, doi:10.1175/JCLI-D-12-00150.1, 2013.

Fay, A. R. and McKinley, G. A.: Global open-ocean biomes: mean and temporal variability, Earth Syst. Sci. Data, 6, 273–284, doi:10.5194/essd-6-273-2014, 2014.

Garcia, H. E. and Keeling, R. F.: On the global oxygen anomaly and air-sea flux, J. Geophys. Res., 106, 31155–31166, 2001.

Gruber, N. and Sarmiento, J. L.: Large-scale biogeochemical-physical interactions in elemental cycles, in: The Sea, Volume 12, edited by: Robinson, A. R., McCarthy, J. J., and Rothschild, B. J., John Wiley & Sons, Inc., New York, 337–399, 2002.

Gurney, K. R., Law, R., Denning, A. S., Rayner, P., Baker, D., Bousquet, P., Bruhwiler, L., Chen, Y.-H., Ciais, P., Fan, S., Fung, I., Gloor, M., Heimann, M., Higuchi, K., John, J., Kowalczyk, E., Maki, T., Maksyutov, S., Peylin, P., Prather, M., Pak, B., Sarmiento, J., Taguchi, S., Takahashi, T., and Yuen, C.-W.: Transcom 3 CO_2 inversion intercomparison: 1. Annual mean control results and sensitivity to transport and prior flux information, Tellus, 55, 555–579, 2003.

Gurney, K. R., Law, R., Denning, A. S., Rayner, P., Pak, B., Baker, D., Bousquet, P., Bruhwiler, L., Chen, Y.-H., Ciais, P., Fan, S., Fung, I., Heimann, M., John, J., Maki, T., Maksyutov, S., Peylin, P., Prather, M., and Taguchi, S.: Transcom 3 inversion intercomparison: Model mean results for the estimation of seasonal carbon sources and sinks, Global Biogeochem. Cy., 18, GB1010, doi:10.1029/2003GB002111, 2004.

Henson, S., Cole, H., Beaulieu, C., and Yool, A.: The impact of global warming on seasonality of ocean primary production,

Biogeosciences, 10, 4357–4369, doi:10.5194/bg-10-4357-2013, 2013.

Ilyina, T., Six, K. D., Segschneider, J., Maier-Reimer, E., Li, H., and Nuñez-Riboni, I.: Global ocean biogeochemistry model HAMOCC: Model architecture and performance as component of the MPI-Earth system model in different CMIP5 experimental realizations, J. Adv. Model. Earth Syst., 5, 287–315, doi:10.1029/2012MS000178, 2013.

Jin, X., Najjar, R. G., Louanchi, F., and Doney, S. C.: A modeling study of the seasonal oxygen budget of the global ocean, J. Geophys. Res., 112, C05017, doi:10.1029/2006JC003731, 2007.

Kahru, M. and Mitchell, B. G.: Blending of ocean colour algorithms applied to the Southern Ocean, Remote Sens. Lett., 1, 119–124, doi:10.1080/01431160903547940, 2010.

Kahru, M., Kudela, R., Manzano-Sarabia M., and Mitchell, B. G.: Trends in primary production in the California Current detected with satellite data, J. Geophys. Res., 114, C02004, doi:10.1029/2008JC004979, 2009.

Keeling, R. F., Najjar, R. P., Bender, M. L., and Tans, P. P.: What atmospheric oxygen measurements can tell us about the global carbon cycle, Global Biogeochem. Cy., 7, 37–67, 1993.

Keeling, R. F., Stephens, B. B., Najjar, R. G., Doney, S. C., Archer, D., and Heimann, M.: Seasonal variations in the atmospheric O_2 / N_2 ration in relation to the kinetics of air-sea gas exchange, Global Biogeochem. Cy., 12, 141–163, 1998.

Keeling, R. F., Koertzinger, A., and Gruber, N.: Ocean Deoxygenation in warming world, Annu. Mar. Rev. Sci., 2, 463–493, 2010.

Khatiwala, S., Primeau, F., and Hall, T.: Reconstruction of the history of anthropogenic CO_2 concentrations in the ocean, Nature, 462, 346–350, 2009.

Laws, E. A.: Export flux and stability as regulators of community composition in pelagic marine biological communities: Implications for regime shifts, Progr. Oceanogr., 60, 343–353, 2004.

Laws, E. A., Falkowski, P. G., Smith, W. O., Jr., Ducklow, H. W., and McCarthy, J. J.: Temperature effects on export production in the open ocean, Global Biogeochem. Cy., 14 1231–1246, 2000.

Lee, Z., Lance, V., Shang, S., Vaillancourt, R., Freeman, S., Lubac, B., Hargreaves, B., Del Castillo, C., Miller, R., Twardowski, M., and Wei, G.: : An assessment of optical properties and primary production derived from remote sensing in the Southern Ocean (SO GasEx), J. Geophys. Res., 116, C00F03, doi:10.1029/2010JC006747, 2011.

Lenton, A., Tilbrook, B., Law, R. M., Bakker, D., Doney, S. C., Gruber, N., Ishii, M., Hoppema, M., Lovenduski, N. S., Matear, R. J., McNeil, B. I., Metzl, N., Mikaloff Fletcher, S. E., Monteiro, P. M. S., Rödenbeck, C., Sweeney, C., and Takahashi, T.: Sea-air CO_2 fluxes in the Southern Ocean for the period 1990–2009, Biogeosciences, 10, 4037–4054, doi:10.5194/bg-10-4037-2013, 2013.

Le Quéré, C., Roedenbeck, C., Buitenhuis, E. T., Conway, T. J., Langenfelds R., Gomez, A., Labuschagne, C., Ramonet, M., Nakazawa, T., Metzl, N., Gillett, N., and Heimann, M: Saturation of the Southern Ocean CO_2 sink due to recent climate change, Science, 316, 1735–1738, 2007.

Manizza, M., Keeling, R. F., and Nevison, C. D.: On the processes controlling the seasonal cycles of the air-sea fluxes of O_2 and N_2O: A modeling study, Tellus B, 64, 18429, doi:10.3402/tellusb.v64i0.18429, 2012.

Manning, A. C. and Keeling, R. F.: Global oceanic and land biotic carbon sinks from the Scripps atmospheric oxygen flask sampling network, Tellus B, 58B, 95–116, 2006.

Mitchell, B. G.: Predictive bio-optical relationships for polar oceans and marginal ice zones, J. Mar. Sys., 3, 91–105, 1992.

Mitchell, B. G. and Kahru, M.: Bio-optical algorithms for ADEOS-2 GLI, J. Remote Sens. Soc., 29, 1, 80–85, 2009.

Moore, J. K., Doney, S. C., Kleypas, J. C., Glover, D. M., and Fung, I. Y.: An intermediate complexity marine ecosystem model for the global domain, Deep-Sea Res.-Pt. II, 49, 403–462, 2002.

Moore, J. K., Doney, S. C., and Lindsay, K.: Upper ocean ecosystem dynamics and iron cycling in a global 3-D model, Global Biogeochem. Cy., 18, GB4028, doi:10.1029/2004GB002220, 2004.

Moore, J., Lindsay, K., Doney, S., Long, M., and Misumi, K.: Marine Ecosystem Dynamics and Biogeochemical Cycling in the Community Earth System Model [CESM1(BGC)]: Comparison of the 1990s with the 2090s under the RCP4.5 and RCP8.5 Scenarios, J. Climate, 26, 9291–9312, doi:10.1175/JCLI-D-12-00566.1, 2013.

Naegler, T., Ciais, P., Orr, J., Aumont, O., and Roedenbeck, C.: On evaluating ocean models with atmospheric potential oxygen, Tellus, 59B, 138–156, 2007.

Najjar, R. G.,Jin, X, Louanchi, F., Aumont, O., Caldeira, K., Doney, S., Dutay, J. C., Follows, M., Gruber, N., Joos, F., Lindsay, K., Maier-Reimer, E., Matear, R., Matsumoto, K., Monfray, P., Mouchet, A., Orr, J., Plattner, G. K., Sarmiento, J., Schlitzer, R., Slater, R., Weirig, M.-F., Yamanaka, Y., and Yool, A: Impact of circulation on export production, dissolved organic matter, and dissolved oxygen in the ocean: Results from Phase II of the Ocean Carbon-cycle Model Intercomparison Project (OCMIP-2), Global Biogeochem. Cy., 21, GB3007, doi:10.1029/2006GB002857, 2007.

Nevison, C. D., Mahowald, N. M., Doney, S. C., Lima, I. D., and Cassar, N.: Impact of variable air-sea O_2 and CO_2 fluxes on atmospheric potential oxygen (APO) and land-ocean carbon sink partitioning, Biogeosciences, 5, 875–889, doi:10.5194/bg-5-875-2008, 2008.

Nevison, C. D., Keeling, R. F., Kahru, M., Manizza, M., Mitchell, B. G., and Cassar, N.: Estimating net community production in the Southern Ocean based on atmospheric potential oxygen and satellite ocean color data, Global Biogeochem. Cy., 26, GB1020, doi:10.1029/2011GB004040, 2012a.

Nevison, C. D., Baker, D. F., and Gurney, K. R.: A methodology for estimating seasonal cycles of atmospheric CO_2 resulting from terrestrial net ecosystem exchange (NEE) fluxes using the Transcom T3L2 pulse-response functions, Geosci. Model Dev. Discuss., 5, 2789–2809, doi:10.5194/gmdd-5-2789-2012, 2012b.

Peylin, P., Law, R. M., Gurney, K. R., Chevallier, F., Jacobson, A. R., Maki, T., Niwa, Y., Patra, P. K., Peters, W., Rayner, P. J., Rödenbeck, C., van der Laan-Luijkx, I. T., and Zhang, X.: Global atmospheric carbon budget: results from an ensemble of atmospheric CO_2 inversions, Biogeosciences, 10, 6699–6720, doi:10.5194/bg-10-6699-2013, 2013.

Sauer, M. J., Roesler, C. S., Werdell, J. P., and Barnard, A.: Under the hood of satellite empirical chlorophyll algorithms: revealing the dependencies of maximum band ratio algorithms on inherent optical properties, Optics Express, 20, 20920–20933, 2012.

Schneider, B., Bopp, L., Gehlen, M., Segschneider, J., Frölicher, T. L., Cadule, P., Friedlingstein, P., Doney, S. C., Behrenfeld,

M. J., and Joos, F.: Climate-induced interannual variability of marine primary and export production in three global coupled climate carbon cycle models, Biogeosciences, 5, 597–614, doi:10.5194/bg-5-597-2008, 2008.

Siegel, D. A., Maritorena, S., Nelson, N. B., and Behrenfeld , M. J.: Independence and interdependencies among global ocean color properties: Reassessing the bio-optical assumption, J. Geophys. Res., 110, C07011, doi:10.1029/2004JC002527, 2005.

Sigman, D. M., Hain, M. P., and Haug, G. H.: The polar ocean and glacial cycles in atmospheric CO_2, Nature, 466, 47–55, 2010.

Steinacher, M., Joos, F., Frölicher, T. L., Bopp, L., Cadule, P., Cocco, V., Doney, S. C., Gehlen, M., Lindsay, K., Moore, J. K., Schneider, B., and Segschneider, J.: Projected 21st century decrease in marine productivity: a multi-model analysis, Biogeosciences, 7, 979–1005, doi:10.5194/bg-7-979-2010, 2010.

Stephens, B. B., Keeling, R. F., Heimann, M., Six, K. D., Murnane, R., and Caldeira, K. Testing global ocean carbon cycle models using measurements of atmospheric O_2 and CO_2 concentration, Global Biogeochem. Cy., 12, 213–230, 1998.

Stephens, B. B., Gurney, K. R., Tans, P., Sweeney, C., Peters, W., Bruhwiler, L., Ciais, P., Ramonet, M., Bousquet, P., Nakazawa, T. Aoki, S., Machida, T., Inoue, G., Vinnichenko, N., Lloyd, J., Jordan, A., Heimann, M., Shibistova, O., Langenfelds, R., Steele, L., Francey, R., and Denning, A. S.: Weak northern and strong tropical land carbon uptake from vertical profiles of atmospheric CO_2, Science, 316, 1732–1735, doi:10.1126/science.1137004, 2007.

Tjiputra, J. F., Roelandt, C., Bentsen, M., Lawrence, D. M., Lorentzen, T., Schwinger, J., Seland, Ø., and Heinze, C.: Evaluation of the carbon cycle components in the Norwegian Earth System Model (NorESM), Geosci. Model Dev., 6, 301–325, doi:10.5194/gmd-6-301-2013, 2013.

Tohjima, Y., Mukai, H., Machida, T., Nojiri, Y., and Gloor ,M.: First measurements of the latitudinal atmospheric O_2 and CO_2 distributions across the western Pacific, Geophys. Res. Lett., 32, L17805, doi:10.1029/2005GL023311, 2005.

Tohjima, Y., Minejima, C., Mukai, H., Machida, T., Yamagishi, H., and Nojiri, Y.: Analysis of seasonality and annual mean distribution of atmospheric potential oxygen (APO) in the Pacific region, Global Biogeochem. Cy., 26, GB004110, doi:10.1029/2011GB004110, 2012.

Uitz, J., Claustre, H., Gentili, B., and Stramski, D.: Phytoplankton class-specific primary production in the world's oceans: Seasonal and interannual variability from satellite observations, Global Biogeochem. Cy., 24, GB3016, doi:10.1029/2009GB003680, 2010.

Wofsy, S. C.: the HIPPO Science Team and Cooperating Modellers and Satellite Teams: HIAPER Pole-to-Pole Observations (HIPPO): Fine grained, global scale measurements for determining rates for transport, surface emissions, and removal of climatically important atmospheric gases and aerosols, Phil. Trans. Roy. Soc., 369, 2073–2086,doi:10.1098/rsta.2010.0313, 2011.

Permissions

All chapters in this book were first published in BD, by Copernicus Publications; hereby published with permission under the Creative Commons Attribution License or equivalent. Every chapter published in this book has been scrutinized by our experts. Their significance has been extensively debated. The topics covered herein carry significant findings which will fuel the growth of the discipline. They may even be implemented as practical applications or may be referred to as a beginning point for another development.

The contributors of this book come from diverse backgrounds, making this book a truly international effort. This book will bring forth new frontiers with its revolutionizing research information and detailed analysis of the nascent developments around the world.

We would like to thank all the contributing authors for lending their expertise to make the book truly unique. They have played a crucial role in the development of this book. Without their invaluable contributions this book wouldn't have been possible. They have made vital efforts to compile up to date information on the varied aspects of this subject to make this book a valuable addition to the collection of many professionals and students.

This book was conceptualized with the vision of imparting up-to-date information and advanced data in this field. To ensure the same, a matchless editorial board was set up. Every individual on the board went through rigorous rounds of assessment to prove their worth. After which they invested a large part of their time researching and compiling the most relevant data for our readers.

The editorial board has been involved in producing this book since its inception. They have spent rigorous hours researching and exploring the diverse topics which have resulted in the successful publishing of this book. They have passed on their knowledge of decades through this book. To expedite this challenging task, the publisher supported the team at every step. A small team of assistant editors was also appointed to further simplify the editing procedure and attain best results for the readers.

Apart from the editorial board, the designing team has also invested a significant amount of their time in understanding the subject and creating the most relevant covers. They scrutinized every image to scout for the most suitable representation of the subject and create an appropriate cover for the book.

The publishing team has been an ardent support to the editorial, designing and production team. Their endless efforts to recruit the best for this project, has resulted in the accomplishment of this book. They are a veteran in the field of academics and their pool of knowledge is as vast as their experience in printing. Their expertise and guidance has proved useful at every step. Their uncompromising quality standards have made this book an exceptional effort. Their encouragement from time to time has been an inspiration for everyone.

The publisher and the editorial board hope that this book will prove to be a valuable piece of knowledge for researchers, students, practitioners and scholars across the globe.

List of Contributors

C. Lin
Department of Forestry and Natural Resources, National Chiayi University, Chiayi, Taiwan

S. C. Popescu
Department of Ecosystem Science and Management, Texas A&M University, College Station, Texas, USA

S. C. Huang
Department of Forestry and Natural Resources, National Chiayi University, Chiayi, Taiwan

P. T. Chang
Department of Horticultural Science, National Chiayi University, Chiayi, Taiwan

H. L. Wen
Department of Forestry and Natural Resources, National Chiayi University, Chiayi, Taiwan

S. Comeau
Department of Biology, California State University, 18111 Nordhoff Street, Northridge, CA 91330-8303, USA

R. C. Carpenter
Department of Biology, California State University, 18111 Nordhoff Street, Northridge, CA 91330-8303, USA

C. A. Lantz
Department of Biology, California State University, 18111 Nordhoff Street, Northridge, CA 91330-8303, USA

P. J. Edmunds
Department of Biology, California State University, 18111 Nordhoff Street, Northridge, CA 91330-8303, USA

Ü. Rannik
Department of Physics, P.O. Box 48, 00014 University of Helsinki, Helsinki, Finland

S. Haapanal
Department of Physics, P.O. Box 48, 00014 University of Helsinki, Helsinki, Finland

N. J. Shurpali
Department of Environmental Science, University of Eastern Finland, Kuopio, Finland

I. Mammarella
Department of Physics, P.O. Box 48, 00014 University of Helsinki, Helsinki, Finland

S. Lind
Department of Environmental Science, University of Eastern Finland, Kuopio, Finland

N. Hyvönen
Department of Environmental Science, University of Eastern Finland, Kuopio, Finland

O. Peltola
Department of Physics, P.O. Box 48, 00014 University of Helsinki, Helsinki, Finland

M. Zahniser
Center for Atmospheric and Environmental Chemistry, Aerodyne Research Inc., Billerica, MA, USA

P. J. Martikainen
Department of Environmental Science, University of Eastern Finland, Kuopio, Finland

T. Vesala
Department of Physics, P.O. Box 48, 00014 University of Helsinki, Helsinki, Finland

B. ˇTupek
Department of Forest Sciences, P.O. Box 27, 00014 University of Helsinki, Finland

K. Minkkinen
Department of Forest Sciences, P.O. Box 27, 00014 University of Helsinki, Finland

J. Pumpanen
Department of Forest Sciences, P.O. Box 27, 00014 University of Helsinki, Finland

E. Nikinmaa
Department of Forest Sciences, P.O. Box 27, 00014 University of Helsinki, Finland

N. Meskhidze
Marine, Earth, and Atmospheric Science, North Carolina State University, Raleigh, NC, USA

A. Sabolis
Marine, Earth, and Atmospheric Science, North Carolina State University, Raleigh, NC, USA

R. Reed
Center for Applied Aquatic Ecology, North Carolina State University, Raleigh, NC 27606, USA

D. Kamykowski
Marine, Earth, and Atmospheric Science, North Carolina State University, Raleigh, NC, USA

P. Brandt
GEOMAR Helmholtz-Zentrum für Ozeanforschung Kiel, Kiel, Germany

H. W. Bange
GEOMAR Helmholtz-Zentrum für Ozeanforschung Kiel, Kiel, Germany

D. Banyte
GEOMAR Helmholtz-Zentrum für Ozeanforschung Kiel, Kiel, Germany

M. Dengler
GEOMAR Helmholtz-Zentrum für Ozeanforschung Kiel, Kiel, Germany

S.-H. Didwischus
GEOMAR Helmholtz-Zentrum für Ozeanforschung Kiel, Kiel, Germany

T. Fischer
GEOMAR Helmholtz-Zentrum für Ozeanforschung Kiel, Kiel, Germany

R. J. Greatbatch
GEOMAR Helmholtz-Zentrum für Ozeanforschung Kiel, Kiel, Germany

J. Hahn
GEOMAR Helmholtz-Zentrum für Ozeanforschung Kiel, Kiel, Germany

T. Kanzow
GEOMAR Helmholtz-Zentrum für Ozeanforschung Kiel, Kiel, Germany

J. Karstensen
GEOMAR Helmholtz-Zentrum für Ozeanforschung Kiel, Kiel, Germany

A. Körtzinger
GEOMAR Helmholtz-Zentrum für Ozeanforschung Kiel, Kiel, Germany

G. Krahmann
GEOMAR Helmholtz-Zentrum für Ozeanforschung Kiel, Kiel, Germany

S. Schmidtko
GEOMAR Helmholtz-Zentrum für Ozeanforschung Kiel, Kiel, Germany

L. Stramma
GEOMAR Helmholtz-Zentrum für Ozeanforschung Kiel, Kiel, Germany

T. Tanhua
GEOMAR Helmholtz-Zentrum für Ozeanforschung Kiel, Kiel, Germany

M. Visbeck
GEOMAR Helmholtz-Zentrum für Ozeanforschung Kiel, Kiel, Germany

L. Merbold
Institute of Agricultural Sciences, ETH Zurich, 8092 Zurich, Switzerland

G. Wohlfahrt
Institute of Ecology, University of Innsbruck, 6020 Innsbruck, Austria European Academy of Bolzano, 39100 Bolzano, Italy

K. Butterbach-Bahl
Karlsruhe Institute of Technology, Institute for Meteorology and Climate Research, Atmospheric Environmental Research (IMK-IFU), 82467 Garmisch-Partenkirchen, Germany
International Livestock Research Institute, Old Naivasha Road, Nairobi 00100, Kenya

K. Pilegaard
Center for Ecosystems and Sustainability, Department of Chemical and Biochemical Engineering, Technical University of Denmark, 4000 Roskilde, Denmark

T. DelSontro
Institute of Biogeochemistry and Pollutant Dynamics, ETH Zurich, 8092 Zurich, Switzerland
Department of Biological Sciences, University of Quebec at Montreal, Montreal, QC, H2J 2K9, Canada

P. Stoy
Department of Land Resources and Environmental Sciences, Montana State University, Bozeman, MT 59717-3120, USA

D. Zona
Department of Animal and Plant Science, University of Sheffield, Western Bank, Sheffield, S10 2TN, UK
Department of Biology, San Diego State University, 5500 Campanile Drive, San Diego, CA 92182, USA

K. Mallick
Environmental Research and Innovation (ERIN), Luxembourg Institute of Science and Technology (LIST), L4422, Belvaux, Luxembourg

A. Jarvis
Lancaster Environment Centre, Lancaster University, Lancaster LA1 4YQ, UK

G. Wohlfahrt
Institute of Ecology, University of Innsbruck, 6020 Innsbruck, Austria

G. Kiely
Hydrometeorology Research Group, Department of Civil and Environmental Engineering, University College Cork, Ireland

T. Hirano
Division of Environmental Resources, Research Faculty of Agriculture, Hokkaido University, Hokkaido, Japan

A. Miyata
National Institute for Agro-Environmental Sciences, Tsukuba, Japan

S. Yamamoto
Graduate School of Environmental Science, Okayama University Tsushimanaka3-1-1, Okayama 700-8530, Japan

L. Hoffmann
Environmental Research and Innovation (ERIN), Luxembourg Institute of Science and Technology (LIST), L4422, Belvaux, Luxembourg

S. Sitch
University of Exeter, Exeter EX4 4QF, UK

P. Friedlingstein
University of Exeter, Exeter EX4 4QF, UK

N. Gruber
Institute of Biogeochemistry and Pollutant Dynamics, ETH Zurich, Zurich, Switzerland

S. D. Jones
Tyndall Centre for Climate Change Research, University of East Anglia, Norwich NR4 7TJ, UK

G. Murray-Tortarolo
University of Exeter, Exeter EX4 4QF, UK

A. Ahlström
Lund University, Department of Physical Geography and Ecosystem Science, Sölvegatan 12, 223 62 Lund, Sweden

S. C. Doney
Marine Chemistry and Geochemistry Department, Woods Hole Oceanographic Institution, 266 Woods Hole Road, Woods Hole, MA 02543, USA

H. Graven
Department of Physics and Grantham Institute for Climate Change, Imperial College London, London SW7 2AZ, UK

C. Heinze
Geophysical Institute, University of Bergen, Bergen, Norway
Bjerknes Centre for Climate Research, Bergen, Norway
Uni Climate, Uni Research AS, Bergen, Norway

C. Huntingford
Centre for Ecology and Hydrology, Benson Lane, Wallingford OX10 8BB, UK

S. Levis
National Center for Atmospheric Research, Boulder, Colorado, USA

P. E. Levy
Centre for Ecology and Hydrology, Bush Estate, Penicuik, Midlothian EH26 0QB, UK

M. Lomas
Department of Animal & Plant Sciences, University of Sheffield, Sheffield S10 2TN, UK

B. Poulter
Institute on Ecosystems and Department of Ecology, Montana State University, Bozeman, MT 59717, USA

N. Viovy
Laboratoire des Sciences du Climat et de l'Environnement, CEA CNRS UVSQ, 91191 Gif-sur-Yvette, France

S. Zaehle
Biogeochemical Integration Department, Max Planck Institute for Biogeochemistry, P.O. Box 10 01 64, 07701 Jena, Germany

N. Zeng
Department of Atmospheric and Oceanic Science, University of Maryland, College Park, MD 20740, USA

A. Arneth
Karlsruhe Institute of Technology, Garmisch-Partenkirchen, Germany

G. Bonan
National Center for Atmospheric Research, Boulder, Colorado, USA

L. Bopp
Laboratoire des Sciences du Climat et de l'Environnement, CEA CNRS UVSQ, 91191 Gif-sur-Yvette, France

J. G. Canadell
Global Carbon Project, CSIRO Oceans and Atmosphere Flagship, Canberra, Australia

F. Chevallier
Laboratoire des Sciences du Climat et de l'Environnement, CEA CNRS UVSQ, 91191 Gif-sur-Yvette, France

P. Ciais
Laboratoire des Sciences du Climat et de l'Environnement, CEA CNRS UVSQ, 91191 Gif-sur-Yvette, France

R. Ellis
Centre for Ecology and Hydrology, Benson Lane, Wallingford OX10 8BB, UK

M. Gloor
University of Leeds, School of Geography, Woodhouse Lane, Leeds LS9 2JT, UK

P. Peylin
Laboratoire des Sciences du Climat et de l'Environnement, CEA CNRS UVSQ, 91191 Gif-sur-Yvette, France

S. L. Piao
College of Urban and Environmental Sciences, Peking University, Beijing 100871, China

C. Le Quéré
Tyndall Centre for Climate Change Research, University of East Anglia, Norwich NR4 7TJ, UK

B. Smith
Lund University, Department of Physical Geography and Ecosystem Science, Sölvegatan 12, 223 62 Lund, Sweden

Z. Zhu
State Key Laboratory of Remote Sensing Science, Institute of Remote Sensing and Digital Earth, Chinese Academy of Sciences, Beijing 100101, China
Center for Applications of Spatial Information Technologies in Public Health, Beijing 100101, China

R. Myneni
Department of Geography and Environment, Boston University, 675 Commonwealth Avenue, Boston, MA 02215, USA

C. L. Faiola
Laboratory for Atmospheric Research, Department of Civil and Environmental Engineering, Washington State University, Pullman, Washington, USA

B. T. Jobson
Laboratory for Atmospheric Research, Department of Civil and Environmental Engineering, Washington State University, Pullman, Washington, USA

T. M. VanReken
Laboratory for Atmospheric Research, Department of Civil and Environmental Engineering, Washington State University, Pullman, Washington, USA

A. W. King
Environmental Sciences Division and Climate Change Science Institute, Oak Ridge National Laboratory, Oak Ridge, Tennessee, USA

R. J. Andres
Environmental Sciences Division and Climate Change Science Institute, Oak Ridge National Laboratory, Oak Ridge, Tennessee, USA

K. J. Davis
Department of Meteorology, The Pennsylvania State University, University Park, Pennsylvania, USA

M. Hafer
Canadian Forest Service, Natural Resources Canada, Victoria, British Columbia, Canada

D. J. Hayes
Environmental Sciences Division and Climate Change Science Institute, Oak Ridge National Laboratory, Oak Ridge, Tennessee, USA

D. N. Huntzinger
School of Earth Sciences and Environmental Sustainability, Northern Arizona University, Arizona, USA

B. de Jong
El Colegio de la Frontera Sur, Unidad Campeche, Campeche, Mexico

W. A. Kurz
Canadian Forest Service, Natural Resources Canada, Victoria, British Columbia, Canada

A. D. McGuire
US Geological Survey, Alaska Cooperative Fish and Wildlife Research Unit, University of Alaska, Fairbanks, Alaska, USA

R. Vargas
Department of Plant and Soil Sciences, University of Delaware, Newark, Delaware, USA

Y. Wei
Environmental Sciences Division and Climate Change Science Institute, Oak Ridge National Laboratory, Oak Ridge, Tennessee, USA

T. O. West
Joint Global Change Research Institute, Pacific Northwest National Laboratory, College Park, Maryland, USA

C. W. Woodall
Northern Research Station, USDA Forest Service, Saint Paul, Minnesota, USA

D. Plake
Max Planck Institute for Chemistry, Biogeochemistry Department, P.O. Box 3060, 55020 Mainz, Germany

M. Sörgel
Max Planck Institute for Chemistry, Biogeochemistry Department, P.O. Box 3060, 55020 Mainz, Germany

P. Stella
Max Planck Institute for Chemistry, Biogeochemistry Department, P.O. Box 3060, 55020 Mainz, Germany

A. Held
University of Bayreuth, Atmospheric Chemistry, Bayreuth, Germany

I. Trebs
Max Planck Institute for Chemistry, Biogeochemistry Department, P.O. Box 3060, 55020 Mainz, Germany

J. Palmiéri
LSCE/IPSL, Laboratoire des Sciences du Climat et de l'Environnement, CEA-CNRS-UVSQ, Gif-sur-Yvette, France

J. C. Orr
LSCE/IPSL, Laboratoire des Sciences du Climat et de l'Environnement, CEA-CNRS-UVSQ, Gif-sur-Yvette, France

J.-C. Dutay
LSCE/IPSL, Laboratoire des Sciences du Climat et de l'Environnement, CEA-CNRS-UVSQ, Gif-sur-Yvette, France

K. Béranger
ENSTA-ParisTech, Palaiseau, France

A. Schneider
GEOMAR; Helmholtz-Zentrum für Ozeanforschung Kiel, Germany

J. Beuvier
Mercator Ocean, Ramonville Saint-Agne, France CNRM/Météo-France, Toulouse, France

S. Somot
CNRM/Météo-France, Toulouse, France

S. Huang
Department of Civil and Environmental Engineering, Princeton University, Princeton, NJ 08544, USA

P. R. Jaffé
Department of Civil and Environmental Engineering, Princeton University, Princeton, NJ 08544, USA

S. E. Hartman
National Oceanography Centre, Southampton, SO14 3ZH, UK

Z.-P. Jiang
National Oceanography Centre, Southampton, SO14 3ZH, UK

D. Turk
Dalhousie University, Halifax, B3H 4R2, Canada
Lamont-Doherty Earth Observatory, Palisades, NY 10964, USA

R. S. Lampitt
National Oceanography Centre, Southampton, SO14 3ZH, UK

H. Frigstad
Norwegian Environment Agency, 0663 Oslo, Norway

C. Ostle
Sir Alister Hardy Foundation for Ocean Science, Plymouth, PL1 2PB, UK
School of Environmental Sciences, University of East Anglia, Norwich, NR4 7TJ, UK

U. Schuster
University of Exeter, Exeter, EX4 4PS, UK

C. D. Nevison
University of Colorado, Boulder, Institute for Arctic and Alpine Research, Boulder, Colorado, USA

M. Manizza
Scripps Institution of Oceanography, La Jolla, California, USA

R. F. Keeling
Scripps Institution of Oceanography, La Jolla, California, USA

M. Kahru
Scripps Institution of Oceanography, La Jolla, California, USA

L. Bopp
IPSL/LSCE, UMR8212, CNRS-CEA-UVSQ, Gif sur Yvette, France

J. Dunne
National Oceanic and Atmospheric Administration/ Geophysical Fluid Dynamics Laboratory, Princeton, New Jersey, USA

J. Tiputra
Uni Climate, Uni Research and Bjerknes Centre for Climate Research, Bergen, Norway

T. Ilyina
Max Planck Institute for Meteorology, Hamburg, Germany

B. G. Mitchell
Scripps Institution of Oceanography, La Jolla, California, USA